Cowen's History of Life

# Cowen's History of Life

*Edited by Michael J. Benton*

*Written by Michael J. Benton, John Cunningham,
Tom Davies, Philip C. J. Donoghue, Andy Fraass, Christine Janis,
Davide Pisani, Emily Rayfield, Daniela Schmidt, Jakob Vinther,
and Tom Williams*

*Palaeobiology Research Group
University of Bristol
UK*

Sixth Edition

*Registered Office(s)*
John Wiley & Sons, Inc., 111 River Street, Hoboken, NJ 07030, USA
John Wiley & Sons Ltd, The Atrium, Southern Gate, Chichester, West Sussex, PO19 8SQ, UK

*Editorial Office*
The Atrium, Southern Gate, Chichester, West Sussex, PO19 8SQ, UK

For details of our global editorial offices, customer services, and more information about Wiley products visit us at www.wiley.com.

Wiley also publishes its books in a variety of electronic formats and by print-on-demand. Some content that appears in standard print versions of this book may not be available in other formats.

*Library of Congress Cataloging-in-Publication Data*

Names: Benton, M. J. (Michael J.), author. | Cowen, Richard, 1940– History of life.
Title: Cowen's history of life / edited by Michael J. Benton ; written by
 Michael J. Benton, John Cunningham, Tom Davies, Philip C. J. Donoghue, Andy
 Fraass, Christine Janis, Davide Pisani, Emily Rayfield, Daniela Schmidt,
 Jakob Vinther, and Tom Williams, Palaeobiology Research Group, University of Bristol.
Other titles: History of life
Description: Sixth edition. | Hoboken, NJ : Wiley-Blackwell, [2019] | Includes index. |
Identifiers: LCCN 2019015116 (print) | LCCN 2019017612 (ebook) | ISBN
 9781119482208 (Adobe PDF) | ISBN 9781119482222 (ePub) | ISBN 9781119482215
 (paperback)
Subjects: LCSH: Paleontology.
Classification: LCC QE711.2 (ebook) | LCC QE711.2 .C68 2019 (print) |
 DDC560–dc23
LC record available at https://lccn.loc.gov/2019015116

Cover Design: Wiley
Cover Image: © Sergey Krasovskiy/Getty Images

Set in 10/12pt Warnock by SPi Global, Pondicherry, India

10 9 8 7 6 5 4 3 2 1

# Contents

# Preface

## Authorship

This book is based on the classic work through five editions by Richard Cowen, and many of his words remain in this new edition. Among the team of revisers, Mike Benton led the revision, contributing more or less to all chapters, supported by Tom Williams (Chapters 1–3), John Cunningham (Chapters 4 and 5), Davide Pisani (Chapters 3–5), Phil Donoghue (Chapters 4, 5, 7 and 8), Daniela Schmidt and Andy Fraass (Chapter 17), Christine Janis and Emily Rayfield (Chapters 18–21 and 23), and Tom Davies (Chapter 22).

## For Everyone

Richard Cowen's *History of Life* has run through five editions since it first appeared in 1990. He based the book on his long experience as an instructor at the University of California, Davis, teaching a course called "History of Life" for 40 years. As he said in his Preface to the fifth edition, the book "is meant not just for students, but for everyone interested in the history of life on our planet. Fortunately, paleontology (= paleobiology) is accessible to the average person without deep scientific training. My aim is ambitious: I try to take you to the edges of our knowledge in paleontology, showing you how life has evolved on Earth, and how we have reconstructed the history of that evolution from the record of rocks and fossils."

The story of the history of life is of profound interest to many because it addresses core questions in philosophy and knowledge, especially about origins, crises, and environmental change. Since the times of the earliest thinkers in China, India, and Greece, people have asked questions about where humans came from and how we relate to the plants and animals we see around us. During the development of modern science in Europe, great philosophers have debated what fossils are and what the rocks show us.

As an example, Leonardo da Vinci (1452–1519), as well as being one of the greatest artists of all time, wrote extensively about scientific topics. In his day, people found fossil shells in the limestone mountains of central Italy, and they wondered how they got there. Had the Roman soldiers dumped the shells after eating their lunch? Leonardo realized, correctly, that these limestones had been deposited in the sea, and that the mountains had subsequently been uplifted. He realized the Earth was ancient, and the animals that had inhabited the shells truly had once lived on the seabed, and it took millions of years for them to reach their present location, many miles from the sea.

Great debates in science in later times focused around rocks and fossils – could extinct species of often unfamiliar appearance even exist? How ancient is the Earth? Has life evolved or simply been created in an instant? Are humans merely naked apes or are they in some way special? How do environmental changes such as rising sea levels, rising temperatures, and ocean acidification affect life? The paleontologist has (some of) the answers.

Since 1990, when *History of Life* first appeared, our knowledge of paleontology has changed enormously – think of the thousands of extraordinary new discoveries from China: the early animals from Chengjiang, the Silurian fishes from Yunnan, the feathered dinosaurs and birds from northern China, and many more. Every week, exciting new finds are made, adding records of the very oldest traces of life on Earth, early microbes, plants, and animals that tell us about the origins of modern biodiversity. A new dinosaur species is named every two weeks.

In a way, more important than the new fossil finds are the changes in *methods*. Richard Cowen has lived through several revolutions in paleobiology. First came cladistics, a set of methods to draw up evolutionary trees according to rules, so that the trees are testable hypotheses rather than merely guesses. Then came phylogenomics, a wholly new set of methods to draw up evolutionary trees using gene sequences from DNA and RNA, to provide an independent test of the fossil-based trees. Then came massive improvements in stratigraphy, with fine tuning of methods such as radiometric dating and astrochronology, so that events can now be dated with much greater precision than was imagined in the 1980s. These three fields of science provide the essential backdrop for paleontology – a well-dated evolutionary tree.

Then there were revolutions in the way paleontologists study macroevolution. First came the explosion of interest in mass extinctions, whether caused by asteroid impact or massive volcanic eruption – this is a field that attracts huge interest from multidisciplinary teams of paleontologists, geologists, geochemists, astronomers, and ecological modelers. New methods of phylogenetic comparative methods (PCM) allowed paleobiologists to explore diversifications, times when new groups of plants or animals radiate rapidly. The PCM provide numerous approaches to test rates of evolution on well-dated evolutionary trees to identify times of unusually high rates of evolution, responses of life to external events, and key characters that might contribute to a group's success.

Finally, in paleobiology itself, the science of bringing fossils to life, new engineering methods can be applied to ancient shells and bones to test their strength characteristics and capabilities. You want to know the bite force of *Tyrannosaurus rex*? Paleobiologists can calculate that. How fast did *Brontosaurus* run? We can test that too. Could an ammonite shell withstand 100 pounds of compression force from a predator? We can work that one out.

What has happened since 1990 is that paleobiology has become a testable science in all its aspects – knowing the shape of evolution, dating fossils, describing ancient crises and their effects, identifying whether a group radiated explosively and why, and testing what extinct organisms could do. No longer can we simply sit around speculating … "I think *T. rex* was purple with green spots," "No, surely it was blue with yellow stripes," "I think the dinosaurs died out because they were just too big and stupid." When you hear these kinds of statements, the answer is, "How can we test that?" If it cannot be tested, it's best left aside for the moment until we can find a way to determine color or reasons for extinction.

As Richard Cowen noted in his introduction to the fifth edition, "The challenge of teaching paleontology, and the challenge of writing a book like this, is to present a complex story in a way that is simple enough to grasp, yet true enough to real events that it paints a reasonable picture of what happened and why. I believe it can be done and done so that you can learn enough to appreciate what's going on in current research projects."

Combining the excitement of the story – all the new fossils, the exotic lands and fossil sites, the big events they tell us about – with a true feel of how that knowledge has been debated and tested is a tricky task. We have attempted to tell the story while also giving a flavor of the hard work behind each discovery.

The book is aimed at students and we hope some will be inspired to find out more, and perhaps sign up for further courses on paleontology, paleobiology, historical geology, macroevolution, and related topics. It's a great life as a paleontologist, even though jobs are thinly spread.

In this new edition, we have kept the basic structure and have substantially rewritten chapters on the origins of eukaryotes, the Cambrian explosion, the terrestrialization of plants and animals, the Triassic recovery of life, the origin of birds, the end-Cretaceous mass extinction, and human evolution. We have added three new chapters, one on the Mesozoic Marine Revolution, all the great events in the oceans as life "speeded up," one on the Cretaceous Terrestrial Revolution, when the diversification of flowering plants changed ecosystems on land forever, and one on the evolution of oceans and climates. This means the book can be used for introductory classes, but the theme can be either "the major steps in evolution from origins to humans" or "history of the diversity of life." The new chapters allow us to keep the strong focus on vertebrate evolution, but to strengthen the more geological and ecological themes of the evolution of oceans, climates, and marine life.

The genesis of this edition may require a brief comment. Richard Cowen has labored valiantly on his own, presenting five editions of the *History of Life* from 1990 to 2013. When the call came from Wiley for a sixth edition, he turned to the paleontology group at the University of Bristol, UK, to help. Richard's idea is that we keep the shape and flavor of the book but freely update it. Mike Benton took the lead and his colleagues in Bristol, Tom Williams, John Cunningham, Davide Pisani, Jakob Vinther, Phil Donoghue, Daniela Schmidt, Andy Fraass, Christine Janis, Emily Rayfield, and Tom Davies (listed roughly in order of their chapters), accepted the challenge with enthusiasm. Adrian Lister at the Natural History Museum, London, provided extensive help with the Ice Ages chapter. We decided collectively also to pool any earnings from the sales of the book and donate them to the Bob Savage Memorial Fund which supports student research – some of which appears in this book!

## To Our Teaching Colleagues

The course for which this book was written serves four audiences at the same time: it is an introduction to paleontology; it is a "general education" course to introduce nonspecialists to science and scientific thought; it provides an overview of the interactions of environmental change and biodiversity; and it can serve as an introduction to the history of life to biologists who know a lot about the present and little about the past. Therefore, the style and language of this book are aimed at accessibility.

We do not use scientific jargon unless it is useful. We have tried to show how we reason out our conclusions – how we choose between bad ideas and good ones. In short, we have aimed this book at the intelligent nonspecialist.

We have not covered the fossil record evenly. We have tried to write compact essays on the most important events and processes that have molded the history of life. They illustrate the most important ways we go about reconstructing the life of the past. We have used case studies from vertebrates more than from other groups simply because those are the animals with which paleontologists and the general reader are most familiar. Most fossils are marine invertebrates, and most paleontologists are invertebrate specialists.

Here, we have introduced more on the history of life in the oceans and impacts of ocean and atmosphere on life than in previous editions. The instructor who has used the book as a basis for their courses of lectures may care to revise the sequence and content of lectures to reflect these additions, or individual chapters can be omitted.

The book is highly illustrated, and we have sought to update illustrations to use the best that are readily available. Most illustrations are photographs of fossils or life reconstructions – these are readily intelligible. But we have also included diagrams to show internal structures of organisms, cladograms, time scales and events, and paleogeographic maps. All images are available in digital format for use in visual presentations through the publisher's website.

The references are a careful mixture of important books, primary literature, news reports, and review articles that bring the latest work into this edition as it went to press. We have tried to choose items that can be readily found through web search engines and published in accessible journals.

If this book contained nothing controversial, it would be very dull and far from representing the state of paleontology as it stands today. We have tried to present arguments for and against particular ideas in case studies that are presented in some detail, such as major extinctions and major evolutionary innovations. Often, however, space or conviction has led us to present only one side of an argument. Please share your dissatisfaction and/or more complete knowledge with your students and tell them why our treatment is one-sided or just plain wrong. That way everyone wins by exposure to the give and take of scientific argument as it ought to be practiced between colleagues.

## To Students

Several thousand people like you have voted with their comments, questions, body language, and formal written evaluations on the content of our courses. As Richard Cowen said about the fifth edition, "Students had more influence on the style and content of this book than anyone else. So, you and your peers at the University of California, Davis, can take whatever credit is due for the style in which the material is presented."

After all the thanks, however, we do have another point to make. You do not have to take any of the interpretations in this book at face value. Facts are facts but ideas are only suggestions. If you can come up with a better idea than one of those we have included here, then work on it, starting with the literature references. It would make a great term paper and (more important) you might be right. The 1960s slogan "Question Authority!" is still valid. Your suggestion would not be the first time that a student found new and better evidence for interpreting the fossil record.

Why should you bother with the past? If we do not understand the past, how can we deal intelligently with the present? We and our environment are reaching such a state of crisis that we need all the help we can get. Nature has run a series of experiments over the last 3.5 billion years on this planet, changing climate and geography, and introducing new kinds of organisms. If we can read the results of those experiments from the fossil record, we can perhaps define the limits to which we can stretch our present biosphere before a biological disaster happens.

The real pay-off from paleontology for all the authors of this book is the fun involved in reconstructing extinct organisms and ancient communities, but if one needs a concrete reason for looking at the fossil record, the future of the human race is surely important enough for anyone.

## Further Reading

We have tried to list widely sold paperbacks and articles in journals such as *Nature, Science, Discover, Scientific American, National Geographic Magazine,* and *American Scientist,* perhaps the six most widely distributed journals that deal with all aspects of science. We also list books and articles in specialized journals; generally, the writing is more detailed and more technical in such journals.

Important earlier work is often summarized in more recent articles we have selected. Always, however, you should be able to work quickly backward to older papers from the references in recent articles. Of course, when reading an older, "classic" piece of writing, you should also use standard search tools such as Google Scholar to track forward in time to read the latest articles on the topic.

## Thanks

We thank all those reviewers who have given careful and calm advice over the years. We also thank the people at Wiley, and formerly at Blackwells, who have encouraged and helped Richard Cowen over the years, and now this new crop of authors. For this edition, it has been a delight to *work* with Athira Menon and Vimali Joseph.

November 2018                                          *Bristol*

## About the Companion Website

Don't forget to visit the companion website for this book:

www.wiley.com/go/cowen/historyoflife

There you will find valuable material designed to enhance your learning, including:

- Slide sets for all the illustrations
- Updates and mini-essays on topics that have arisen after the book went to press

1

The Origin of Life on Earth

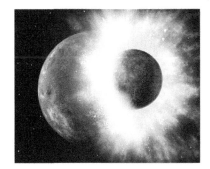

| In This Chapter |
|---|

First, we describe what geology and paleontology aim to study. But in dealing with the history of life, we have to face the most difficult question first: where did Earth's life come from? Astronomers find that organic compounds exist almost everywhere in space, yet we only know of life on one planet: Earth. We discuss the planets and moons of our solar system, and there are good reasons why none of them (except Earth) have life. Life exists in cells, so we discuss at length how complex organic molecules might have come together inside cells which survived, reproduced, and evolved on the early Earth. Laboratory experiments have already mimicked many of the steps in that process in the laboratory, but there is still a lot of work to be done.

## How Geology Works

Geology is the study of the Earth we live on. It's about fundamental ideas, such as origins and deep time, but also has immediate, practical aspects, such as guiding the commercial search for oil and useful minerals, as well as being core to understanding current and future climate change. Geology has methods and principles of its own, but also draws from many other sciences: physics, chemistry, biology, mathematics, and statistics are just a few. Geologists cannot be narrow specialists, because geology is a broad science that works best for people who think broadly. So, geologists cannot be successful if they are geeks (although a few seem to manage it). Above all, geology deals with the reality of the Earth: its rocks, minerals, its rivers, lakes, and oceans, its surface and its deep structure.

Some geologists deal with the Earth as it is now: they do not need to look at the past. Deep Earth history does not matter much to a geologist trying to deal with ecological repair to an abandoned gold mine. But many geologists do study Earth history, and they find that our planet has changed, at all scales of space and time, and sometimes in the most surprising ways. For 200 years, fossils have provided direct and solid proof of change through time. Life began and evolved on a planet that is changing too. Fossils often provide insight into Earth's environmental changes, whether or not they survived those changes. Paleontology is not just a fascinating side branch of geology, but a vital component of it.

As they run their life processes, organisms take in, alter, and release chemicals. Given enough organisms and enough time, biological processes can change the chemical and physical world. Photosynthesis, which provides the oxygen in our atmosphere, is only one of these processes. In turn, physical processes of the Earth such as continental movement, volcanism, and climate change affect organisms, influencing their evolution, and, in turn, affecting the way they can make changes to the physical Earth. This gigantic interaction, or *feedback mechanism*, has been going on since life evolved on Earth. Paleontologists and geologists who ignore this interaction are likely to get the wrong answers as they try to reconstruct the past.

## How Paleontology Works

Traces of Earth's ancient life have been preserved in rocks as fossils. Paleontology is the science of studying these fossils. Paleontology aims to understand fossils as once-living organisms, living, breeding, and dying in a

*Cowen's History of Life*, Sixth Edition. Edited by Michael J. Benton.
© 2020 John Wiley & Sons Ltd. Published 2020 by John Wiley & Sons Ltd.

real environment on a real but past Earth that we can no longer touch, smell, or see directly.

Most paleontologists do not study fossils for their intrinsic interest, although some of us do. Their greater value lies in what they tell us about ourselves and our background. We care about our future, which is a continuation of our past. One good reason for trying to understand ancient life is to manage better the biology of our planet today, so we need to use some kind of reasonable logic for clear interpretation of the life of the past.

Some basic problems of paleontology are much like those of archaeology and history: how do we know we have found the right explanation for some past event? How do we know we are not just making up a story?

Anything we suggest about the biology of ancient organisms should make sense in terms of what we know about the biology of living organisms, unless there is very good evidence to the contrary. This rule applies throughout biology, from cell biochemistry to genetics, physiology, ecology, behavior, and evolution.

There are three levels of paleontological interpretation. First, there are *inevitable conclusions* for which there are no possible alternatives. For example, there's no doubt that extinct ichthyosaurs were swimming marine reptiles; look at their body shape (Figure 1.1) and compare it with a dolphin or a shark, and there is also the fact their skeletons are always found in marine sediments.

At the next level, there are *likely interpretations.* There may be alternatives, but a large body of evidence supports one leading idea. For example, there is good evidence that suggests ichthyosaurs gave birth to live young rather than laying eggs. The evidence includes dozens of fossils of mother ichthyosaurs carrying well-developed babies inside their rib cages. Almost all paleontologists view this as the best hypothesis and would be surprised if contrary evidence turned up.

Then there are *speculations.* They may be right, but there is not much real evidence one way or another. Paleontologists, and indeed all other scientists, can accept speculations as tentative ideas to work with and to test carefully, but they should not be surprised or upset to find them wrong. For example, it seems reasonable that ichthyosaurs were warm-blooded, but it's a speculative idea because it's difficult to test. If new evidence showed that the idea was unlikely, people might be disappointed but they would not be distressed scientifically.

It's important for students of paleontology not to fall into the apologetic trap of saying, "Well, it all happened millions of years ago, so we can only really guess about stuff." Not true! The geologic time scale is well established based on multiple dating evidence, and we know the past geography of the Earth as the continents moved, again based on multiple lines of evidence that agree.

This is the first point: we can often get at information about the past through multiple routes, and so we confirm our hypotheses with different, independent datasets. We have powerful tools to reconstruct exact patterns of evolution, as you will read in this and later chapters, and we know about times of crisis – mass extinctions – again through massive amounts of data and multiple lines of evidence. These basics cannot be denied, only modified as knowledge becomes more precise.

The other kind of analysis in paleontology is where we can *test ideas numerically.* For example, paleobiologists can calculate dinosaur running speeds using ancient trackways – you just need the stride length and the leg length and it's a simple calculation that works for all modern animals, and so must also work for dinosaurs. Paleobiologists can also calculate the bite forces of dinosaurs by constructing three-dimensional digital models

 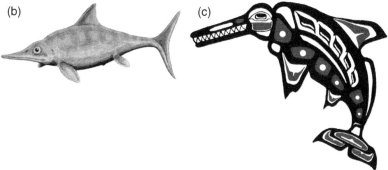

**Figure 1.1** Guesses about ichthyosaur color patterns. (a) Ichthyosaur painting by Heinrich Harder 1916. (b) Art by Nobu Tamura, with muted colors (Wikimedia). (c) Stylistic artwork. *Source:* © Danny Anduza, used by permission. See more of Danny's work at www.cafepress.com/dannysdinosaurs.

from computed tomography (CT) scan data. The digital models are assigned material properties of bones, teeth, etc. and are then exact replicas of the living flesh and bone – calculations of stresses and strains are real and they bring the past to life! Like all scientific hypotheses, these numerical approaches, which are very new, can be tested and accepted or rejected.

As an example of when speculation becomes reality, we can look at the color of ancient organisms. For example, a few years ago it might have been argued that the color we assign to ichthyosaurs must always be a speculation, even a guess. But in fact, color has become scientific recently because of the recognition that fundamental color-bearing organelles, called **melanosomes**, are frequently found in the fossil skin of ichthyosaurs, as well as the feathers of fossil birds and dinosaurs. In the case of ichthyosaurs, the melanosomes, which bore the pigment **melanin**, show the animals were black over much of the body, and even the belly, although more specimens are needed to test whether some species were counter-shaded (dark above, pale belly; Figure 1.1a) or uniformly colored (Figure 1.1b). Some artistic interpretations are very pleasing to the eye (Figure 1.1c), but most unlikely!

The fossil record gradually gets poorer as we go back in time, for two reasons. Biologically, there were fewer types of organisms in the past. Geologically, relatively few rocks (and fossils) have survived from older times, and those that have survived have often suffered heating, deformation, and other changes, all of which tend to destroy fossils. Earth's early life was certainly microscopic and soft-bodied, a very unpromising combination for fossilization. So direct evidence about early life on Earth is very scanty, although speculation and guesses are abundant.

In the absence of direct evidence, paleobiologists have turned to genetic data from modern organisms to inform their reasoning about the distant past. Comparing the genomes of living organisms provides an estimate of the length of time that has passed since they shared a common ancestor, and can identify likely ancestral character states, from behavior to metabolism. These molecular techniques have allowed paleontologists to broaden their window on the past, and to extend their study to times in evolutionary history when the fossil record is sparse or nonexistent.

## The Origin of Life

The fact of observation is that there is no evidence of life, let alone evidence of intelligence or civilization, anywhere in the universe except on our planet, Earth (for example, Smith 2011). This fact endures in the face of strenuous efforts by science fiction writers, tabloid magazines, movie directors, and NASA publicists to persuade us otherwise (Figure 1.2). However, we have to face up to its implications. Most important, it implies (but does not prove) that Earth's life evolved here on Earth. How difficult would that have been?

We can test the idea that life evolved here on Earth, from nonliving chemicals, by observation and experiment. Geologists and astronomers look for evidence from the Earth, Moon, and other planets to reconstruct conditions in the early solar system. Chemists and biochemists determine how complex organic molecules could have formed in such environments. Geologists try to find out when life appeared on Earth, and biologists design experiments to test whether these facts fit with ideas of the evolution of life from nonliving chemicals.

Complex organic molecules have been found in interstellar space, in the dust clouds around newly forming

Figure 1.2 Edgar Rice Burroughs published the first in his series of Martian stories, *A Princess of Mars*, as a book in 1917. *Source:* Cover art by Frank E. Schoonover (Wikimedia).

stars, on comets and asteroids and interplanetary dust, and on the meteorites that hit Earth from time to time. These compounds form naturally in space, generated as gas clouds, dust particles, and cometary and meteorite surfaces are bathed in cosmic and stellar radiation. Laboratory experiments designed to mimic such conditions in space have yielded organic molecules. Probably any solid surface near any star in the universe received organic molecules at some point in its history (Ciesla and Sandford 2012). Analyses of meteorites that have hit the Earth show they were carrying many of the basic organic molecules needed in the evolution of life.

But life as we know it is not just made of organic compounds: life consists of cells, composed mostly of liquid water that is vital to life. It is almost impossible to imagine the formation of any kind of water-laden cell in outer space; that can only happen on a planet that has oceans and therefore an atmosphere.

Planets have organic compounds delivered to them from space, especially from comets or meteorites, but this process by itself is unlikely to lead to the evolution of life. For example, organic molecules must have been delivered everywhere in the solar system, including Mercury, Mars, Venus, and the Moon, only to be destroyed by inhospitable conditions on those lifeless planets.

If conditions on a planet's surface were mild enough to allow organic molecules to survive after they arrived on comets, it is very likely that organic molecules were also forming naturally on that planet. Space-borne molecules may have added to the supply on a planetary surface, but they are unlikely to have been the only source of organic molecules there.

## Planets in Our Solar System

Scientists reconstructed the process of star and planet formation long before we could check it by observing stars forming out in the universe. Stars form from collapsing clouds of dust and gas, and in the process, planets and smaller bodies often form in orbit around the new stars. Now that we have telescopes powerful enough, the theories have been confirmed. In 2010, a spectacular new star, surrounded by dust and gas, was discovered in the process of forming in the constellation Centaurus (Figure 1.3). Astronomers have now found hundreds of planets around other stars, most of them large ones because they are easier to detect.

Our star the Sun formed with Earth, which is one of four terrestrial (rocky) planets in the inner part of our solar system. Venus and Earth are about the same size, and Mars and Mercury are significantly smaller. They all formed from dust and gas in the same way, about 4570 Ma (million years ago) (Lin 2008).

Most likely, all the planets were largely complete by 4500 Ma, although they were bombarded heavily for hundreds of millions of years afterwards as stray asteroids struck their surfaces. The heat energy released as the planets formed would have made them partly or totally molten. Earth in particular was struck by a huge Mars-sized body late in its formation. That impact probably melted the entire Earth, while most of the debris collected close to Earth to form the Moon (Figure 1.4). If life originated on Earth, it likely evolved after this point.

All the inner planets melted deeply enough to have hot surfaces that gave off gases to form atmospheres. But

Figure 1.3 A new star forms in the constellation Centaurus. (a) A bright new star (left side of the image) with a dust cloud around it. *Source:* NASA/JPL-Caltech/ESO/ S. Kraus image. (b) Artist's impression of the new star. *Source:* NASA/JPL-Caltech/R. Hurt (SSC) image.

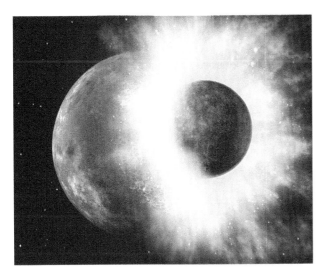

Figure 1.4 The early Earth was hit by a Mars-sized asteroid, and the debris that was blasted into space quickly collected to form the Moon. *Source:* NASA/JPL-Caltech image.

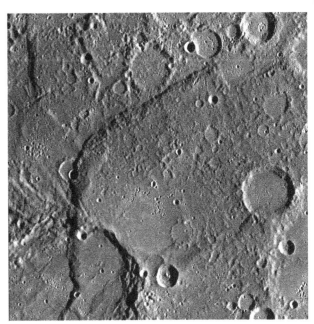

Figure 1.5 Image of the surface of Mercury showing scarps and craters. *Source:* NASA image.

Figure 1.6 Topographic map of the far side of the Moon: airless and lifeless. *Source:* USGS image.

there the similarity ended, and each inner planet has had its own later history.

Once a planet cools, conditions on its surface are largely controlled by its distance from the Sun and by any volcanic gases that erupt into its atmosphere from its interior. From this point onward, the geology of a planet greatly affects the chances that life might evolve on it.

Liquid water is vital for life as we know it, so surface temperature is perhaps the single most important feature of a young planet. Surface temperature is mainly determined by distance from the Sun: too far, and water freezes to ice; too close, and water evaporates to form water vapor.

But distance from the Sun is not the only factor that affects surface temperature. A planet with an early atmosphere that contained gases such as methane, carbon dioxide, and water vapor would trap solar radiation in the "greenhouse" effect, and would be warmer than an astronomer would predict just from its distance from the Sun.

In addition, distance from the Sun alone does not determine whether a planet has water, otherwise the Moon would have oceans like Earth's. The size of the planet is important, because gases escape into space from the weak gravitational field of a small planet. Gas molecules such as water vapor are lost faster from a small planet than from a larger one, and heavier gases as well as light ones are lost from a small planet. Thus, Mars has only a thin atmosphere, and Mercury (Figure 1.5) and the Moon (Figure 1.6) have practically none.

Gases may be absorbed out of an atmosphere if they react chemically with the surface rocks of the planet. As they do so, they become part of the planet's geology, but may be released again if those rocks are melted in volcanic activity. But a small planet cools faster than a large one, so any volcanic activity quickly stops as its interior freezes. After that, no more eruptions can return or add gases to the atmosphere. Therefore, a small planet quickly evolves to have a very thin atmosphere or no atmosphere at all, and no chance of gaining one.

Volcanoes typically erupt large amounts of water vapor and $CO_2$, and these are both powerful greenhouse gases. Earth would have been frozen for most of its history without volcanic $CO_2$ and water vapor in its atmosphere. Together, they add perhaps 33 °C to Earth's average temperature.

With these principles in mind, let us look at the prospects for life on other planets of our solar system. The brief story is that there is none. Both Mercury and the Moon had active volcanic eruptions early in their history, but they are small. They cooled quickly and are now solid throughout. Their atmospheric gases either escaped quickly to space from their weak gravitational fields or were blown off by major impacts. Today, Mercury and the Moon are airless and lifeless.

Venus is larger than the Moon or Mercury, almost the same size as Earth. Volcanic rocks cover most of its surface. Like Earth, Venus has had a long and active geological history, with a continuing supply of volcanic gases for its atmosphere, and it has a strong gravitational field that can hold most gases. But Venus is closer to the Sun than Earth is, and the larger amount of solar radiation hitting the planet was trapped so effectively by water vapor and $CO_2$ that water molecules may never have been able to condense to become liquid water. Instead, water remained as vapor in the atmosphere until most of it was dissociated, broken up into hydrogen ($H_2$), which was lost to space, and oxygen ($O_2$), which was taken up chemically by reacting with hot surface rocks (Figure 1.7).

Today, Venus has a dense, massive atmosphere made largely of $CO_2$. Volcanic gases react in the atmosphere to make tiny droplets of sulfuric acid ($H_2SO_4$), forming thick clouds that hide the planetary surface. Water vapor has vanished completely. Although the sulfuric acid clouds reflect 80% of solar radiation, $CO_2$ traps the rest, so the surface temperature is about 450 °C (850 °F). We can be sure that there is no life on the grim surface of Venus under its toxic clouds.

Mars is much more interesting than Venus from a biological point of view. It is smaller than Earth (Figure 1.8) and farther from the Sun but it is large enough to have held on to a thin atmosphere, mainly composed of $CO_2$. Mars today is cold, dry, and windswept; dust storms sometimes cover half the planet.

No organic material can survive now on the surface of Mars. There is almost no liquid water, and the soil is highly oxidizing. But while Mars was still young and was actively erupting volcanic gases from a hot interior, the planet may have had a thicker atmosphere with substantial amounts of water vapor. The crust still contains ice that could be set free as water if large impacts heated the

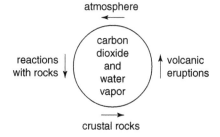

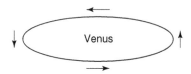

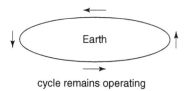

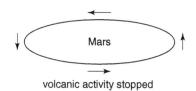

Figure 1.7 An idealized rocky planet, with surface reactions. Earth is like this, but Venus and Mars are not. This has made all the difference in their history. Mars is frozen and dead, Venus is hot and toxic. *Source:* See text for details.

surface rocks deeply enough to melt it, or if climatic changes were to melt it briefly.

So Mars does have water, but it is ice, frozen as part of the ice caps or under the surface sediment, where it is shielded from the sun. Ice can sublimate off the Martian surface, changing directly into water vapor. This blows around, sometimes being lost to space, sometimes freezing out again in the Martian winter.

Mars occasionally had surface water in the distant past. Canyons, channels, and plains look as if they were shaped by huge floods (Figure 1.9), and other features look like ancient sandbars, islands, and lake beds. Ancient craters on Mars, especially in the lowland plains, have been eroded by gullies, and sheets of sediment lap around and inside the old craters, sometimes reducing them to ghostly rims sticking out of the flat surface.

Mars was too small to sustain geological activity for long. As the little planet cooled, its volcanic activity

Figure 1.8 Earth and Mars at the same scale. North poles at left. *Source:* NASA/JPL-Caltech image.

Figure 1.10 Olympus Mons, an enormous but long-extinct volcano on Mars, standing 27 km (17 miles) higher than the average crust of Mars, and over 600 km (370 miles) across at the base. *Source:* NASA image.

stopped (Figures 1.7 and 1.10). Its atmosphere was largely lost, blasted off by impacts or by slow leakage to space, or by chemical reactions with the rocks and soil. There may never have been oceans, and even lakes would have lasted a very short time. The surface is a dry frozen waste, and likely has been for well over 3 billion years. Even floods generated by a large meteorite impact would drain away or evaporate very quickly – they could not have lasted long enough to sustain life. In short, Mars is a lifeless ice ball, and has been for billions of years.

In 1996, researchers reported they had found fossil bacteria in a meteorite that originated on Mars. (It was blasted into space by an asteroid impact and fell on to Earth's Antarctic ice cap after spending thousands of years in space.) The researchers suggested that the bacteria were Martian. However, the report has been discounted; the objects are not bacteria and they are not evidence for life on Mars.

The asteroid belt lies outside the orbit of Mars. Some asteroids have had a complex geological history, but there is no question of life in the asteroid belt now. Outside the asteroid belt, Jupiter and Saturn have ice-rich moons but no planet or moon outside the orbit of Mars could trap enough solar radiation to form liquid water on its surface to provide the basis for life. Complex hydrocarbon compounds can accumulate and survive on asteroids, or in the atmospheres of the outer planets or on some of their satellites, but those bodies are frigid and lifeless.

Looking further afield, there is absolutely no evidence of life anywhere else in the universe. Many scientists argue that the universe is so vast that there must be other life out there, but that is speculation, not science.

Figure 1.9 Ancient channels on the surface of Mars. *Source:* NASA image.

As we discover more planets around other stars, we find that many of them are in orbits that would make life impossible.

## The Early Earth

So we return to Earth as the only known site of life. Gases released by eruptions and impacts formed a thick atmosphere around the early Earth, consisting mainly of $CO_2$ but with small amounts of nitrogen, water vapor, and sulfur gases. By about 4.4 billion years ago (4400 Ma or 4.4 Ga), Earth's surface was cool enough to have a solid crust, and liquid water accumulated on it to form oceans. Ocean water in turn helped to dissolve $CO_2$ out of the atmosphere and deposit it into carbonate rocks on the sea floor. This absorbed so much $CO_2$ that Earth did not develop runaway greenhouse heating as Venus did (Figure 1.8). Large shallow oceans probably covered most of Earth, with a few crater rims and volcanoes sticking out as islands. The evidence for a cool watery Earth early in its history comes from a few zircon crystals that survived as recycled grains in later rocks. Some of the zircon crystals are dated close to 4.4 Ga.

We know from crater impacts and lunar samples that the Earth and Moon suffered a heavy late bombardment of asteroids around 3900 Ma, and the same event probably affected all the inner planets (Bottke and Norman 2017). Those catastrophic impacts must have destroyed almost all geological evidence of the early Earth's structure. Earth must have been hit by 100 or more giant asteroids and many smaller ones. At the same time, huge craters and basins filled with basalt lava were formed on the Moon (Figure 1.11). The incoming asteroids seem to have been dislodged from their original orbits by changes in the orbits of Jupiter and perhaps Saturn as well, as those giant planets went through final gravitational adjustments in the complex dynamics of the solar system.

As the great bombardment died away, small late impacts may have encouraged the evolution of life on Earth. All comets and a few meteorites carry organic molecules, and comets in particular are largely made of ice. These bodies could have delivered organic chemicals and water to Earth. But Earth already had water, and processes here on Earth also formed organic chemicals. Intense ultraviolet (UV) radiation from the young Sun acted on the atmosphere to form small amounts of very many gases. Most of these dissolved easily in water, and fell out in rain, making Earth's surface water rich in carbon compounds. The compounds included ammonia ($NH_3$), methane ($CH_4$), carbon monoxide (CO), ethane ($C_2H_6$), and formaldehyde ($CH_2O$). They could have formed at a rate of millions of tons a year. Nitrates built up in water as photochemical smog and nitric acid

Figure 1.11 The Late Heavy Bombardment hits the Moon (*top*), leaving scars that are still visible today (*bottom*). The effect on Earth would have been even greater because of Earth's greater mass. *Source:* Image by Tim Wetherell of the Australian National University (Wikimedia).

from lightning strikes also rained out. But the most important chemical of all may have been cyanide (HCN). It would have formed easily in the upper atmosphere from solar radiation and meteorite impact, then dissolved in raindrops. Today, it is broken down almost at once by oxygen, but early in Earth's history it built up at low concentrations in lakes and oceans. Cyanide is a basic building block for more complex organic molecules such as amino acids and nucleic acid bases. Life probably evolved in chemical conditions that would kill us instantly!

We have a good idea of the conditions of the early Earth, and of the many possible organic molecules that might have been present in its atmosphere and ocean. But how did that result in the evolution of life? First, we look at the biology and the laboratory experiments that help us to solve the question, and then we look at real-world environments to help us to work out where it happened.

## Life Exists in Cells

The simplest cell alive today is very complex; after all, its ancestors have evolved through many billions of generations. We must try to strip away these complexities as we wonder what the first living cell might have looked like and how it worked.

A living thing has several properties: it has organized structure and the capacity to reproduce (replicate itself) and to store information; and it has behavior and energy flow (metabolism). Mineral crystals have the first two but not the last two.

A living thing has a boundary that separates it from the environment. It operates its own chemical reactions, and if it did not have a boundary, those reactions would be unable to work; they would be diluted by outside water or compromised by outside contaminants. So a living "cell" has some sort of protective membrane around it.

A cell, like a computer, has hardware, software, and a protective case, all working well together. The case, or **cell membrane**, is made from molecules called **lipids**. The software that contains the information for running a cell is coded on **nucleic acids** (DNA and RNA), which use a four-character code rather than the two-character code (0 and 1) that our computers all use. The hardware consists largely of **proteins**, long molecules made from strings of **amino acids**. All those components had to become parts of a functioning organism.

A living thing can grow and it can **replicate**; that is, it can make another structure just like itself. Both processes require complex chemistry. Growth and replication use materials that must be brought in from outside, through the cell wall.

A living thing interacts with its environment in an active way: it has **behavior**. The simplest behavior is the chemical flow of substances in and out of the cell, which can be turned on and off. The chemical flow will change the immediate environment, and the presence or absence of the desired chemicals will decide whether the cell turns the flow on or off. Temperature and other outside conditions also affect the behavior of even the simplest cell.

The chemical activity of the cell includes an energy flow that is called **metabolism** in living things. The cell must make molecules from simpler precursors or break down complex molecules into simpler ones. If a cell grows or reproduces, it builds complex organic molecules, and those reactions need energy. The cell obtains that energy from outside, in the form of radiation or "food" molecules that it breaks down.

These attributes of a living cell are not different things; they are all intertwined, connected with gathering and processing energy and material into new chemical compounds (tissues) and continuing those processes into new cells. Any reconstruction of the evolution of life, as opposed to its creation by a Divine Being, must include a period of time during which lifeless molecules evolved the characteristics listed above and thereby became living. The phrase for this process is **chemical evolution**. We have to be able to argue that every step in the process could reasonably have happened on the early Earth in a natural, spontaneous way. It's easy to see that a protocell could grow effectively, given the right conditions. The critical turning point that defines life comes when relatively accurate replication evolves.

Even with a time machine, it would be very difficult to pick out the first living thing from the mass of growing organic blobs that must have surrounded it. But that cell survived and replicated accurately, and as time went by, its descendant cells that were more efficient remained alive and replicated, while those that were less efficient died or replicated more slowly. So as living things slowly emerged, chemical evolution slowly changed into *biological* evolution as we understand it today, subject to natural selection and extinction. Some lines of cells flourished, others became extinct. So living cells today do not exactly have the same genetic and biochemical machinery their ancestors had; they have long had major upgrades of their original software.

That brings one other concept into our discussion: *improvement* or *progress*. There is no question that the simplest living cells today are more efficient than their distant ancestors. Arguments rage about the politically correct word to use to describe this. The fossil record shows many examples of improved performance that can be analyzed mechanically. Living horses and living humans run far more efficiently, living whales swim more efficiently, and living birds fly more efficiently than their ancestors did. No doubt similar trends have occurred in physiology, biochemistry, reproduction, and so on.

But progress is not constant or one-way, because environments on Earth change all the time and so evolution may follow zig-zag routes. Also, of course, there is no evidence that life has a future aim, such as the production of *Tyrannosaurus rex* or humans; remember, the Earth is still dominated by Bacteria and Archaea like the very earliest living things. The history of life has been progressive, although in the addition of new forms and the long-term increase in maximum size, from microscopic to truly huge, such as dinosaurs, whales, and giant redwoods.

We turn now to experiments that help us to see how life evolved from nonliving chemicals. The only life we know is on Earth, so we are testing the hypothesis that ingredients for the first cells were available on Earth, and that the first cells could have evolved along reasonable pathways.

The first stages in reconstructing the evolution of life were experiments in making the different necessary chemical components in plausible conditions. Now, with success in that first stage, research has moved on to determine how the components were successfully assembled into working units, getting closer to objects we might call "protocells."

## Making Organic Molecules

In 1953 Stanley Miller, a young graduate student at the University of Chicago, passed energy (electric sparks) through a mixture of hydrogen, ammonia, and methane in an attempt to simulate likely conditions on the early Earth (Figure 1.12). Any chemical products fell out into a protected flask. Among these products, which included cyanide and formaldehyde, were amino acids. This result was surprising at the time because amino acids are complex compounds and are also vital components of all living cells.

The experiment that Miller published used a rather unlikely mixture of starting gases, but he also did a number of other experiments that gave similar results. Some were not published at the time, but Miller stored all his lab notes and experimental vials. When they were discovered after his death and analyzed with twenty-first century techniques, it turns out that the best results came when Miller added volcanic gases to his mixtures (McCollom 2013).

It is now clear that almost all the amino acids found in living cells today could have formed naturally on the early Earth, from a wide range of ingredients, over a wide range of conditions. They form readily from mixtures that include the gases of Earth's early atmosphere. The same amino acids that form most easily in laboratory experiments are also the most common in living cells today. The only major condition is that amino acids do not form if oxygen is present.

Miller's experiments made amino acids in sterile glass flasks but in later experiments, it was found that amino acids form even more easily on the surfaces of clay particles. Clay minerals are abundant in nature, have a long linear crystal structure, and are very good at attracting and adsorbing organic substances; cat litter is made from a natural clay and works on this principle.

People used to talk about "primordial soup," with the idea that interesting organic molecules would have been present throughout Earth's oceans. Everyone recognizes now that for the later stages of complex organic chemistry, organic molecules need to be concentrated, which allows them to react faster and more efficiently. Life may have begun in a rather unusual local environment.

For example, linking sequences of amino acid molecules into chains to form protein-like molecules involves the loss of water, so scientists have tried evaporation experiments in simulated early Earth conditions. Four natural concentration mechanisms are evaporation, freezing, being enclosed inside membranes in scums, droplets, or bubbles, and concentration by being absorbed onto the surfaces of mineral grains. High temperatures help evaporation but organic molecules tend to break down if they are heated too much. The longer the molecule, the more vulnerable it is to heat damage. However, experiments at

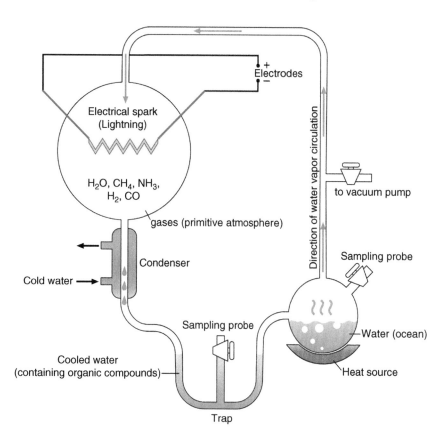

Figure 1.12 Stanley Miller's classic 1953 experiment, designed to simulate conditions on the early Earth. An atmosphere largely of water vapor, methane, and ammonia was subjected to lightning discharges. The reaction products cooled, condensed, and rained out to collect in the ocean. Those reaction products included amino acids. *Source:* Diagram by Yassine Mrabet (Wikimedia).

low temperature form large molecules rather well. As water freezes into ice, other chemicals present are greatly concentrated. If they react to form larger organic molecules, the new molecules survive well.

Nucleic acids (RNA and DNA) have structures made of nucleic acid bases, or **nucleobases**, sugars, and phosphates. All the nucleobases have now been made in reasonable laboratory experiments. Sugars form in experiments that simulate water flow from hot springs over clay beds. Sugars and nucleobases could have formed in reactions powered by lightning. Naturally occurring phosphate minerals are associated with volcanic activity. Thus, all the ingredients for nucleic acids were present on the early Earth, and the cell fuel adenosine triphosphate (ATP) could also have formed easily.

Linking sugars, phosphates, and nucleobases to form fragments of nucleic acid called **nucleotides** also involves the loss of water molecules, and the phosphates themselves can act as catalysts here. Long nucleotides form much more easily on phosphate or clay surfaces than they do in suspension in water.

Many organic membranes are made of sheets of molecules called lipids. A lipid molecule has one end that attracts water and one end that repels water. Lipid molecules line up naturally with heads and tails always facing in opposite directions (Figure 1.13); a bilayer sheet of lipid molecules therefore repels water. If a single or double sheet of lipids happens to fold around to meet itself, it forms globular waterproof membranes (micelles) or hollow pills (liposomes or **vesicles**). Such shapes form spontaneously in lipid mixtures. Whipping up an egg in the kitchen produces lipid globules as the contents are frothed around. In the real world, lipid foams can form in the scum on wave surfaces (Figure 1.14).

A breakthrough came when David Deamer's research group found that fatty acid molecules occur in the Murchison meteorite (Figure 1.15), which fell in Australia in 1969. Those fatty acids could be extracted and formed into lipid vesicles by drying them out and then rewetting them (Figure 1.16). Vesicles can also form from mixtures of molecules that would have been present on the early Earth. Deamer shook mixtures of lipids, amino acids, and nucleic acids, and found that they formed spontaneously into many vesicles with organic molecules trapped inside them. They became tiny reaction chambers, inside which complex chemical changes could and did happen.

Nature has done experiments on making organic molecules. The meteorites and comets that strike Earth often carry organic compounds, and we can analyze them knowing that they formed somewhere in space. The most common organic compounds in meteorites are also the most abundant in experiments that try to simulate chemistry in space. Many forms of amino acids, sugars, and nucleobases are found in meteorites, and so are fatty

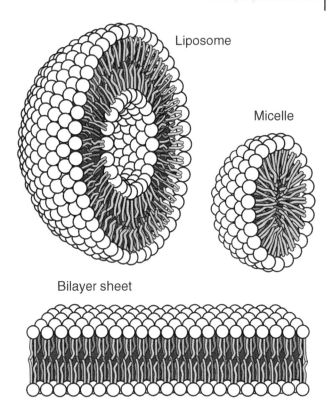

**Figure 1.13** The different shapes that lipid layers can form. Liposomes are also called vesicles. Vesicles can enclose mixtures of chemicals in a central cavity, and are very important in origin-of-life experiments. *Source:* Image by Lady of Hats, Mariana Ruiz Villarreal (Wikimedia).

**Figure 1.14** Sea foam, formed by waves on a South Australian beach. The dog is for scale. *Source:* Photo taken by Bahudhara (Wikimedia).

acids that easily form lipid membranes. Thousands of different organic compounds could have been supplied to the early Earth (Schmitt-Kopplin et al. 2010). We do not know how much organic matter was formed in natural processes on Earth and how much was delivered on comets and meteorites before and after the Late

Figure 1.15 A fragment of the Murchison meteorite yielded fatty acids that readily form into vesicles. *Source:* Image from the US Department of Energy.

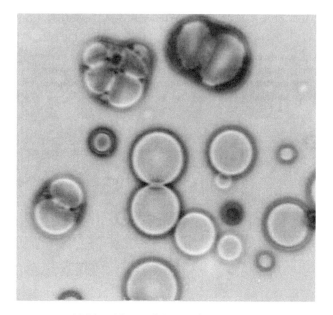

Figure 1.16 Lipid vesicles made in David Deamer's laboratory from fatty acids extracted from the Murchison meteorite. *Source:* NASA image.

Bombardment. Either way, the right materials were present on the early Earth to encourage further reactions.

## Toward the First Living Cell

How did basic organic molecules evolve into a cell that could reproduce itself? Deamer's early experiments began a new style of prebiotic experiments, using vesicles rather than test tubes. After all, vesicles with cell-like contents could have formed in great numbers as waves thrashed around lipids on water surfaces (Figure 1.14), or as lipid scums washed up on a muddy shore with clays in the water, or in the turbulent convection in and around hot springs. These vesicles would have had very variable contents (some with amino acids, primitive forms of nucleic acid, and so on). The "best" ones would have operated chemical reactions much more efficiently than the "worst." They would have done this because they had "better" nucleic acids, coded to produce "better" sets of protein enzymes to run efficient reactions.

Researchers have now found that vesicles can form 100 times as fast as usual if clay is added to the experimental mixtures. Some vesicles can take in substances from outside, through the lipid walls, and use them to build new walls and new contents; that is, they can grow. Irene Chen found that an active vesicle can "steal" (attract and absorb) part of the membrane from a less active neighbor and use it to grow! Vesicles can display a kind of "reproduction" in the sense that a large vesicle may divide into two, each keeping some of the original vesicle contents (Pressman et al. 2015; Saha and Chen 2015).

So we can imagine some watery environment where vesicles were growing and dividing more and more efficiently as their nucleic acids, their proteins, and their vesicle walls came to work well together.

In living cells today, information for making proteins is coded on long sequences of nucleic acid. The molecules of DNA that specify these protein structures are difficult to replicate, and replication requires many proteins to act as enzymes to catalyze the reactions. In living cells today, protein synthesis and DNA replication are interwoven: they depend on one another. So how could DNA and proteins have been formed independently, then evolved to depend on each other?

The answer lies with the simpler nucleic acid, RNA. Some RNA sequences called **ribozymes** can act as enzymes and make more RNA, even when no proteins are present. Other RNA sequences speed up the assembly of proteins. Perhaps the first living things were efficient vesicles that contained ribozymes with the right structure to replicate themselves accurately. Ribozymes would also have coded for the proteins needed to grow the vesicle and divide. In theory, RNA ribozymes on the early Earth could have replicated themselves with minimal proteins, in vesicles that we can now call **protocells**. Increasingly successful protocells would very quickly have outcompeted their neighbors. At some point, a successful protocell became the ancestor of all later life on Earth. The scenario that begins with ribozymes in an RNA world (Higgs and Lehman 2015; Pressman et al. 2015) is currently the best hypothesis for the origin of life on Earth.

## Where Did Life Evolve?

Most theories of the origin of life suggest surface or shoreline habitats in lakes, lagoons, or oceans. But it's unlikely that life evolved in the open sea. Complex organic molecules are vulnerable to damage from the sodium and chlorine in seawater. Most likely, life evolved in lakes or in seashore lagoons that were well supplied with river water. We have come to think of lagoons as tropical; the very name conjures up blue water and palm trees. Warm temperatures promote chemical reactions, and an early tropical island would most likely have been volcanic and therefore likely to have interesting minerals. But RNA bases are increasingly unstable as temperatures rise; normal tropical water, at 25 °C, is about as warm at it could be for the origin of life.

So perhaps lakes or lagoons on cold volcanic islands were the best environments favoring organic reactions on the early Earth. In the laboratory, cyanide and formaldehyde reactions occur readily in half-frozen mixtures. Volcanic eruptions often generate lightning storms (Figure 1.17), so eruptions, lightning, fresh clays, and near-freezing temperatures (ice, snow, hailstones) could all have been present on the shore of a cold volcanic island (Figure 1.18). Note that if this environment is the correct one, there had to have been land and sea when life evolved; fresh water can only occur on Earth if it is physically separated from the ocean.

Figure 1.18 A volcanic island set in a cold climate: Onekotan, in the Kurile Islands on the Russian east coast. The southern volcano, on an island in a large crater, is Krenitzyn Peak. *Source:* Image from NASA Earth Observatory.

Solar radiation or lightning are likely energy sources for the reactions leading toward life. But deep in the oceans are places where intense geothermal heating generates hot springs on the sea floor. Most of these lie on the midocean ridges, long underwater rifts where the sea floor is tearing apart and forming new oceanic crust. Enormous quantities of heat are released in the process, much of it through hot water vents, and myriad bacteria flourish in the hot water. Perhaps life began nowhere near the ocean surface but deep below it, at these **hydrothermal vents** (Figure 1.19).

Laboratory experiments have implied that amino acids and other important molecules can form in such conditions, even linking into short protein-like molecules, and currently the deep-sea hypothesis is popular. But if life evolved by way of **naked genes** (genes that are not enclosed in a cell), then it did not do so in hot springs. RNA and DNA are unstable at such high temperatures. Naked genes could not have existed (for long enough) in hot springs.

If black smokers (the narrow rocky tubes that release sulfur and other chemicals as steam deep in the midocean ridge systems) are too hot, there are also cool, alkaline hydrothermal vents that might be more plausible sites for the origin of life (Lane 2016). Today, anaerobic reactions are reported from alkaline hydrothermal vents in the ocean floor. Natural proton gradients are observed

Figure 1.17 Volcanic lightning in an eruption cloud, at Rinjani volcano in Indonesia, 1995. *Source:* Photograph by Oliver Spalt (Wikimedia).

Figure 1.19 Hydrothermal vents on the Pacific Ocean floor. *Source:* Image from NOAA.

associated with these vents, and this provides a possible means to harness energy; an energy-releasing reaction is coupled to the pumping of protons (positive ions) across a membrane from the inside of the cell to the outside.

The deep-sea hypothesis has led to speculation that life might have evolved deep under the surface layers of other planets or satellites. (For example, Jupiter's moon Europa probably has liquid water under its icy crust, and Saturn's moon Enceladus has been seen to erupt water vapor "geysers.") The speculation helps to generate money for NASA's planetary probes. But the internal energy of such planets and moons is very low, and water-borne organic reactions are much less likely to work deep under the icy crust of Europa or Enceladus than in Earth's oceans. In any case, the under-ice oceans of icy moons are salty (that's how they were detected), so an origin of life is very unlikely in such environments.

## Energy Sources for the First Life

Living things use energy. Much of biology consists of studying metabolism and ecology: how living things acquire and use the energy they need to grow and reproduce.

Living organisms take in outside energy in two ways: **autotrophy** and **heterotrophy**. Autotrophs make their own food molecules, paying the cost of building them by harnessing energy from outside. Heterotrophs obtain their metabolic energy by breaking down organic molecules they obtain from the environment: hummingbirds sip nectar and humans eat doughnuts. Heterotrophs do not pay the cost of building the organic molecules. They simply have to operate the reactions that break them down. But they must live where they can find "food" molecules.

So, were the first lifeforms autotrophs or heterotrophs? On the early Earth, the most abundant source of carbon for building organic molecules was $CO_2$ gas, and complex, digestible organic molecules to support heterotrophy were in short supply. So, while we do not have a definitive answer to this question, the earliest lifeforms were probably autotrophs. Heterotrophs came later, once a primary ecology based on autotrophy had evolved to support them.

Modern autotrophs generate their own energy in two completely different ways. Some extract chemical energy from inorganic molecules (**lithotrophy**), while others gain energy by trapping solar radiation (**photosynthesis**). Lithotrophy was likely the first autotrophic metabolism. The most ancient biochemical pathway for lithotrophy is the acetyl-CoA pathway, which both fixes carbon and captures energy via the reduction of $CO_2$ with electrons from $H_2$ – both of which were abundant gases in the early oceans (Weiss et al. 2018).

Today, this pathway is still used for carbon fixation and energy generation by a diverse range of microorganisms from both prokaryotic domains of life, the Archaea and Bacteria (see Chapter 2). For example, Archaea called **methanogens** gain energy from lithotrophy by breaking up carbon dioxide and transferring the oxygen to hydrogen, forming water and methane as by-products:

$$4H_2 + CO_2 \rightarrow CH_4 + 2H_2O + energy$$

If lithotrophy evolved very early, it may have been the first time (but not the last) that living things modified Earth's chemistry and climate. By replacing the greenhouse gas carbon dioxide with the even more powerful greenhouse gas methane, the activity of methanogens might have warmed the early Earth (see Chapter 2).

### Heterotrophy

The simplest reaction used by cells to break down organic molecules is **fermentation**, to break down sugars such as glucose. This is what early heterotrophs must have done. Glucose is often called the universal cellular fuel for living organisms, and it was probably the most abundant sugar available on the early Earth. (Today, humans use fermenting microorganisms to produce beer, cheese, vinegar, wine, tea, and yogurt, and to break down much of our sewage.)

As heterotrophs used up the molecules that were easiest to break down, there would have been intense competition among them to break down more complex ones. One can imagine a huge advantage for cells that evolved enzymes to break down molecules that their competitors could not use (remember that, for them, fermentation is feeding). New sets of fermentation reactions would quickly have evolved, and different lineages of heterotrophic cells would have come to be specialists in their chemistry.

In becoming more efficient heterotrophs, some early cells found a way to import energy to make their internal chemistry run faster at no extra cost. In recent years, microbiologists have found that billions of heterotrophic microbes living in the world's shallow waters, in seas and lakes, and even in the ice around Antarctica can absorb light energy and use it to help their internal chemical reactions. The molecules that can absorb light in this fashion are called **rhodopsins**.

We and many other creatures now use rhodopsins in our eye cells as light sensors. Light hitting a rhodopsin molecule activates it, and after a cascade of reactions, a nerve impulse is sent to the brain. Rhodopsins are the universal molecules in biological visual systems, allowing bacteria and fungi as well as humans to detect and react to light.

But the first rhodopsin molecules probably did something else. Rhodopsin is triggered by light to add electric charges to protons, and those protons can then be taken off to power chemical reactions inside the cell. Light-powered chemistry thus gives an advantage to rhodopsin-bearing heterotrophs over their competitors. Much of the biology in the ocean's surface waters is powered by rhodopsin reactions, and we knew nothing about them until recently! This system is called **phototrophy** ("feeding by light") because the rhodopsin reactions help to break down molecules, but do not build them up. Rhodopsin reactions aid heterotrophs, not autotrophs.

The first rhodopsin systems probably evolved only once, in some lucky mutant cell. The genes that code for rhodopsin are not large, and they seem to have passed easily from one cell to another, so that now, after billions of years, many different lineages of heterotrophic cells use rhodopsin to save energy. Of course, rhodopsin is useful only in water that is shallow enough to receive sunlight. Heterotrophs living in dark environments must run at lower energy levels.

## Photosynthesis

Photosynthesis is simple in concept: energy from light is absorbed into specific molecules called **chlorophylls**. The process is biochemically more complex than lithotrophy or phototrophy. Chlorophylls (and the genes that code for making them) seem to have evolved only once.

The evolution of photosynthesis produced major ecological changes on Earth. Light energy trapped by chlorophyll was used to build more *biomass* (biological substance), giving photosynthetic cells an energy store, a buffer against times of low food supply, that could be used when needed. It's easy to see how such cells could come to depend almost entirely on photosynthesis for energy. In doing so, they did not have to compete directly with heterotrophs. In addition, as photosynthesizers died and their cell contents were released into the environment, they inadvertently provided a dramatic new source of nutrition for heterotrophs. Photosynthesis greatly increased the energy flow in Earth's biological systems, and for the first time considerable amounts of energy were being transferred from organism to organism, in Earth's first true ecosystem.

The earliest photosynthetic cells probably used hydrogen from $H_2$ or $H_2S$. For example, the reaction:

$$H_2S + CO_2 + light \rightarrow (CH_2O) + 2S$$

released sulfur into the environment as a by-product of photosynthesis. Later, photosynthetic bacteria began to break up the strong hydrogen bonds of the water molecule. Bacteria that successfully broke down $H_2O$ rather than $H_2S$, like this:

$$2H_2O + CO_2 + light \rightarrow (CH_2O) + 2O$$

immediately gained access to a much more plentiful resource. There was a penalty, however. The waste product of $H_2S$ photosynthesis is sulfur (S), which is easily disposed of. The waste product of $H_2O$ photosynthesis is an oxygen radical, monatomic oxygen (O), which is a deadly poison to a cell because it can break down vital organic molecules by oxidizing them. Even for humans, it is dangerous to breathe pure oxygen or ozone-polluted air for long periods.

Cells needed a natural antidote to this oxygen poison before they could operate the new photosynthesis consistently inside their cells. **Cyanobacteria** were the organisms that made the first breakthrough to oxygen photosynthesis using water. A lucky mutation allowed them to make a powerful antioxidant enzyme called **superoxide dismutase** to prevent O from damaging them; essentially, the enzyme packaged up the O into less dangerous $O_2$ that was ejected out of the cell wall into the environment.

From then on, we can imagine early communities of microorganisms made up of autotrophs and heterotrophs, each group evolving improved ways of gathering or making food molecules.

Photosynthesizers need nutrients such as phosphorus and nitrogen to build up their cells, as well as light and $CO_2$. In most habitats, the nutrient supply varies with the seasons, as winds and currents change during the year. Light, too, varies with the seasons, especially in high latitudes. Since light is required for photosynthesis, great seasonal fluctuations in the primary productivity of the natural world began with photosynthesis. Seasonal cycles still dominate our modern world, among wild creatures and in agriculture and fisheries.

We can now envisage a world with a considerable biological energy budget and large populations of microorganisms: Archaea, photosynthetic Bacteria, and heterotrophic Bacteria. So there is at least a chance that a paleontologist might find evidence of very early life as fossils in the rock record. In Chapter 2, we will look at geology, rocks, and fossils, instead of relying on reasonable but speculative arguments about Earth's early history and life.

## References

Bottke, W.F. and Norman, M. (2017). The late heavy bombardment. *Annual Review of Earth and Planetary Sciences* 45: 619–647.

Ciesla, F.J. and Sandford, S.A. (2012). Organic synthesis via irradiation and warming of ice grains in the solar nebula. *Science* 336: 452–454.

Higgs, P.G. and Lehman, N. (2015). The RNA world: molecular cooperation at the origins of life. *Nature Reviews Genetics* 16: 7–17.

Lane, N. (2016). *The Vital Question: Energy, Evolution and the Origins of Complex Life*. New York: W.W. Norton & Co.

Lin, D.N.C. (2008). The genesis of planets. *Scientific American* 298 (5): 50–59.

McCollom, T.M. (2013). Miller-Urey and beyond: what have we learned about prebiotic organic synthesis reactions in the past 60 years? *Annual Review of Earth and Planetary Sciences* 41: 207–229.

Pressman, A., Blanco, C., and Chen, I.A. (2015). The RNA world as a model system to study the origin of life. *Current Biology* 25: R953–R963.

Saha, R. and Chen, I.A. (2015). Origin of life: protocells red in tooth and claw. *Current Biology* 25: R1175–R1177.

Schmitt-Kopplin, P., Gabelica, Z., Gougeon, R.D. et al. (2010). High molecular diversity of extraterrestrial organic matter in Murchison meteorite revealed 40 years after its fall. *Proceedings of the National Academy of Sciences of the United States of America* 107: 2763–2768.

Smith, H.A. (2011). Alone in the universe. *American Scientist* 99: 320–328.

Weiss, M.C., M, P., Xavier, J.C. et al. (2018). The last universal common ancestor between ancient earth chemistry and the onset of genetics. *PLoS Genetics* 14 (8): e1007518.

## Further Reading

Carroll, S. (2017). *The Big Picture: On the Origins of Life, Meaning, and the Universe Itself*. London: Oneworld Publications.

Deamer, D. (2011). *First Life: Discovering the Connections Between Stars, Cells, and How Life Formed*. Berkeley: University of California Press.

Knoll, A.H. (2015). *Life on a Young Planet: The First Three Billion Years of Evolution on Earth*, updated edition. Princeton: Princeton University Press.

Ricardo, A. and Szostak, J. (2009). Life on earth. *Scientific American* 301 (3): 54–61.

Szostak, J.W. (2017). The origin of life on Earth and the design of alternative life forms. *Molecular Frontiers Journal* 1: 121–131.

## Questions for Thought, Study, and Discussion

1 It is clear that after Earth had cooled, comets and meteorites added important ingredients to its surface: ice (= water) and a great variety of organic molecules. Many scientists think that this "late accretion" gave Earth the ingredients for the formation of life. However, the same ingredients must have been added to Mars and Venus and the Moon also, with no sign that they ever evolved life. So why did Earth evolve life while the others did not?

2 Many movies have portrayed extinct animals. Suppose I said to you that none of the portrayals were scientific. Give a careful response to this assertion.

3 Where on Earth did life first evolve? When you decide where it was, give a careful summary of the evidence that helped you to come to your answer.

2

## The Earliest Life on Earth

**In This Chapter**

We turn now to geological, paleontological, and phylogenetic evidence for Earth's early life. First, we explain what fossils are and how we can find out how old they are. Since organisms run chemical reactions, they change Earth's chemistry, particularly in the ocean surface, on land, and in the atmosphere, and as they do so, they leave clues about ancient life processes as chemical traces in ancient rocks. Ancient rocks may carry subtle chemical markers of ancient life but in very special circumstances, they can carry traces of ancient cells. As life expanded, its chemical influence on Earth's processes widened, and we see ancient iron-bearing rocks that mark a transition from an anoxic atmosphere to the oxygen-bearing atmosphere we breathe today.

## Introduction

When we move from astronomy and the laboratory to the Earth itself to search for evidence about early life, we look for **fossils**. A fossil is the remnant of an organism preserved in the geological record. There are three kinds of fossils: body fossils, trace fossils, and chemical fossils. We are most familiar with body fossils, in which part or all of an organism is preserved. If an organism had body parts that were made of resistant materials, such as shells, bones, or wood, it is much more likely than a "soft-bodied" creature to be preserved in the geological record. Such fossils may look more or less unchanged after death. Minerals may crystallize out of ground water to fill up large or small cracks, crevices, and cavities in the original substance, so body fossils may be denser and harder than they were in life. Sometimes, the original shell or bone may be replaced by another mineral, making the fossil easier to recognize or to extract from the rock (Figure 2.1).

Obviously, the hard parts of an organism are far more likely to be preserved than more fragile parts. But occasionally, soft parts may leave an impression on soft sediment before they rot. Even more rarely, a complete organism may be encased in soft sediment that later hardens into a rock. Bees, ants, flies, frogs, and even a dinosaur tail have been preserved as fossils in amber (fossilized tree resin) (Figure 2.2a), and individual cells have been preserved in chert, a rock formed from silica gel that impregnated the cells and retained their shapes in three dimensions.

A **trace fossil** is not part of an organism at all, but was made by an organism and therefore may tell us something about that creature. Trace fossils may be marks left by active organisms (footprints, trails, or burrows; Figure 2.2b) or fecal masses (Figure 2.3), or even a spider web. Trace fossils may give us insight into behavior that would not be available from a body fossil. For example, although dinosaur skeletons suggest that they could have run, trace fossils of dinosaur footprints tell us that they certainly did run (see Chapter 13).

**Chemical fossils** are compounds produced by organisms and preserved in the rock record. They may be molecules that were originally part of the organism or that were produced in the metabolic processes the organism operated. They may provide information about the organisms that produced them. In special cases, where an organism absorbs one isotope of an atom over another in its food or water, the chemical fossils of these isotopes can be used to gain information too, as described later in this chapter.

All kinds of agents may destroy or damage organisms beyond recognition before they can become fossils or while they are fossils. After death, the soft parts of

*Cowen's History of Life*, Sixth Edition. Edited by Michael J. Benton.
© 2020 John Wiley & Sons Ltd. Published 2020 by John Wiley & Sons Ltd.

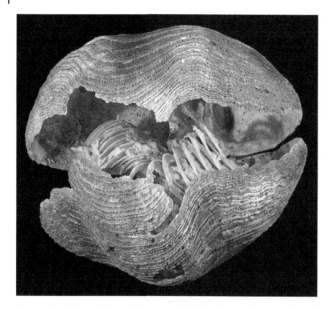

Figure 2.3 A trace fossil: a coprolite or fossil dung ball from a carnivorous dinosaur. This is a trace fossil because it was not part of the original organism but is evidence that it once existed (during the late Cretaceous in Saskatchewan, Canada). Scale is 15 cm (about 6 in.). *Source:* Image from the United States Geological Survey.

Figure 2.1 A brachiopod whose original calcite shell was replaced by silica. This made it fairly easy to dissolve the shell out of rock for study. This is the brachiopod *Spiriferina*, from the early Jurassic of France. In life, the spiral structure supported soft tissue that filtered sea water for plankton and oxygen. *Source:* Photograph by Didier Descouens (Wikimedia).

organisms may rot or be eaten. Any hard parts may be dissolved by water or broken or crushed and scattered by scavengers or by storms, floods, wind, and frost. Remains must be buried to become part of a rock, but a fossil may be cracked or crushed as it is buried. After burial, groundwater seeping through the sediment may dissolve bones and shells. Earth movements may smear or crush the fossils beyond recognition or may heat them too much. Even if a fossil survives and is eventually

exposed at the Earth's surface, it is very unlikely to be found and collected before it is destroyed by weathering and erosion.

Even when they are studied carefully, fossils are a very biased sample of ancient life. Fossils are much more likely to be preserved on the sea floor than on land. Even on land, animals and plants living or dying by a river or lake are more likely to be preserved than those in mountains or deserts. Different parts of a single skeleton have different chances of being preserved. Animal teeth, for example, are much more common in the fossil record than are tail bones and toe bones. Teeth are usually the only part of sharks to be fossilized. Large fossils are usually tougher than small ones and are more easily seen in the rock. Spectacular fossils are much more likely to be collected than apparently ordinary ones. Even if a fossil is

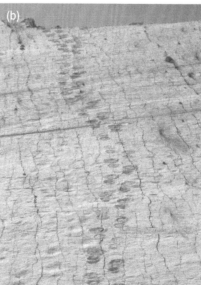

Figure 2.2 (a) A fossil ant, preserved in the famous amber found on the shores of the Baltic Sea. *Source:* Image by Anders Damgaard (Wikimedia). See www.amber-inclusions.dk for more of his images. (b) The trackway of a dinosaur with very big feet, preserved on a tilted rock face in Bolivia. *Source:* Image by Jerry Daykin (Wikimedia).

collected by a professional paleontologist and sent to an expert for examination, it may never be studied. All the major museums in the world have crates of fossils lying unopened in the basement or the attic.

When we look at museum display cases, it seems that we have a good idea of the history of life but most of the creatures that were living at any time are not in a museum. They were microscopic or soft-bodied, or both, or they were rare or fragile and were not preserved, or they have not been discovered. We do have enough evidence to begin to put together a story but that story is always changing as we discover new fossils and look more closely at the fossils we have found already.

## How to Find the Age of a Fossil

Fossils are found in rocks, and usually geologists try to establish the age of the containing rock or a layer of rock that is not far under or over the fossil (so might be close to it in age). The age of rocks is measured in two different ways, known as relative and absolute dating.

Age dating of rocks can only work if one identifies components of the rocks that change with time or are in some way characteristic of the time at which the rocks formed. The same principles are used in dating archaeological objects. Coins may bear a date in years (**absolute dating**), and one can be certain that a piece of jewelry containing a gold coin could not have been made before the date stamped on the coin. The age of waste dumps can be gauged by the type of container thrown into them: bottles with various shapes and tops, steel cans, aluminum cans, and so on.

Absolute geological ages can be determined because newly formed mineral crystals sometimes contain unstable, radioactive atoms. Radioactive isotopes break down at a rate that no known physical or chemical agent can alter (Figure 2.4), and as they do so, they may change into other elements. For example, potassium-40, $^{40}$K, breaks down to form argon-40, $^{40}$Ar. By measuring the amount of radioactive decay in a mineral crystal, one can calculate the time since it was newly formed, just as one reads the date from a coin. The principle is simple, although the techniques are often laborious. For example, $^{40}$K breaks down to form $^{40}$Ar at a rate such that half of it has gone in about 1.3 billion years (Figure 2.4). If we measure the $^{40}$Ar in a potassium feldspar crystal today and find that half the original amount of $^{40}$K has gone, then the age of the crystal is 1300 Ma. Other dating methods use this same principle. (By convention, absolute ages in millions of years are given in megayears [Ma], while time periods or intervals are expressed in millions of years [m.y.].) Ages in billions of years are gigayears [Ga].)

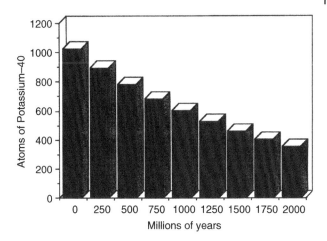

Figure 2.4 The radioactive decay of an isotope proceeds on a logarithmic time-table that is constant under all known conditions. If the decay is recorded in a rock or mineral, we can infer the date when the decay began. Often, but not always, that tells us the age of the rock. This graph shows the atoms of potassium-40 remaining in a crystal on a time scale measured in millions of years, compared to a starting value of 1024 atoms at time zero. *Source:* For more details, see Hazen (2010).

Absolute dating must be done carefully. Crystals may have been reheated or even recrystallized, resetting their radioactive clocks back to zero well after the time the rock originally formed. Chemical alteration of the rock may have removed some of the newly produced element, also giving a date younger than the true age. Geologists are familiar with these problems, and go to immense trouble to find and use fresh clean crystals.

Most elements used for radioactive age dating are not used by animals to build shells or bones, so usually we cannot date fossils directly. Instead, we have to measure the age of a lava flow or volcanic ash layer as close to the fossil-bearing bed as possible (Figure 2.5), which does contain crystal we can use.

Paleontologists more often deal with a **relative time scale**, in which one says "Fossil A is older than Fossil B" (as in Figure 2.5) without specifying the age in absolute years. This is much the same way that archaeologists date Egyptian artifacts. We know which pharaoh followed which, although we do not know the calendar years for some earlier dynasties. So Egyptian history is scaled according to the reigns of individual pharaohs, rather than recorded in absolute years. We can work this way with fossils, because it is a fact of observation that fossils preserved in the rock record at particular times are almost always different from those preserved at other times. These principles have been firmly established over the past two centuries by geologists working in rock sequences to define successive layers, each layer lying on and thus being younger than the one underneath.

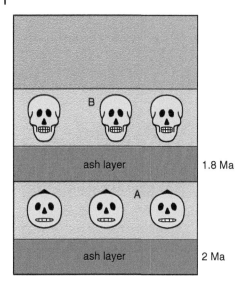

Figure 2.5 The skulls of these two fossil hominids do not contain any radioactive isotopes, but they lie close to two layers of volcanic ash that do. By using relative dating methods, one can say that hominid A is older than hominid B. Using absolute dating methods, the age of hominid A can be fixed closely between 1.8 Ma and 2.0 Ma because there are dated ash layers above and below it. All we know about the age of hominid B from this situation is that it is younger than 1.8 Ma.

With the occasional check from absolute methods, the geological record has been arranged into a standard sequence: the **geological time scale** (Figure 2.6). The time scale is divided into a hierarchy of units for easy reference, with the divisions between major units often corresponding to important changes in the history of life on Earth. The names of the eras and periods are often unfamiliar and have bizarre historical roots. For example, the Permian period was named when British geologist Sir Roderick Murchison and French paleontologist Edouard de Verneuil took a stagecoach tour of Russia in 1841 and discovered unfamiliar new rocks near the city of Perm.

Murchison's dream in 1841 was to show that these major divisions of geological time would apply not just in Britain or Russia but worldwide, and he was right. The time scale (Figure 2.6) is now the formal, international reference that is constantly refined and sharpened, and applies to rocks on every continent, whether laid down in the sea or on land.

## Life Alters a Planet

For too long, paleontologists thought of life as a set of passengers on a planet that had a certain geology, chemistry, and climate. Evolution took place as organisms interacted with each other or as a response to the physical environment. But we know now that biological processes dramatically affect the physical Earth, in a mutual interaction that has complex patterns. Scientists can no longer study any component of the Earth system on its own because the interplay is so important. This may make life difficult for Earth scientists but we do our humble best.

For example, the fact that Earth's atmosphere today has 21% oxygen reflects the continuous production of oxygen by photosynthesizers on land and in surface waters. Without life, oxygen cannot be present at more than a few parts per million. Chemically, 21% oxygen provides enough $O_2$ to form an ozone ($O_3$) layer in the high atmosphere, helping to shield the surface and its life from ultraviolet (UV) radiation. Physically, 21% oxygen affects the chemistry of seawater (iron will not dissolve in it, for example), and it affects the reactions by which rocks break down on the surface and turn into sediment. The oxygen is extracted from $CO_2$, which reduces the concentration of that greenhouse gas in the atmosphere and ocean, and cools the climate. There is nothing magic about a level of 21% oxygen; that level, and Earth's surface temperature, has fluctuated (moderately) for hundreds of millions of years.

In another example, methane is a greenhouse gas, so at times when methanogens were globally important autotrophs, their methane release may have warmed the Earth.

As we follow the history of life, we shall see that major biological changes led to major environmental changes, which in turn led to further biological events, and so on. And in reverse, major physical changes led to biological changes, and so on. It is the dynamic interplay which is important. Life has such an important effect on a planet that NASA is working out strategies for detecting the presence of life on extrasolar planets by searching for its chemical signature. We could use the same strategy in trying to work out when life arose on Earth, and what form it took.

## Isotope Evidence for Biology

Most chemical elements have two or more isotopes; that is, their atoms may have slightly different masses. Thus, most carbon atoms weigh 12 amu (the nucleus has six protons and six neutrons). But a few carbon atoms have an extra neutron, so they weigh 13 units, and are called carbon-13 or $^{13}C$.

The extra mass does not affect the chemistry but it has physical effects. The heavier carbon atom moves a little more slowly than the lighter. In the molecule $CO_2$, for example, molecules with $^{13}C$ move more slowly than molecules with $^{12}C$. Photosynthesizers, in air or in water, take in $CO_2$ and break it up, building the carbon into their

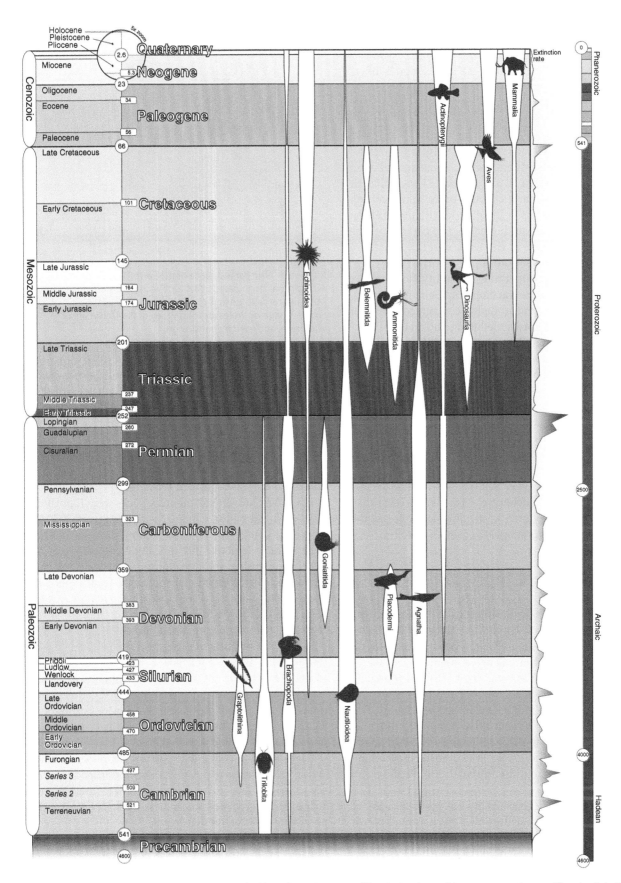

Figure 2.6 The international geological time scale. The column is arranged in relative time, with youngest at the top. The absolute time scale is on the right. Obviously, the depths of the divisions are not to scale. *Source:* Diagram by Frederik Lerouge (Wikimedia).

tissues. But since they take in $^{12}C$ molecules more easily than $^{13}C$, carbon that has gone through photosynthesis contains more $^{12}C$; it has a ratio of $^{13}C$ to $^{12}C$ that is different from the ratio in the $CO_2$ it came from, skewed toward the lighter carbon isotope. The difference is called isotope fractionation and can be measured in a mass spectrometer (at around $25 per sample). The isotope fractionation is expressed typically in parts per thousand (or "per mil"). Photosynthetic carbon in the ocean has an isotope fractionation, or $\partial^{13}C$, of about –20 per mil: the negative sign means that it contains more lighter carbon than "normal." (This has nothing to do with radiocarbon dating, which is based on the radioactive carbon isotope $^{14}C$. The isotopes used in the work described here are nonradioactive or stable isotopes.)

Different organisms operating different reactions may cause a different isotope fractionation. So methanogens, which split $CO_2$ and make methane, produce a $\partial^{13}C$ of about –60; in other words, methanogenic methane contains very light carbon. Bacteria that are lithotrophs, oxidizing iron and making an energetic profit, fractionate the iron isotopes they process, thus leaving a chemical trace of their activity in the sediments where that iron oxide is deposited.

Nitrogen isotopes are used to help us to assess ancient diets. If a heterotroph eats the tissue of another creature, it will digest, absorb, and perhaps lay down that nitrogen in organic components of bones or teeth. During that digestion, nitrogen isotopes $^{15}N$ and $^{14}N$ are fractionated by about +3 per mil. If that heterotroph is eaten by yet another, $\partial^{15}N$ increases by another +3 per mil. One can assess the ecology of some extinct animals by using N isotopes.

## Earth's Oldest Rocks

The first one-third of Earth's history is called the **Archaean** (Figure 2.6), a time when the early Earth was very different from today's planet. There was little or no oxygen in the atmosphere. There was much less life in the seas and none on the land. The Earth was young; its interior was hotter and its internal energy was greater. Volcanic activity was much greater, but we have no idea whether it was more violent or just more continuous. Very large asteroid impacts smashed into the Earth every 40 million years or so (Kerr 2011), so reconstructing "normal" conditions on the early Earth is difficult.

Rocks older than 3.5 Ga (3500 Ma) are very rare. The oldest minerals on Earth are zircon crystals dated at 4400 Ma (Figure 2.7), but they have been eroded out of their original rocks and deposited as fragments in younger rocks. These grains do contain evidence that there were patches of continental crust on Earth at or before 4400 Ma (Valley et al. 2014).

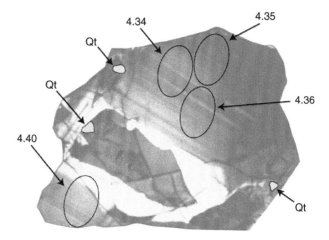

**Figure 2.7** "The earliest piece of the Earth." This zircon crystal from Australia has been dated at 4.4 Ga. The crystal is not really blue, but it fluoresces blue in the radiation used to examine it.
*Source:* Image by Professor John Valley of the University of Wisconsin, and used by permission.

This is important because continental crust, dominated by granite, is only found on Earth. Its chemistry is very different from ocean crust and includes important minerals that release phosphorus and potassium as they are broken down in weathering at the surface. Phosphorus in particular is vital for life (see Chapter 1). Continental crust may be yet another unique feature of Earth that encouraged the evolution of life here.

The oldest known sedimentary rocks were reported in 2017 from Labrador, Canada, with an age of 3950 Ma. These are pelites, metamorphosed mudstones that have been folded, faulted, and reheated, but they can still tell us something about conditions on the early Earth when they were formed. The pelites are associated with conglomerates, rocks composed of recycled older rocks, so in fact there were older rocks but it may be hard to identify sedimentary rocks older than, say, 4000 Ma because temperatures on the surface of the Earth were too hot and there would have been no water. Water is essential for the formation of most sedimentary rocks, so water, sediments, and of course life go together. These rocks tell us that conditions on Earth were hospitable to life by 3950 Ma at the latest, including the fact that there was land as well as ocean (see Chapter 1).

The Labrador rocks have carbonate rocks in them, and several groups of scientists have examined that carbon. On the face of it, carbon isotope fractionations indicate the activity of life processing that carbon, either photosynthetic or methanogenic. There is a difference of up to 25 parts per thousand in $\partial^{13}C$ between the carbonate carbon and organic carbon values, which has been interpreted by Tashiro et al. (2017) as evidence for the existence of life already by that time. The question is still being debated but in younger rocks, there would be no argument; the fractionations would be accepted as traces of biological processes because the alternative hypothesis is more complex.

If these conclusions are correct, then different lineages of cells were flourishing by 3950 Ma when we see the first reasonably well-preserved sedimentary rocks on Earth. No cells are preserved; only their chemical traces reveal they were there. What were these early lifeforms, and how did they give rise to the diversity of extant life we see around us today? To answer those questions, we turn to an evolutionary tree of the main groups of extant organisms: Bacteria, Archaea, and eukaryotes. Tracing the origins of these modern groups in deep time helps us to interpret the earliest traces of life in the rock record.

## Last Universal Common Ancestor (LUCA) and the Early Evolution of Cells

Phylogenetic trees are branching diagrams that depict the evolutionary relationships among organisms (Figure 2.8). For organisms that are alive today, phylogenetic trees can be inferred based on genetic data; organisms are grouped together on the tree according to the similarity of their DNA sequences. This approach does not work for organisms that went extinct millions of years ago – like dinosaurs, trilobites, or ancient microbes – because DNA breaks down after a few hundred thousand years. The oldest DNA that has been successfully isolated and sequenced is about 560 000 years old. This "ancient DNA" is very useful indeed for studies of the recent past, including the origin of humans (see Chapter 22), but is of no help in understanding earlier periods in the history of life. For organisms that fossilize, like animals and plants, we can learn a lot by studying those fossils and interpreting them in the light of modern relatives. But the earliest life forms were microbes, and their fossil record is sparse and extremely difficult to interpret.

So how can we investigate the earliest events in the history of life, given that we lack direct evidence of the organisms themselves? Part of the answer lies in the fact that all modern lifeforms, from humans to bacteria, share a common ancestor that lived early in the history of the Earth, perhaps more than 4 billion years ago (Figure 2.8). We know that all life has a common origin because of fundamental similarities in the way that all cells work, particularly in how they replicate their genome and make proteins from genes by means of a universal genetic code. The simplest explanation for these shared, complex features is that all life is descended from a LUCA that already possessed these traits. By working out the deepest relationships

among all modern life, therefore, we can work back toward the LUCA and learn much about early life on Earth. So, even though we cannot find a fossil of the LUCA, we can say a lot about it from comparisons of living organisms.

Phylogenetic trees built from the core genes shared by all lifeforms (Figure 2.8) suggest that the deepest split in the tree of life lies between two groups of relatively simple cells: Bacteria and Archaea. In terms of body plan, Bacteria and Archaea are prokaryotes; they are mostly small (~1 μm in diameter) and single-celled, and contain a compact and efficiently organized DNA genome. Genes with similar functions – such as the different genes required for a particular metabolic pathway – are often found side by side on the genome, so that they can be turned on and off as a unit by the same set of regulatory sequences. When genes are turned on, they are copied into an RNA message which is directly translated into protein by ribosomes. Since these features are shared by Bacteria and Archaea, it is likely that they were also possessed by the LUCA. By looking for other traits shared by Bacteria and Archaea, it is possible to reconstruct a sketch of the kind of organism the LUCA might have been; it was probably an autotroph, making energy and fixing carbon using the acetyl-CoA pathway in much the same way as modern methanogens (Archaea) and acetogens (Bacteria) do. Some of the reactions of the acetyl-CoA pathway can occur without any biology, in the presence of transition metal catalysts. This means that the earliest lifeforms may have coopted and then refined chemical reactions that were already taking place to power their growth and reproduction.

Fossil evidence and time-calibrated phylogenetic trees indicate that Bacteria and Archaea are the oldest lineages of organisms, and that fully the first half of life's history was entirely prokaryotic. Eukaryotes – the cell lineage to which we belong – evolved about 2 billion years ago, resulting from an evolutionary merger between an archaeon and a bacterium (see Chapter 3). This does not in any way mean that prokaryotes are "lower" than or inferior to eukaryotes! Modern Bacteria and Archaea far exceed eukaryotes in terms of abundance, biodiversity, and evolutionary rate, and are key drivers of the planetary biogeochemical cycles that recycle nutrients and maintain a relatively stable environment for macroscopic life.

The same analyses that date eukaryotes to ~2 billion years ago suggest that the LUCA may be about twice as old, evolving before 4 Ga. If so, then the earliest traces of life in the fossil record ~3.5–3.8 Ga likely postdate the LUCA, and so belong to early groups of Bacteria and Archaea. With this framework in mind, we now turn to the oldest rocks that document the deep origins of some modern prokaryotic groups.

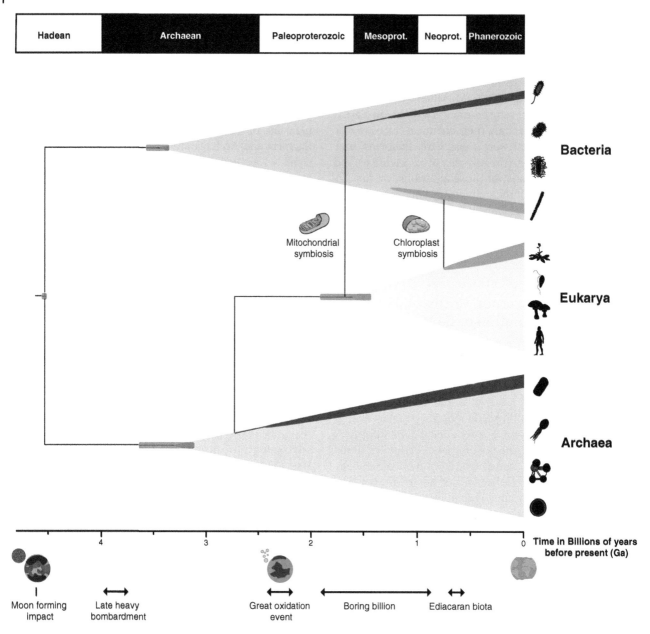

Figure 2.8 **A dated time scale of early life, based on genes and fossils (Betts et al. 2018). The last universal common ancestor (LUCA) of modern life lived at least 4 billion years ago. Bacteria and Archaea were the first of the extant lineages to evolve, before 3 billion years ago. Eukaryotes – our lineage – evolved much later, in an endosymbiosis between an archaeon and a bacterial endosymbiont that evolved to become the mitochondria of modern eukaryotic cells.**

## Stromatolites

The Pilbara is a remote region of Northwest Australia. Pilbara rocks include the Warrawoona Series, dated to about 3300–3550 Ma. The Warrawoona rocks are mainly volcanic lavas erupted in shallow water or nearby on shore, but there are sedimentary rocks too. The sediments include storm-disturbed mudflakes, wave-washed sands, and minerals formed by evaporation in very shallow pools. The rocks have not been tilted, folded, or heated very much, and the environment can be reconstructed accurately. The rocks formed along shore-lines that we can interpret clearly because we can match them to modern environments.

The Warrawoona rocks contain structures called **stromatolites**, which are low mounds or domes of finely laminated sediment (Figure 2.9). Fossil stromatolites are known from many different geological periods, and cut specimens can show in detail how they built up as a series of thin layers (Figure 2.10).

We know what stromatolites are because they are still forming today in a few places. Thus, we can study the living forms to try to understand the fossil structures. Stromatolites are formed by mat-like masses of abundant microbes, usually including photosynthetic cyanobacteria. Stromatolites live today in Shark Bay, Western Australia, in warm salty waters in long shallow inlets along a desert coast (Figures 2.11 and 2.12). They form from the highest tide level down to subtidal levels, but the higher ones close to shore have been better studied (sea snakes, not sharks, are the problem).

The cyanobacteria that grow and photosynthesize in Shark Bay so luxuriantly thrive in water that is too salty for grazing animals such as snails and sea urchins that would otherwise eat them. Like most bacteria, they secrete slime, and can also move a little in a gliding motion. Sediment thrown up in the waves may stick to the slime and cover up some of the bacteria. But they quickly slide and grow through the sediment back into the light, trapping sediment as they do so. As the cycle repeats itself, sediment is built up under the growing mats. Eventually, the mats grow as high as the highest tide, but cannot grow higher without becoming too hot and dry. Some mats harden because the photosynthetic activity of the bacteria helps carbonate to precipitate from seawater, binding the sediment into a rock-like consistency that resists wave action (Figure 2.11). However, sediment stabilization in stromatolites today works best in the light. Stromatolites placed experimentally in the dark lose stability (Paterson et al. 2008). This implies that all stromatolites, ancient and modern, formed and built rock-like trace fossils through photosynthesis.

Some cyanobacterial mats are so dense that light may penetrate only 1 mm. The topmost layer of cyanobacteria absorbs about 95% of the blue and green light, but just underneath is a zone where light is dimmer but exposure to UV radiation and heat is also less. Green and purple bacteria live here in huge numbers and also contribute to the growth of the mat. Deeper still in the mat, light is too low for photosynthesis, and there heterotrophic bacteria

Figure 2.9 The oldest trace fossils on Earth. Cone-shaped stromatolites from the Warrawoona rock sequence in Western Australia. They are found with wave-affected sediments, and therefore formed in very shallow water. Comparing these structures with those forming today (see Figure 2.11), they can be interpreted as having been formed by cyanobacterial mats 3.5 billion years ago. The image shows an eroded surface cutting through the cones. *Source:* From Macquarie University at http://pilbara.mq.edu.au/wiki.

Figure 2.10 Stromatolites in the Soeginina Beds (Paadla Formation, Ludlow, Silurian) near Kübassaare, Saaremaa, Estonia, showing the internal structure of multiple thin layers, building up over long spans of time. *Source:* Photo by Wilson44691 (Wikimedia).

Figure 2.11  The famous modern stromatolites from shallow seas at Hamelin Pool Marine Nature Reserve, Shark Bay, Western Australia. *Source:* Photo by Paul Harrison (Wikimedia).

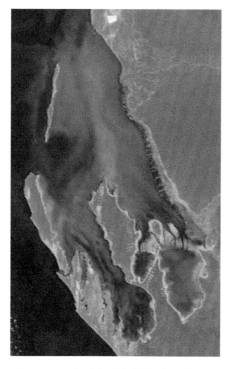

Figure 2.12  Shark Bay World Heritage Site is a series of shallow bays on the coast of Western Australia. Stromatolites form today in the warm salty water close to shore. *Source:* NASA image taken from space.

absorb and process the dying and dead remains of the bacteria above them. Oxygen diffuses down into the mat from above, and sulfide diffuses upward from the zone below, creating an extraordinary zone where chemistry can change within minutes and within millimeters.

Night follows day, of course, and photosynthesis stops at night. The oxygen in the top layers of the stromatolite is quickly lost. Sulfide dominates the nighttime hours, oxygen dominates the daylight hours, and all the bacteria must be able to adjust quickly to the daily change. The internal chemistry in stromatolites is as complex as the mix of bacteria. There is no reason to suppose that ancient stromatolites were any different.

## Ancient Stromatolites

Stromatolites are trace fossils. They are formed by the action of living cells, even if those cells are hardly ever preserved in them as fossils. Because they are large, and because their distinctive structure makes them easy to recognize, stromatolites are the most conspicuous fossils for 3 billion years of Earth history, from about 3500 Ma to the end of the Proterozoic at about 550 Ma. They are rare in Archaean rocks, probably because there were few clear, shallow-water shelf environments suitable for stromatolite growth at the time. The few Archaean land masses were volcanically active, generating high rates of sedimentation that probably inhibited mat growth in many shoreline environments. Even so, stromatolites flourished in Australia and South Africa around 3430 Ma (Figure 2.9), and locally covered miles of shoreline.

Solar UV radiation was intense in Archaean time, with no oxygen (or ozone layer) in the atmosphere. A shield of perhaps 10 m (30+ feet) of water might have been needed to prevent damage to a normal early cell by UV radiation. However, the early evolution of the stromatolitic way of life by cyanobacteria may have been a response to UV radiation. With light (and UV) penetrating only a little way into the mat, bacteria were able to live essentially at the food-rich water surface without damage from UV. Cyanobacteria were not just existing at Warrawoona; they were already modifying their microenvironment for survival and success. Stromatolites were not just the sites of simple microbial mats but were complex miniature ecosystems teeming with life.

Newer work reminds us that care is needed in interpreting stromatolites. The evidence for the importance of photosynthesis is clear, but rare examples of modern stromatolites have been found growing well below the photic zone, such as some examples in 731 m water depth in the Arabian Sea (Himmler et al. 2018). These grow by chemotrophic processes in methane seeps, where the microbes metabolize the methane and sulfur, and such settings could explain the formation of some ancient stromatolites.

Banded Iron Formations (BIF)

## Identifying Fossil Cells in Ancient Rocks

Chert is a rock formed of microscopic silica particles ($SiO_2$). It does not form easily today because all kinds of organisms, including sponges, take silica from seawater to make their skeletons. But silica-using organisms had not evolved in Archaean times, so cherts are often abundant in Archaean rocks. As chert forms from a gel-like goo on the sea floor, it may surround cells and impregnate them with silica, preserving them in exquisite detail as the silica hardens into chert. Once it hardens, chert is watertight, so percolating water does not easily contaminate the fossil cells.

Several processes can generate inorganic blobs of chemicals in rocks, and blobs in chert have often been mistaken for fossil cells. But cell-like structures in the Apex Chert, a formation in the Warrawoona rocks dated at 3465 Ma, are genuine cells (Figure 2.13). This is hardly surprising, given that locally huge areas of the Warrawoona coastline were covered with stromatolites at about this time (Figures 2.9 and 2.12).

These earliest stromatolites formed around 3430 Ma (Figure 2.9). By 3.1 Ga, there were two distinctly different styles of bacterial mat: by 2.9 Ga, bacterial mats were forming on soft silty sea floors, and by 2800 Ma stromatolites are known from salt lake environments as well as oceanic shorelines. A diverse set of cells is known from cherts in the Pilbara at 3000 Ma. Bacterial mats were abundant and varied by Late Archaean time.

There were important geological changes at the end of the Archaean, which is dated at 2500 Ma. The Earth had cooled internally to some extent, and the crust was thicker and stronger. The thicker crust affected tectonic patterns: the way the crust moves, buckles, and cracks under stress. Continents became larger and more stable in Early Proterozoic times, with wide shallow continental shelves that favored the growth and preservation of stromatolites. Most Proterozoic carbonate rocks include stromatolites, some of them enormous in extent. Proterozoic stromatolites evolved new and complex shapes as bacterial communities became richer and expanded into more environments.

## Banded Iron Formations (BIF)

From the beginning of the Archaean (around 3800 Ma), we find increasing accumulations of a peculiar rock type. **Banded iron formations** are sedimentary rocks found mainly in sequences older than 1800 Ma. The bands are alternations of iron oxide and chert (Figure 2.14), sometimes repeated millions of times in microscopic bands (Figure 2.15). No iron deposits like this are forming now, but we can make intelligent deductions about the conditions in which BIF were laid down.

The chemistry of seawater on an Earth without oxygen differed greatly from today's situation. Today, there is practically no dissolved iron in the ocean, but iron dissolves readily in water without oxygen. Even today, in oxygen-poor water on the floor of the Red Sea, iron is enriched 5000 times above normal levels. So Archaean oceans must have contained a great deal of dissolved iron as well as silica.

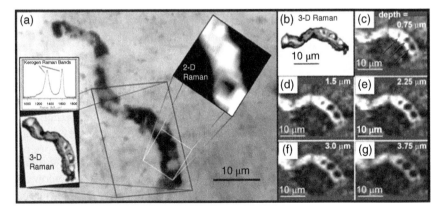

Figure 2.13 This fossil is *Primaevifilum*, from the Apex Chert in the Warrawoona rocks of Western Australia, dated about 3465 Ma. A photograph down a microscope (a) shows a rather blurry outline, which caused unnecessary vicious criticism suggesting that it was not a cell and not a fossil. However, new techniques for imaging the fossil show its internal structure and 3D shape, confirming it as the earliest fossil cell so far found. It looks like a cyanobacterium, but that is hard to prove with our current technology. *Source:* Image from Schopf and Kudryavtsev (2010); courtesy of Professor J. W. Schopf.

Figure 2.14  Block of banded iron. *Source:* Image by André Karwath (Wikimedia).

Figure 2.15  Close-up of banding in BIF specimen from the Proterozoic of Michigan. Scale bar is 5 mm. *Source:* Photograph by Mark Wilson of the College of Wooster (Wikimedia).

Silica would have been depositing more or less continuously on an Archaean sea floor to form chert beds, especially in areas that did not receive much silt and sand from the land. But iron oxide can only have precipitated out of seawater in such massive amounts by a chemical reaction that included oxygen.

Therefore, to form the iron oxide layers in BIF, there must have been occasional or regular oxidation events to produce iron ore, against a background of regular chert formation. Between oxidation events, dissolved iron was replenished from erosion down rivers or from deep sea volcanic vents. What were these oxidation events, and what started them? The most likely hypothesis to explain BIF formation calls on seasonal changes in sunlight and temperature that in turn affect bacterial action and mineral deposition.

Lake Matano, on the Indonesian island of Sulawesi, gives us an idea of how an Archaean ocean may have worked (Busigny et al. 2014). The lake is small but very deep. The tropical climate means that the surface waters are always warm, so they never sink or mix with the deeper water below. The surface waters are well lit, and floating cyanobacteria photosynthesize there. But the lake water has few nutrients, so these surface bacteria are not important, except that they keep the surface layers of the lake oxygenated.

Below the surface layer is water with no oxygen, rich in dissolved iron, just as we imagine the Archaean ocean to have been. Sunlight reaches the top layer of this deeper water, but the light is too dim for cyanobacteria to photosynthesize. Instead, huge numbers of green sulfur bacteria, with a variety of chlorophyll that works better in dim light, operate lithotrophy (see Chapter 1) about 120 m down in Lake Matano. They break down water, use the oxygen to oxidize the dissolved iron, and make a metabolic profit from the reaction. Oxidized iron sinks to the lake floor in large quantities.

Laboratory experiments suggest that in this situation, iron oxidation would work fastest in a narrow temperature range, with silica depositing faster at higher or lower temperatures. Using the evidence from Lake Matano, it looks as if sulfur bacteria could also have formed the alternating mineral bands in Archaean BIF. Although they involve oxidation, the reactions occur in environments without oxygen, and they do not produce any. Since sulfur bacteria are very ancient, there is no problem in suggesting that they were involved in producing the first BIF in the Isua rocks in Greenland, at 3800 Ma, some of the oldest rocks on Earth.

In an extensive Archaean ocean, rather than a small tropical lake like Lake Matano, we would expect that iron oxidation would occur over a large area. Indeed, BIF were often deposited in bands that can be traced for hundreds of kilometers.

Today, we probably see only a small fraction of the BIF that once formed on Archaean sea floors, because most ocean crust has since been recycled back into the Earth. But even the amounts remaining are staggering. BIF make up thousands of meters of rocks in some areas and they contain by far the greatest deposits of iron ore on Earth. At least 640 billion tonnes of BIF were laid down in the early Proterozoic between 2500 Ma and 2000 Ma (that's an average of half a million tonnes of iron per year). (The metric tonne that is used internationally is 1000 kg, very close to an American ton.) The Hamersley Iron Province in Western Australia alone contains 20 billion tonnes of iron ore, with 55% iron content. At times, iron was dropping out in that basin at 30 million tonnes a year. Most modern steel industries are based on iron ores laid down in BIF during that time (Figure 2.16).

Figure 2.16 An enormous iron mine in BIFs of the Pilbara region of Western Australia, abandoned in 2008. *Source:* Photograph by Philipist (Wikimedia).

## Banded Iron Formations, Stromatolites and Oxygen

The current best hypothesis for forming BIF requires oxygen-free ocean water with dissolved iron, at least below the surface. However, cyanobacteria in stromatolites were producing free oxygen in the Archaean, starting with the first major stromatolites around 3500 Ma.

Cyanobacteria evolved oxygen dismutase as an antidote to oxygen poisoning (see Chapter 1), and that gave them the opportunity to control and then use that oxygen in a new process, **respiration** (biological oxidation). Fermenting sugars leave by-products such as lactic acid that still have energy bonded within them. By using oxygen to break those by-products all the way down to carbon dioxide and water, a cell can release up to 18 times more energy from a sugar molecule by respiration than it can by fermentation.

Cyanobacteria can photosynthesize in light and respire in the dark. To do this, they must be able to store oxygen in a stable, nontoxic state for hours at a time. Most likely, they began to use oxygen in respiration very early; the energy advantages are astounding. The early success of cyanobacteria probably reflects their access to an abundant and reliable energy supply in two different ways: first, in photosynthesis, and second, by breaking down food molecules by respiration rather than fermentation.

However, that success poses a problem. If cyanobacteria made such an important biochemical breakthrough, using a process that produces oxygen as a by-product, why did the ocean and atmosphere not become oxygenated quickly? Stomatolites were locally abundant by 3500 Ma, producing "whiffs of oxygen" (as one researcher has written). However, stromatolites lived only along shallow shorelines, and early continents were small. So any effects of free oxygen would at first have been local rather than global.

Nevertheless, even whiffs of oxygen can be detected by skilful geochemists. Molecules called steranes can only be produced in reactions that use free molecules of oxygen, and steranes have been detected in rocks at 2720 Ma (Gold et al. 2017).

Stromatolites increased dramatically at about 2500 Ma, along with the formation of larger continents and more shallow-water habitat. BIF production reached a peak around that time, too, coinciding with massive volcanic eruptions that must have vastly enriched dissolved oxygen supplies in seawater.

By 2500 Ma, the official beginning of the Proterozoic, Earth's surface chemistry had been changed by life for a billion years. Cyanobacteria in stromatolites, and probably in surface waters everywhere, were producing waves of oxygen large enough to oxygenate large areas of shallow ocean waters, especially along continental shores. Green sulfur bacteria were forming huge masses of BIF in anoxic iron-rich waters in the oceans, and Archaea were producing methane. Yet the atmosphere had practically no oxygen until about 2300 Ma.

There are many ways in which oxygen can be used up before and after it is produced in photosynthesis. Today, only about 5% of all the oxygen produced in photosynthesis reaches the atmosphere and ocean. The other 95% reacts with iron and sulfur compounds to form iron oxides and sulfates. These and similar reactions must have used up almost every oxygen molecule produced on the early Earth, too, and any surplus oxygen would have used up to oxidize organic molecules in the water. It could easily have taken a billion years before free oxygen began to accumulate in air and water on a global scale.

In addition, photosynthesis (and oxygen production) may have been a lot slower than we might imagine. Photosynthesis does not just need light, water, and carbon dioxide; the plants or bacteria that operate it need nutrients as well. Phosphorus is a limiting nutrient in many environments even today (many of our fertilizers contain phosphorus).

Phosphorus comes largely from continental crust, and Archaean continents were small. So phosphorus supplies may have been limited, especially out in the vast expanses of Archaean oceans. In addition, iron minerals absorb some phosphorus as they form, so forming BIF would have used up (and locked up) a lot of phosphorus as well as a lot of iron. Lack of phosphorus would then have slowed cyanobacterial growth, which would have slowed photosynthesis and oxygen production, until renewed weathering and erosion brought new iron and phosphorus supplies to ocean water. Altogether, this would have dramatically slowed the oxygenation of Earth.

In the end, however, the supply of oxygen became large enough that free oxygen began to accumulate in oceans and atmosphere, setting the stage for dramatic changes in Earth's surface chemistry, and its life.

## The Great Oxidation Event

Banded iron formation production rose to a peak around 2500 Ma, then declined, with only occasional bursts of activity over the next billion years. Other geological evidence confirms that the ocean surface waters and the atmosphere were oxygenated early in the Proterozoic. The uranium mineral **uraninite** cannot exist for long if it is exposed to oxygen, and it is not found in rocks younger than about 2300 Ma. The sulfur isotopes in rocks suggest that sulfate levels rose in the ocean, lowering methane production by Archaea, while the methane that they continued to produce was quickly broken down by free oxygen. More complex indicators of ancient oxygen levels come from isotope changes in a number of metals, but there is general agreement on the timing. The great change in surface oxygen levels occurred between 2400 and 2000 Ma, with periods of slow change and periods of rapid change. Oxygen levels rose from perhaps one-millionth of their present atmospheric level, to 1% by 2000 Ma (Lyons et al. 2014). In turn, the drop in methane levels in the atmosphere cooled the Earth, and there is evidence of a very large ice age between 2400 and 2200 Ma.

Once oxygen was part of the atmosphere, it would have rusted any iron minerals exposed on the land surface by weathering. Rivers would have run red over the Earth's surface before vegetation invaded the land. On land and in shallow seas, **red beds**, sediments bearing iron oxides, date from about 2300 Ma (Figures 2.17 and 2.18).

Photosynthesis produces oxygen only in surface waters, because that is as far as useable light penetrates water. Surface waters are warmer than deeper layers, so are less dense and tend to stay at the surface. The deep waters of the ocean receive no oxygen directly. In today's oceans, oxygen-rich surface waters can sink but only if they are unusually dense: if they are very cold, for example, or if they are very salty, or both. Examples are polar seas – in the North Atlantic and around Antarctica – or hot shallow salty tropical seas such as the Persian Gulf.

In some seas today, the surface waters do not sink so there is no oxygen, and little life, below the surface layers. The Black Sea is the best-known example, but the Red Sea also has deep basins that lack oxygen.

Today, enough surface water sinks to carry oxygen to most of the world's oceans. But it would have been different in the Proterozoic. Clearly, the surface waters would have become oxygen rich before the bulk of the ocean did. And the atmosphere, which can exchange

Figure 2.17 Image taken by an alien spacecraft of the Earth's Proterozoic land surface shortly after the Great Oxidation Event. I tell a lie: alien spacecraft do not exist. This is actually an image of red sand dunes in the Namib Sand Sea of Namibia, taken from the International Space Station. *Source:* NASA Earth Observatory photograph.

Figure 2.18 Red beds have been common rocks on the Earth since the Great Oxidation Event around 2300 Ma. This splendid outcrop of red sandstone in Namibia is called Lion's Head. *Source:* Photograph by Violet Gottrop (Wikimedia).

gases with the surface waters, would also have become oxygen bearing before the deep ocean did. We can imagine a Proterozoic world that had free oxygen only in surface waters and the atmosphere. The deep ocean would still have been **anoxic**, rich in dissolved iron and silica and sulfide, and inhabited by bacteria and methanogens, while the surface waters had photosynthesizers and oxygen-tolerant microbes.

Banded iron formations are rare after 2300 Ma, as oxygen levels in the surface waters extended deeper, driving iron-rich water deeper and sulfur bacteria so deep that they had no light to form BIF. BIF could form only in rare isolated basins, like the famous ones in Michigan, or at times of crisis when iron-rich waters extended upward into shallow depths.

The term that summarizes all the chemical, geological, and biological changes around 2330 Ma is the **Great Oxidation Event** (Luo et al. 2016). The surface chemistry of Earth's air, land, and water changed forever, and one indirect result was vital for the further evolution of living things. Solar UV radiation acts on any free oxygen high in the atmosphere to produce ozone, which is $O_3$ rather than $O_2$. Even a very thin layer of ozone can block most UV radiation. Earth's surface, land and water, has been protected from massive UV radiation ever since free oxygen entered the atmosphere. It then became possible for organisms to evolve that were larger and had more stringent energy requirements than Bacteria or Archaea: the eukaryotes (see Chapter 3).

## References

Betts, H.C., Puttick, M.N., Clark, J. et al. (2018). Integrated genomic and fossil evidence illuminates life's early evolution and eukaryote origin. *Nature Ecology and Evolution* 2: 1556–1562.

Busigny, V., Planavsky, N.V., Jézéquel, D. et al. (2014). Iron isotopes in an Archean ocean analogue. *Geochimica et Cosmochimica Acta* 133: 443–462.

Gold, D.A., Caron, A.M., Fournier, G. et al. (2017). Paleoproterozoic sterol biosynthesis and the rise of oxygen. *Nature* 543: 420–423.

Hazen, R.M. (2010). How old is Earth, and how do we know? *Evolution: Education and Outreach* 3: 198–205.

Himmler, T., Smrzka, D., Zwicker, J. et al. (2018). Stromatolites below the photic zone in the northern Arabian Sea formed by calcifying chemotrophic microbial mats. *Geology* 46: 339–342.

Kerr, R.A. (2011). Asteroid model shows early life suffered a billion-year battering. *Science* 332: 302–303.

Luo, G., Ono, S., Beukes, N.J. et al. (2016). Rapid oxygenation of Earth's atmosphere 2.33 billion years ago. *Science Advances* 2: e1600134.

Lyons, T.W., Reinhard, C.T., and Planavsky, N.J. (2014). The rise of oxygen in Earth's early ocean and atmosphere. *Nature* 506: 307–315.

Paterson, D.M., Aspden, R.J., Visscher, P.T. et al. (2008). Light-dependent biostabilisation of sediments by stromatolite assemblages. *PLoS One* 3 (9): e3176.

Schopf, J.W. and Kudryavtsev, A.B. (2010). Biogenicity of Earth's earliest fossils: a resolution of the controversy. *Gondwana Research* 22: 761–771.

Tashiro, T., Ishida, A., Hori, M. et al. (2017). Early trace of life from 3.95 Ga sedimentary rocks in Labrador, Canada. *Nature* 549: 516–518.

Valley, J.W., Cavosie, A.J., Ushikubo, T. et al. (2014). Hadean age for a post-magma-ocean zircon confirmed by atom-probe tomography. *Nature Geoscience* 7: 219–223.

## Further Reading

Arndt, N.T. and Nisbet, E.G. (2012). Processes on the young Earth and the habitats of early life. *Annual Review of Earth & Planetary Sciences* 40: 521–549.

Craddock, P.R. and Dauphas, N. (2011). Iron and carbon isotope evidence for microbial iron respiration throughout the Archean. *Earth and Planetary Science Letters* 303: 121–132. [Microbes were respiring iron at 3.8 Ga].

Mills, D.B. and Canfield, D.E. (2014). Oxygen and animal evolution: did a rise of atmospheric oxygen "trigger" the origin of animals? *BioEssays* 36: 1145–1155.

Olson, S.L., Kump, L.R., and Kasting, J.F. (2013). Quantifying the areal extent and dissolved oxygen concentrations of Archean oxygen oases. *Chemical Geology* 362: 3543.

Sugitani, K., Mimura, K., Takeuchi, M. et al. (2015). Early evolution of large micro-organisms with cytological complexity revealed by microanalyses of 3.4 Ga organic-walled microfossils. *Geobiology* 13: 507–521.

## Questions for Thought, Study, and Discussion

1 Explain the difficulty of finding the age of a rock bed by the fossils in it, and at the same time finding its age by radiometric methods.

2 If you were able to watch the Earth from a spacecraft during Earth's first 2 billion years of existence, how would you tell that life had evolved on the planet? This is not necessarily the time when life evolved: it's the time when an observer in space could detect it. The spacecraft has reasonable instruments for remote sensing.

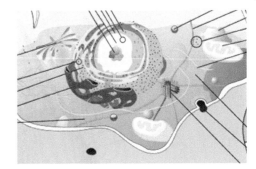

# 3

# The Origin of Eukaryotes

| In This Chapter |
|---|

For the first half of Earth's history, life consisted of prokaryotes (Archaea and Bacteria), but the evolution of eukaryotes (cells with nuclei) changed the biological world forever. Eukaryotes have complex cells resulting from a symbiosis, or stable evolutionary partnership, between an archaeal host cell and one or more bacterial endosymbionts. These are major steps in evolution and we discuss how they happened. While most eukaryotes are single-celled microbes, multicellularity evolved several times and gave rise to the most familiar eukaryotic groups, including animals, plants, and fungi. In this chapter, we also explore how biologists and paleontologists make sense of the huge diversity of species on Earth – the secret has been the revolution in cladistics and genomics in reconstructing the tree of life.

## Single-Celled Life

The microbes that were Earth's first life evolved into two different major groups or domains: Archaea and Bacteria. They shared much the same body plan, however, and we group them together as prokaryotes (Figure 3.1). Prokaryotes were and are very successful in an incredible range of habitats, from stinking swamps to the hindgut of termites and from hot springs in the deep sea to the ice desert of Antarctica, and deep in rocks underground. They occur in numbers averaging 500 million/l in surface ocean waters, 1 billion/l in fresh water, and about 300 million on the skin of the average human. The evolutionary success of prokaryotes is due to their metabolic versatility, resilience to a broad range of environmental conditions, and large population sizes. With such large populations, adaptive evolution to new nutrient sources and environmental challenges typically proceeds much faster in prokaryotes than in eukaryotes – the cellular lineage to which we belong.

Eukaryotes today are larger and much more complex than prokaryotes (Box 3.1). Most of their DNA is contained in a nucleus with a membrane around it, and the eukaryotic cell also contains organelles, each one wrapped in a membrane, that perform functions in the cell and can contain a small circular genome. The **genome** is the sum total of the genetic code of a single individual or species. It is clear from their structure and the fact that they contain their own, independently replicating DNA that some of these organelles were once cells in their own right, but are now part of the eukaryote. In this chapter, we will investigate how the ancestor of these organelles came to live inside eukaryotes, and how the structural complexity of eukaryotic cells first evolved.

## Symbiosis and Endosymbiosis

Symbiosis is a relationship in which two different organisms live together. Often, they both get some benefit from the arrangement. Examples range from the symbiosis between humans and dogs to bizarre relationships such as that of acacia plants, which house and feed ant colonies that in turn protect the acacia against herbivorous animals and insects. Sometimes, symbiosis evolves into endosymbiosis, in which one organism lives inside its partner. Animals as varied as termites, sea turtles, and cattle can live on plant material because they contain

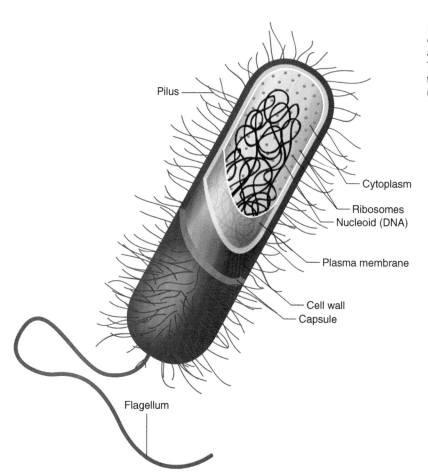

Pilus

Cytoplasm

Ribosomes
Nucleoid (DNA)

Plasma membrane

Cell wall
Capsule

Flagellum

**Figure 3.1** A prokaryotic cell. The DNA (*blue cords*) is twisted and folded to fit into the cell and floats free in the cell cytoplasm (*orange*). This prokaryote is mobile and propelled by a flagellum. *Source:* Image by Ali Zifan (Wikimedia).

---

**Box 3.1 Differences Between Prokaryotes and Eukaryotes**

1) Eukaryotes have their DNA contained inside a membrane, and under the microscope this package forms a distinct body called the nucleus. Prokaryotes have their strands of DNA loose in the cell cytoplasm.

2) Prokaryotes generally have no internal subdivisions of the cell, but almost all living eukaryotes have organelles as well as a nucleus, and there is evidence that eukaryotes without organelles have lost them through evolutionary processes leading to simplified morphologies (generally associated with parasitic lifestyles – as in the case of the intestinal parasite *Giardia*). Organelles are subunits of the cell that are bounded by membranes and they perform some specific function in the cell. Some organelles, plastids, and mitochondria have their own independent genome and evolved through processes of endosymbiosis. Plastids perform photosynthesis inside the cell, harvesting solar energy that they transform into food molecules (chemical energy) and releasing oxygen as a waste product. Mitochondria contain the respiratory enzymes of the cell. Food molecules are first fermented in the cytoplasm, then passed to the mitochondria for respiration. Mitochondria generate adenosine triphosphate (ATP) as they break food molecules down to water and $CO_2$, and they pass energy and waste products to the rest of the cell. They also make steroids, which help to form cell membranes in eukaryotes and give them much more flexibility than prokaryote membranes.

3) Eukaryotes can perform sexual reproduction, in which the DNA of two cells is shuffled and redealt into new combinations.

4) Prokaryotes generally have rather inflexible cell walls, so they cannot easily engulf other cells. The flexibility of eukaryotic cell membranes allows them to engulf large particles (phagocytosis), to form cell vacuoles, and to move freely. Plant cells, armored by cellulose, are the only eukaryotes that have given up a flexible outer cell wall for most of their lives.

5) Eukaryotes have a well-organized system for duplicating their DNA exactly into two copies during cell division. This process, mitosis, is much more complex than the simple splitting found in prokaryotes.

6) Eukaryotes are almost always much larger than prokaryotes. A eukaryote is typically 10 times larger in diameter, which means that it has about 1000 times the volume of a prokaryote.

7) Eukaryotes have perhaps 1000 times as much DNA as prokaryotes. They have multiple copies of their DNA, with much repetition of sequences. The DNA content of prokaryotes is small, and they have only one copy of it. Therefore, genetic regulation is not well developed in prokaryotes, which means that they cannot produce the differentiated cells that we and other eukaryotes can. Multicellular colonies of bacteria are generally made up of the same cell type, repeated many times in a clone. Therefore, any species of bacterium is very good at one thing but cannot do others; its range of functions is narrow.

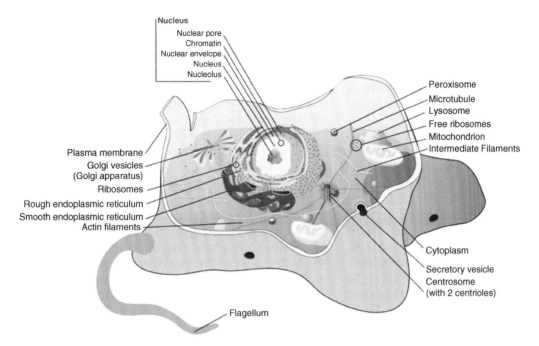

**Nucleus**
- Nuclear pore
- Chromatin
- Nuclear envelope
- Nucleus
- Nucleolus

Plasma membrane
Golgi vesicles
(Golgi apparatus)
Ribosomes
Rough endoplasmic reticulum
Smooth endoplasmic reticulum
Actin filaments

Peroxisome
Microtubule
Lysosome
Free ribosomes
Mitochondrion
Intermediate Filaments

Cytoplasm
Secretory vesicle
Centrosome
(with 2 centrioles)

Flagellum

Figure 3.2 A eukaryotic cell (more specifically, a protist). A cell membrane (*light blue*) surrounds the cell's cytoplasm (*darker blue*). Among the cell contents are pill-shaped mitochondria (*lime green*), and a nuclear membrane (*orange*) surrounding the nucleus itself, which contains DNA. This motile cell is driven by a flagellum which has a different structure from its prokaryotic equivalent. *Source:* Image by Mariana Ruiz Villareal, Lady of Hats (Wikimedia).

bacteria in their digestive system with the enzymes to break down the cellulose that is unaffected by the host's own digestive juices. Many tropical reef organisms have symbiotic partners in the form of photosynthesizing microorganisms. Living inside the tissues of corals or giant clams, these symbiotic partners have a safe place to live. In turn, the host receives a share of their photosynthetic production.

It is now clear that endosymbiosis was a critical step in the evolution of eukaryotes. A prokaryotic host cell that was part of the archaeal domain made a dramatic evolutionary breakthrough: it entered into symbiosis with a bacterium that later evolved to become the first mitochondrion (plural, mitochondria), the organelles that generate energy and carry out other key biochemical conversions in modern eukaryotic cells. Later, one descendant branch of the eukaryotic lineage acquired another endosymbiotic bacterium which evolved to become the first plastid of plants and other photosynthetic eukaryotes. Plastids contain all the chlorophyll in the cell, perform photosynthesis inside what we now call plant cells, turning light (solar energy) into food for the cell.

Once free-living bacteria, these organelles are now so closely integrated into a host cell that they are to all effects part of it (Figure 3.2). At least five major pieces of evidence show that organelles (and therefore eukaryotes) originated by endosymbiosis (Box 3.2).

---

**Box 3.2 Evidence for Organelle/Eukaryote Symbiosis**

1) The DNA in mitochondria and plastids is not the same as the DNA in the eukaryotic cell nucleus.
2) Mitochondria and plastids are separated from the rest of the eukaryotic cell by membranes; thus, they are really "outside" the cell. The cell itself makes the membrane, but inside it is a second membrane made by the organelle.
3) Plastids, mitochondria, and prokaryotes make proteins by similar biochemical pathways, which differ from those in the cytoplasm of eukaryotes.
4) Mitochondria and plastids are susceptible to antibiotics such as streptomycin and tetracycline, like prokaryotes; eukaryotic cytoplasm is not affected by these drugs.
5) Mitochondria and plastids can multiply only by dividing; they cannot be made by the eukaryotic cell. Thus, organelles have their own independent reproductive mechanism. A cell that loses its mitochondria or plastids cannot make any more.

---

## The Prokaryotic Ancestors of Eukaryotic Cells

Several plausible scenarios for the origin of eukaryotes have been proposed but testing them is difficult because of the enormous time scales involved and the limited available data. The key players were microbes that fossilize poorly or not at all, and so most of the ideas are based on considerations of how modern prokaryotes engage in symbiosis, along with evolutionary reconstructions of ancient cells based on the genomes of bacteria and archaea that are alive today.

While the evidence for symbiosis between an archaeon and a bacterium is strong, the selective benefit to the two partners remains debated, and there are several possible ideas. Modern symbioses between prokaryotes often involve cross-feeding, where one partner consumes the waste products of the other, and so this seems a reasonable starting point; one influential hypothesis is that the initial interaction was based on molecular hydrogen, provided by the bacterium and used as source of electrons by the archaeon (Martin and Muller 1998). To maximize the surface area for the exchange of hydrogen or other materials, physical cell-to-cell contact between the two partners would have become increasingly intimate, until the bacterium ended up being entirely contained within the archaeon. (Note here the uses of words: the word "bacteria" refers to many bacteria, and a single one is a "bacterium." For Archaea, a single one is an "archaeon," and we can use adjectival forms such as "bacterial" and "archaeal" to describe processes and anatomy, such as "bacterial cell." It's unfortunate that the first division of geological time is the Archaean, as we saw in Chapter 1.)

The bacteria were protected from external stresses, and as they respired food molecules they released some energy to their archaeal hosts. Their capacity for aerobic respiration might also have helped to consume, and therefore detoxify, environmental oxygen, which may have been increasing in atmospheric concentration at the time due to the evolution of cyanobacteria and oxygenic photosynthesis. With time, the bacteria lost their cell walls and became organelles (mitochondria), and the host was no longer a simple archaeal prokaryote but a true eukaryotic cell. This symbiotic organism would have had (and still has) two genomes (three in the case of photosynthetic eukaryotes) – one from each symbiont – and the nuclear membrane might have evolved to keep them distinct.

This reorganization of two individual organisms into a single, larger and interdependent whole would have radically changed the energy budget of the first eukaryotic cells, as well as the mechanisms that were required for growth and replication. It would also have opened up new ecological opportunities and changed the evolutionary pressures faced by the cell and probably enabled the evolution of increasing intracellular complexity.

These early eukaryotes now received so much energy from the respiration of their mitochondria that they came to depend entirely on them to provide them with ATP. The number of mitochondria had to be matched closely to the needs of the host cell, so most of the genes that controlled mitochondrial reproduction and development were transferred away from the mitochondria and packaged into the host's DNA, inside the nucleus. Now as eukaryotic cells grew and flourished, so did their mitochondria, and as a eukaryote divided, each daughter cell took some mitochondria with it.

The modern relatives of the archaeal and bacterial partners that merged into the first eukaryote can be traced by comparing genes found in the eukaryotic nuclear and mitochondrial genomes to modern, free-living Archaea and Bacteria. Our understanding of the modern diversity of prokaryotes is improving rapidly, through the development of techniques that enable microbiologists to sequence genomes directly from environmental samples, without first needing to grow the microorganisms in the lab. This ever-expanding diversity has allowed us to identify the closest modern relatives of the archaeal and bacterial cells that gave rise to eukaryotes, and to study their genomes to understand the origins of eukaryotic cellular complexity (Williams et al. 2013).

Among prokaryotes, the Asgard Archaea (Eme et al. 2017) are the modern lineage that is most closely related to the host cell; eukaryote nuclei contain a trace of this archaeal ancestry in the slow-evolving genes that encode core components of the cell, such as the machinery for DNA replication and protein production (translation). Similar analyses of mitochondrial DNA suggest that mitochondria are related to the Alpha-proteobacteria, a major group of bacteria. The genetic similarities among all eukaryotes, and all mitochondria, suggest that modern eukaryotes all descend from a single endosymbiotic event.

This scenario produces protists (Figure 3.2), single-celled eukaryotes capable of moving and feeding by engulfing other organisms. Their food is fermented in the cytoplasm and oxidized in mitochondria. The same process occurs in our cells today.

In another major evolution of cell symbiosis that also happened only once, an early protist took in cyanobacteria as symbiotic partners which became plastids (Figure 3.3). The cyanobacteria benefited more from nutrients in the host's wastes than they would as independent cells. In time, the protist came to rely so much on the photosynthesis of its partners that it gave up hunting and engulfing other cells, gave up locomotion, grew a strong cellulose cell wall for protection, settled or floated in well-lit waters, and took on the way of life that we now associate with the word *plant*. Plant

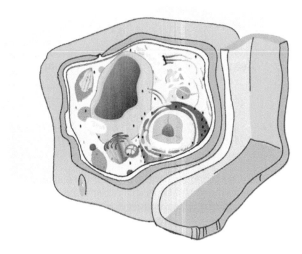

Figure 3.3 A plant cell. Here the cell membrane (*green*) surrounds the cell cytoplasm (*blue*), which contains not only the nucleus (*red*) and mitochondria (pill-shaped, *lime green*), but also chloroplasts, organelles that once were free-living cyanobacteria (*green*). *Source:* Image by Mariana Ruiz Villareal, Lady of Hats (Wikimedia).

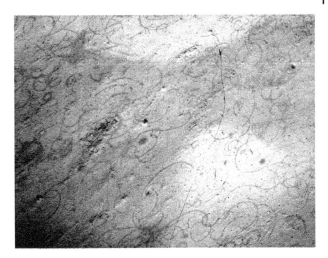

Figure 3.4 A block of rock showing fossils of *Grypania spiralis*, from the Negaunee Iron Formation of Michigan. *Grypania* is probably the earliest multicellular alga, with fronds about 1 mm across. *Source:* Photograph by Xvazquez (Wikimedia).

photosynthesis is not performed in the cell cytoplasm, but only in the plastids.

The word **symbiogenesis** is used to describe the appearance of a dramatically new biological or ecological ability by symbiosis rather than simple mutation. It is a useful term, because we are discovering more and more examples in living ecosystems. For example, many plants are successful because they have symbiotic fungi in or around their roots, which help to break down soil debris and make it available to their plant partners; in turn, they take some nutrition from the plant roots. Nevertheless, although symbiogenesis seems dramatic, it is a normal part of evolution. Species must acquire the mutations that allow them to take part in symbioses. Each species in the symbiosis continues to evolve under natural selection, and individuals that take part in symbioses do so because they reproduce more effectively than those that do not. That is as true for the partners in a eukaryote cell as it is in the later examples of symbiosis.

## Eukaryotes in the Fossil Record

The endosymbiotic theory for the origin of eukaryotes is based on biological and molecular evidence (see Box 3.2) but it is difficult to identify the first fossil eukaryotes (Fig. 3.4). Most fossil cells are small spherical objects with no distinguishing features. Most eukaryotes are much larger than prokaryotes, but at least one living prokaryote approaches normal eukaryotic size. Experiments to make artificial fossils from rotting prokaryotes have shown that it is almost impossible to distinguish them from eukaryotes after death. After death, the cell contents of prokaryotes can form blobs or dark spots that look like fossilized nuclei or organelles. Rotting colonies of cyanobacteria can look like multicellular eukaryotes, and filamentous bacteria can look like fungal hyphae. And finally, early eukaryotes were probably small and thin-walled, and therefore are most unlikely to be preserved as fossils.

We have to interpret the geological record as best we can. First of all, eukaryotes could not have evolved before oxygen became a permanent component of seawater. But that could have happened in early stromatolites, if "oxygen oases" formed round patches of stromatolites while the rest of the world was anoxic.

The oldest eukaryotes are acritarchs from the Changcheng Formation in North China. These are discriminated from prokaryotes by their large size (40–250 μm) and complex wall structure, including striations, longitudinal ruptures, and a trilaminar organization. However, it can only be said that these are eukaryotes, but they cannot be assigned to any modern group of eukaryotes. But this places a minimum fossil date on the origin of Eukaryota of 1.62 Ga, the age of the Changcheng Formation, and so it is likely that they had evolved by 1.8 Ga, according to recent studies of genomic data and fossils (Betts et al. 2018). *Tappania* may not be the earliest eukaryote, but it is among the most beautiful. The specimen in Figure 3.5 is likely a resting stage or cyst, and it dates from about 1.4 Ga.

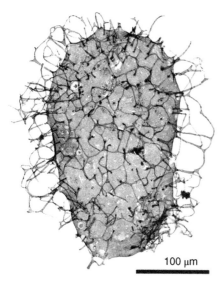

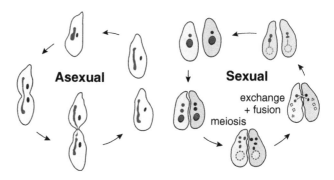

Figure 3.6 Comparing the basics of asexual and sexual reproduction, where cells simply split or clone (asexual, *left*), or males and females combine, exchange genetic material, and the genetic material is redistributed during meiosis. *Source:* IHMC image, Creative Commons.

100 μm

Figure 3.5 *Tappania*, an undoubted eukaryote. This specimen dates from about 1400 Ma. It is called an "acritarch" to signify that despite its excellent preservation, we are not sure which group of eukaryotes it belongs to. *Source:* Image courtesy of Nicholas Butterfield of Cambridge University, who thinks *Tappania* is a fungal cell (Butterfield 2015).

## The Evolution of Sex

Eukaryotes have sexual reproduction, but prokaryotes do not. Essentially, every prokaryote is its own lineage, either dying, budding off, or splitting into daughter cells that are clones of the parent, in the process of asexual reproduction. Nonetheless, prokaryotes do obtain new genes (and improved versions of existing genes) by means of **horizontal gene transfer**, a process whereby viruses and other prokaryotes inject parts of their genome into another organism, which then incorporates genes into their own genome.

Cell division is simple for prokaryotes. There are no mates to find, no organelles to organize. Daughter cells are clones, with the same DNA as the parent cell (except for mutations happening when the prokaryotic DNA is duplicated), so they are already well adapted to their microenvironment. However, because prokaryotic populations are large, prokaryotes still generate a significant amount of genetic diversity at each generation. They use this diversity to gamble against changes in the environment; if a change occurs that kills an individual, that change would stand a good chance of wiping out all that individual's clones too. However, mutations in prokaryotic populations allow for the fast emergence of genetic variants that can survive in an ever-changing environounment. Another name for cloning is **asexual reproduction**, because reproduction happens without having males and females and exchange of genetic material (Figure 3.6).

In eukaryotes, all the DNA of two individuals taking part in **sexual reproduction** is shuffled and redealt to their offspring, and in eukaryotic populations the genetic diversity of the next generation does not depend exclusively on newly emerged mutations. Offspring are similar but not identical to their parents: in fact, there is an impossibly low chance that any two sexually reproduced individuals are genetically identical, unless they developed from the same egg, as identical twins do.

The offspring of sexual reproduction resemble their parents in all major features but are unique in their combination of minor characters. Sexually reproducing species have built-in genetic variability that is often lacking in clones of bacteria. Nevertheless, in both asexual and sexual organisms, individuals vary in the characters of their bodies, which often means that some individuals are better fitted to the environment than others, so stand a better chance of reproducing. The genomes of those individuals are thus differentially represented in future populations.

In organisms that reproduce by cloning, a favorable mutation can spread successfully over many cycles of cloning if it occurred in an individual that divided faster than its competitors. In contrast, a mutation in a sexually reproducing individual is shuffled into a different combination in each of its offspring. For example, a mutant oyster might find her mutation being tested in different combinations in each of her 100 000 eggs. Natural selection could then operate on 100 000 prototypes, not just one. Favorable combinations of genes can be passed on effectively. An important advantage of sex is that sexually reproducing populations can evolve rapidly and smoothly in changing environments.

At the same time, sexual reproduction is conservative. Extreme mutations, good or bad, can be diluted out at each generation by recombination with normal genes. The genes may not disappear from the population but may lurk as recessives, likely to reappear at unpredictable times as recombination shuffles them around.

In eukaryotes, physical, chemical, and behavioral mechanisms usually ensure that sex is attempted only by individuals that share much the same DNA. Such a set of organisms forms a species, defined as a set of individuals that are potentially or actually interbreeding. The composite total of genes that are found in a species is called the gene pool.

Sexual reproduction has two major costs. First, the male or female sexual individual passes on only half of its DNA to the offspring, with the other half coming from the partner. Therefore, to pass on all its genes, a sexual individual has to invest double the effort of an asexual individual. Second, the offspring of sexual parents are not identical. Sets of incompatible genes may be shuffled together into the DNA of an unfortunate individual, which may die early or fail to reproduce. At every generation, then, some reproductive "wastage" occurs.

Many eukaryotes can also reproduce by simple fission, cloning identical copies of themselves. An amoeba is perhaps the most familiar example, but corals, strawberries, Bermuda grass, and aphids often use this method too. But it's likely that sex evolved once, in the earliest eukaryotes, and only a few lineages now resort to asexual reproduction. It appears that, in general, the benefits of sex in changing environments outweigh its costs under many other conditions.

## The Classification of Eukaryotes

Eukaryotes occur in the natural world in ecological and evolutionary units called species. As we have seen, species are groups of individuals whose genetic material is drawn from the same gene pool but is almost always incompatible with that of another gene pool. Members of the same species, therefore, can potentially interbreed to produce viable offspring. They tend to share more physical, behavioral, and biochemical features (characters) with one another than they do with members of other species. Defining and comparing such characters allows us to distinguish between species of organisms. A species is not an arbitrary group of organisms but a real, or natural, unit.

Biologists use the Linnean system of naming species, after the Swedish biologist Carl von Linné who invented it in the eighteenth century. A species is given a unique name (a specific name) by which we can refer to it unambiguously. Linné gave the specific name *noctua* to the European little owl (Figure 3.7) because it flies at night. Species that share a large number of characters are gathered together into groups called genera (the singular is genus) and given unique generic names. Linné gave the little owl the generic name *Athene*. Athena is the Greek goddess of wisdom, and

Figure 3.7 *Athene noctua*, the little owl of Europe, was named by Carl von Linné. *Source:* Photograph by Trebol-a (Wikimedia).

the little owl is the symbol of the city of Athens, stamped on its ancient coins. However, taxonomic names do not have to carry a message, even though a simple and appropriate name is easier to remember. (One must be careful about names, too: *Puffinus puffinus* is not a puffin but a shearwater, and *Pinguinus* is not a penguin but the extinct great auk!) Thus, Linnean names are only a convenience, but a very valuable one. The bird that the British call the tawny owl, the Germans the wood owl, and the Swedes the cat owl is *Strix aluco* among international scientists.

Genera may be grouped together into higher categories (Figure 3.8). For example, *Athene* and *Strix* and many other owls are grouped together to form the Family Strigidae, named after *Strix*. Families may be grouped into superfamilies, and then into orders, classes, and phyla.

Any division or subdivision that is used to group organisms is called a taxon (plural, taxa). Biologists who try to recognize, describe, name, define, and classify organisms are taxonomists or systematists, and the practice is called **taxonomy** or **systematics** or **classification**. Slightly different ranks of categories are used for different kingdoms of organisms, but the basic units of classification recognized by all biologists remain the species and genus. Taxonomy is an international endeavor, governed by agreed rules for naming plants, animals, and microorganisms. These rules set standards of good practice (e.g., how to describe your new species adequately; making sure the first name to be given to a species historically has precedence).

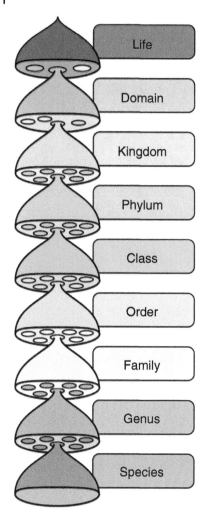

Figure 3.8 A simple diagram of the Linnean hierarchy of taxa as it is used today. *Source:* Drawn by Peter Halasz (Wikimedia).

Charles Darwin in 1859 showed why organisms can be classified into definite groups according to their evolution. For example, everyone agrees what are cats and cat-like (members of Family Felidae) and what are dogs and dog-like (members of the Family Canidae). Paleontologists have shown that there was once an ancestral cat-like species that gave rise to modern lions, tigers, cheetahs and domestic cats, and an ancestral dog that gave rise to modern wolves, foxes, hunting dogs, and domestic dogs. Generally, species fit in their categories and do not randomly hop about, and this is because of their evolutionary history. But how are we to understand the evolutionary history of such categories and ultimately of life?

## Cladistics

The solution to discovering the tree of life came about because of two revolutions in the 1960s. The first was the **molecular clock**, which we shall explore further in

Chapter 5, and the second was **cladistics**, a method of turning taxonomy into a science rather than an art. Up to the 1960s, taxonomists had used ideas about general resemblance to classify things, and it usually worked. After all, everyone agrees that lions are in the cat family and wolves are in the dog family. So, without the molecular clock and cladistics, biologists and paleontologists mainly got it right – but they could not explain exactly why, nor could their ideas be tested. This is the basis of science – hypotheses (= claims about nature) should be testable.

The word *cladistics* describes an approach to classifying species that uses the fact that every species began by branching from another (*klados* is the Greek word for a branch). So, every species is a **clade**, if it is defined accurately. And the same is true of every genus, family, order, class, and phylum – they are all clades, and each one is defined in such a way that it shares a single common ancestor and includes all descendants of that common ancestor. So, groups such as Dinosauria, Mammalia, Canidae, Mollusca, and Homo are clades with a point of origin, a particular history, and usually a set of distinguishing or diagnostic characters.

The idea of **diagnostic characters** is at the core of the cladistic method. We must look at all the features of an organism and decide which ones really matter; which ones mark a point in evolution. Some familiar cases might make this clearer. For example, everyone knows that birds have feathers, and so we might choose the character "Possession of feathers" to be diagnostic of a clade. We think about this carefully – could feathers have evolved many times, and so are we being hoodwinked? Probably not, because feathers are highly complex in all sorts of details, and we can study their development in baby birds to adults, and it's the same every time. Feathers are different from hairs (which diagnose mammals) and scales, which are seen in various forms in fishes and reptiles.

But which clade is identified by the diagnostic character of feathers? If we only look at living animals, it's birds, or more formally the Class Aves. But when we include fossils, the clade gets larger. Up to 1990, the oldest bird was identified as *Archaeopteryx* (see Chapter 14), and the fossils are so remarkably preserved that they show *Archaeopteryx* had feathers, even 150 million years ago in the Late Jurassic. Now we know that many dinosaurs more basal in the evolutionary tree than *Archaeopteryx* also had feathers, thanks to the amazing new fossils from China. We could then perhaps limit the definition to "Possession of pennaceous feathers," meaning feathers with a central quill and branching barbs on either side (excluding whiskers, bristles, protofeathers, down feathers, etc.). Pennaceous feathers are diagnostic of a large clade called Coelurosauria, including many dinosaurs as well as all birds (Figure 3.9).

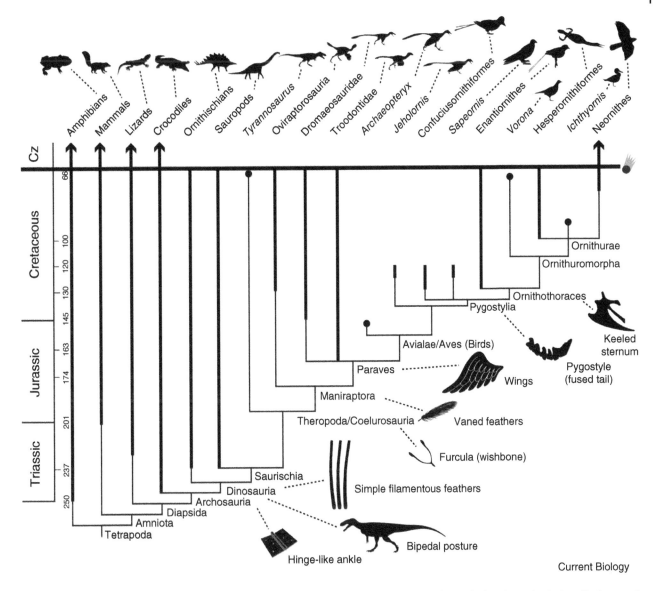

Figure 3.9 An evolutionary tree of tetrapods, showing how they are related and when the splits took place in geologic time. Each named group along the top, and climbing up the tree (horizontal terms, from Tetrapoda to Ornithurae) are clades, and in the case of the line from dinosaurs to birds some diagnostic anatomical characters are indicated, including ankle type, posture, various feather types, and aspects of the skeleton. *Source:* After Brusatte et al. (2015, *Current Biology*), with permission.

In this example (Figure 3.9), within Coelurosauria we can identify a whole series of included clades, among which the wider clade Aves or Avialae (i.e., *Archaeopteryx* plus the sparrow plus all their common ancestors) is characterized by a diagnostic character such as "Wing capable of flapping flight," and modern birds, the clade Neornithes, are characterized by the diagnostic characters of teeth absent, dentaries fused anteriorly, and 11 or more sacral vertebrae.

But how do we identify these important diagnostic characters? Why not say that birds should be diagnosed by the fact that they all have eyes or tails? It probably does not take long to answer that – nearly all animals

have eyes and all vertebrates have tails. The characters eyes and tails are not diagnostic for birds alone, but occur much more widely. So, the core of cladistics is to consider the possible arrangements of any groups of species of interest and try to identify those useful characters that might be diagnostic or count as evolutionary **novelties**, literally "something new." Feathers were novelties when they emerged first, whether in the Triassic or Jurassic, and they are also diagnostic.

Let us look at a simple example. Three living species, A, B, and C, could be related to each other in three different ways (Figure 3.10). Which is correct? Which two of the three species are most closely linked? Two species

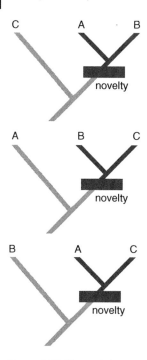

Figure 3.10 Three cladograms that show all the possible relationships between the three species A, B, and C, that had a common ancestor. Two of them may share a newly evolved character that the third does not have: the two species with the derived character are sister species. The cladograms reflect the evolutionary changes that occurred within the group.

may look very similar because they share similar characters, but if those are shared ancestral characters that were also present in a common ancestor, they cannot tell us anything about evolution within the group, because they have not changed within that history. The useful characters for solving the problem are the novelties, because they define how the species have changed since they shared the same ancestral characters in their common ancestor (Figure 3.9).

Figure 3.9 shows three **cladograms** which display the distribution of characters in a visual form. The cladogram that requires the simplest and fewest evolutionary changes is assumed to show the most likely history of the species. A cladogram therefore expresses a hypothesis about the phylogeny of a group. Two species are most closely linked and form a sister group. In turn, the third species becomes their sister group in a larger clade.

Care is required in choosing diagnostic characters. For example, bats and birds both have wings, and in each group the wing is a derived character that has been modified from some other structure. But bats and birds share very few other derived characters, and even their wings have a different basic structure. The weight of evidence

suggests that birds are a clade, bats are probably a clade, but [bats + birds] is not a clade. Bats have hair, mammary glands, complex teeth and all the other diagnostic characters of mammals, and the weight of this evidence, plus the fact that their wings are very different in detail, immediately highlights the mistake.

In 1995, a complete reading of the genome of a living organism, the gram-negative bacterium *Haemophilus influenzae*, was completed (see Chapter 5). The first sequencing of a complete genome was a momentous scientific breakthrough that catapulted biology into the *"genomic era."* To date, genomic data for 10s of 1000s of living species across the tree of life, and several key extinct species (like our closest cousin, *Homo sapiens neanderthalensis*) have been obtained. It quickly emerged that genomic data are ideally suited for cladistic and molecular clock analysis, ushering in a new understanding of the evolution of life. Cladograms (hypotheses of phylogenetic relationships) derived using morphological data could now be tested using the information in our genomes.

Being able to cross-compare results from two broadly independent approaches (cladistic analysis of morphology and genomic analysis) has given real power to studies in comparative paleontology and biology. The cladogram or genomic tree can be converted into a **time tree** (see Figure 3.9 for an example) by adding a time scale and paying close attention to the ages of fossils.

Unexpected patterns sometimes emerge from cladistic and genomic analyses. We are all used to thinking about living fishes, amphibians, reptiles, birds, and mammals as classes of vertebrates, equal in rank to one another (Figure 3.11a). But this is not a cladistic classification. Tetrapods are actually a clade within fishes, derived from them by acquiring some novel characters, including feet, and amphibians are a clade within tetrapods. Reptiles, mammals, and birds are also clades of derived tetrapods.

If we classify all living reptiles as one group and draw a cladogram of vertebrates (Figure 3.11b), we display the well-known fact that living reptiles and birds are more alike than either is to mammals. The cladogram also carries other information. It shows that warm blood, a derived character that living birds and mammals share, must have evolved independently at least twice, unless living reptiles have lost warm blood.

As we consider smaller subgroups of living and fossil reptiles, we find that this neat picture of reptile classification breaks down, so we must revise our ideas about tetrapod evolution. Figure 3.11c shows that "living reptiles" is not a clade. We could define a clade called "reptiles" but we would have to include birds in it. In the same way, humans are derived fishes, derived amniotes, and derived primates, all at the same time.

**Figure 3.11** (a) A traditional classification of the five classes of living vertebrates, showing them as equal rank. (b) A cladogram of those five classes, showing that cladistically they are not equal in rank; for example, living mammals are the sister group of (living reptiles + living birds). Warm blood (W) was independently evolved as a derived character in both mammals and birds, according to the hypothesis expressed in this cladogram. (c) When we look more deeply into "living reptiles" (*shown in blue*) and we add the extinct dinosaurs to the cladogram, we must change our assessment of vertebrate evolution. Reptiles are not a clade unless they also include birds, and "living reptiles" are not a clade either. Dinosaurs are the sister group of birds, which brings up the question of where warm blood (W) evolved in that lineage. Is it a derived character of birds or is it a derived character shared by both birds and dinosaurs? Drawing a cladogram forces us to look at such evolutionary questions!

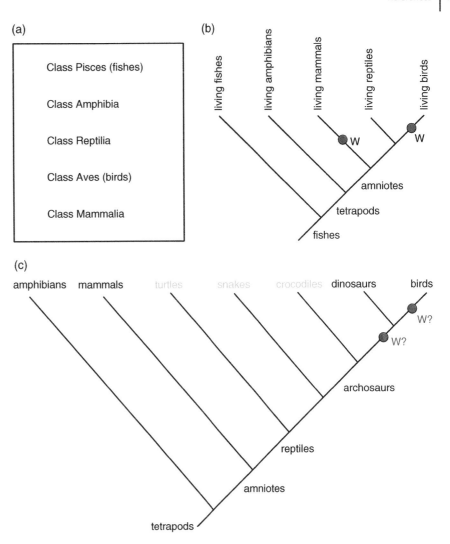

We have fulfilled the vision Charles Darwin had that all of life originated in the deep, distant past in "some warm little pond" – well, maybe in the ocean. But the point is that all organisms, from humans to bacteria and viruses to oak trees, are all part of a single great evolutionary tree, and we now have forensic, scientific ways such as cladistics and genomics to discover the exact shape of that tree.

## References

Betts, H.C., Puttick, M.N., Clark, J. et al. (2018). Integrated genomic and fossil evidence illuminates life's early evolution and eukaryote origin. *Nature Ecology and Evolution* 2: 1556–1562.

Brusatte, S.L., O'Connor, J.K., and Jarvis, E.D. (2015). The origin and diversification of birds. *Current Biology* 25: R888–R898.

Butterfield, N.J. (2015). Early evolution of the Eukaryota. *Palaeontology* 58: 5–17.

Eme, L., Spang, A., Lombard, J. et al. (2017). Archaea and the origin of eukaryotes. *Nature Reviews Microbiology* 15: 711–723.

Martin, W. and Muller, M. (1998). The hydrogen hypothesis for the first eukaryote. *Nature* 392: 37–41.

Williams, T.A., Foster, P.G., Cox, C.J., and Embley, M. (2013). An archaeal origin of eukaryotes supports only two primary domains of life. *Nature* 504: 231–236.

## Further Reading

Cotton, J.A. and McInerney, J.O. (2010). Eukaryotic genes of archaebacterial origin are more important than the more numerous eubacterial genes, irrespective of function. *Proceedings of the National Academy of Sciences of the United States of America* 107: 17252–17255.

Gross, J. and Bhattacharya, D. (2010). Uniting sex and eukaryote origins in an emerging oxygenic world. *Biology Direct* 5: 53.

Longsdon, J.R. (2010). Eukaryotic evolution: the importance of being archaebacterial. *Current Biology* 20: 1078–1079.

## Question for Thought, Study, and Discussion

If sexual reproduction is so inefficient and wasteful, why is it that we do not simply evolve virgin birth? Many species of insects, spiders, snails, sharks, and lizards do this, having only females. Women would give birth to babies identical to themselves, obviously completely capable of living successful lives.

4

## The Evolution of Metazoans

In This Chapter

In this chapter, we follow the evolution of single-celled eukaryotes into multicellular organisms, and then into multicellular animals with complex structures that include different organ systems. We start with a new discovery about the evolution of multicellular life in the laboratory. But different ways of life evolve in specific environments on the Earth, so we discuss how climate change on Earth affected major breakthroughs in evolution, during a cold period called Snowball or Slushball Earth. Then as climate changed again, the animals evolved. There has been heated debate around when this happened in relation to the dramatic explosion of animal groups in Earth's seas, which is recorded in the fossil record from around the start of the Cambrian period 541 million years ago. We look at the evidence for the timing of animal origins from both the fossil record and from information encoded in the genetic material of living animals. Taken together, these suggest that the first animals evolved long before the start of the Cambrian, but that there was then a major diversification in the Cambrian, when many of the modern animal groups evolved.

## Proterozoic Microbes

For much of the Proterozoic, we see little change in ocean chemistry and slow biological evolution, and this has led scientists to call the interval between 1800 and 800 Ma the "boring billion." The Proterozoic ocean was anoxic, rich in dissolved iron and silica and sulfide. It probably had low quantities of some biologically important elements, which may have limited eukaryote activity and slowed their diversification. The deep waters must have been largely inhabited by bacteria and methanogens, because eukaryotes require some oxygen to run their mitochondria. Only a zone of surface waters had photosynthesizers, and even then the oxygen levels were low and probably varied from time to time and from place to place. Oxygen-tolerant microbes and single-celled eukaryotes were probably confined to this surface zone.

Microfossils occur in these Proterozoic rocks, but they are not abundant or diverse. Beginning about 1800 Ma, we find **acritarchs**, spherical microfossils with thick and complex organic walls (see Chapter 3, Figure 3.5). They were organisms, most likely eukaryotes of various kinds (Figure 4.1), that grew thick organic walls (cysts) in a resting stage of their life cycle but spent the rest of their lives floating among the plankton, the organisms that live in the surface waters of oceans and lakes. For the next 800 million years, we see microfossils that are difficult to interpret but probably reflect a slow diversification of eukaryote protists. Multicellular eukaryotes including algae and fungi evolved during this time and are found in the fossil record.

No automatic barrier forces protists to be single-celled. A protist that divided with its daughter cells remaining together could form a multicelled colony in only a few cycles of division. A process like this in an algal cell could have produced multicellular *Grypania* (see Chapter 3). However, there was a breakthrough on this question when a team led by Will Ratcliff (Ratcliff et al. 2012) applied artificial selection to a unicellular strain of brewer's yeast. They selected out all yeast that sank slowly in water, allowing the fastest 5% to keep on reproducing. This encouraged the yeast to form small "colonies" of cells that stayed together after dividing, because larger cell clusters sink faster than single cells. After 60 days, these authors had yeast cultures that were entirely

*Cowen's History of Life*, Sixth Edition. Edited by Michael J. Benton.
© 2020 John Wiley & Sons Ltd. Published 2020 by John Wiley & Sons Ltd.

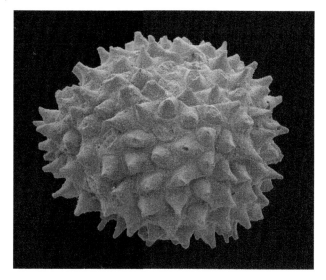

Figure 4.1 An acritarch from the Ediacaran Doushantuo Formation. *Source:* Image by Kelly Vargas used previously by Cunningham et al. (2017b).

multicellular, little spherical colonies that looked like snowflakes. The snowflakes reproduced like multicellular organisms, breaking apart the parent snowflake into two. And when the sinking selection was removed, none of the cultures reverted back to a single-celled existence. They had become, biologically and functionally, multicellular organisms (and their DNA had changed, so the switch to multicellularity was an evolutionary process). What this means for paleontologists is that we can stop worrying about how multicellular organisms could evolve – it was (comparatively) easy!

## Snowball or Slushball Earth

In the late Proterozoic, starting about 800 Ma, increased oxygen levels seem to have gradually extended the oxygen-bearing zone deeper into the ocean, cutting down methane production in deep waters. There was a notable diversification of eukaryotes then, and more of the shallow sea floor would have become inhabitable by eukaryotic protists as well as Bacteria and Archaea. For the first time in global history, we can envisage sea floors with successful populations of protists.

After this, the Earth went through a series of dramatic cold periods late in the Proterozoic, in a time period called the **Cryogenian**, between 720 Ma and 635 Ma. These were likely brought on by the break-up of the supercontinent Rodinia and the associated eruption of flood basalts. The continental rearrangement is thought to have led to increased rainfall and therefore rapid weathering of the continents, including the flood basalts.

This weathering would have removed $CO_2$ from the atmosphere and so led to a dramatic cooling.

Many deposits from this period contain glacial debris, and many of them occur in regions that are reliably reconstructed near the Equator at the time. These deposits imply episodes of massive and widespread glaciation, much more extensive than any glaciations that have occurred since. One scenario that attempts to explain their wide distribution is "Snowball Earth."

Immediately after the Cryogenian, we see a radiation of new multicellular organisms, including possible animals, and so many researchers have speculated on a link between the dramatic physical events of Snowball Earth and the evolutionary emergence of metazoans.

The Snowball Earth model (Hoffman et al. 2017) proposes that the ocean surface was frozen all the way to the Equator. Surface temperatures dropped to about –40 °C. As the ice spread, photosynthesis was choked off and most life in the oceans died off. The only surviving life would have been around sea floor hot vents and (perhaps) in surface ice. Solar radiation would be reflected back into space, so you would think Earth would be locked permanently into a snowball state.

However, the model proposes that volcanoes continued to erupt, putting carbon dioxide back into the atmosphere until there was once again enough carbon dioxide to trap solar heat and melt the ice. Calculations suggest it would have taken an enormous amount of carbon dioxide to break the grip of Snowball Earth. The ice cover would not have melted until volcanoes had erupted around 350 times more carbon dioxide than there is in our present atmosphere.

It is argued that the ice then melted quickly, but the enormous reservoir of carbon dioxide in the atmosphere rocketed the whole Earth directly into a "greenhouse" hot period, with temperatures averaging around 50 °C (over 120 °F). Deluges of acid rain would then have acted on the sterilized continents, flooding the ocean with carbonate, and thick limestones formed very quickly on top of the glacial deposits. Finally, weathering and photosynthesis would have brought down carbon dioxide levels, and the world recovered biologically. However, the geographic set-up that had begun the Snowball Earth cycle was still present, so the cycle then repeated itself three times, twice in the Cryogenian and once in the succeeding Ediacaran.

However, there is good evidence that the Snowball Earth scenario is too extreme (Figure 4.2). The glacial sediments include dropstones, rocks which fall from floating icebergs into soft sea-floor sediment. Those icebergs must have been floating freely, in open water. A "Slushball Earth" concept, which calls for less than a

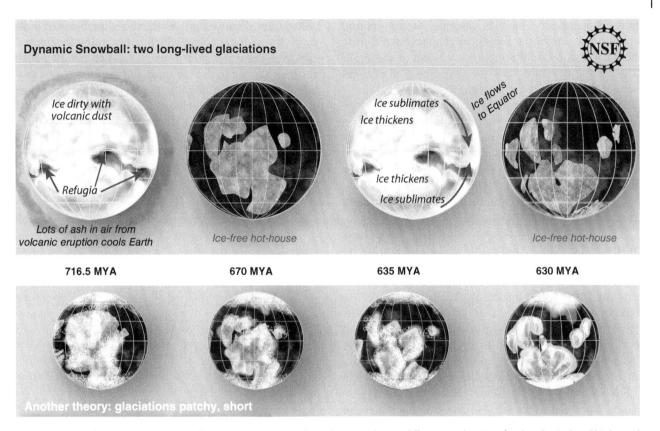

**Figure 4.2** Snowball Earth and Slushball Earth: two alternative hypotheses with very different implications for the physical and biological Earth. *Source:* Graphic by Zena Deretsky: courtesy National Science Foundation.

completely frozen Earth, is supported by computer models that suggest a stable climate, with low-latitude continental ice sheets and seasonal floating sea ice over much of the world's oceans. These models project open tropical waters at cool to mild temperatures (up to 10 °C at the Equator). The model atmosphere has only 2.5 times today's carbon dioxide, and the stability of the model implies that any small rise in those levels will easily revert conditions to Earth normal, without the extreme greenhouse called for by the Snowball Earth idea.

The deglaciations that took place at this time are thought to have played an important role in shaping the evolution of life. Brocks et al. (2017) extracted biochemical compounds, called **biomarkers**, that indicate bacteria and algae. They showed that cyanobacteria had dominated the oceans before the Cryogenian and during the first major glaciation. Eukaryotic algae became ecologically important during the deglaciation event between the two major Cryogenian glaciations, probably because of a surge in nutrients caused by extreme weathering. Further biomarkers that may have been produced by sponges are found at the end of the Cryogenian, suggesting a potential link between this event and the origin of animals.

## The Ediacaran Period

The end of the great glaciations of the Cryogenian around 635 Ma begins a major change in Earth's physical and biological evolution. There was a biological revolution as metazoans became major players in the oceans, Earth's atmosphere became more oxygen rich, and its climate was moderated so that there have never again been glaciations of Cryogenian magnitude. A variety of potential, but controversial, animal fossils are known from this time. These include possible animal embryos from around 609 Ma. Soon after 565 Ma, in rocks found worldwide from Canada to Russia to Australia, we find soft-bodied animals that make up the Ediacaran fauna, named after rocks found in the Ediacara Gorge in the Flinders Ranges near Adelaide. These fossils lived during a newly recognized period of geological time, the Ediacaran Period, between the end of the Cryogenian at 635 Ma and the base of the Cambrian at 538 Ma.

The Ediacaran represents a major shift in the history of life; before this interval, life was dominated by stable microbial ecosystems, afterwards by dynamic macroevolution of large and complex eukaryotes. The environment also changed markedly, with the oceans becoming more oxygenated. There is ongoing debate around whether this drove the biological changes seen during the Ediacaran by facilitating the evolution of metabolically active organisms, or if the environmental changes were, in fact, mediated by changes in biology.

Eukaryotic algae are generally larger than cyanobacteria that had dominated previously. Lenton and colleagues (2014) argued that this made them sink more rapidly through the water column. Oxygen is usually used up when organic tissue decays but if organic matter falls unoxidized to the sea floor, it reduces the oxygen consumption in the upper waters. This allows oxygen to reach deeper levels in the oceans.

The evolution of metazoans with guts would have enhanced the effect (Butterfield 2018). They process their food and produce carbon-rich waste in compact fecal pellets. Those fecal pellets drop quickly through the water and if they are buried quickly, that will increase oxygen levels in the sea and in the atmosphere. So, the rise of metazoans big enough to produce quantities of fecal pellets could have led to a rise in oxygen. In turn, higher oxygen levels may have permitted larger, more active metazoans to evolve, and so on.

The evolution of larger metazoans may have prevented the recurrence of the extreme glaciations of Snowball or Slushball Earth. Surface productivity could no longer draw down carbon dioxide to critically low levels because primary producers were eaten back by new planktonic predators (small or larval metazoans). Stronger metazoan burrowers dug up buried carbon from sea-floor muds and recycled it into carbon dioxide. And complex populations in near-shore waters intercepted nutrients before they reached the oceanic sea surface.

All this is another example of the continuous interplay between life and Earth's physical environment. The evolution of metazoans was made possible by the increase in oxygen levels that resulted from increased photosynthesis, which in turn resulted from increased nutrients released by major glaciations. But the evolution of metazoans also acted to moderate the dramatic shifts in Earth's climate.

Next, we will examine how animals might have first evolved, before looking at the timing and fossil record of their early evolution.

## Making a Metazoan

A flagellate protist is a single cell with a lashing filament, a **flagellum** (plural, flagella), that moves it through the water. Some flagellates called **choanoflagellates** are close relatives of the animals. Choanoflagellates build a conical collar around the flagellum (*choana* is the Greek word for collar) that has pores through it, rather like a

coffee filter (Figure 4.3a). As the flagellum beats, it pulls water through the collar, which collects tiny food particles from the water such as bacteria. A choanoflagellate can also anchor to the sea floor and feed using this system. Instead of dividing into two independent daughter cells, some choanoflagellates form **colonies**: they bud off new individuals which then stay together to form a group (Figure 4.4a). The flagella of all the members

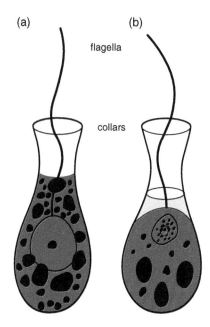

**Figure 4.3** From eukaryote to metazoan. (a) A choanoflagellate that collects food as the flagellum pulls water through the collar. (b) A collar cell or chanocyte from a sponge. Here the collar cell is firmly anchored in the body of the sponge. *Source:* Adapted from Barnes et al. (2001). © Blackwell Science.

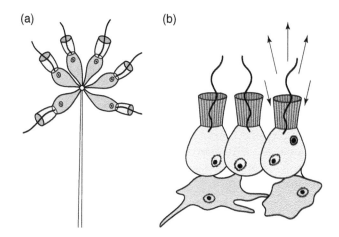

**Figure 4.4** (a) A colonial choanoflagellate. It has divided several times but the daughter cells have stayed together, and generate a powerful feeding current for the colony as a whole. (b) A group of choanocyte cells from a sponge. They are embedded in other tissues in this metazoan animal. *Source:* Both images from Barnes et al. (2001). © Blackwell Science.

of the colony now beat together to generate a powerful water current that makes a very efficient filtering system.

Metazoans evolved just once. They all originally had one cilium or flagellum per cell, for example. They also share the same kind of early development. They quickly form into folded balls of internal cells which are often free to move, and are covered by outer sheets of cells that form an external skin-like coating for the young animal. Sponges probably branched off first from the ancestral metazoan, by extending the choanoflagellate way of life to large size and sophisticated packaging.

Metazoans are not just multicellular. They have different kinds of cells that perform different functions. **Sponges** are the simplest metazoans living today. They contain many flagellated cells called **choanocytes** (Figure 4.3b), which are arranged so that they generate efficient feeding currents (Figure 4.4b). In turn, efficient groups of choanocytes pump water (and the oxygen and bacteria they capture from it) through the sponge, in internal filtering modules (Figure 4.5).

Sponges surely evolved from choanoflagellates (Figures 4.3 and 4.4), but they are much more complex because they also have other specialized cells. One breakthrough was to link cells firmly together to form a body wall, using a gene complex that is also found in all later metazoans. The sponge body wall can (very slowly) contract the sponge as a defense mechanism, even though it has no muscle cells. Other cells digest and distribute the food that the choanocytes collect, and yet others construct a stiffening framework, often made of mineral, that allows sponges to become large without collapsing into a heap of jelly (Figure 4.6).

**Cnidarians**, including sea anemones, jellyfish, and corals, are built mostly of *sheets* of cells, and they exploit the large surface area of the sheets in sophisticated ways to make a living. The cnidarian sheet of tissue has cells on each surface and a layer of jelly-like substance in the middle. The sheet is shaped into a bag-like form to define an outer and an inner surface (Figure 4.7). A cnidarian thus contains a lot of seawater in a largely enclosed cavity lined by the inner surface of the sheet. The neck of the bag forms a mouth, which can be closed by muscles that act like a drawstring. A network of nerve cells runs through the tissue sheet to coordinate the actions of the animal.

In most cnidarians, the outer surface of the sheet is simply a protective skin. The inner surface is mainly digestive and absorbs food molecules from the water in the enclosed cavity. Because cnidarians are built only of thin sheets of tissue, they weigh very little and can exist on small amounts of food. They can absorb all the oxygen they need from the water that surrounds them.

Figure 4.6 The skeleton of a deep-sea glass sponge, made of silica spicules. *Source:* Photograph by Randolph Femmer for the United States Geological Survey.

Figure 4.5 Sponges build modular filtering units that can reach high complexity. In these three examples of sponge structure, sets of choanocytes are in red, the outer skin of the sponge is yellow, and the outgoing water current of water that has been filtered is blue-green. *Source:* Diagram created by Philcha (Wikimedia).

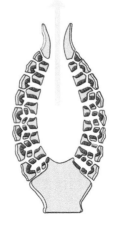
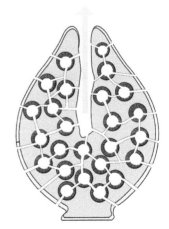

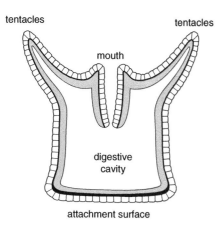

Figure 4.7 Basic structure of a cnidarian. A two-sided sheet of tissue defines the inside and outside surfaces of a bag-shaped digestive cavity. *Source:* From Boardman et al. (1987). © Blackwell Scientific.

Cnidarians have **nematocysts** or stinging cells set into the outer skin surface. The toxins of some nematocysts are powerful enough to kill fish, and people have died after being stung by swarms of jellyfish. Nematocysts are usually concentrated on the surfaces and the ends of tentacles, which form a ring around the mouth. They provide an effective defense for the cnidarian, but they are also powerful weapons for catching and killing prey, which the tentacles then push through the mouth into the digestive cavity (Figure 4.8). The tissues of the prey are then broken down by powerful enzymes and the food molecules are absorbed through the cells of the inner lining of the cavity. A cnidarian without jaws or a real gut can thus eat prey.

Hardly any sponges can tackle food particles larger than a bacterium, although there are a few exceptions. Yet living cnidarians routinely trap, kill, and digest creatures that outweigh them many times by using their nematocysts. However, there is no guarantee that the first cnidarians had nematocysts. They may simply have absorbed dissolved organic nutrients from seawater.

The third and most complex metazoan group contains all the other metazoans, including vertebrates. These are the Bilateria or **bilaterians**, metazoans with a distinct bilateral symmetry that influences their biology

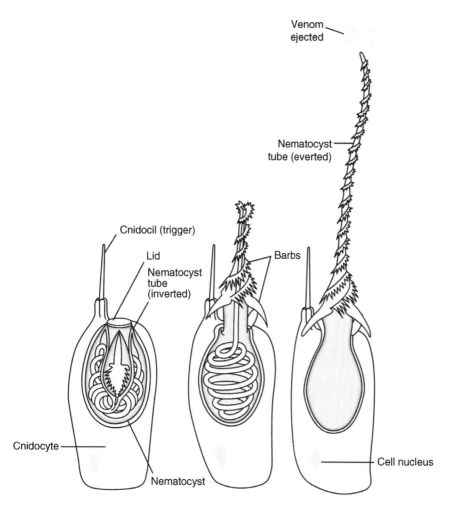

Figure 4.8 How a cnidarian fires a stinging netamocyst. *Source:* Reprinted from the Exploring Our Fluid Earth website at https://manoa. hawaii.edu/exploringourfluidearth/. © University of Hawaii. Reprinted with permission from the Curriculum Research & Development Group. The Exploring Our Fluid Earth program is a product of the Curriculum Research & Development Group (CRDG) of the University of Hawaii.

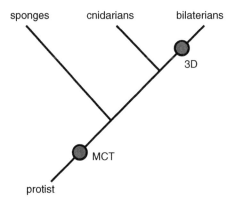

Figure 4.9 Simple phylogram of metazoans. MCT indicates the evolutionary point at which multicellular tissues evolved. 3D indicates a 3D structure of the body.

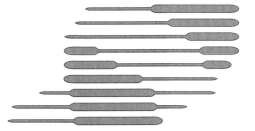

Figure 4.10 This worm-like bilaterian uses its coelom to burrow from right to left. It squeezes fluid forward to push out the front end, then makes it into a bulb. The back can then be pulled forward, and the cycle repeats.

enormously. They consist basically of a double sheet of tissue that is folded around with the inner surfaces largely joining to form a three-dimensional animal. In contrast to sponges and cnidarians, they have complex organ systems made from specialized cells, and those organ systems are built as the animal grows by special regulatory mechanisms coded in the genes (Figure 4.9). Worms are simple bilaterians.

All sponges and most cnidarians are attached to the sea floor as adults, and depend on trapping food from the water. But many bilaterians were (and are) mostly free-living animals, making a living as mobile scavengers and predators. The bilateral symmetry is undoubtedly linked with mobility; any other shape would lead to an animal that could not move efficiently.

The first bilaterians would have been worm-like. Worms creep along the sea floor on their ventral (lower) surface, which may be different from the dorsal (upper) surface. They prefer to move in one direction, and a head at the (front) end contains major nerve centers associated with sensing the environment. A well-developed nervous system coordinates muscles so that a worm can react quickly and efficiently to external stimuli. The mobility of early bilaterians on the sea floor probably led to the differentiation of the body into anterior and posterior (head and tail) and into dorsal and ventral surfaces, as the various parts of the animal encountered different stimuli and had to be able to react to them.

All but the simplest bilaterians have an internal fluid-filled cavity called a **coelom**, which may be highly modified in living forms. In humans, for example, the coelom is the sac containing all the internal organs. The coelom may have evolved as a useful hydraulic device. Liquid is incompressible, and a bilaterian with a coelom (a coelomate) can squeeze this internal reservoir by body muscles. Such squeezing pokes out the body wall at its weakest point, which is usually an end (Figure 4.10). Such a hydraulic extension of the body can be used as a power drill for burrowing into the sediment to find food or safety.

The coelom could have provided another great advantage for bilaterians. Oxygen must reach all the cells in the body for respiration and metabolism. Single-celled organisms can usually get all the oxygen they need because it simply diffuses through the cell wall into their tiny bodies. Sponges pump water throughout their bodies as they feed, and cnidarians and flatworms are at most two sheets of tissue thick. But larger animals with thicker tissues cannot supply all the oxygen they need by diffusion. Oxygen supply to the innermost tissues becomes a genuine problem with any increase in body thickness or complexity. If the animal evolved some exchange system so that its coelomic fluid was oxygenated, the coelom could then become a large store of reserve oxygen. Eventually, the animal could evolve pumps and branches and circuits connected with the coelom to form an efficient circulatory system. The coelom is core to bilaterian anatomy (Figure. 4.11). The front end of bilaterians usually features the food intake, a mouth through which food is passed into and along a specialized one-way internal digestive tract instead of being digested in a simple seawater cavity. No sponge cell or cnidarian cell is very far away from a food-absorbing (digestive) cell, so these creatures have no specialized internal transport system. But the digestive system of bilaterians needs an oxygen supply, and the nutrients absorbed there have to be transported to the rest of the body. Bilaterians therefore have a circulation system, and the larger and more three-dimensional they are, the better the circulation system must be.

Many advanced bilaterians have **segments**: their bodies are divided by septa that separate the coelom into separate chambers connected by valves. This arrangement is more efficient for burrowing than a simple, single coelomic cavity (Figures 4.10, 4.11). The segmentation of many animals, including earthworms, may be derived from this invention on the Precambrian sea floor.

Respiration problems probably prevented early coelomates from burrowing for food in rich organic sediments, which are very low in oxygen. But a coelomate burrowing for protection might have evolved some special organs to obtain oxygen from the overlying seawater at one end while the main body remained safely below the surface. Many coelomates that live in shallow

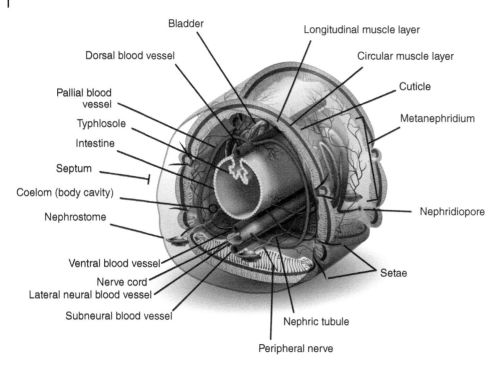

Bladder
Longitudinal muscle layer
Dorsal blood vessel
Circular muscle layer
Pallial blood vessel
Cuticle
Typhlosole
Metanephridium
Intestine
Septum
Coelom (body cavity)
Nephrostome
Nephridiopore
Ventral blood vessel
Nerve cord
Setae
Lateral neural blood vessel
Subneural blood vessel
Nephric tubule
Peripheral nerve

Figure 4.11 Cross section of an earthworm showing the gut (pale green) in the center, surrounded by the coelom, which contains many of the basic organs and the blood vessels (red), and then surrounded by the external muscles and skin (pink and orange). *Source:* artwork by KDS444 (Wikimedia).

burrows have various kinds of tentacles, filaments, and gills that they extend into the water as respiratory organs. It is a very short step from here to the point where a coelomate collects food as well as oxygen from the water by filter feeding (Figure 4.12), as in all bryozoans and brachiopods, in some molluscs, worms, and echinoderms, and in simple chordates.

## The Doushantuo Formation

Fossils found soon after the start of the Cambrian can be unequivocally assigned to a wide variety of the major groups of modern animals, which are called **phyla** (singular = phylum) as we shall see in Chapter 5. In contrast, the Ediacaran fossil record also includes potential animal fossils, but they are controversial and difficult to interpret.

Early in the Ediacaran, the Doushantuo Formation was laid down as a set of rocks in South China, very soon after the last major glaciation. It contains some exquisite fossils that are the key to understanding the life of this time. The Doushantuo rocks are so rich in phosphate that they are mined for fertilizer, so they have been well studied. One layer, dated about 609 Ma, contains tiny fossils that were preserved so soon after death that phosphate replaced the individual cells, preserving them in 3D. Structures within the cells are even preserved, including yolk granules and nuclei.

Figure 4.12 Some worms build tubes to make their own burrows. With the body safe inside the tube, they extend tentacles to collect food and oxygen from the water. This is the Christmas tree worm *Spirobranchus* from the tropical Pacific. *Source:* Photograph by Nick Hobgood (Wikimedia).

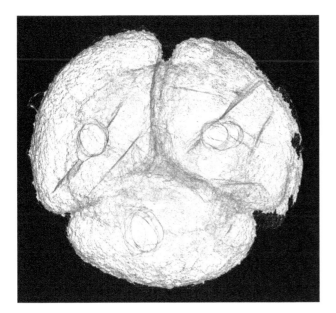

Figure 4.13 An embryo-like fossil from the Doushantuo Formation. This is a three-dimensional reconstruction showing possible nuclei in turquoise. *Source:* From Cunningham et al. (2017a).

Figure 4.14 The rangeomorph *Avalofructus* from the Mistaken Point Formation in Newfoundland. Reconstruction of a large specimen showing branches, frondlets, and the basal holdfast. *Source:* Image from Narbonne et al. (2009). © Guy Narbonne and The Paleontological Society, used by permission.

Thousands of these Doushantuo microfossils have been studied. Figure 4.13 gives an idea of the quality of the preservation. Enthusiastic accounts of their astounding variety identified protists and algae, tiny metazoans (sponges and cnidarians), and eggs and embryos of yet more metazoans.

However, many small creatures build cysts (see Chapter 3, Figure 3.5) as resting stages while conditions are bad, and cysts are more likely to be preserved as fossils than other stages of the life cycle. The cyst stages of some protists have much the same structures as the Doushantuo microfossils. They can reproduce by cloning dozens, hundreds, or thousands of identical individuals before releasing them. Metazoan embryos do not do that. Each cell, programmed differently, is destined to become a separate individual cell in the metazoan body. This is a controversial topic, but the weight of evidence now suggests that the Doushantuo fossils are not true metazoans.

This statement could change overnight, of course. There may be very early metazoans in the Doushantuo fossils. A group of big acritarchs in the early Ediacaran is not found in the later Ediacaran (Figure 4.1). They seem large to be single-celled, so they may be micrometazoan cysts.

## Large Ediacaran Organisms

Many Ediacaran fossils belong to an extinct group called **rangeomorphs**, but there are possible Ediacaran sponges, cnidarians, and bilaterians, too. Rangeomorphs became extinct at the end of the Ediacaran, at or before

541 Ma, but the others could have been the ancestors of the Cambrian animals that followed.

All rangeomorphs were soft-bodied. It is only when their corpses were colonized after death by layers of bacteria that we see them at all, typically as "ghost" outlines where biofilms of bacteria compacted the sediment. These Ediacaran organisms colonized the sea floor, from shallow water to well below the well-lit surface zone.

In the Mistaken Point Formation in Newfoundland, Canada, we find the earliest large organized Ediacaran organisms, from about 565 Ma. Here masses of rangeomorphs (and a few other organisms) were killed and buried by very fine-grained volcanic ash falling through the water. The animals are preserved in great detail in three dimensions, giving us a unique opportunity to interpret their mode of life.

Rangeomorphs are built from small blade-shaped units ("frondlets") about 1 cm long. Young forms have only a few frondlets, but larger ones have multiple branching supports, each one bearing multiple frondlets, and growing up to a meter long. The animal is fixed to the sea floor by a circular disk or holdfast (Figure 4.14).

There are no openings in the rangeomorph body wall, and the simplest hypothesis for their biology is **osmotrophy**:

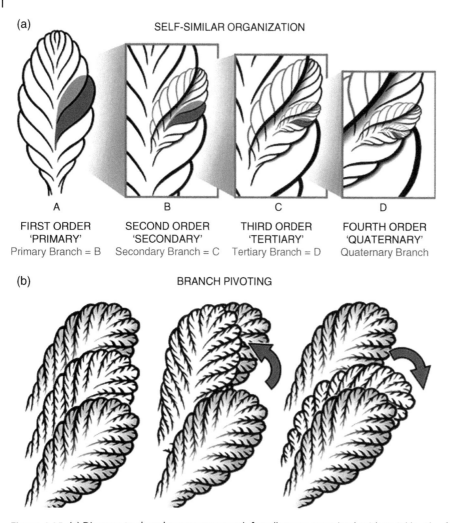

**Figure 4.15** (a) Diagram to show how rangeomorph frondlets are organized, with serial levels of complex branching. (b) Diagram showing how rangeomorphs were able to pivot branches, presumably at all levels, to best intercept nutrient-laden water currents. *Source:* Both images from Narbonne et al. (2009). © Guy Narbonne and The Paleontological Society, used by permission.

taking up dissolved nutrients from the water directly through the skin by osmosis. Each frondlet thus obtains its own nutrition, but clearly there must be some nutrient transport through the body to grow the nonfeeding holdfast and the supporting tissues. The fractal arrangement of branches and frondlets approaches a mathematical optimum for an array of osmotrophic collectors (Figure 4.15).

Many marine invertebrates get some nutrition this way, through skin, gills, or tentacles; jellyfish are just one example. Even a vertebrate, the ghastly hagfish, can burrow inside a whale carcass and absorb dissolved nutrients from the rotting flesh, through its gills and its skin. However, nutrients are not concentrated enough in most environments today to feed larger animals entirely by osmotrophy. Probably Ediacaran sea floors had more dissolved nutrients because there were few organisms eating plankton at the surface, or intercepting and eating dead and dying plankton before they decayed to release nutrients. Rangeomorphs may have died out as larger

metazoans radiated at the base of the Cambrian and depleted their nutrient supplies.

There is a growing consensus that at least some of the large Ediacaran organisms were animals. *Dickinsonia* (Figure 4.16) is flat and large, and seems to have fed by osmotrophy through its lower side as it moved across the sea floor. This, as well as the way that its body segments were added through growth, suggests that *Dickinsonia* was an animal. What is more, biomarkers of probable animal origin have been isolated from *Dickinsonia* specimens. *Kimberella* (Figure 4.17) is the strongest candidate for a bilaterian animal from the Ediacaran. It may have been evolving toward a slug-like early mollusk, and is found with scrape-marks that suggest that it grazed on algal mats. Some Ediacaran fossils resist interpretation.

How could these large metazoans survive if Ediacaran environments had low oxygen conditions, as would certainly have been the case for the rangeomorphs at Mistaken Point? Osmotrophic animals today have very

Figure 4.16 *Dickinsonia*, a bilaterian from the Ediacaran. Scale in cm. *Source:* Image by Merikanto (Wikimedia).

Figure 4.17 *Kimberella*, a bilaterian from the Ediacaran of Russia. About 1 cm long. *Source:* Image by Aleksey Nagovitsyn (Wikimedia).

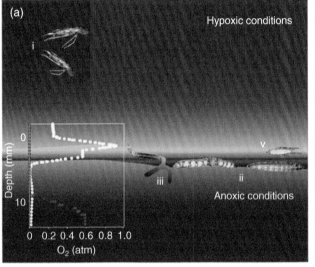

Figure 4.18 Oxygen miners. (a) Results found today at Las Roques, Venezuela. An oxygen-rich algal mat covers the surface under oxygen-poor lagoon water and above sea floor sediment with no oxygen. (Daytime oxygen levels are shown as yellow dots, sulfur levels as red dots.) Small swimming metazoans make forays into the water, but spend a lot of time on the bottom. Others crawl on or burrow into or under the algal mats, mining them for oxygen and food. (b) Analogous conditions inferred for some Ediacaran sea floors, with small metazoans living and feeding close to oxygen-rich mats. *Source:* From Gingras et al. (2011), used by permission.

low metabolic rates, so the Ediacaran rangeomorphs, and *Dickinsonia*, probably had the same low oxygen requirements.

However, some Ediacaran animals left trace fossils of their burrowing activity in and on the surface, and some of these, found from 555 Ma onwards, presumably were bilaterians using a coelom to move through the sediment – a relatively high-energy way to move about.

A large coral reef complex lies off the north coast of Venezuela, around the Las Roques islands. Some shallow lagoons are warm and very salty, so that normal marine animals do not live there. Cyanobacterial mats flourish in very shallow water and produce oxygen by day under the tropical sun. The water immediately around the mats can contain up to four times normal oxygen levels. At night, photosynthesis stops and oxygen levels drop sharply toward zero, except for oxygen bubbles trapped in and under the mats. Thus, the shallow water of the lagoon has generally low oxygen levels but the mats form "oxygen oases" in an "oxygen desert." A few metazoans flourish around the mats. Some of them graze the mats, others eat organic sediment. All of them live close to the mats, burrowing into them or under them, "mining" oxygen to support their metazoan metabolism (Figure 4.18, (a)).

Figure 4.19 Trace fossil of an Ediacaran oxygen miner from Australia. This trace was made by a burrowing metazoan in a horizontal plane immediately under a fossil bacterial mat, probably mining it for oxygen and food. *Source:* From Gingras et al. (2011), used by permission.

These studies (Gingras et al. 2011) are important because metazoans left fossil burrows and trails in Ediacaran rocks. Many Ediacaran burrows are very like the burrows in the mats at Las Roques (Figure 4.19), so Gingras et al. confidently infer that Ediacaran metazoans survived low oxygen conditions by oxygen mining in and around bacterial mats (Figure 4.18, (b)). In this sense, Ediacaran bacterial mats were forcing-houses of metazoan evolution, in the same way that stromatolites were forcing-houses of eukaryote evolution. In each case, biological events in just a tiny portion of the global ecosystem took on an importance much greater than the area involved.

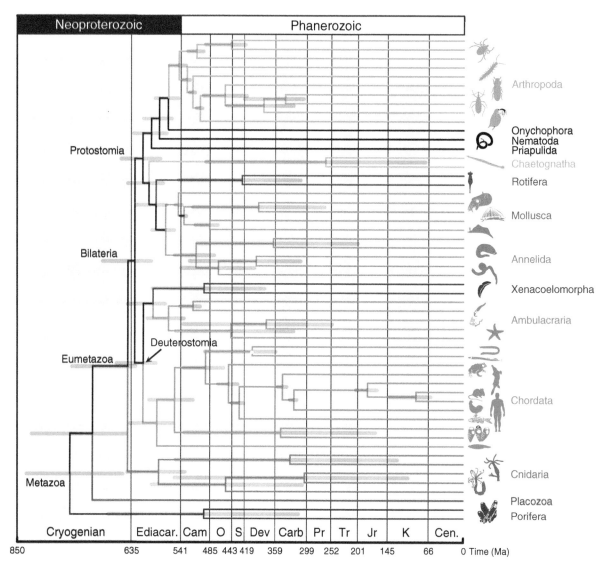

Figure 4.20 Molecular clock time scale for the emergence of metazoan lineages. The gray bars on the cladogram reflect uncertainties in the estimates of clade age; most of this uncertainty is a result of uncertainties in the early animal fossil record. *Source:* After dos Reis et al. (2015).

## Molecular Clocks

We can now read the genetic code of many living organisms: all it takes is time and money, although costs are coming down. Information encoded in these genetic sequences can help us understand not only the evolutionary relationships and genetic distances between living species, but also the timing of early animal evolution. This is called the **molecular clock** hypothesis which relies on four very simple assumptions:

1) mutations in gene sequences happen as a result of random copy errors, and they accumulate over time
2) if you compare the same genes from two species, the number of genetic differences in their coding sequence reflects the amount of time that has gone by since they last shared an ancestor
3) if you have fossil evidence for the age of that ancestor, you can work out the rate of mutation
4) using that rate, you can calculate the age of the ancestors shared with other species based on the number of differences in their gene sequences.

In reality, the rate of mutation varies from branch to branch, and Bayesian statistical methods are used to model uncertainties in the fossil evidence, genetic distances, mutation rate, etc. The analysis millions of times, considering all the uncertainties, all the time seeking the best evolutionary time scale that explains all the data. This approach is used to study the time scale of various events in the history of life; for example, in Chapter 8 we discuss how it can be used to study early plant evolution.

When these methods are applied to early animal evolution, molecular clocks consistently estimate that animals evolved during the Cryogenian or perhaps even earlier (dos Reis et al. 2015; Figure 4.20). At first glance, this seems to be at odds with the fossil record, with animals evolving over 100 million years before the undisputed animal fossils of the Cambrian. On closer inspection, though, the mismatch is not so great. There is convincing trace fossil evidence for animals by 565 Ma and bilaterians by 555 Ma, with probable animal body fossils from a similar age. While members of the living bilaterian phyla may be absent, modern molecular clock studies suggest that there was a lag between the origin of animals and the rapid diversification of animal phyla that occurred around the start of the Cambrian period and this will be the focus of our next chapter.

## References

Barnes, R., Callow, P., Olive, P. et al. (2001). *The Invertebrates: A Synthesis*, 3e. Oxford: Blackwell Science.

Boardman, R., Cheetham, A., and Rowell, A. (1987). *Fossil Invertebrates*. Oxford: Blackwell Science.

Brocks, J.J., Jarrett, A.J.M., Sirantoine, E. et al. (2017). The rise of algae in Cryogenian oceans and the emergence of animals. *Nature* 548: 578–581.

Butterfield, N.J. (2018). Oxygen, animals and aquatic bioturbation: an updated account. *Geobiology* 16: 3–16.

Cunningham, J.A., Liu, A.G., Bengtson, S. et al. (2017a). The origin of animals: can molecular clocks and the fossil record be reconciled? *BioEssays* 39: 1600120.

Cunningham, J.A., Vargas, K., Yin, Z. et al. (2017b). The Weng'an Biota (Doushantuo formation): an Ediacaran window on soft-bodied and multicellular microorganisms. *Journal of the Geological Society, London* 174: 793–802.

dos Reis, M., Thawornwattana, Y., Angelis, K. et al. (2015). Uncertainty in the timing of origin of animals and the limits of precision in molecular timescales. *Current Biology* 25: 1–12.

Gingras, M., Hagadorn, J.W., Seilacher, A. et al. (2011). Possible evolution of mobile animals in association with microbial mats. *Nature Geoscience* 4: 372–375.

Hoffman, P.F., Abbott, D.S., Ashkenazy, Y. et al. (2017). Snowball earth climate dynamics and Cryogenian geology-geobiology. *Science Advances* 3 (11): e1600983.

Lenton, T.M., Boyle, R.A., Poulton, S.W. et al. (2014). Co-evolution of eukaryotes and ocean oxygenation in the Neoproterozoic era. *Nature Geoscience* 7: 257–265.

Narbonne, G.M., La Flamme, M., Greentree, C., and Trusler, P. (2009). Reconstructing a lost world: Ediacaran rangeomorphs from Spaniard's Bay, Newfoundland. *Journal of Paleontology* 83: 503–523.

## Further Reading

Bobrovskiy, I., Hope, J.M., Ivantsov, A. et al. (2018). Ancient steroids establish the Ediacaran fossil *Dickinsonia* as one of the earliest animals. *Science* 361: 1246–1249.

Cohen, P.A. and Macdonald, F.A. (2015). The Proterozoic record of eukaryotes. *Paleobiology* 41: 610–632.

Droser, M.L., Tarhan, L.G., and Gehling, J.G. (2017). The rise of animals in a changing environment: global ecological innovation in the late Ediacaran. *Annual Review of Earth and Planetary Sciences* 45: 593–617.

Dunn, F.S., Liu, A.G., and Donoghue, P.C.J. (2018). Ediacaran developmental biology. *Biological Reviews* 94: 913–932.

Erwin, D.H. and Valentine, J.W. (2013). *The Cambrian Explosion: the Construction of Animal Biodiversity*. Greenwood Village: Roberts.

Hoekzema, R.S., Brasier, M.D., Dunn, F.S. et al. (2017). Quantitative study of developmental biology confirms *Dickinsonia* as a metazoan. *Proceedings of the Royal Society B: Biological Sciences* 284: 1862.

Lyons, T.W., Reinhard, C.T., and Planavsky, N.J. (2014). The rise of oxygen in Earth's early ocean and atmosphere. *Nature* 506: 307–315.

Laflamme, M., Xiao, S., and Kowalewski, M. (2009). From the Cover: Osmotrophy in modular Ediacara organisms. *Proceedings of the National Academy of Sciences of the United States of America* 106: 14438–14443.

## Question for Thought, Study, and Discussion

Briefly describe the rangeomorphs and their unusual body construction. What is the best explanation (at the moment) of their way of life? Why do so few animals today live like this?

5

The Cambrian Explosion

## The Cambrian Explosion

The waves of evolutionary novelty that appeared in the seas during the Early Cambrian have few parallels in the history of life. As we saw in Chapter 4, animals originated well before this time but in the Cambrian, we find their skeletons in abundance. It is still debated to what extent this is an explosion of diversity and to what extent it is an explosion of fossils. In other words, skeletons originated at about this time and we have to ask whether animal life boomed as a result of the skeleton (a real explosion) or whether the different animal groups were already abundant but soft-bodied, and so not found as fossils until they had evolved skeletons.

For the moment, we report what we see. In the Cambrian, we find animals with all kinds of innovations that allowed them to see, crawl, swim, burrow, prey on other animals and defend themselves from predators. We will first look at the variety of animal life, before studying the major evolutionary steps and the fossils that help us to understand them.

## The Variety of Metazoans

When one animal group is radically different from another and is also considered to be a clade that evolved from some single ancestral species (see Chapter 3), it is a **phylum**, defined by its own particular body structure, ecology, and evolutionary history. Mollusca and Arthropoda are familiar phyla. They must once have had a common bilaterian metazoan ancestor, but that ancestor wasn't a mollusk or an arthropod (by definition as well as common sense). There are arguments about the number of phyla among living metazoans, mostly because there is a bewildering variety of worm-like organisms, but most people would count about 30 phyla. Because only creatures with hard parts are easy to recognize as fossils, only nine or 10 phyla are or have ever been important in paleontology (Box 5.1).

It is stunning to realize that all these phyla are known from Cambrian rocks, but none of them are known for certain in rocks older than Cambrian. That could mean one of two things: first, that there was an "explosive" burst of evolution at the beginning of the Cambrian and, second, that we have no fossil record of the metazoan evolution that gave rise to the phyla that we recognize today. However, as we have seen in Chapter 4, evidence from molecular clocks suggests that animals were indeed present well before the Cambrian. Many animal phyla include largely soft-bodied forms, so we cannot be sure that the fossil record would capture them.

As we have seen, two basal metazoan phyla are sponges and cnidarians. The bilaterians are all complex and three-dimensional, and they have Hox genes controlling the placement of structures along their axis of symmetry. Genomic evidence confirms that advanced bilaterians form three major clades, each including a cluster of phyla: Ecdysozoa, Lophotrochozoa, and Deuterostomia.

**Ecdysozoa** are animals that molt off their outer skins as they grow. This can be an important way of getting rid of unwanted external parasites. Molting is characteristic of the Arthropoda, for example, and for many of them it is a major evolutionary burden as well as an advantage. Crabs and lobsters must molt many times over the years of their (natural) lives, and each time they do so they are very vulnerable to predators and must spend a considerable time hiding while their new shell hardens. Some ecdysozoans have found a way to avoid this evolutionary constraint. For example, insects do not molt as adults. However, insects can only do this by having a very short adult life, with all their growth taking place in earlier life stages (as larvae). (A short adult life is an extreme but successful way of avoiding a major evolutionary constraint!) Arthropoda often have hard shells and are the dominant members of the Ecdysozoa in the fossil record. Other ecdysozoans include gloriously obscure phyla, most of which are microscopic: Nematoda (round worms), Nematomorpha (horsehair worms), Kinorhyncha (mud dragons), Lorifinera, and Priapulida (penis worms).

**Lophotrochozoa** are animals with a cute, fuzzy little floating larva and a way of life that originally involved filter feeding from the water. The Mollusca are the best-known phylum of Lophotrochozoa, well fossilized, well understood, and very varied in their anatomy and ecology. Brachiopoda and Bryozoa are also important fossil groups, while Annelida (worms) has a comparatively poor fossil record. Other lophotrochozoan phyla (Chaetognatha, Rotifera, Acanthocephala, Platyhelminthes, Gastrotricha, Mesozoa, Cycliophora, Nemertea) lack a material fossil record.

**Deuterostomia** also seem to have been originally filter feeders with floating larvae, but their larvae are so different from those of lophotrochozoans that they cannot belong to the same clade. Deuterostomes include Hemichordata (acorn worms and the extinct graptolites), Echinodermata (sea-stars and relatives), Chordata (tunicates, fish, and tetrapods – including ourselves).

Much of the mismatch between the fossil record and molecular clock estimates for the antiquity of metazoan clades can be accounted for by problems in interpreting Ediacaran and Cryogenian physical, trace, and biochemical fossil evidence, the paucity of Ediacaran rocks in which to search for fossils (in comparison to the Cambrian), and the challenges of identifying primitive animals (Cunningham et al. 2017). The anatomy of the ancestral animal is difficult to discern, while the first bilaterians may have resembled modern flatworms and evidence of their activity is manifest by 555 Ma, in the form of burrows in sea-floor sediments. These trace fossils exhibit increasing complexity and three-dimensionality with proximity to the Cambrian.

## The Evolution of Skeletons: Small Shelly Fossils

One of the most important events in the history of life was the evolution of mineralized hard parts in animals. Starting rather suddenly at the beginning of the Cambrian, the fossil record contains skeletons: shells and other pieces of mineral that were formed biochemically by animals. Humans have an internal skeleton or endoskeleton, where the mineralization is internal and the soft tissues

---

**Box 5.1  The Major Phyla of Fossil Invertebrates**

- Porifera or sponges (includes †Archaeocyatha)
- Cnidaria
- Bryozoa
- Brachiopoda
- Mollusca
- Arthropoda
- Echinodermata
- Hemichordata (with †graptolites)
- Chordata (including vertebrates)

(† indicates an extinct group)

lie outside. Most animals have the reverse arrangement, with a mineralized exoskeleton on the outside and soft tissues inside, as in most mollusks and in arthropods (Figure 5.1). The shell or test of an echinoderm is technically internal but usually lies so close to the surface that it is external for all practical purposes. The hard parts laid down by corals are external, but underneath the body, so that the soft parts lie on top of the hard parts and seem comparatively unprotected by them. Sponge skeletons are simply networks of tiny spicules that form a largely internal framework. There is incredible variety in the type, function, arrangement, chemistry, and formation of animal skeletons: biomineralization is a whole science in itself.

With the evolution of hard parts, the fossil record became much richer, because hard parts resist the destructive agents that affect the soft parts of bodies. The evolution of hard parts marks the beginning of a new eon in Earth history, the **Phanerozoic**, a new era, the **Paleozoic**, and its oldest subdivision, the **Cambrian** Period. Many different ways of life, using many different hard parts in many different body plans, seem to have been explored as soon as animals evolved the biochemical pathways for making hard parts. Thus, the Cambrian "explosion" was genetic as well as ecological, and dramatic indeed to paleontologists collecting their fossils. The new animals evolved the features that allow us to identify most of them as members of the metazoan phyla that survive today.

In Siberia and China, rocks in the very latest Ediacaran contain some small shells in the form of tiny cones and tubes (Figure 5.2). In the Early Cambrian, there are diverse small shelly fossils (so called because they are small, they are shelly, and they are fossils). Some are the shells of small animals, including mollusks. Others are disarticulated fragments of larger animals, a fact that was revealed by discoveries of exceptionally preserved fossils that are fully articulated. For example, *Microdictyon* was originally known as isolated mineralized net-like plates, but these are now known to have been the elaborate shoulder pads of an ancient Cambrian ecdysozoan (Figure 5.3). At this time, we also find archaeocyathid sponges forming large reef patches (Figure 5.4). The next stage of the Cambrian saw the appearance of more abundant and more complex creatures, worldwide, in a few million years after 520 Ma. Dominant among these animals were trilobites, brachiopods, and echinoderms.

## Larger Cambrian Animals

Trilobites are arthropods, complex creatures with thick jointed armor covering them from head to tail (Figure 5.5). They had antennae and large eyes, they were mobile on the

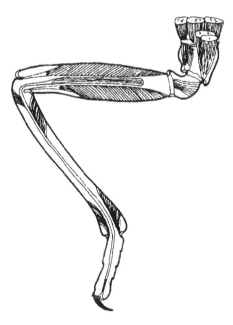

Figure 5.1 An arthropod leg. Arthropods have jointed exoskeletons operated from inside by muscles and ligaments. *Source:* From Barnes et al. (2001). © 2001 Blackwell Science.

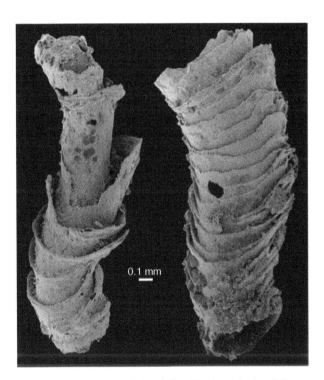

Figure 5.2 The small shelly fossil *Cloudina*, from the late Ediacaran of China. The animal, whatever it was, grew a tube-like shell about 0.5 mm in diameter. It presumably collected food from the water. But notice the trace fossil: a hole bored through the shell, presumably by an unknown predator. *Source:* Image by Stefan Bengtson, Swedish Museum of Natural History, from Bengtson & Yue (1992).

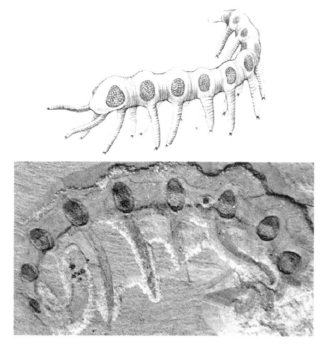

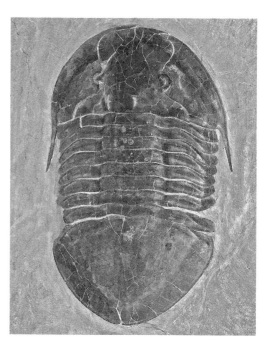

Figure 5.3 The ecdysozoan worm *Microdictyon*. Note the net-like plates that had previously been found as disarticulated small shelly fossils. *Source:* Image from Hou et al. (2007).

Figure 5.5 A trilobite, *Megalaspides*, from the Ordovician of Ohio. Two prominent eyes are set on the head shield, with lines of weakness running past them to make molting easier. *Source:* Image by Llez (H. Zell) (Wikimedia).

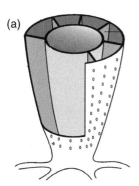

Figure 5.4 (a) Diagram of a Cambrian archaeocyathid sponge. Water was taken in through the side of the colony, filtered through a compartmented body, and expelled into a central exit cone. *Source:* Image by Muriel Gottrop (Wikimedia). (b) An Early Cambrian archaeocyathid reef exposed by erosion on the coast of southern Labrador. Undergraduate researcher Hannah Clemente of Smith College shown for scale. *Source:* Image © Dr Sara Pruss of Smith College, used by permission.

sea floor using long jointed legs, and they were something like crustaceans and horseshoe crabs in structure. They did not have the complex mouth parts of living crustaceans, so their diet may have been restricted to sediment or very small or soft prey. They burrowed actively, leaving traces of their activities in the sediment, and they are by far the most numerous fossils in Cambrian rocks. The number of fossils they left behind was increased by the fact that they molted their armor as they grew, like living crustaceans. Thus, a large adult trilobite could have contributed 20 or more suits of armor to the fossil record before its final death. Even allowing for this bias of the fossil record, it is clear that Cambrian sea floors were dominated by trilobites. Other large arthropods are also known from Early Cambrian rocks, although they are much less common.

Brachiopods are relatively abundant Cambrian fossils, creatures that had two shells protecting a small body and a large water-filled cavity where food was filtered from seawater pumped in and out of the shell (see Chapter 2, Figure 2.1). Cambrian brachiopods lived on the sediment surface or burrowed just under it and gathered food from the seawater.

These animals are large, and they are easily assigned to living phyla. For the first time, the sea floor would have looked reasonably familiar to a marine ecologist.

## Soft-Bodied Cambrian Animals

We have so far discussed the "Cambrian explosion" as if it related entirely to the evolution of skeletons. While this is basically true in terms of fossil abundance, there was also dramatic evolution at the same time among animal groups with little or no skeleton. Trace fossils – tracks, trails, and burrows – increase in abundance at the beginning of the Cambrian, and soft-bodied animals appeared with some amazingly sophisticated body plans.

Many soft-bodied animals were preserved by quirks of the environment in early Cambrian rocks in South China (the Chengjiang Fauna), and in middle Cambrian rocks in the Canadian Rockies (in the Burgess Shale). Similar fossils are now known from Cambrian rocks in several other places. Here, we will call them all the "Burgess Fauna."

More than half the Burgess animals burrowed in or lived freely on the sea floor, and most of these were deposit feeders. Arthropods (such as *Marrella*, Figure 5.6a) and worms dominate the Burgess Fauna. Only about 30% of the species were fixed to the sea floor or lived stationary lives on it, and these were probably filter feeders, mainly sponges and worms. Thus, the dom-inance of most Cambrian fossil collections by bottom-dwelling, deposit-feeding arthropods is not a bias of the preservation of hard parts; it occurs among soft-bodied communities too. Trilobites are fair representatives of Cambrian animals and Cambrian ecology.

The main delights of the Burgess Fauna are the unu-sual animals which have provided fun and headaches for paleontologists. *Aysheaia* is a lobopod; it looks like a caterpillar, with thick soft legs (Figure 5.6b). It has stubby little appendages near its head that may be slime glands for entangling prey. *Hallucigenia*, named for its bizarre appearance, is a lobopod with spines – related to *Microdictyon*, which we met earlier in the chapter (Figure 5.3). *Opabinia* was a highly evolved arthropod relative, long and slim, with a vertical tail fin. It had five eyes and only one large grasping claw on the front of its head (Figure 5.7).

There are predators in the Burgess Fauna. Priapulid worms today live in shallow burrows and capture soft-bodied prey by plunging a hooked proboscis into them as they crawl by. The Burgess priapulid *Ottoia* (Figure 5.8) probably did the same.

Anomalocarids are the most spectacular Cambrian predators. They are an extinct group of animals related to arthropods; pieces of Burgess animals suggest that they could have been a meter long! Anomalocarids have been difficult to reconstruct because they are usually found as pieces that have to be fitted together. They were swimmers, and most of them had very large grasping

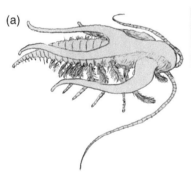
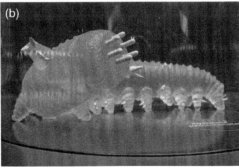

Figure 5.6 (a) A complex arthropod, *Marrella*, from the Burgess Shale. Drawing by Ghedoghedo (Wikimedia). (b) A compelling glass-fiber model of the lobopod *Aysheaia*. *Source:* Image by Eduard Solà Vázquez (Wikimedia).

Figure 5.7 *Opabinia* from the Burgess Shale, about 6 cm long. It is clearly an anomalocarid, but has only one large appendage and multiple eyes. (a) Photograph by Jstuby (Wikimedia). (b) Reconstruction by Nobu Tamura (Wikimedia).

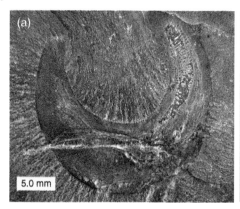

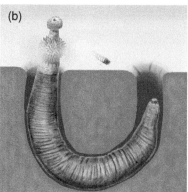

Figure 5.8 The priapulid worm *Ottoia* from the Burgess Shale. A typical *Ottoia* is about 30 mm long (stretched out!). (a) Photograph by Dr Mark Wilson of the College of Wooster (Wikimedia). (b) Reconstruction of an *Ottoia* in its burrow. *Source:* Image by Smokeybjb (Wikimedia).

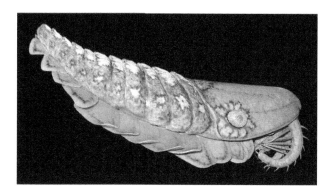

Figure 5.9 A model reconstruction of the anomalocarid *Peytoia* from the Burgess Shale. The grasping appendages are curled around the mouth, and the animal is posed in swimming position. *Peytoia* was typically about 60 cm (2 ft) long. (The blue eyes are appealing, but the color is speculation!) Model by Espen Horn. *Source:* Photograph by Llez (H. Zell) (Wikimedia).

Figure 5.10 *Wiwaxia* from the Burgess Shale, about 5 cm long. *Source:* Photograph by Jstuby (Wikimedia).

appendages, and a mouth with scraping or piercing saw-like edges. Although they have a very lightly built outer skeleton, they would have been powerful predators, especially on equally thin-skinned prey. *Anomalocaris* itself was the largest, at a meter long, and *Peytoia* is the best known (Figure 5.9). Fossils from Greenland show that one anomalocarid, *Tamisiocaris*, evolved from being a predator to become a large and actively swimming suspension feeder occupying a similar niche to modern basking sharks and baleen whales.

The eyes of *Anomalocaris* are astounding. Eyes discovered in Early Cambrian rocks of South Australia are preserved in such detail that one can estimate that each eye had 16 000 little lenses (Paterson et al. 2011). This would have given *Anomalocaris* a finely detailed image of its surroundings and its potential prey. The discovery adds to the picture of a highly effective predator in the Early Cambrian. Do not forget, too, that an eye with 16 000 lenses must have had a complex system of visual receptors and nerve networks to transmit the images from each eye to the brain. And since arthropods molt their outer covering in order to

grow, every anomalocarid would have molted off its eye lenses with each growth stage, and then would have had to grow a new lens system, precisely coordinating with the exposed visual system under it. A meter-long anomalocarid could easily have molted 20 times during its life, growing a new and larger lens system each time.

*Wiwaxia* (Figure 5.10) is a flat creature that crept along the sea floor under a cover of tiny scales that were interspersed with tall strong spines. Halkieriids, best known from the Burgess Fauna of Greenland, look like flattened worms, with perhaps 2000 spines forming a protective coating embedded into the dorsal surface. Yet two distinct subcircular shells are embedded in the upper surface close to each end. These creatures are probably mollusks.

The Burgess animals also include worm-like creatures that are identified as early chordates and vertebrates – in other words, the remote ancestors of ourselves and all other vertebrates (see Chapter 7).

Altogether, the Burgess Fauna give us a good idea of the exciting but extinct soft-bodied creatures that may always have lived alongside the trilobites but were hardly ever preserved. They show that the variety of Cambrian life on and near the sea floor was greater than the typical fossil collection would suggest.

## What Caused the Cambrian Explosion?

Many factors have been proposed as possible "triggers" for the explosive diversification of animals in the Cambrian. These fall into three broad categories: intrinsic biological factors (e.g., genetic or developmental innovations), extrinsic environmental changes, and ecological processes. We will discuss the factors in this order, although in reality complex interactions between many of these may well have been responsible.

### Evolution and Development

We can now recognize certain strings of DNA as genes, and understand what many of them do within the living organism. For example, the entire genome (genetic code) of the human parasite *Mycoplasma genitalium*, one of the smallest genomes so far discovered, contains only 580 070 base pairs (humans have 3 billion). Sorting, slicing, and dicing this genome, geneticists have concluded that *M. genitalium* has 525 genes, of which only 382 code for proteins and so are essential for biological functions. The emphasis on proteins is because they perform so many cell functions: building lipids for the cell membrane, transporting phosphate, breaking down glucose, and so on.

But it is more complex to grow a viable metazoan than a single-celled protist. The genome must contain the information to build many kinds of cells rather than just one, and the information to grow them in the right place at the right time, and to develop the control mechanisms, sensory systems, transport systems, and whole-body biochemical reactions that operate in a metazoan.

The genetic programming that builds a metazoan from a single cell need not specify individual cells one by one. Like a well-written computer program, there can be tricks that promote efficiency. For example, one could program a computer to draw a flower, specifying the size, shape, and position of each petal. But the petals of any given flower are typically much alike, so one can use the same shape and size for each petal, and simply tell the computer to move the pen to the right place and draw the same petal each time.

In the same way, metazoans have **structural genes** to contribute to building components of the animal, and **regulatory genes** that make sure the piece is built in the right place at the right time. These are linked together in a **gene regulatory network** of positive and negative interactions between genes and the proteins they encode, and by cell signaling molecules that effect coordinated development between cells. For example, a network of regulatory genes could be used in combination with a network of "segment" genes to build all the segments along a growing worm. The same sort of regulatory gene networks could easily be used to build legs on, say, a millipede or a crab, by calling on "leg" network the appropriate number of times instead of "segment" network. Slight modifications of the leg network during development could result in an animal whose legs were different along its length (as in insects), or build a vertebrate with different bones along the length of a backbone. For example, embryonic snakes have limb buds that show us where once there were legs on ancestral snakes. Today, those buds do not develop into legs because the ancestral regulatory network inherited from limbed ancestors has been modified by evolution. There can be no doubt that the origin and diversification of animal body plans are in part a consequence of the evolution of gene regulatory networks.

### Oxygen Levels

The environment in which these evolutionary changes were taking place is also vitally important. As we have seen in Chapter 4, the early evolution of animals was intimately linked to rising oxygen levels. The evolution of large bodies and skeletons was probably made possible, or encouraged, by high oxygen concentrations. Shells and thick tissues prevent the free diffusion of oxygen into a body, so they could not have evolved unless there was a high enough oxygen level to push oxygen into the body through the few remaining areas of exposed tissue: through gills, for example. This also cannot be the whole story, because sponges and cnidarians could have evolved their skeletons (which do not inhibit respiration) in low oxygen conditions. Again, oxygen levels could explain much of the Cambrian explosion.

### Seawater Chemistry

The unusual seawater chemistry in the Early Cambrian has been proposed as a driver for the Cambrian explosion. In the Neoproterozoic, large-scale erosion took place, leaving the continents low-lying and covered in soils resulting from weathered crystalline rocks. In the Cambrian there was a major sea level rise, which flooded the continents and remobilized the soils. Peters and Gaines (2012) showed that this led to high concentrations of elements such as calcium and phosphorus in the oceans and they argued that this might have driven the evolution of mineralized skeletons in multiple animal groups.

## Mineralization

Many groups of fossils appeared quite suddenly in the fossil record, thanks to their evolution of skeletons (Figure 5.11), sometimes at comparatively large body size. Given the Ediacaran legacy of metazoans and relatively high oxygen levels, however, it is most likely that the Cambrian explosion simply records the invention and exploitation of skeletons for many good reasons associated with locomotion (walking, digging, and swimming), size, support, defense, and other functions, made even more complex by the fact that animals interact ecologically with other species as they evolve.

A skeleton may support soft tissue, from the inside or from the outside, and simply allow an animal to grow larger. Therefore, sponges could grow larger and higher after they evolved supporting structures of protein or mineral (see Figure 4.6), and they could reach further into the water take advantage of currents and to gather food. Large size also protects animals from predators large and small. A large animal is less likely to be totally consumed, and in an animal like a sponge that has little organization, damage can eventually be repaired if even a part of the animal survives attack. As skeletons evolved, even for other reasons, they helped animals to survive because of their defensive value.

For some animals, skeletons provided a box that gave organs a controlled environment in which to work. Filters were less exposed to currents, so perhaps they would not clog so easily from silt and mud (see Figure 2.1). A box-like skeleton would also have given an advantage against predation. Mollusks and brachiopods may have evolved skeletons for these reasons.

In other animals, hard parts may have performed more specific functions. We have already seen that worms tend to burrow head first in sediment. But after penetrating the sediment, they squirm through it (see Figure 4.10). A worm that evolved a hardened head covering could use a different and perhaps better technique, shoveling sediment aside like a bulldozer. Richard Fortey suggested that the large head shield of trilobites was evolved and used in this fashion.

But arthropods, and especially trilobites, are strongly armored all over their dorsal surfaces, not just in the head region. Most likely, their armor served for the attachment of strong muscles. Muscles pull and cannot push. Worms move by using internal hydraulic systems, as we have seen. On the other hand, walking and digging demand that limbs push on the sediment, and that is very unrewarding if the other end of the leg is unbraced. Arthropods evolved a large, strong dorsal skeleton against which their jointed legs were firmly braced, allowing them to move much more efficiently than worms or lobopods do.

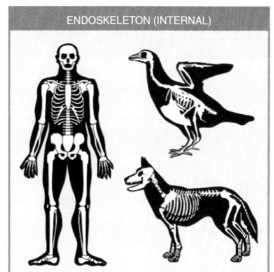 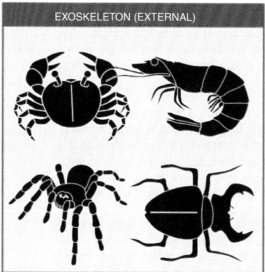

Figure 5.11 The contrast between internal and external skeletons, the endoskeleton and exoskeleton. The contrast is a little more complex than suggested here, because, for example the vertebrate endoskeleton indeed comprises portions that derive from deep tissue sources, but also others, such as most of the skull and shoulder girdle, that derive from sources in the skin. *Source:* artwork by diffzi.

But despite all the discussion of skeletons, the Burgess animals show that dramatic evolution took place also in animals that did not have strong skeletons. Many of these animals had outer coverings that were tough but lightly mineralized; the Burgess arthropods are particularly good examples.

## Predation

The predation theory has two aspects. The first is a general ecological argument. The ecologist Robert Paine removed the top predator (a starfish) from rocky shore communities on the Washington coast and found that diversity dropped. In the absence of the starfish, mussels took over all available rocky surfaces and smothered all their competitors. Paine suggested that a major ecological principle was at work: effective predators maintain diversity in a community. If a prey species becomes dominant and numerous, the top predator eats it back, maintaining diversity by keeping space available for other species.

Steven Stanley used Paine's work to suggest that the evolution of predation triggered the Cambrian radiation. Stanley made an intellectual jump to suggest that predators can cause additional diversity in their prey. He argued that if predators first appeared in the Early Cambrian, they may have caused the increase in diversity at that time. Perhaps predators also encouraged the evolution of many different types of skeletonized animals.

Geerat Vermeij supported Stanley's idea, suggesting how new predators might indeed cause diversification among prey (at any time). In response to new predators, prey creatures might evolve large size, or hard coverings made from any available biochemical substance, or powerful toxins, or changes in life style or behavior (such as deeper burrowing), or any combination of these, all to become more predator proof. And as the new predators in turn evolve more sophisticated ways of attacking prey, the responses and counterresponses might well add up to a significant burst of evolutionary change.

The rules of the predator/prey game probably changed radically as large metazoans evolved. Many Early Cambrian fossils have hard parts that look defensive. Some sharp little conical shells called sclerites may have been spines that were carried pointing outwards on the dorsal and lateral sides of animals, to fend off predators. There are armored and spined Early Cambrian animals, and some Early Cambrian trilobites have healed injuries that may indicate damage by a predator. Defensive structures made of hard parts could therefore have contributed to the increase in the number of fossils in Early Cambrian rocks.

So predation played an important part in generating the Cambrian event. The only major predators we have discovered are the anomalocarids, but they are certainly impressive animals. However, predation alone does not explain the timing of the Cambrian explosion: why not 100 m.y. earlier or later? And predation alone cannot account for all the variety of skeletons that we see.

## Exploiting the Sea Floor

Cambrian animals, whether they had skeletons or not, lived in and close to the sea floor. There was likely a rich supply of organic matter there. Ediacaran sea floors and Cambrian sea floors were ecologically different, so that people sometimes refer to the "Cambrian substrate revolution" (Figure 5.12), marked by larger animals that dug and burrowed deeper and more powerfully than before, leaving trace fossils and sediment disturbance to document their life style.

Jack Sepkoski's "Cambrian Fauna" (see Chapter 6) is dominated by trilobites, and the dominant disturbers of Cambrian se-afloor sediment were trilobites. Their multiple limbs were effective at walking, digging, and stirring up the organic-rich Cambrian sea-floor mud. They had small mouths and rather ineffective appendages round the mouth, so they probably ate mud – a lot of it. These "ecosystem engineers" were also "evolutionary engineers," because their activity helped to underpin the "Cambrian substrate revolution." This vivid phraseology helps to describe one of the fundamental changes in the Earth system at the beginning of the Cambrian. Butterfield (2018) gives a concise and convincing summary.

After all these specific suggestions for the Cambrian explosion, we come back to the general overall idea of the synergistic relationship between metazoans and oxygen levels as the root cause of the dramatic change in the physical and biological worlds (giving the timing), with predation as a major accelerant for the dramatic changes in body plans, especially in hard parts of the skeleton.

After the dramatic changes in the Early Cambrian, the continued increase in numbers and diversity of fossils later in the period seems anticlimactic. Cambrian fossil collections are not complex ecologically; they are dominated by trilobites, most of which lived on the sea floor and were deposit feeders. Filtering organisms are very much secondary, and large carnivores are represented only by anomalocarids.

The Cambrian explosion is spectacular but it is not unique; the great diversification of the diapsid reptiles, especially archosaurs, in the Triassic is an analogous case (see Chapter 12), as is the diversification of the flowering plants and insects in the Cretaceous (Chapter 15), and of modern mammals in the Paleocene (Chapter 17). These radiations stand out from "normal" evolutionary events just as "mass extinctions" stand out from the rest (see Chapter 6). On a real planet inhabited by real organisms, evolutionary rates are likely to vary in time and space,

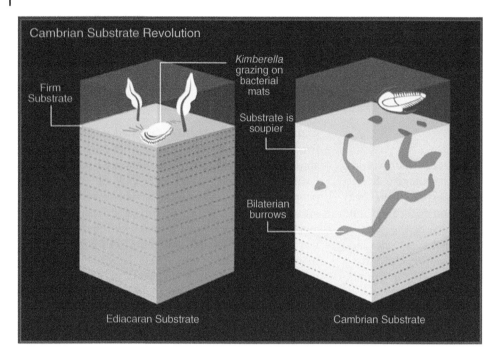

**Figure 5.12** The Cambrian substrate revolution. (*Left*) An Ediacaran sea floor, with oxygen miners concentrated immediately around a surface bacterial mat. (*Right*) A Cambrian sea floor, with powerful burrowers and diggers. *Source:* With permission of the Royal Ontario Museum © ROM.

and evolutionary events are likely to vary in magnitude, duration, and frequency. We should not expect that ideal rules we might propose for an ideal planet would be followed by the natural world; instead, we have to find out from that natural world what the rules actually were.

Given a healthy fossil record, we can explore from the Cambrian onward how life varied through time, along with the physical changes on Earth. So, the next chapter deals with changing life on a changing planet, with much more evidence to help us.

## References

Barnes, R., Callow, P., Olive, P. et al. (2001). *The Invertebrates: A Synthesis*, 3e. Oxford: Blackwell Science.

Butterfield, N.J. (2018). Oxygen, animals and aquatic bioturbation: an updated account. *Geobiology* 16: 3–16.

Cunningham, J.A., Liu, A.G., Bengtson, S. et al. (2017). The origin of animals: can molecular clocks and the fossil record be reconciled? *BioEssays* 39: 1600120.

Hou, X.-G., Siveter, D.J., Siveter, D.J. et al. (2007). *The Cambrian Fossils of Chengjiang, China: The Flowering of Early Animal Life*, 2e. Oxford: Wiley-Blackwell.

Paterson, J.R., García-Bellido, D.C., Lee, M.S.Y. et al. (2011). Acute vision in the giant Cambrian predator *Anomalocaris* and the origin of compound eyes. *Nature* 480: 237–240.

Peters, S.E. and Gaines, R.R. (2012). Formation of the 'Great Unconformity' as a trigger for the Cambrian explosion. *Nature* 484: 363–366.

## Further Reading

Bengtson, S. and Yue, Z. (1992). Predatorial borings in Late Precambrian mineralized exoskeletons. *Science* 257: 367–369.

Briggs, D.E.G., Collier, F.J., and Erwin, D.H. (1995). *The Fossils of the Burgess Shale*. Washington, DC: Smithsonian Books.

Buatois, L.A. and Mángano, M.G. (2018). The other biodiversity record: innovations in animal-substrate interactions through geologic time. *GSA Today* 28: 4–10.

Deline, B., Greenwood, J.M., Clark, J.W. et al. (2018). Evolution of metazoan morphological disparity.

*Proceedings of the National Academy of Sciences of the United States of America* 115: E8909–E8918.

Erwin, D.H., Laflamme, M., Tweedt, S.M. et al. (2011). The Cambrian conundrum: early divergence and later ecological success in the early history of animals. *Science* 334: 1091–1097.

Erwin, D.H. and Valentine, J.W. (2013). *The Cambrian Explosion: The Construction of Animal Biodiversity*. Greenwood Village: Roberts.

Kouchinsky, A., Bengtson, S., Runnegar, B. et al. (2012). Chronology of early Cambrian biomineralisation. *Geological Magazine* 149: 221–251.

Murdock, D.J.E. and Donoghue, P.C.J. (2011). Evolutionary origins of animal skeletal biomineralization. *Cells, Tissue and Organs* 194: 98–102.

Sebe-Pedros, A., Degnan, B.M., and Ruiz-Trillo, I. (2017). The origin of Metazoa: a unicellular perspective. *Nature Reviews Genetics* 18: 498–512.

Telford, M.J., Budd, G.E., and Philippe, H. (2015). Phylogenomic insights into animal evolution. *Current Biology* 25: R876–R887.

Vinther, J. (2015). The origin of molluscs. *Palaeontology* 58: 19–34.

Wood, R. (2018). Exploring the drivers of early biomineralization. *Emerging Topics in Life Sciences* 2: 201–212.

## Questions for Thought, Study, and Discussion

1. Choose one of the stranger Burgess animals (that is, not a crustacean or a sponge or other familiar animal). Describe its body and explain how it might have lived.

2. How can predators encourage animal diversity to increase? After all, they are eating them!

some of it ends up in sea-floor sediment, where it powers benthic communities and then may be buried.

The global ecosystem is thus powered by energy flows between one part of the system and another. Although energy seems to flow mainly downward, physical mechanisms such as upwelling in the ocean reverse that to some extent. But we also have to recognize that the system is largely solar powered by photosynthesis, so the downward flow is based on a real factor.

**Ecosystems**, the whole network of microbes, plants, animals and their interactions, can be described at a single spot, in a region, or globally. Energy flows through an ecosystem from bottom to top, forming a so-called **trophic pyramid** (Figure 6.1a). In most ecosystems, energy from the Sun is captured by photosynthesizing organisms such as cyanobacteria, diatoms, or green plants, and is then passed up through plant-eaters to carnivores. The pyramid comprises huge masses of primary producers at the bottom to a small number of predators at the top. The energy flow can also be visualized in the form of a **food web** (Figure 6.1b), a diagram that summarizes who eats what.

Energy flows and total global biodiversity have jumped upwards at different points in the history of life, coinciding with the origin of oxygen, the origin of eukaryotes and the origin of animals (see Chapters 2–4). Then, the move of life onto land (see Chapters 8 and 9) expanded the ability of life to capture more energy and indeed total **productivity**. Productivity is the ability of primary consumers to capture energy to power the whole food pyramid. Further jumps in global productivity and biodiversity occurred with the Mesozoic Marine Revolution (Chapter 11) and the Cretaceous Terrestrial Revolution (Chapter 15).

Before 500 million years ago, with no important life on land, there would have been very little soil. Every rain would have caused a flash flood, leaving mostly bare, unweathered rock on the continents. Chemical and organic nutrient flow to the ocean would have been small, so marine productivity must have been some unknown but small fraction of today's.

There were other "feedback loops" too. Lower productivity means less carbon being caught up into organic tissues and sediments. With less carbon being buried, oxygen levels in the oceans and atmosphere would not only have been lower than today's but would have been more vulnerable to severe swings. With more frequent and more dangerous changes in the global environment, ecosystems would have been less stable, more prone to disruption or even destruction.

Groups of microbes, plants, and animals that live and interact together form a **community**. For example, the intertidal rocky shore community in New Zealand has its ecological equivalent in British Columbia, even though the families and genera of animals are quite different in the two communities. Each community of organisms lives together in a particular **habitat** – rocky shore communities, mudflat communities, and so on. Within each community and habitat, each species occupies a **niche**, which describes its ecological "address" and might specify its preferred temperature and humidity range, its preferred food, its enemies (predators, disease organisms), and other aspects of its daily life.

Energy and ecology are crucial in the history of life, but so too is geography.

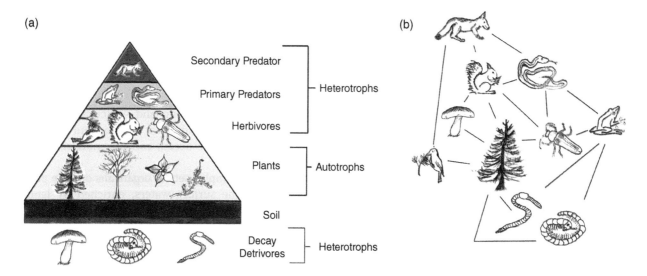

Figure 6.1 A trophic pyramid, showing energy flow from bottom to top (a), and a simple food web of a temperate woodland (b). *Source:* Art by Thompsma (Wikimedia).

# Biogeography and Paleobiogeography

The ways in which organisms occupy the Earth and their distributions are their **biogeography**, and the history of the changing distributions of organisms is referred to as their **paleobiogeography**. As the surface of the Earth and the oceans have been remodeled, as sea level rose and fell, and as groups originated and disappeared, paleobiogeographic provinces have kept changing. The history of life has been much affected by changing geography and changing climates.

## Provinces

Moving from ecology to geography, we have geographic **provinces**. For example, coastal communities of the world can be arranged into geographically separate provinces (Figure 6.2), with each province containing its own set of communities, such as the Oregonian and Californian provinces of western North America.

Provinces are real phenomena, not artifacts of a human tendency to classify things. There are natural ecological breaks on the Earth's surface, usually at places where geographic or climatic gradients are sharp and divide one environmental regime from another in a short distance. A classic example is at Point Conception on the California coast. Here, the ocean circulation patterns cause a sharp gradient in water temperature. In human terms, Point Conception marks the northern limit of West Coast beaches for surfing without a wetsuit, but marine creatures surely feel that difference too. The communities on each side of Point Conception are very different, so a provincial boundary is drawn here, with the Oregonian province grading very sharply into the Californian province (Figure 6.2).

As provinces are identified around the coasts of the world, it seems that the number of species in common between neighboring provinces is usually 20% or less. About 30 provinces have been defined along the world's coasts, mostly on the basis of mollusks, which are obvious, abundant, and easily identified members of coastal communities. Some provinces are very large because they inhabit long coastlines that lie in the same climatic belt (the Indo-Pacific, Antarctic, and Arctic provinces); some are small, like the Zealandian province, which includes only the communities around the coasts of New Zealand (Figure 6.2).

The total diversity of the world's shallow marine fauna directly reflects the number of provinces, which in turn reflects climate and geography. But if tectonic

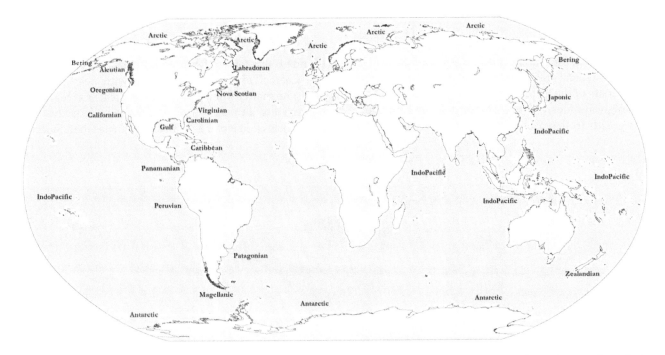

Figure 6.2  Today's marine biosphere includes 30 or so biological provinces along the coasts of the world. This figure does not show all of them but includes all the provinces around the Americas, to stress the differences between east and west coasts and the strong latitudinal gradient that sets many provinces along north–south coastlines. In contrast, the Arctic and Indo-Pacific provinces are very large, because organisms migrate easily around these oceans. The Zealandian province is small because it can occupy only a restricted area of shallow shelf. Land masses, yellow; ocean, blue; low-lying land and shallow coastal waters, white.

movements were to change Earth's geography enough, they would also alter the number of provinces of organisms, and that in turn would increase or decrease world biodiversity. Other things being equal, a world with widely split continents would have greater lengths of shoreline, scattered around the world, giving many marine provinces and high diversity of life.

### Poles and Tropics

The Equator has a fairly uniform climate, and the same applies to the broad tropical zone on Earth, which lies between 23.5° north and south. The Sun is always strong in the tropics and the temperature variation between seasons is small. The general result is that food supply is stable, available at about the same level all year round. Therefore, a species can rely on one or two particular food sources that are always available. As each species comes to depend on a narrow range of food sources, it adapts so well to harvesting them that it cannot easily switch to alternatives. This can generate high levels of biodiversity.

In the tropics, a great variety of specialized species may evolve, competing only marginally with one another, at least for food. For example, on the Serengeti plains of East Africa, several species of vultures are all scavengers on carcasses. But the lappet-faced vulture has a head and beak capable of tearing through the tough hide of a fresh carcass, the white-backed vulture can eat the soft insides from an opened carcass, and the slim-beaked hooded vulture is adept at cleaning bones and gleaning scraps (Figure 6.3). In the sea, the tremendous diversity of life in and around coral reefs is a major contributor to the overall diversity of the tropics.

In high latitudes, on the other hand, food supplies may vary greatly from season to season and from year to year.

The total amount of food supply may be high. Tundra vegetation blooms in spectacular fashion in the spring. There are rich plankton blooms in polar waters in spring and summer, and millions of seabirds and thousands of whales migrate there to share in the abundant food that is produced. Antarctic waters teem with millions of tons of tiny crustaceans (krill) that eat plankton and in turn are fed on by fish, seabirds, penguins, whales, and seals. Yet for organisms that live all year in polar regions, spring abundance contrasts with winter famine. Plants will not grow in winter darkness. Food variability is a major problem. The Arctic tern migrates almost from pole to pole, timing its stay at each end of the world to coincide with abundant food supply.

Where food supplies vary, animals cannot be specialists on only one food source; they must be versatile generalists. Generalists share some food sources and probably compete more than specialists do. If so, fewer generalists than specialists can coexist on the same food resources. In seasonal or variable environments, where organisms must be generalists, diversity is lower. So, there is a rather dramatic global **diversity gradient**, with high diversity at the Equator and low diversity at the pole (Figure 6.4).

Tectonic movements can move continents around the globe. When many continents are in the tropics, global diversity may be higher than when many continents are in high latitudes.

### Islands, Continents, and Climates

Island groups tend to have milder climates – "maritime" or "oceanic" climates – compared with nearby continents, no matter whether they are tropical or at high latitudes. Thus, although they lie at the same latitudes, the British Isles and Japan have milder climates than Siberia;

**Figure 6.3** Serengeti vultures. (a) The lappet faced vulture, with its huge beak, can tear through tough hides to scavenge a carcass. (b) The white-backed vulture can feed on a carcass once it is opened. (c) The hooded vulture, with its slim weak beak, is well adapted to gleaning scraps and cleaning bones. *Source:* Images by Lip Kee Yap, Frank Wouters, and Atamari (Wikimedia).

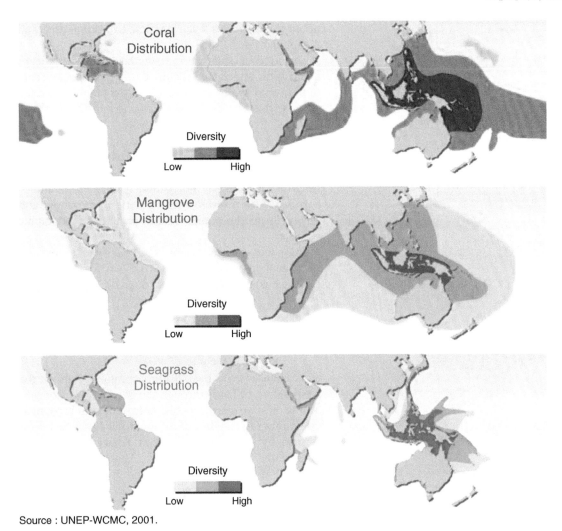

Source : UNEP-WCMC, 2001.

Figure 6.4 Distribution of major elements of shallow marine water biodiversity today – corals, mangroves, and seagrass. Note the high biodiversity in the tropics, tailing off to the poles. *Source:* From United Nations Environment Program –World Conservation Monitoring Center (2001).

the West Indies have milder climates than Mexico; and Indonesia has a milder climate than Indochina.

Large continental areas have especially severe climates for their latitudes. Asia, for example, is so large that extreme heat builds up in its interior in the northern summer, forming an intense low-pressure area. Eventually, the low pressure draws in a giant inflow of air from the ocean, the **summer monsoon**, that brings a wet season to areas all along the south and east edges of the continent, from China to Pakistan. In winter, the interior of Asia becomes very cold, a high-pressure system is set up and an outflow of air, the **winter monsoon**, brings very chilly weather to India, Pakistan, China, and Korea (Figure 6.5).

Land organisms respond to the great seasonality of the monsoon climate, and organisms in the shallow coastal waters are affected strongly too. As nutrient-poor water is blown in from the surface of the open ocean in the summer monsoon, food becomes scarce; as water is blown offshore in the winter monsoon, deeper water is sucked to the surface and brings nutrients and high food levels. As a result, the diversity of marine creatures along the coasts of India is far less than it is in the Philippines and Indonesia, which are far enough away from the center of Asia that they feel the effects of the monsoons much less strongly (Figure 6.5). Reefs are scarce and poor in diversity along the Asian mainland coast but are rich and diverse in a great arc from the Philippines to the Australian Barrier Reef.

## Geographic Isolation and Evolution

Geography is an important driver of evolution. Charles Darwin noticed this as he traveled round the world in the

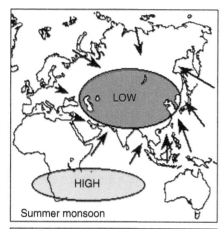

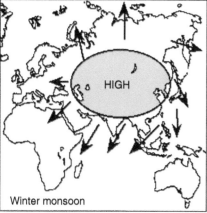

Figure 6.5 The monsoons of Asia. In summer, heat builds up over the continent and generates low pressure that draws in moist air from the surrounding oceans. In winter, high pressure over the continent generates cold winds that blow offshore. As a result, southern Asia is more seasonal than most regions that lie on or near the Equator.

1830s. Everywhere he went, there were different species of plants and animals, some of them occurring in only very restricted areas. For example, he saw that the plants and animals of Australia were quite different from any-where else – they were **endemic**, meaning they only occurred in that one place. He surmised that the unique kangaroos and wombats of Australia had evolved there for many millions of years in isolation.

He also noted many examples of groups of related species that lived close together. For example, he studied the turtles and "finches" on the Galápagos Islands. These islands, lying on the Equator 906 km (563 miles) west of Ecuador, were formed from volcanoes. So all the plants and animals on the islands had arrived across the ocean, from the nearest mainland, which is Ecuador. Birds could fly in and seeds could be blown across but what about the giant land tortoises? In any case, giant tortoises are present on many of the Galápagos islands, and Darwin

noted three species, each restricted to one island, and distinguishable by shell shape.

This kind of observation is the basis of the model of **geographic speciation** (Figure 6.6). Many species seem to have split in this way because of geographic barriers. These tortoises got to their specific islands thousands or millions of years ago, and have evolved in isolation until they are distinct species. The same can happen between neighboring valleys or mountain tops, where plants or animals become isolated and evolve independently. Darwin's ideas, based on natural history observations, have since been tested using genetic information, and this confirms the origin of many species by geographic isolation.

These ideas combine aspects of space (geography) and time (geology), and we now turn to the fossil record to explore evidence of diversity through time.

## Diversity Patterns in the Fossil Record

Jack Sepkoski spent over 20 years compiling data on the fossil record of Earth through the Phanerozoic, concentrating most on marine fossils. At first, he simply counted the number of families of marine fossils that had been defined by paleontologists from Ediacaran to recent times; later he compiled genera. These data on global (marine) diversity show clear and reasonably simple trends (Figure 6.7). Few families of marine animals existed in Ediacaran times but the beginning of the Cambrian saw a dramatic increase that followed a steep curve to a Late Cambrian level. A new, dramatic rise at the beginning of the Ordovician raised the total to a high level that remained comparatively stable through the rest of the Paleozoic. In the Late Permian, there was a dramatic diversity drop in a very large extinction that marks the end of the Paleozoic Era. A steady rise that began in the Triassic has continued to the present, with a small and short-lived reversal (extinction) at the end of the Triassic, and a steeper and deeper extinction at the end of the Cretaceous, which also marks the end of the Mesozoic Era.

This general pattern has been familiar to paleontologists for a long time, and indeed the apparently sharp changes from Paleozoic to Mesozoic and Mesozoic to Cenozoic marine faunas marked the divisions between those eras, when they were named about 1840. They are also now recognized to mark two of the great mass extinctions, at the end of the Permian and Cretaceous.

It's easy to think of possible problems with Sepkoski's approach. For example, only some parts of the world have been thoroughly searched for fossils; some parts of the geological record have been searched more carefully

Figure 6.6 Geographic isolation and speciation. Three species of giant land tortoise have evolved on the Galápagos Islands. When Charles Darwin visited in the 1830s, local people told him they could easily distinguish the three species by the shell shape. These species had arisen at different times in the past 2 million years. *Source:* Image from Slideserve/Prentice Hall.

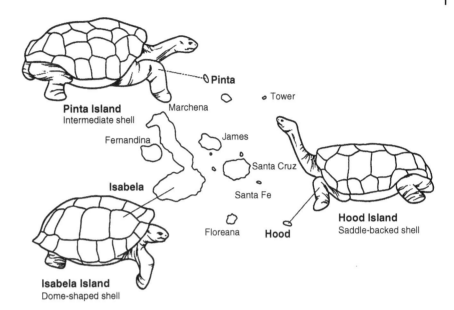

Figure 6.7 Jack Sepkoski's compilation of all marine fossil families for the Phanerozoic. The geological periods are indicated along the x-axis. Major extinctions occurred at the ends of the Paleozoic and Mesozoic eras (vertical lines).

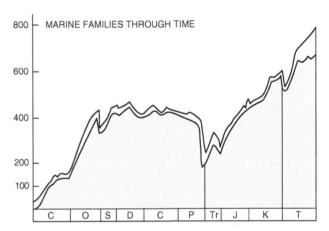

than others; older rocks have been preferentially destroyed or covered over by normal processes such as erosion and deposition. Different researchers mean different things as they define species, genera, and families in the group they study.

There have been quite heated discussions recently about whether we can believe anything in the raw data (Figure 6.7) – could this graph represent reality or is it so compromised by incomplete preservation and incomplete sampling by paleontologists as to be meaningless? Here, we will follow the general assumption that the plot reflects reality more than bias and error. Evidence in favor of this view is that the order of fossils in the rocks is roughly the same as predicted from phylogenomic trees (see Chapter 3) – the fossils are in the right order, and so we can trust that the broad pattern, while full of error, is not positively misleading.

Paleontologists have been searching the world for fossils for 200 years. The best-sampled fossil communities are shelly faunas that lived on shallow marine shelves,

and our estimate of their diversity through time is likely to be a fair sample of the diversity of all life through time. Larger groups of animals are harder to miss than smaller groups, so we have probably discovered all the phyla of shallow marine animals with hard skeletons. Perhaps we have only found a few percent of the species in the fossil record but we have probably discovered many of the families. In any case, if the search for fossils has been roughly random (and there's no reason to doubt it), the shallow marine fossil record as we now know it is a fair sample of the marine fossil record as a whole. So we can now ask what influenced the patterns that Sepkoski documented. Were they affected by the changing geography of Earth and if so, what were the causal connections?

Following Jack Sepkoski's lone efforts, John Alroy launched a major community initiative in 1998, the Paleobiology Database, which now contains millions of pieces of information and can be used as a research and education tool to plot diversity-through-time curves and paleogeographic maps of fossil distributions, and to explore different habitats and possible sources of error in the data.

## Global Tectonics and Global Diversity

We have known for over 40 years that the Earth's crust is made up of great rigid *plates* that move about under the influence of the convection of the Earth's hot interior. As they move, the plates affect one another along their edges, with results that alter the geography of the Earth's surface in major ways. Two plates can separate to split continents apart, to form new oceans, or to enlarge existing oceans by forming new crust in giant rifts in the ocean floor. Two plates can slide past one another, forming long *transform faults* such as the San Andreas Fault of California. Plates can converge and collide, forming chains of volcanic islands and deep trenches in the ocean, volcanic mountain belts along coasts, or giant belts of folded mountains between continental masses. At times, the Earth has had widely separated continents; at other times, the continental crust has largely been gathered into just one or two "supercontinents." These movements and their physical consequences are studied in the branch of geology called **plate tectonics.**

Plate tectonic movements affect the geography of continents and oceans, which can in turn affect food supply, climate, and the diversity of life. In other words, the tectonic history of Earth should have been a first-order influence on the diversity of the fossil record. Do we see a correlation? The brief answer is yes.

In the Early Cambrian, most continental pieces were more or less close together in a great belt across the South Pole. They split progressively during the Cambrian and Ordovician to form several small continents that were generally more scattered and in lower latitudes (Figure 6.8). This dispersion of continents coincides with the great diversity rise of the Cambrian and Ordovician. There were several continental collisions from the Middle Paleozoic through the Permian, and larger land masses were formed (Figure 6.9a). The great extinction at the end of the Permian coincides with the final merger of the continents into a giant global supercontinent, **Pangea** (Figure 6.9b), composed of a large northern land mass, **Laurasia**, and a southern land mass, **Gondwana**.

The rise in diversity that began in the Triassic and continued into the Cenozoic coincides very well with the progressive break-up of Pangea. The break-up was under way by the Jurassic and reached a climax in the Cretaceous (Figure 6.10). The continental fragments have continued to drift and today, the continents are perhaps as well separated as one could ever expect, even in a random world, with diversity at an all-time high level.

Thus, the tectonic events that affected Earth over the past 550 million years are reflected in the diversity curve. What are the connecting factors?

In an oceanic world, with continents small and widely separated, so that there are many provinces, each community in a province tends to have stable food supplies and high diversity. Therefore, the more the continents are fragmented into smaller units, the more oceanic the world's climate becomes and the more diverse its total biota.

The other extreme occurs when all the world's continents are together in a supercontinent, such as Pangea: not only are there fewer provinces but each province has low-diversity communities. This is true for life in the sea as well, or at least the richest life of

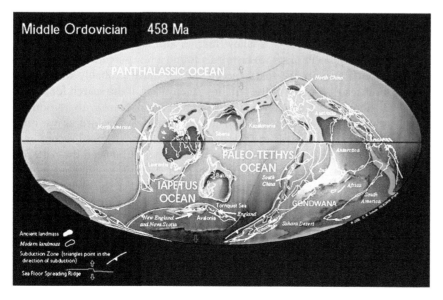

Figure 6.8 Ordovician paleogeography. There are several separate continents in the tropical regions, and the southern continents make the supercontinent Gondwana. *Source:* Paleogeographic map by C.R. Scotese © 2012, PALEOMAP Project (www.scotese.com).

(a)

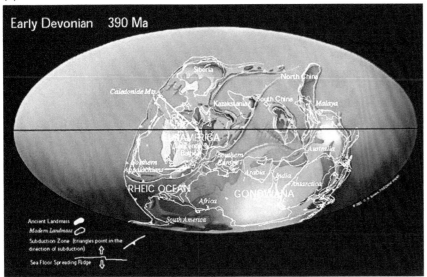

(b)

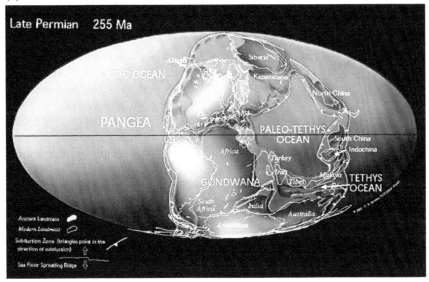

Figure 6.9 Late Paleozoic paleogeography. (a) By the Early Devonian, 390 Ma, Gondwana is drifting north into warmer latitudes and other continents are converging together. (b) By the Late Permian (255 Ma), a complete Laurasia has united with Gondwana to form the global supercontinent Pangea. *Source:* Paleogeographic maps by C.R. Scotese © 2012, PALEOMAP Project (www.scotese.com).

Figure 6.10 Cretaceous paleogeography. By the end of the Cretaceous, Gondwana and Laurasia have split into pieces, with Australia just leaving Antarctica. *Source:* Paleogeographic map by C.R. Scotese © 2012, PALEOMAP Project (www.scotese. com).

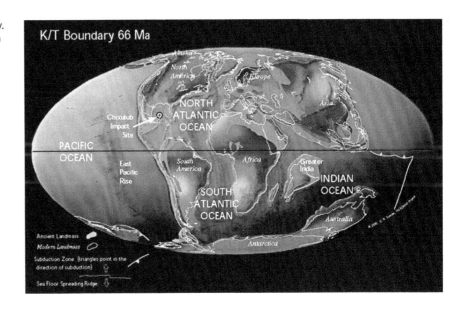

the continental shelves – a single supercontinent means more connections between shallow marine faunas worldwide.

The overall pattern of diversity data through time does receive a first-order explanation from plate tectonic effects. But that cannot be the whole story, for several reasons.

1) **Changing faunas through time.** If plate tectonics were the only control on diversity, much the same groups of animals should rise and fall with the changes in global geography. Instead, we see dramatic changes in different animal groups that succeed one another in time. As we shall see, some of these new groups were able to reach much higher levels of diversity in the same space.

2) **Increase in global diversity.** The overall increase in global diversity from Ediacaran to recent times is not predicted on plate tectonic grounds. This is especially true of the rapid diversity increase in the past 100 million years.

3) **Mass extinctions.** At one time, it was thought that mass extinctions were caused by plate tectonic movements. For example, the end-Permian mass extinction coincided with the supercontinent Pangea, and it was suggested that greater uniformity worldwide reduced provinces and so reduced global biodiversity. However, as we shall see, the event was sudden and driven by volcanic catastrophe. Mass extinctions such as these had a huge influence on the subsequent recovery of life.

## Changing Faunas Through Time

### Three Great Faunas

Jack Sepkoski sorted his data on marine families through time to see if there were subsets of organisms that shared similar patterns of diversity. He distinguished three great divisions of marine life through time, which explain about 90% of the data (Figures 6.11 and 6.12). Sepkoski called them the Cambrian Fauna, the Paleozoic Fauna, and the Modern Fauna. These "faunas" overlap in time and the names are only for convenience but they do reflect the fact that different sets of organisms have had very different histories.

The **Cambrian Fauna** includes the groups of organisms, particularly trilobites, that were largely responsible for the Cambrian increase in diversity. But after a Late Cambrian diversity peak, the Cambrian Fauna declined in diversity in the Ordovician and afterward.

The success of the **Paleozoic Fauna** was almost entirely responsible for the great rise in diversity in the Ordovician,

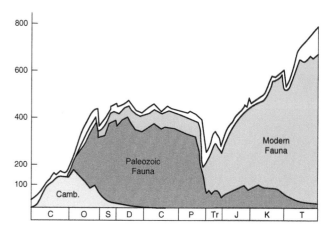

**Figure 6.11** The three great faunas defined by Jack Sepkoski in his analysis of the marine fossil record through time. They are subsets of the data shown in Figure 6.4. *Source:* Scale in numbers of families. Data from Sepkoski (1981, 1984).

and it slowly declined afterward. The Paleozoic Fauna suffered severely in the Late Permian extinction, and its recovery afterward was insignificant compared with the dramatic diversification of the **Modern Fauna**.

Figure 6.12 also shows the major animals making up the three faunas. Their definition is approximate because Sepkoski tried to make his analysis simple by using animal groups at the level of classes or subphyla. The analysis could be repeated by subdividing the marine animals into groups that would give even sharper divisions between the three faunas: for example, one could separate Paleozoic corals from later ones. There is no zoological affinity connecting the members of the three faunas but they do have ecological meaning.

### Explaining the Three Great Faunas

The diversity patterns imply that ecological opportunities in the world's oceans changed through time to favor one ecological assemblage and then to allow the diversification of others. The diversity patterns have been known in outline for some time, so some of the explanations predate Sepkoski's analysis.

In the 1970s, James Valentine pointed to the different ways of life that are encouraged under different types of food supply. In the Cambrian, he argued, the continents were not widely separated, food supplies were variable, and the most favored way of life would have been deposit feeding; there is always some nutrition in sea-floor mud. Thus, Cambrian animals, wrote Valentine, are "plain, even grubby." The Burgess fossils may not be plain but many of them were certainly mud grubbers. Even among soft-bodied animals, arthropods dominate Cambrian

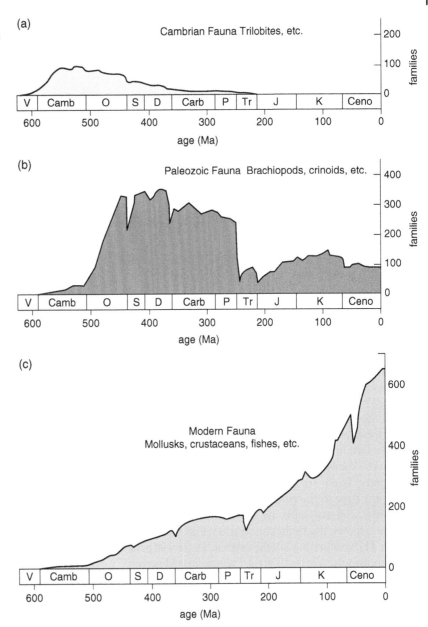

Figure 6.12 The individual histories through time of the three great faunas defined by Jack Sepkoski. In particular, note the difference between them at the end of the Permian: the Paleozoic Fauna suffers a tremendous extinction while the others are hit less hard. *Source:* Based on Sepkoski (1981, 1984).

faunas in numbers and diversity, and most of them were deposit feeders.

The Paleozoic Fauna lived in more tightly defined communities, with a more complex ecological structure. Indeed, filter feeders reached higher in the water and fed at different levels, and there was more burrowing in the sediment and to greater depths than in the Cambrian. The development of reef-like structures gave opportunities for many groups of fixed filter-feeding animals to diversify. This larger food supply in the form of stationary benthic filter feeders allowed slow-moving carnivores to become more diverse. This is especially true of cephalopods and fishes which diversified hugely, with new kinds of swimming forms. The overall trend was to add new ways of life, or **guilds**, to marine faunas. Altogether, Paleozoic animals seem to have subdivided their ways of life more finely over time.

The mass extinction at the end of the Permian wiped out most Paleozoic groups, and it seems to have related more to shock changes in oceans and atmospheres than any long-term processes. Continents may have been amalgamating as Pangea, but the old idea that this was a driver of extinction has been largely discarded. It's more the ocean warming, acidification, and loss of oxygen from sea floors that hit the corals, brachiopods, cephalopods, bryozoans, and crinoids hard.

Further evidence for the minimal role of continental movements in driving large-scale evolution of ocean life comes in the rise of the Modern Fauna. This should have witnessed the reemergence of Paleozoic-style predators and filter feeders. Most of the Mesozoic diversification was achieved by other groups that stand out in Sepkoski's analysis as the Modern Fauna. These new groups included new guilds, especially more mobile animals, more infaunal burrowing animals, and new predators, including lobsters and crabs, gastropods, and fishes.

These new groups were all parts of the Mesozoic Marine Revolution, as we shall see in Chapter 11. At first, paleontologists focused on changes in marine ecosystems in the Cretaceous, when new predators were more effective than their predecessors at attacking animals on the sea floor. Modern gastropods evolved, capable of attacking shells with drilling radulae backed with acid secretions and poisons. Advanced shell-crushing crustaceans became abundant, and so did bony fishes with effective shell-crushing teeth. The filter feeders of the Paleozoic were replaced by animals that could hide in burrows or escape fast. In fact, current evidence suggests the Mesozoic Marine Revolution began in the Triassic, marking the rise of the Modern Fauna, when burrowing bivalves and burrowing echinoids make up important components of the Modern Fauna, together with effective, wide-roaming predators such as lobsters, gastropods, and fishes.

The evolution of new modes of life – new niches and new guilds – was associated with overall expansions in biodiversity. There are more species in the oceans today than at any time in the past. Indeed, some of the groups within the Modern Fauna are enormously species rich. Think of teleosts, the dominant modern bony fishes (see Chapter 11), comprising 30 000 species. There was probably nothing quite like them in the Paleozoic or Mesozoic.

Similar huge hikes in global biodiversity are even more apparent when we look at life on land. At one time, there was no life on land, then microbes, then green plants, then bugs and spiders, then tetrapods, then forests emerged, flying animals, flowering plants … Each step, including terrestrialization in the Paleozoic (see Chapters 8 and 9) and the Cretaceous Terrestrial Explosion (see Chapter 15), mark major increases in biodiversity. There are more than a million species of insects today; once there were no species of insects. Life has burgeoned in the oceans, and even more so on land.

## Mass Extinctions

Extinction happens all the time. Martha, the last passenger pigeon left in the world, died of old age in the Cincinnati Zoo in 1914 (Figure 6.13). This officially made

Figure 6.13 An icon of extinction: Martha the passenger pigeon, last of her species, who died in 1914 at the Cincinnati Zoo. *Source:* Photograph by Enno Meyer (Wikimedia).

her species, *Ectopistes migratorius*, extinct although of course the species had been ecologically doomed as soon as the last breeding birds were gone. It is estimated that hundreds of species of plants and animals are driven to extinction each year by habitat destruction and other human activities.

The current biodiversity crisis is exceptional because of the rate of species loss. However, extinction occurs naturally, although generally at a much slower rate. Extinction occurs on all scales, from local to global, and it happens at different rates at different times and in different regions. Some species have small populations that depend on a particularly narrow range of food or habitat, and are vulnerable to even small-scale ecological disturbance. So, there must be a steady leakage of species, through extinction, out of the global biosphere. Occasionally, by bad luck perhaps, one of these species will be the last of its family, and the loss of that family would become visible in a compilation like Sepkoski's.

Sooner or later, every species has some chance of becoming extinct: extinction is the expected fate of species, not a rarity. In a world with steady diversity through time, existing species (and families) would become extinct about as often as new species (and families) evolved. With overall rising diversity (Figure 6.7), new species and families emerge more often than matching extinctions. But there are times when extinctions exceed originations, and these may be regionally or ecologically restricted, or global. An example of an ecologically restricted and regional **extinction event** is the loss of large mammals 11 000 years ago as the northern ice sheets retreated – the time when mammoths, mastodons, and wooly rhinos disappeared from northern parts of North America, Europe, and Asia (see Chapter 23).

**Mass extinctions** are the larger, global-scale events, when species of all kinds, marine and terrestrial, died out. Paleontologists had long noted a series of five big events, and these were codified through statistical study of Sepkoski's databases and the Paleobiology Database. The "big five" mass extinctions are, in geological order:

- at the end of the Ordovician
- at the end of the Frasnian stage of the Late Devonian (F–F)
- at the end of the Permian (Permo-Triassic or P–T)
- at the end of the Triassic
- at the end of the Cretaceous (Cretaceous-Paleogene or K–Pg)

A mass extinction is defined as a time when many species, of diverse ecology, went extinct worldwide in a geologically short span of time. This definition cannot be quantified further because each event, it seems, is unique, and some might have happened literally overnight – perhaps the K–Pg event, which resulted from an asteroid strike (see Chapter 16) – whereas others were driven by volcanic eruptions or severe climate change and so might have lasted as long as a million years.

The definition is not always purely numerical in any case. Ecology matters, and mass extinctions can be classified in terms of the seriousness of their impact on ecosystems. George McGhee and colleagues identified the end-Permian mass extinction as the most severe, and it wins both in terms of numbers of species lost and ecological impact, but the K–Pg comes a close second even though it was numerically on a par with the other three events. This is because the disappearance of all the large dinosaurs at the end of the Cretaceous, together with the large marine reptiles and ammonites, damaged huge parts of ecosystems on land and sea. This would have made the K–Pg mass extinction much more of a global ecological disaster than you would expect from simply counting extinct reptiles.

It is already clear that mass extinctions did not all have the same cause. In fact, each one may have had a unique cause or combination of causes. That becomes important if we are to assess the ongoing human-induced mass extinction currently affecting the globe; it looks as if it will come into the category of a sixth mass extinction (see Chapter 23).

## Explaining Mass Extinction

Mass extinctions were global phenomena, so they have to be explained by global processes. The first that comes to mind is plate tectonics. However, tectonic changes are relatively slow in geological terms, so if tectonic extinctions were to happen, they would be slow. But mass extinctions are relatively sudden, so we would have to suggest something else to make a case for a tectonic extinction.

Some plausible agents for global extinctions are:

- a failure of normal ocean circulation, affecting ocean chemistry enough to cause global changes in climate and atmosphere
- a rapid change in sea level affecting global ecology and climate
- an enormous volcanic eruption affecting global ecology and climate
- an extraterrestrial impact by an asteroid affecting global ecology and climate.

Of these possible agents, enormous volcanic eruptions leave behind great masses of volcanic rocks so they are relatively easy to detect in the geological record. Further, new geological observations and Earth system modeling have identified a plausible sequence of consequences of massive volcanic eruption, including sharp global warming, acid rain, destruction of vegetation and wash-off of soils into the ocean, with acidification, warming, stagnation, and anoxia in the oceans. Global changes in sea level would change the distribution of sediments laid down on the Earth's surface; as long as the sea-level change lasts long enough to leave behind this kind of evidence, we will be able to find it. But a failure of ocean circulation typically would be expected to leave behind only subtle chemical evidence, and it is likely to be short-lived, so evidence might be difficult to find and interpret.

Extraterrestrial impact is an instantaneous event that might leave behind only a thin layer of evidence. Unless we find a crater or some unique piece of evidence that only an impact can produce, it may be very difficult to identify an impact, especially in more ancient rock. Three major indicators are:

- a defined layer or spike of the element iridium (Ir), which occurs in greater abundance in meteorites than in Earth's crust

- tektites, which are tiny glass blobs (spherules) formed as a meteorite or asteroid splashes molten drops of rock at high speed into the atmosphere
- shocked quartz: quartz crystals with characteristic damage that can only be caused by intense shock waves.

At present, the leading hypotheses for the causes of the largest six extinctions are:

- end-Ordovician climate (a short-lived ice age)
- Late Devonian (F–F) oceanic crisis
- end-Permian (P–Tr) giant eruptions, and their climatic consequences
- end-Triassic very large eruptions, and their climatic consequences
- end-Cretaceous (K–Pg) huge asteroid impact, plus giant eruptions.

We discuss the K–Pg event in Chapter 16. Here we discuss the four others but concentrate on the largest of them all, at the Permo-Triassic boundary.

## The Ordovician Mass Extinction

The mass extinction at or near the end of the Ordovician seems to be closely linked with a major climatic change. A first pulse of extinction happened as a big ice age began, and the second occurred as it ended. This "mass extinction" included the loss of a lot of shelly fossils, including bivalves, echinoderms, bryozoans, and corals, but ecologically it was a comparatively minor event. There was minimal life on land at the time, so the Ordovician extinction is purely a marine event.

## The Late Devonian (F–F) Mass Extinction

A mass extinction took place, possibly in several separate events, at the boundary between the last two stages of the Devonian: the Frasnian and Famennian (the F–F boundary). There was a major worldwide extinction of coral reefs and their associated faunas, and many other groups of animals and plants were severely affected. The land plants in wet lowland areas and the first amphibians that lived at the water's edge do not seem to have been affected by this extinction. However, the event marked a major extinction of various armored fish groups and ushered in new fish faunas in the Carboniferous.

The causes of the multiple Late Devonian mass extinctions seem to have been climatic, with major changes in

sea level and ocean chemistry at the same time. Carbon isotope shifts indicate that global organic productivity changed rapidly before the boundary. There was widespread anoxia on the seabed and this was associated with massive extinction.

Some have tried to link these physical and chemical changes in the oceans to asteroid impact but the evidence is limited. Geologists have an impressive list of indicators of impact and these have all been identified for the K–Pg event (see Chapter 16), but very few of them have been found in the Late Devonian. It may be that all these effects followed from a series of major volcanic events, where great volumes of lava were spewed out over parts of Russia, and volcanic gases led to global warming, anoxia, and the other devastating killing effects. If so, then this event falls in line with a common style of volcanic model that may have driven other mass extinction events, including the Permian–Triassic.

## The Permian–Triassic (P–Tr) Extinction

The extinction at 252 Ma, at the end of the Permian, is the largest of all time, numerically and ecologically. By Sepkoski's count, an estimated 57% of all families and 95% of all species of marine animals became extinct (Figure 6.11). The Paleozoic Fauna was very hard hit (Figure 6.12), losing very many suspension feeders and carnivores and almost all the reef dwellers. During the other "big five" mass extinctions, the losses were much lower, roughly 15–30% of families and 50% of species.

The P–Tr extinction was rapid, taking place in much less than a million years. It was much more severe in the ocean but it affected terrestrial ecosystems too. Overall, the P–Tr extinction is a major watershed in the history of life on Earth – it triggered the replacement of the Paleozoic marine fauna by the Modern Fauna, as we have seen, and on land it triggered the rise of many remarkable new groups (see Chapter 12). Study of the killing models for the P–Tr mass extinction has drawn attention to a new model for catastrophe, times when global warming caused a cascade of catastrophic consequences, called **hyperthermals**, meaning "high temperatures."

### Plume Volcanism

The Permian extinction coincides with the largest known volcanic eruption in Earth history: one of a few giant **plume eruptions**. Occasionally, an event at the boundary between the Earth's core and mantle sets a

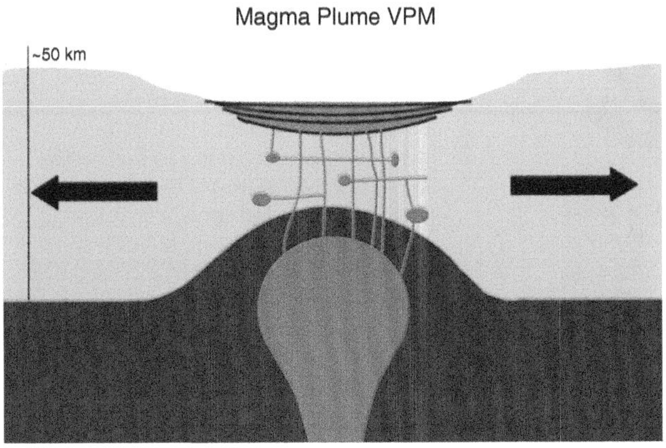

Magma Plume VPM

Figure 6.14 Operation of a mantle plume. Magma reaches the surface through radiating sills and dikes, forming basalt flows, as well as deep and shallow magma chambers below the surface. The crust gradually thins due to thermal subsidence, and basalt flows accumulate in thick layers along the volcanic passive margin (VPM). *Source:* Art by Graham Mills (Wikimedia).

giant pulse of heat rising toward the surface as a plume (Figure 6.14). As it approaches the surface, the plume melts or distorts the crust to develop a flat head of molten magma that can be 1000 km across and 100 km thick. Melting the crust, the plume generates enormous volcanic eruptions that pour millions of cubic kilometers of basalt – **flood basalts** – out on to the surface. If a plume erupts through a continent, it blasts material into the atmosphere as well. After the head of the plume has erupted, the much narrower tail may continue to erupt for 100 m.y. or more, but now its effects are more local, affecting only 100 km or so of terrain as it forms a long-lasting **hot spot** of volcanic activity.

Plume events are rare: there have been only eight enormous plume eruptions in the last 250 m.y. The most recent is the Yellowstone plume; at about 17 Ma, it burned through the crust to form enormous lava fields that are now known as the Columbia Plateau basalts of Oregon and Washington, best seen in the Columbia River gorge. North America drifted westward over this "hot spot," which continued to erupt to form the volcanic rocks of the Snake River plain in Idaho (Valley of the Moon and so on), and it now sits under Yellowstone National Park. The hot spot is in a quiet period now, with geyser activity rather than active eruption, but it produced enormous volcanic explosions about 500 000 years ago that blasted ash over most of the mountain states and into Canada. Even so, it was not large enough to cause a mass extinction.

At 252 Ma, a massive plume burned through the continental crust in what is now western Siberia to form the Siberian Traps, gigantic flood basalts about 5 million $km^2$ in area (Figure 6.15) and perhaps 3 million $km^3$ in volume. The eruptions lasted at full intensity for only about half a million years; these are the largest known, most intense eruptions in the history of the Earth, and are dated exactly at the P–Tr boundary, 252 Ma. Can this be a coincidence?

### The Hyperthermal Killing Model

The answer is, "they are linked." In the past 20 years, intensive studies of remarkable rock sections that span the Permian–Triassic boundary, in central Europe, the Middle East, Pakistan, Canada, Spitsbergen, Greenland, and especially in South China, have revealed nearly every detail of what happened. Not only can it be shown that the event coincides with the volcanic activity, but there were two pulses of extinction, separated by as little as 60 000 years. This is an amazing piece of dating precision! The new work has also helped to solidify the details of how times of major eruptions, hyperthermals, can change environments and kill life in a huge way.

The geological clues are firm evidence that these things happened in the sea: loss of oxygen from seabed sediments (black, carbon-rich muds and limestones, rich in pyrite); sharp warming by 15 °C (oxygen isotopes); massive loss of life (change from rich communities of reefs, shell-beds, and burrowers to limited communities of four or five species). On land, there is evidence from terrestrial successions in Russia and South Africa showing these things: sharp warming by 15 °C (oxygen isotopes); massive loss of forests and wash of sediment into the oceans (sedimentological shift from meandering to braided rivers, plus peak of sand and organic plant matter in oceans); massive loss of life (reduction of complex communities of plants and tetrapods to very few species).

The fact of the massive Siberian Traps eruptions, coupled with the repeated evidence from terrestrial and marine rock sections, has allowed geologists to put together a flow chart (Figure 6.16) that nicely summarizes the key consequences of the eruptions. It's all to do with the gases poured out of the volcano, not so much the lava. Remember, this is a rift-type eruption where lava leaks out of a crack in the crust, as on Iceland today, not an explosive eruption from a pyramidal volcano. The carbon dioxide and methane pouring out of the eruptive fissures are **greenhouse gases**, meaning they absorb and emit radiant energy. Their sudden injection into the atmosphere drove the rapid rise in temperature, and the rise was so sharp it stressed animals and plants on land and in surface seas and caused sudden extinction of many species.

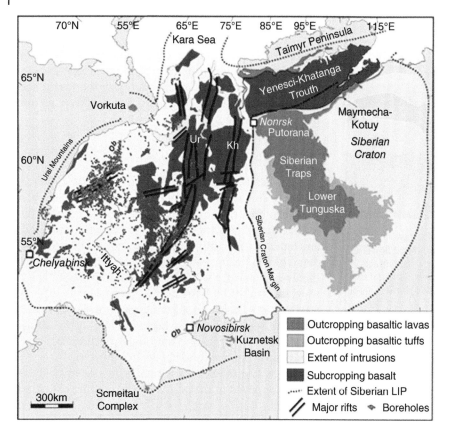

Figure 6.15 A map of the total extent of the Siberian Traps. Dark and medium green: surface outcrops of lava. Light green: intrusions of lava into the continental sediments. Red: lava discovered underground by drilling the crust. The total extent of the igneous rock is shown by dotted lines. *Source:* Image from Andy Saunders and Marc Reichow at the University of Leicester, and used by permission of Professor Saunders.

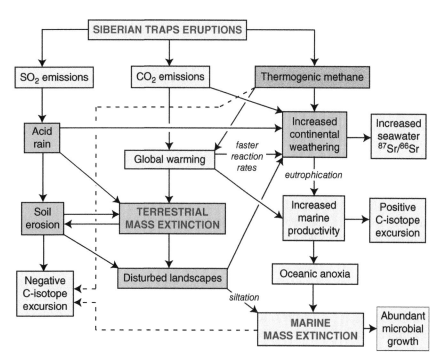

Figure 6.16 Model of likely environmental effects of the Siberian Traps eruptions, showing the flows of consequences of global warming and acid rain. Causal links are indicated by solid arrows and possible second-order controls on the negative carbonate carbon-isotope excursion at the P–Tr mass extinction are indicated by dashed lines. *Source:* From Algeo et al. (2011), by permission of Thomas J. Algeo.

A second effect of the gases was to produce **acid rain** when they combined with rainwater in the atmosphere. The acid rain fell on the Late Permian forests and the trees died and fell over, so normal rainfall could then erode the soil that had been held in place by the tree roots. A massive flood of material was released and the landscapes became rocky and bare. As the soil and plant debris entered the oceans, it blocked the filter-feeding structures of corals and seabed creatures and killed them too. The increased continental weathering also caused changes in strontium isotope ratios and led to a plankton bloom, where surface organisms had a short boom and then died off. The acid rain also caused **ocean acidification**, where the pH of the water dropped and any organism with a shell made from calcium carbonate, such as brachiopods, bivalves, corals, and echinoderms, could not build their skeletons and so died.

The global warming also led to stagnation of ocean currents, meaning that the normal circulation was stopped. In the normal deep ocean circulation, cold bottom waters from the North and South Pole run along the bottom of the oceans and come up at the Equator, absorbing oxygen from the atmosphere, and then cycle back at the surface and down to the bottom. With stagnation, these cycles stopped and oxygen was not delivered to the sea bed and again, life died. Evidence of this **anoxia** (lack of oxygen) is from the black sediments with pyrite – the sediments are black because organic matter (carbon) builds up and cannot be eaten by seabed animals and microbes because of the lack of oxygen. Iron pyrites forms in these anoxic settings – think of the sulfurous stink of a black-leaf stagnant pond.

The summary flowchart (Figure 6.16) shows how all the observations from field geology, paleontology, and isotope geochemistry provide evidence for a coherent killing model. Life wasn't killed by the lava, as people might expect, but by the millions of tons of volcanic gases and their impact on the atmosphere–ocean system. In fact, this core killing model might explain numerous mass extinctions and extinction events, and in recent reviews it has been applied, as we saw, to the Late Devonian events and to the end-Triassic mass extinction, as well as some other extinction events in the Middle Permian, Late Triassic, and Early Jurassic.

The Siberian Traps volcanic eruptions may have triggered other catastrophic consequences. For example, as the magma from the plume broke through the lower crust, it encountered very large oil and gas fields in Ediacaran and Cambrian sediments. As the oil and gas were heated, they formed even more gases to add to those already in the magma. Then, the rising magma reached and heated a giant salt-bearing field in the Cambrian rocks, which added chlorine gases like hydrochloric acid

aerosols to the mix. Even higher in the crust were thick coal-bearing strata, so huge quantities of carbon were added, as carbon dust, carbon dioxide, or methane gas. The cumulative pressure in the trapped gases was great enough to punch hundreds of giant "pipes" through the crust, many of them hundreds of meters across, and gigantic fountains of toxic gases were blasted high into the atmosphere along with volcanic ash.

## Ocean Chemistry Crisis

The ocean went through a chemical crisis at the P–Tr boundary. Many of the marine organisms that went extinct made skeletons of calcium carbonate. They would have been particularly susceptible to very high levels of acid in the seawater because that would inhibit the reactions they used to make their skeletons. As we have seen, ocean acidification was generated by excess carbon dioxide from the eruptions and acid rain. Indeed, human activity that involves burning fossil fuel is acidifying the oceans today. In addition, limestone sediments exposed to ocean water in the late Permian often show signs of being etched by acidic sea water.

Andy Knoll and colleagues suggested that the extinction was caused by a catastrophic overturn of an ocean that was supersaturated in carbon dioxide. Others have suggested hydrogen sulfide and methane as components in the ocean. If a mass of anoxic water in the deep oceans, loaded with dissolved carbon dioxide, methane, and hydrogen sulfide, were brought suddenly to the surface, it would degas violently on a global scale and would likely trigger greenhouse heating and a major and sudden climatic warming. Ultimately, these possible oceanic chemical crises were triggered by the Siberian Traps eruptions.

The idea of a crisis that resulted from suddenly overturning an anoxic ocean has not been well defined, although such a crisis would clearly devastate the oceans and damage ecosystems along the shores. In particular, the Permian ocean has long been understood to have been one giant ocean, **Panthalassa**, as a counterpart to the giant continent Pangea. But it is very difficult to generate a crisis in a world ocean. Global wind patterns circulate it efficiently, at least in the surface waters. Today, water circulates vertically in the ocean when surface waters become dense and sink to the bottom, stirring up deep waters that upwell to the surface. Various ideas about opening and closing seaways that triggered flows of fresh water or chemically perturbed waters have been proposed, but evidence is mixed.

One chemical consequence, however, is that the overall warming driven by the eruptions might have caused further releases of methane. Methane can form from

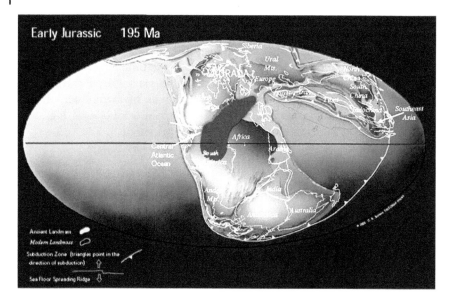

Figure 6.17 The Earth in the Early Jurassic, just a few million years after the end-Triassic extinction. The Central Atlantic Magmatic Province (CAMP) basalts are shown in red. They look bigger than the Siberian Traps but that is because of the map projection used.
*Source:* Paleogeographic map by C.R. Scotese © 2012, PALEOMAP Project (www.scotese.com).

decayed plant material in the frozen tundra and from decayed plankton in the deep oceans. Methane is a gas comprising carbon and hydrogen and it is a very effective greenhouse gas, causing 28 times the amount of global warming as carbon dioxide. Warming atmospheres and oceans at the end of the Permian might have melted frozen stores of methane in the deep ocean, causing it to bubble up and accelerate the warming effects already set in motion by the eruptions.

This warming-leading-to-more-warming is a scary prospect, as it defies the normal kinds of feedback processes on Earth, and it has been termed the "runaway greenhouse." Normally, if temperature rises as a result of excess carbon dioxide, plants mop up the gas as a normal part of photosynthesis and temperatures return to normal. What occurred in the P–Tr mass extinction was happening too fast and the plants were all being killed off anyway by acid rain. We are seeing a runaway greenhouse today, when carbon dioxide is being produced too fast from factory chimneys and cars, and methane is being produced too fast by the melting permafrost and all those factory-farmed cows. Photosynthesis and other natural processes cannot mop up all the greenhouse gases fast enough, so the climate is changing.

## The End-Triassic Extinction

Fifty million years after the P–Tr extinction, at the end of the Triassic about 201 Ma, there were more huge eruptions as Pangea began to split apart and the Atlantic Ocean began to form. Eleven million km$^2$ (4 million mi$^2$) of basalt lava were erupted, most of it within only

half a million years, to form the so-called CAMP (Central Atlantic Magmatic Province) that stretches from France to Brazil (Figure 6.17). Carbon isotopes suggest major changes in ocean productivity, and there were important extinctions in the sea (Figure 6.10, and the Paleozoic Fauna in Figure 6.11). On land, various groups of plants and insects, and some key reptile groups died out.

Here, the eruptions are the prime suspect for the extinction, and the model is very like the hyperthermal model developed for the P–Tr event (Figure 6.16). The huge amounts of lava from the CAMP would have released enormous quantities of carbon dioxide, sulfur gases, and aerosols that could have affected ocean chemistry directly, and probably affected productivity in the surface waters and on several continents. The amount of carbon involved in the carbon isotope changes suggests that very large quantities of methane were released as well.

But this was not another P–Tr event. The Triassic event is much smaller. As we work out the details, it will be very interesting to see which of the special circumstances of the P–Tr might also have occurred during the end-Triassic event. One thing is sure: at all scales, over all geological time, the geology of the Earth at any time is unique. As the ancient Greek philosopher Heraclitus said, "You can never step twice in the same river," meaning everything keeps changing.

## Evolutionary Radiations

New species appear all the time, just as species become extinct all the time. Occasionally, we can look back into the record and see that a particular new species happened to be the first of a very successful group that we define as a family or an even larger group. For example, it would be intriguing to track back to the very first bird or the very first beetle – these species were the first possessors of some pretty nifty specializations that were to give rise to highly successful groups.

The early years of any clade, whether it grows to be highly successful or not, are called **evolutionary radiations**. Documenting the shape of such a radiation is important – did the clade expand slowly or fast, how long did it continue expanding, and did it happen at the same time all over the world? Once the pattern is established, it is important to think about the processes that enabled or drove the radiation. A radiation is a response to an opportunity. So what kind of opportunity would set off a radiation so large that it would show up in a compilation of global diversity? Perhaps there are three, namely the clearing of life by mass extinctions, invasion of a new habitat, and a new biological invention.

## Mass Extinctions

By their very nature, mass extinctions remove many organisms from the biosphere. If the mass extinction was a one-time massive physical disaster (volcanic eruption, asteroid impact), the physical world might recover quickly to "normal" yet have a biology that was missing major components. This situation provides a major opportunity for surviving organisms to evolve to fill those ecological gaps. The newcomers will not have the same anatomy and will not re-evolve the same characters as their extinct predecessors, so we are likely to see a wave of evolutionary novelty wash across the world.

Obvious examples include the radiation of the Modern Fauna after the P–Tr extinction (see Chapters 11 and 12); the radiation of land mammals after the extinction of most dinosaurs at the K–Pg extinction (see Chapter 19); and the radiations of bats and whales after the extinction of most flying and swimming reptiles, again at the K–Pg extinction.

Mass extinctions remove the **incumbent effect**. This powerful metaphor is easily understood by Americans, who live with a political system in which an elected representative to Congress (let us say) is very difficult to remove from office, once elected, even though the election process is entirely open and democratic. The reason is that the incumbent has name recognition, and has a lot of power and access to money, while any prospective challenger typically does not.

The incumbent effect works in biology too. Any species is well adjusted to its normal environment; it evolved in that environment and its adaptations have been honed by natural selection for success there. Any invading species is likely to be less well fitted to that environment. As the incumbent effect pervades communities and provinces as well, ecosystems typically are stable over long time periods. Yet just as hurricanes can smash a local area of forest and allow weeds to flourish, or clean off a low-lying island that can be recolonized, so disasters such as mass extinctions can remove incumbents and allow survivors their place in the sun. A perfect example is the rise of mammals immediately after the K–Pg mass extinction. Mammals arose in the Late Triassic, at about the same time as the dinosaurs, and yet they existed through the Jurassic and Cretaceous mainly at small size and in the shade of the dinosaurs. Once the dinosaurs were gone, the mammals radiated explosively.

## Invading a New Habitat

Evolution works by natural selection, which implies continual testing of new mutations against the environment. Some organisms are always "pushing the envelope" and occasionally a lineage will evolve a body plan that allows it to invade a new habitat that may have been available for a long time but had been unexploited. If successful, that lineage may expand into a radiation as subclades explore the different ways of life that are possible in that new habitat. Obvious global examples include the first land plants and the first land animals (see Chapters 8 and 9) and the first flying animals (see Chapter 14).

This kind of opportunity probably exists at all scales. Land animals reaching a biologically "empty" isolated continent or island may radiate there; obvious examples are the marsupials of Australia and the mammals of South America during the Cenozoic (see Chapter 19), not to mention the reptiles and birds of the Galápagos that influenced Darwin so much.

## New Biological Inventions

Occasionally, a lineage will evolve a body plan that allows it to do things that no organism has done before. If successful (if the timing and the ecology are just right), that lineage may expand into a radiation as new subclades explore the different ways of exploiting that new invention. Obvious examples include the first eukaryotes (see Chapter 3), the early metazoans (see Chapters 4 and 5), and (again) the various groups that evolved the apparatus for flight (see Chapter 14). Both dinosaurs (see Chapter 13) and mammals evolved warm blood and the erect limbs that allowed them a very active life style on land. Bats and whales evolved sonar, hominids invented the capacity to make tools... we could go on for pages, and in fact we do, in the chapters listed above.

## References

Algeo, T.J., Chen, Z.-Q., Fraiser, M.L. et al. (2011). Terrestrial-marine teleconnections in the collapse and rebuilding of Early Triassic marine ecosystems. *Palaeogeography, Palaeoclimatology, Palaeoecology* 308: 1–11.

Sepkoski, J.J. (1981). A factor analytic description of the Phanerozoic marine fossil record. *Paleobiology* 7: 36–53. [Identifying the three great marine faunas].

Sepkoski, J.J. (1984). A kinetic model of Phanerozoic taxonomic diversity. III. Post-Paleozoic families and mass extinctions. *Paleobiology* 10: 246–267. [Putting it all together].

## Further Reading

### Diversity Through Time

Alroy, J., Aberhan, M., Bottjer, D.J. et al. (2008). Phanerozoic trends in the global diversity of marine invertebrates. *Science* 321: 97–100. [Paper using data from the Paleobiology Database, emphasizing bias in the data].

Benton, M.J. (2009). The Red Queen and the Court Jester: species diversity and the role of biotic and abiotic factors through time. *Science* 323: 728–732. [Overview of drivers of large-scale evolution, emphasizing the broad evidence from global paleodiversity curves].

Hannisdal, B. and Peters, S.E. (2011). Phanerozoic earth system evolution and marine biodiversity. *Science* 334: 1121–1124. [Physical drivers of global marine biodiversity].

Paleobiology Database: https://paleobiodb.org [There are some great online tools and R packages to query the database and plot some great graphs, timelines and paleomaps].

### Mass Extinctions

Benton, M.J. (2015). *When Life Nearly Died*, 2e. New York: Thames & Hudson.

Benton, M.J. (2018). Hyperthermal-driven mass extinctions: killing models during the Permian–Triassic mass extinction. *Philosophical Transactions of the Royal Society A – Mathematical Physical and Engineering Sciences* 376: 20170076. [How hyperthermals killed life on land and in the sea].

Bond, D.P.G. and Grasby, S.E. (2017). On the causes of mass extinctions. *Palaeogeography, Palaeoclimatology, Palaeoecology* 478: 3–29. [The common volcanic driver of several mass extinctions].

Burgess, S.D. and Bowring, S.A. (2015). High-precision geochronology confirms voluminous magmatism before, during, and after Earth's most severe extinction. *Science Advances* 1 (7): e1500470. [The link between the Siberian Traps eruptions and mass extinction].

Chen, Z.-Q. and Benton, M.J. (2012). The timing and pattern of biotic recovery following the end-Permian mass extinction. *Nature Geoscience* 5: 375–383.

Davies, J.H.F.L., Marzoli, A., Bertrand, H. et al. (2017). End Triassic mass extinction started by intrusive CAMP activity. *Nature Communications* 8: 15596.

Harper, D.A.T., Hammarlund, E.U., and Rasmussen, C.M.Ø. (2014). End Ordovician extinctions: a coincidence of causes. *Gondwana Research* 25: 1294–1307.

Knoll, A.H., Bambach, R.K., Payne, J.L. et al. (2007). Paleophysiology and end-Permian mass extinction. *Earth and Planetary Science Letters* 256: 295–313.

McGhee, G.R., Sheehan, P.M., Bottjer, D.J. et al. (2004). Ecological ranking of Phanerozoic biodiversity crises: ecological and taxonomic severities are decoupled. *Palaeogeography, Palaeoclimatology, Palaeoecology* 211: 289–297.

Sallan, L.C. and Coates, M.I. (2010). End-Devonian extinction and a bottleneck in the early evolution of modern jawed vertebrates. *Proceedings of the National Academy of Sciences of the United States of America* 107: 10131–10135. [An excellent, statistical study of fish evolution through one of the "big five" mass extinctions].

Saunders, A. and Reichow, M. (2009). The Siberian Traps and the end-Permian extinction: a critical review. *Chinese Science Bulletin* 54: 20–37.

Wignall, P.B. (2017). *The Worst of Times: How Life Survived Eighty Million Years of Extinctions*. Princeton: Princeton University Press. [The hyperthermal killing model applied to a series of five extinction events, from Permian to Jurassic].

## Question for Thought, Study, and Discussion

In the last paragraph of the chapter, we mention new biological inventions. All of them are fascinating but here's the question. Lots of people worry about the possibility of intelligent life somewhere in the universe but intelligent life has to evolve, as it did here. Starting from a world inhabited only by the first cells, list (in order) some of the new biological inventions that were crucial along the way to intelligent life on Earth. For example, was vision one of them? Be prepared to defend your list.

# 7

## The Early Vertebrates

| In This Chapter |
| --- |

We now turn to following the history of vertebrates, partly because we are familiar with them and because they include us. They evolved from soft-bodied invertebrates, which leads to some uncertainty about their earliest members. Even so, we can identify the earliest known vertebrate in Cambrian rocks from China. The vertebrate fossil record improves with the evolution of a mineralized skeleton, first manifest as external dermal plates. A variety of fish groups shows increasing swimming ability. All early fishes lacked jaws but some time in the Ordovician, fishes evolved a jaw from parts of the gill system. The jawed fishes now dominate living fish faunas. Jaws allowed some fishes to become large predators, and the placoderms of the Devonian included some of the largest and most effective predators the seas have ever seen. Sharks are ancient fishes too, and their survival in today's seas reflects their array of adaptations to marine carnivory. Some early fishes evolved to breathe air, paving the way for fishes to leave the water.

## Vertebrates

Vertebrates dominate land, water, and air today in ways of life that combine mobility and large size (more than a few grams). Only arthropods (insects on land and crustaceans in the sea) come close to competing for these ecological niches. As vertebrates ourselves, we have a particular interest in the evolutionary history of our own species and our remote ancestors. It's hardly surprising that vertebrates should receive special treatment in this and almost every other book on the history of life.

It is easier for us to identify with vertebrates than with invertebrates. We can feel how ligaments, muscles, and bones work. We feed by using our jaws and teeth. We have sensory skin and good vision, and we sense vibrations in our ears. We walk, run, and swim. We have bodily sensations as we thermoregulate, and we understand by experience the bizarre system we have for getting oxygen and circulating it around the body. All vertebrates share some of these systems, and many vertebrates have them all. In contrast, most invertebrates have quite different body systems that are more difficult for us to identify with and to understand.

Our familiarity with vertebrate biology helps to make up for the rarity of vertebrate fossils. Vertebrates are rare even today in comparison with arthropods or mollusks, and vertebrate hard parts are held together only by skin, muscles, cartilage, and ligaments that rot easily after death. Mollusks and arthropods, on the other hand, have hard external skeletons (shells) that readily form fossils. Even bones crumble and dissolve rather easily once they lose the organic matter that permeates them in life. Land vertebrates in particular live in habitats that offer few chances for preservation. However, paleontologists have always paid special attention to fossil mammals, dinosaurs, and fishes, and modern methods allow us to extract a great deal of information about how they lived.

## Vertebrate Origin

Vertebrates are members of the chordate phylum, the other members of which are invertebrates. Most vertebrates have a spine, a bony (or cartilaginous) column that contains a nerve canal and a notochord. The **notochord** is a specialized structure that looks like a stiff rod of

*Cowen's History of Life*, Sixth Edition. Edited by Michael J. Benton.
© 2020 John Wiley & Sons Ltd. Published 2020 by John Wiley & Sons Ltd.

dense tissue. It is a more fundamental character than the spine that surrounds it. It is a shared derived character that places vertebrates in the phylum Chordata, together with some soft-bodied creatures that have a notochord but do not have a distinct head or a skeleton.

By using the stiffness of the notochord, a chordate without a spine can give its muscles a firm base to pull against, while retaining enough flexibility to allow a push against the water for efficient swimming. The notochord can store elastic energy that is released at the right moment to help swimming. The evolution of the notochord, with this mechanism for energy storage and release, preceded by a long time the evolution of the skeleton of a typical vertebrate.

Urochordates and cephalochordates are two living groups of soft-bodied chordates that help to show us what a vertebrate ancestor might have looked like. Urochordates include **tunicates** (sea squirts), small bag-like creatures that live as adults in colonies fixed to the sea floor and filter food from the water (Figure 7.1). But tunicate larvae swim actively, using the notochord and muscle fibers in a tail-like structure that is lost soon after they settle as adults. The tunicate *Ciona* has had its genome completely sequenced, and that genome looks much like that of vertebrates but simpler because it has lost many genes seen in other chordates.

**Cephalochordates** (Figure 7.2) are marine creatures that filter small particles from seawater, which is also used for respiration. The notochord runs along the dorsal axis and is surrounded by segmental body muscles arranged in hollow-cone blocks that appear V-shaped in lateral aspect. Alternate contractions of the muscle packs flex the body from side to side in a wave-like pattern that allows the creature to swim. Nerve tissue at the anterior end of the notochord marks the position of a rudimentary brain. In most of these characters, cephalochordates are much like fishes, even to the pattern of V-shaped muscles that is so obvious when one dissects a fish carefully in a laboratory or a restaurant.

*Branchiostoma*, the amphioxus (Figure 7.2), is a typical cephalochordate. It lives and moves between sand grains and in open water, squirming and swimming with its muscle blocks and notochord acting against one another. Its genome has been sequenced and it is more complex than that of *Ciona*, sharing key genes with vertebrates, although often in a single copy where vertebrates have multiple duplicates. These shared genes show shared functions in directing early development and adaptive immunity.

Molecular phylogenetic analysis of these genes shows that urochordates are closer relatives of vertebrates than cephalochordates, which is perhaps a little counterintuitive since amphioxus looks so much like a little fish. It is then possible to consider where some fascinating fossil early chordates fit in the phylogeny.

Possible cephalochordates and tunicates have been described from the Cambrian Chengjiang Lagerstätte of China (see Chapter 4), but more convincing still are fossils of *Haikouichthys* (Figure 7.3), a creature with a

Figure 7.1 A colonial tunicate. This is *Rhopalaea*, from East Timor. Water is drawn into the buccal siphon at the top, oxygen and food particles are extracted, and the water is then expelled through the smaller atrial siphon on the side. *Source:* Image by Nick Hobgood (Wikimedia).

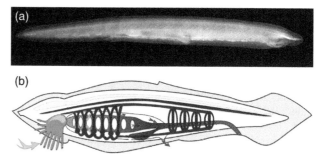

Figure 7.2 Photograph and diagram of a cephalochordate. (a) *Branchiostoma* from the North Sea, about 2 cm long. Note the V-shaped muscle pattern. *Source:* © Hans Hillewaart, used by permission. (b) Diagram of some body parts. Food and water are taken in (*big blue arrow*) through a mouth that bears tentacles (*dark green*). Water passes through a pharynx for respiration and exits at gill slits. Food is filtered from the water and passes through a gut to the anus. Blood (*red*) is circulated round vital organs. The body is stiffened by a notochord (*brown*), which lies next to a nerve chord with a bulge of nervous tissue at the front (*yellow*). *Source:* Based on a diagram by PioM (Piotr Michal Jaworski) (Wikimedia).

Figure 7.3 Reconstruction of the earliest known fish, *Haikouichthys* from the Lower Cambrian Chengjiang Fauna of China. *Source:* Art by Nobu Tamaru (Wikimedia).

distinct head. The soft parts are difficult to interpret. Paired dark spots at the front of the organism have been interpreted as eyes and nasal sacs. A gill skeleton is quite clear, lying immediately under this "head" region, but the position and nature of the mouth are unclear. Some specimens preserve serially repeated collars wrapped around the notochord, perhaps representing cartilaginous precursors of vertebrae. Regardless, the presence of paired sense organs demonstrates that *Haikouichthys* is a vertebrate since these characters are not present in invertebrate chordates.

The notochord probably evolved as a structure that aided swimming. But the physics of hydrodynamics dictates that swimming efficiency increases with body length. As early chordates explored various ways of life, the more actively swimming species probably increased in body size. But there must come a body size for which efficient swimming requires more stiffness than a notochord can give, and a cartilaginous or mineralized skeleton then becomes a cheap way of increasing efficiency. At the moment *Haikouichthys* is the first sign of this breakthrough in mechanical efficiency.

## The Cyclostomes – Vertebrates Without Vertebrae

It might seem clear that the defining character of vertebrates is the presence of vertebrae – but your intuition would be wrong since the most primitive living vertebrates, the eel-like jawless hagfishes and lampreys, lack true vertebrae. They have arcualia – cartilaginous "nubbins" – associated with the notochord, that may (or may not) be evolutionary rudiments of vertebrae, but they have most of the other features shared by other vertebrates, like paired sense organs, a differentiated brain, fin rays, etc.

The hagfishes and lampreys are each other's closest relatives, united in a group called "cyclostomes" named after their round mouths. They mostly live parasitic adult lives, lampreys feeding on the blood of their prey and hagfish eating dead or dying fish. They are perhaps

the best guide we have to the nature of the ancestral vertebrate, but they have surely evolved specialist adaptations not seen in the vertebrate ancestor, and they have certainly lost many features that were once ancestrally present. Only a handful of cyclostome fossils are known, dating from the Devonian and younger. If nothing else, the cyclostomes are an object lesson that the names of taxonomic groupings are a poor guide to the features that unite their membership.

## Ostracoderms

The earliest fishes with hard parts, from the Ordovician, did not have a bony internal skeleton. Instead, they evolved mineralized bony plates that covered some or all of their bodies, adding stiffness and giving rise to the informal group name **ostracoderms** ("plated skin") for them. The plates of ostracoderms would have provided protection too, from predators and from abrasion by sand and rock surfaces. Solving the same problem in a different way, sharks today have an internal bony skeleton made of cartilage rather than bone, and they have a tough skin with strong fibers that stiffen the body considerably. Most other living fishes have a light scaly skin and an internal bony skeleton but they are all descended from ancestral ostracoderms.

There are many distinct groups of ostracoderms which are related by degree to the living gnathostomes. These include the heterostracans, anaspids, thelodonts, galeaspids, osteostracans, and placoderms. *Astraspis* from the Ordovician of Colorado is one of the best-preserved early fishes (Figure 7.4). A headshield protected

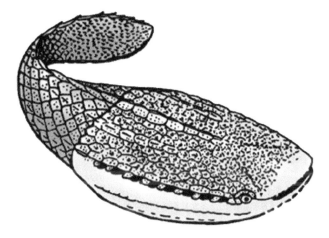

Figure 7.4 Reconstruction of one of the earliest ostracoderms. *Astraspis,* from the Ordovician of Colorado, was about 13 cm (5 in.) long. *Source:* Courtesy of Philippe Janvier and David K. Elliott.

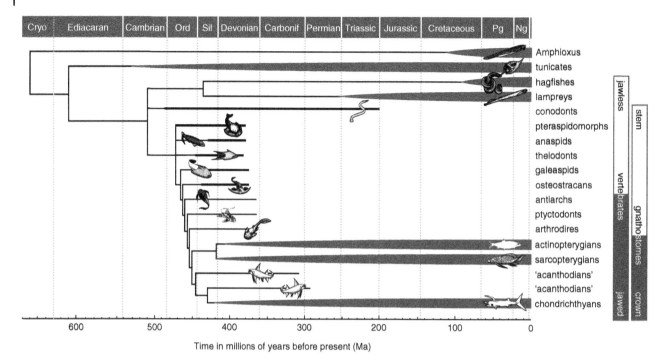

**Figure 7.5** Cladogram of the basal chordates and vertebrates, showing modern forms and the fossil armored fishes. *Source:* from Donoghue and Keating (2014), with permission.

the brain and visceral organs. Behind the eyes were plates with multiple openings to allow water to flow out past the gills. The tail was short, stubby, symmetrical, and small, and these fishes probably swam well but not fast.

It is possible to understand the evolution of the early Paleozoic fishes by first constructing a cladogram of the living forms, and then inserting the fossils according to anatomical features of their brains, sensory systems, and other organs (Figure 7.5). The living forms are the cephalochordates and tunicates, as well as the cyclostomes and gnathostomes. Living gnathostomes all possess jaws (which we will look at shortly) but they are descended from jawless ostracoderms. Also in the cladogram are the conodonts, a group of vertebrates that have long been known from remains of their tiny teeth which formed complex food-grabbing and -processing structures in the mouth or pharynx of a long, cyclostome-like fish.

## Heterostracans

Heterostracans are the diverse grouping to which the astraspids belong, and they are among the earliest abundant fishes (in the Silurian and Devonian). They had flattened headshields with eyes at the side, and they look well adapted for scooping food off the sea floor (Figure 7.6). Some had plates around the mouth

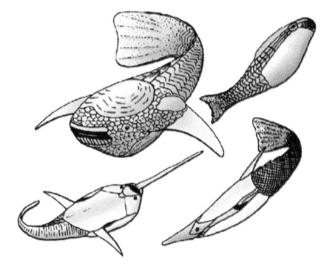

**Figure 7.6** A selection of heterostracans, including a primitive tolypelepid (*top right*). Most heterostracans are pteraspidiforms, such as the pteraspidid (*bottom right*), doryaspidid (*bottom left*) and the huge psammosteid (*top left*), which are the youngest known members of the group. *Source:* Image by Philippe Janvier (Wikimedia).

that could have been extended out into a shoveling scoop. The rigid head and the stiff, heavy-plated body imply that propulsion came mainly from the tail in a simple swimming style, with none of the control surfaces provided by the paired fins of gnathostomes. Even so, the heterostracan way of life was successful, and their fossils are found all across the current northern hemisphere.

Early heterostracans were similar to astraspids. The tolypelepids had broad armor plates above and below, stiff fin-like bony projections at the side. The other heterostracans in Figure 7.6 are pteraspids, including *Pteraspis* itself (bottom right), a doryaspidid (bottom left), and a huge psammosteid (top left). Many pteraspidids such as *Pteraspis* had beautifully streamlined armored headshields, with a sharp nose cone and a smooth curved shape that gave an upward motion to counteract the density of the armor. In some species, a spine projected backward over the lightly plated trunk, partly for protection and partly for hydrodynamic stability. Pteraspids had symmetrical paddle-like forelimbs, aiding propulsion. The mouth lay under the head and a ventral plate covered the gills. Water was taken in through the mouth, and the exit passages were neatly tucked toward the back of the headshield. In some forms, the headshield was very flattened, for gliding through water as a delta-wing aircraft glides through air.

But in all this successful evolution, heterostracans never evolved paired fins. Their swimming power came entirely from the trunk and tail.

## Anaspids

Anaspids are a small clade of heterostracan-like jawless fishes that can be distinguished by the fact that they lacked large cranial plates. Instead, they were covered in hundreds of small lath-like scales. Never more than a few tens of centimeters in length, anaspids were the first vertebrates with a stomach, facilitating the digestion of nutrients. They were among the first vertebrates to possess paired fins although it is not clear that these are homologous to the paired fins of gnathostomes; they occur as a single pair of long fins which some have interpreted as a single evolutionary rudiment of two pairs of fins (pectoral and pelvic) seen in gnathostomes. Anaspids ranged in age from the late Ordovician to the Devonian. They are rare as articulated body fossils but analysis of isolated scales suggests that they were common elements of early Paleozoic vertebrate communities in modern-day North America and Europe.

## Thelodonts

The jawless thelodonts are superficially shark-like, in that their external skeleton comprised thousands of individual scales. Like anaspids, thelodonts possessed a distinct stomach and one or more pairs of fin-like lateral extensions to the body. They exhibit a diversity of body shapes, including both deep-bodied and flattened forms. Thelodonts ranged in size from small *Phlebolepis*, which was about 10 cm in length, to the adult halibut-sized *Turinia*. Thelodonts have a global distribution, indicating that they were less facies restricted than any other ostracoderm group, perhaps because their minute scales allowed them to be better swimmers than their more inflexible armor-encased relatives. Thelodonts were also among the longest-ranging ostracoderm groups, originating in the late Ordovician and extending to the end of the Devonian. Some thelodonts had internal oral and pharyngeal scales arranged like the whorls of replacement teeth in sharks, but rather than reflecting an origin of teeth before jaws, this similarity appears to be a consequence of evolutionary convergence.

## Galeaspids

Galeaspids are a clade of jawless fishes that ranged in size from about 10 cm to many tens of centimeters. They were geographically confined to eastern Asia, especially China and Vietnam. Galeaspids were successful, however, and range from Silurian to Devonian. They are unusual in having a central nostril on the top side of the headshield (Figure 7.7). The mouth and gill openings are on the underside, as usual, and the eyes are small and set wide on the headshield.

A new galeaspid from the Silurian of China called *Shuyu* is so beautifully preserved that the internal structures of the head could be reconstructed from X-ray data (Gai et al. 2011). This study uncovered the gross anatomy of the brain and sense organs of galeaspids, including the bilaterally arranged nasal sacs (Figure 7.7). Galeaspids are the earliest vertebrates in which this gnathostome-like condition is seen, revealing the reorganization of the cranial sense organs in the prelude to the origin of jaws.

Galeaspids possess many features that are considered specific to gnathostomes, including a mineralized braincase and perichondral bone. However, another group, the osteostracans, shares even more gnathostome features, including paired pectoral fins, demonstrating that they are the closest relatives of the jawed vertebrates.

## Osteostracans

Other things being equal, any swimming creature would benefit by evolving powerful swimming and better maneuverability. Heterostracans achieved this with their symmetrical tails and rigid paired spines which may have provided dynamic lift. However, osteostracans are the

(a)

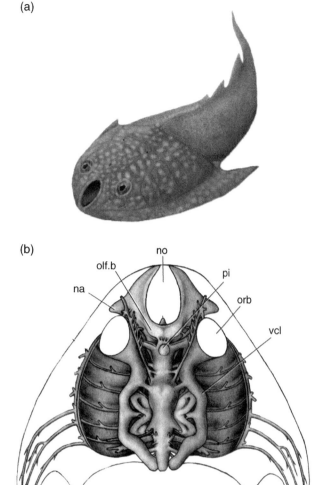

(b)

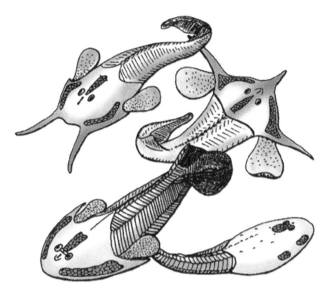

Figure 7.8 A selection of osteostracans, including the primitive zenaspidid *Zenaspis* (*bottom left*), and derived forms with elaborate headshields, the benneviaspidids *Hoelaspis* (*top right*) and *Tauraspis* (*top left*), and the thyestiid *Tremataspis* (*bottom right*). *Source:* Image by Philippe Janvier (Wikimedia).

Figure 7.7 (a) *Shuyu*, a galeaspid from the Silurian of China, about 3 cm (1 in.) long. It has a prominent opening in the center of the headshield. Artwork by Brian Choo, used by permission. (b) Computed tomography scans of the interior (Gai et al. 2011) showed a complex nasal structure inside the headshield. Courtesy of Dr. Zhikun Gai. The important labels are: no, nasal opening; ol.b, olfactory bulb, na, nasal sac, orb, eye. Scale bar, 2 mm.

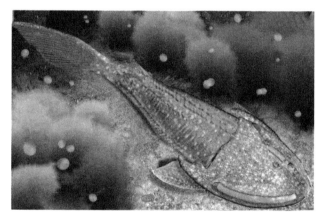

Figure 7.9 Reconstruction of the osteostracan *Cephalaspis* in life, grubbing for food on the bottom of a pond. *Source:* Image by Rod6807 (Wikimedia).

first with a crucial innovation, paired fins, although only in the pectoral position – evolutionary equivalents of our arms. This evolutionary advance was inherited by the jawless vertebrates which also possess pelvic appendages – equivalents of our legs.

Osteostracans were like heterostracans and galeaspids in that they had a more or less fused bony headshield and a comparatively flexible body and tail that provided most of the propulsion. The best known osteostracans resemble *Zenaspis* (Figure 7.8, bottom left), with a rounded front to the headshield. But others, such as *Hoelaspis* (top right) and *Tauraspis* (top left), showed lateral and

anterior horns whose function is a mystery. Unusually, the thyestiid *Tremataspis* (bottom right) has lost the paired fins.

Cephalaspids are the most general type of osteostracans, representing the rootstock from which other osteostracan clades emerged; they thrived from the Late Silurian to the Late Devonian. Their large horseshoe-shaped headshields often had two spines extending backward at each corner (Figure 7.9), surrounding and protecting the armored pectoral fins. There was no distinct shoulder girdle supporting the fins; rather, the fins attached to the back of the braincase which was composed of a large mass of cartilage and/or perichondral bone that lined the inner surface of the dorsal headshield. The body behind the headshield was rounded or laterally compressed, as in most living fishes, and one or more dorsal fins added stability. The cephalaspid tail was more versatile than the heterostracan tail, with an extended upper lobe resembling the heterocercal tail of sharks.

Cephalaspids are conventionally interpreted as bottom swimmers. The mouth was on the flat underside of the headshield. The eyes were small and close together on the top of the headshield (Figure 7.9). In addition, cephalaspids had large sensory areas on each side of the headshield, covered with very small plates. These organs may have served as pressure sensors in murky water, although they may also have sensed electrical fields, as in living sharks.

## The Evolution of Jaws

By the Late Ordovician-Early Silurian, the jawless fishes were quite varied. Without jaws, they were confined to eating small particles, such as plankton from the surface, deposit feeding in sediment on the sea floor or soft, easily swallowed food. In fact, so far as we know, none of these Paleozoic jawless fishes had evolved the flesh ripping or blood sucking seen in some modern hagfishes and lampreys. But somewhere among them were fishes

in the process of a major breakthrough for the feeding ecology of fishes: **evolving jaws**.

Studies of anatomy and embryology suggest that the bones that form the vertebrate jaw evolved originally from the same population of cells that the gill arches of jawless fishes develop from. In living fishes, water is taken in at the mouth and passes backward past the gills, where oxygen and $CO_2$ are exchanged with the blood system (Figure 7.10). Gills are soft, so they must be supported in the water current by a thin skeleton of bone or cartilage called **gill arches** (Figure 7.11). The more water passing the gills, the more oxygen can be absorbed and the higher the energy the fish can generate. Living fishes usually have pumps of some kind to increase and regulate the flow of water passing the gills. Most fishes use a pumping action in which they increase and decrease the volume of the mouth cavity by flexing the jaws. The cyclostomes have a modified gill arch that supports a pumping scroll-like velum, and ostracoderm groups, including the osteostracans and galeaspids, may have had the same.

If jaws evolved through reprogramming the development of a gill arch, the evolution of the jaw might have been associated with respiration rather than feeding. Water flow over the gills of jawless fishes may have been impeded by their small mouths and by a slow flow of water past the gills, so their swimming performance may have been limited by oxygen shortage. Perhaps a joint evolved in the forward gill arch so that it flexed to open the mouth wider, pumping more water backward over the gills (Figure 7.11) and transforming the gill arch into a true jaw.

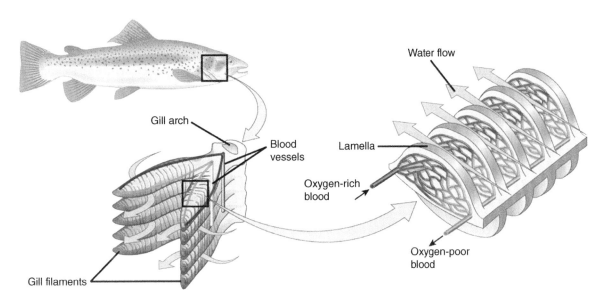

Figure 7.10 How gills work in most living fishes. Gills (*left*) are arrays of thin plate-like structures set in rows supported on a strong axis. Oxygen-poor blood is pumped along a one-way system through each plate-like structure (*right*). Water is pumped the other way, exchanging gasses with the blood by shedding oxygen and taking up carbon dioxide. *Source:* Image by Charles Molnar and Jane Gair, Creative Commons 4.0.

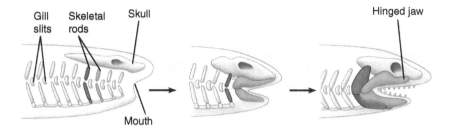

Figure 7.11 Evolving the fish jaw, from an agnathan or jawless fish (*left*), through an early jawed fish (*middle*) to an advanced jawed fish (*right*). At the first step, the first gill arch, the mandibular arch, flexes a little to help water flow into the mouth and out through the first gill opening. Then, the mandibular arch becomes larger and stronger as muscles flex it more strongly to improve the respiration of the fish. In the gnathostome, the mandibular arch has evolved into a jaw, capable of biting down on a food item. At the right, the fully evolved jawed fish, with teeth forming along the structure of the jaws. *Source:* © 2009 Pearson Publishing.

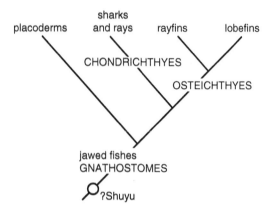

Figure 7.12 A simplified cladogram of early jawed fishes (gnathostomes). Evolving from something like the galeaspid *Shuyu*, the extinct placoderms are the basal group in this scheme. The extinct acanthodians are a "stem group" of various lineages of chondrichthyans. The cartilaginous fishes Chondrichthyes are a clade and so are the bony fishes, the Osteichthyes, which quickly diverged into rayfin fishes and lobefin fishes.

Living jawed fishes comprise the **gnathostomes**, divided into two groups: the chondrichthyans (sharks, rays, chimaeroids) and osteichthyans (bony fishes, and their descendants, the tetrapods – which include us). The extinct jawed placoderms are the nearest extinct relatives of gnathostomes, and the acanthodians are early members of the chondrichthyan lineage (Figure 7.12).

Chondrichthyans are known from the late Ordovician of North America where they occur in the same assemblages as the jawless *Astraspis*. This indicates that early jawed vertebrates lived alongside their jawless relatives for tens of millions of years – there is no evidence that jawed vertebrates drove their jawless relatives to extinction, at least not initially. Instead, during the Late Ordovician-Devonian interval, jawed vertebrates simply added to vertebrate diversity.

The evolution of jaws and a resulting extension of the potential food supply were keys to the tremendous evolutionary success of jawed fishes. But teeth and jaws are only

weapons; they must be applied to targets by a delivery system. The history of fishes since the Devonian has been largely one of increasing effectiveness in mounting and hinging the jaws, in the speed of strike, and in the hydrodynamics of propulsion and maneuverability. All these factors meant that the jawed fishes were able to extend their ecological range into the three-dimensional world of open water, as opposed to the largely bottom-feeding agnathans.

## Placoderms

Placoderms were abundant and globally distributed by the Devonian. Most placoderms had a well-developed headshield made of several plates, jointed to an armored girdle surrounding the front part of the trunk. The rest of the trunk was lightly scaled, and the trunk and the long tail were flexible, for powerful swimming. There were several pairs of fins, providing good control over movement. Placoderms include a range of groups, the best known being the large predatory **arthrodires** and the small box-like **antiarchs**.

Arthrodires were powerful, streamlined fishes but their great weight of armor and their flattened body shapes may have limited their swimming performance. Large pectoral fins aided stability and provided lift for the heavy armored head. The small eyes imply that they probably used other senses to a large extent, just as living sharks do. The jaws vary quite a lot but some advanced placoderms had vicious, sharp-edged, tooth-bearing plates set into the jaw (Rücklin et al. 2012) (Figure 7.13). Others had large crushing tooth plates, perhaps for eating mollusks or arthropods. Ecologically, the rich Late Devonian placoderm fauna included generalists and cutters and crushers. In fact, it is clear that the range of ecology in placoderms is similar in broad outline to that of today's bony fishes (Anderson et al. 2011). Giant arthrodires include *Tityosteus*, the largest known Early Devonian fish, with a length of about 2.5 m (8 ft). But

*Tityosteus* is dwarfed by a Middle Devonian freshwater arthrodire, *Heterosteus*, and by the Late Devonian *Dunkleosteus*, both of which grew to 6 m (20 ft) long (Figure 7.13).

The arthrodires evolved a unique set of joints that operated the jaws, headshield, and trunk armor in a spectacular way. The head could be levered upward while the lower jaw dropped at the same time, quickly opening a wide gape that would have sucked prey toward the jaws just as they closed (Figure 7.14). Models show that the bite of a huge *Dunkleosteus* is one of the most powerful ever evolved (Anderson and Westneat 2009). These were large carnivores up to 6 m (20 ft) long, in their day the largest animals on Earth, and indeed the largest animals ever to have lived on Earth up to that point.

Antiarchs are much more difficult to understand but they were successful worldwide, mostly in freshwater environments. They were small, with headshields up to 50 cm (20 in.) long and a maximum known length of just over a meter (3 ft). Their headshields were flattened against the bottom, with the eyes set close together high on the headshield. The mouth lay just under the snout. The body armor was long. Instead of pectoral fins, antiarchs had long, jointed appendages that were encased in thick armor (Figure 7.15). Antiarchs had small mouths and their feeding is poorly understood. It's clear that they were slow, rather clumsy swimmers.

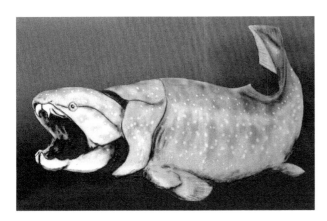

Figure 7.13 The giant arthrodire *Dunkleosteus*, from the Late Devonian of North America, which was up to 6 m (20 ft) long. *Source:* Artwork by Nobu Tamura (Wikimedia).

Figure 7.15 Reconstruction of the little antiarch *Bothriolepis*, from the Devonian of Canada. It is about 30 cm in length (about 1 ft). *Source:* Image by Citron (Wikimedia).

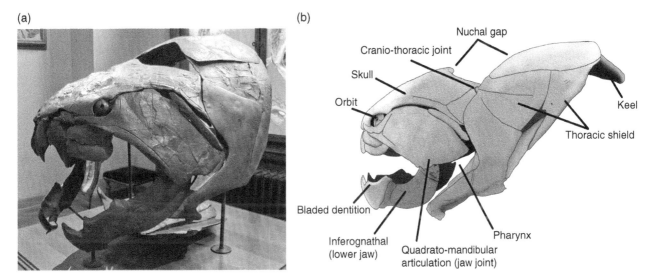

Figure 7.14 The skull of the mighty *Dunkleosteus*, showing the sharpened, bone-lined jaws (a), and detail of the anatomy (b), showing the quadratomandibular (jaw) joint and the additional craniothoracic joint between the skull and shoulder bones. *Source:* Images by Zachi Evenor (a) and Phil Anderson (b) (Wikimedia).

Placoderms had another unusual biological attribute: they gave birth to live young, which also implies internal fertilization. Today's bony fish almost all spawn eggs and sperm into the water, but living sharks and rays have internal fertilization and live birth, and so does the coelacanth (see later in this chapter). The surprise discovery of embryos preserved inside Devonian placoderms was followed quickly by the realization that the sexes were dimorphic, with the males having pelvic fins modified into "claspers" that are important for mating. These features have been reported now from both arthrodires and antiarchs, and so were present in most placoderms (Long et al. 2015).

Placoderms were long thought to be a natural group but this view is disputed, with most researchers now arguing that the different placoderm families are related by degree to the overall gnathostome clade and, perhaps, some were more closely related to the modern bony fishes. Whether they comprise a clade or an evolutionary grade, placoderms became extinct in the Late Devonian, one of the great mass extinctions (see Chapter 6), a time of major turnover in fishy faunas that involved also the loss of osteostracans and other armored ostracoderms. The cartilaginous fishes and the bony fishes, each with their own adaptations, now dominate living fish faunas.

**Figure 7.16** Three acanthodians, *Mesacanthus, Parexus,* and *Ischnacanthus,* from the Early Devonian of Great Britain. *Source:* Images by Jobanbo (Wikimedia).

## Acanthodians

Another extinct group of gnathostomes are the **acanthodians**, known from the Ordovician to the Permian. These generally small fishes were immensely common in some lakes and seas, and they are sometimes represented by hundreds of specimens on a single slab. Their name refers to the bony spines that stood in front of each of the fins, and even lined the underside (Figure 7.16). Acanthodians do not in fact form a natural biological group but are instead related by degree to the living chondrichthyans.

## Cartilaginous Fishes (Sharks and Rays)

Sharks and rays, and all their ancestors, have cartilaginous skeletons rather than bone. This distinction dates back to the Early Devonian, when this group of fishes was just one of the many early successful lines that had recently evolved jaws. However, it is clear that they evolved from ostracoderm ancestors with a more fully developed bony dermal skeleton and endoskeleton.

The fossil record of sharks and rays is poor, because they rarely preserve well as fossils. They have cartilage rather than bone, and minute tooth-like skin denticles rather than heavy scales. They do have formidable teeth, which are often well preserved as fossils, but teeth alone give only a vague insight into the entire fish. Following the Late Devonian mass extinctions of osteostracans and placoderms, sharks proliferated in the Early Carboniferous, and some weird and wonderful forms emerged. For example, the rare find of a complete specimen of *Akmonistion,* preserving its skeleton, teeth, and body outline, shows the hydrodynamic shark-shaped body but also the elaborate, ironing board-shaped bony headcrest, an elaboration of the shoulder girdle, complete with teeth (Figure 7.17). New groups, especially the modern-style sharks, the neoselachians, appeared in the Mesozoic (see Chapter 11).

Sharks have excellent vision and smell and an electrical sense, all of which combine to equip them well for hunting in all kinds of environments. They all have internal fertilization, and some have live birth. Most sharks are top predators in their marine ecosystems but in the past, there were many successful smaller sharks even in fresh waters. We are all familiar with the image of the giant, terrifying man-eater but the largest shark today, the whale shark, filter feeds on plankton.

## Bony Fishes

The bony fishes, or Osteichthyes, are known from fossil remains that date back to the Late Silurian and Early Devonian, but there are few fossils from that time.

Figure 7.17 The fossil shark *Akmonistion*, from the Carboniferous. Its shape is uncannily like that of many living sharks, except for the prominent dorsal structures. *Source:* Artwork by Nobu Tamura (Wikimedia).

Almost as soon as they appeared, they were already diversified into two major groups: the rayfins (Actinopterygii) and the lobefins (Sarcopterygii) (Figure 7.13). The critical fish faunas involved in the very earliest radiation were discovered in South China, which was an isolated minicontinent at the time. Fossils from this region include *Guiyu*, the oldest reasonably complete bony fish (see later in this chapter).

## Air Breathing

We know already that air breathing had evolved in an early lineage of bony fishes. Some primitive rayfin fishes that survive today breathe air with lungs. Living lobefins, lungfishes, do the same. But sharks, rays, and cyclostomes today all breathe through their gills. Air breathing evolved in early bony fishes but has been lost in most rayfins, and in coelacanths.

### The Problem

Why did early bony fishes evolve the ability to breathe air? The answer may be that this was the most efficient way for them to obtain oxygen and that gills required complex adaptation in order to make them adequate for the task. The model here comes from the work of Colleen Farmer, a physiologist at the University of Utah (Farmer 1999).

Animal respiration is fundamental to survival. Animals take in oxygen to burn their food in respiration, and they produce $CO_2$ as a waste product. Carbon dioxide is toxic because it dissolves easily in water to form carbonic acid. Animals can tolerate only a small build-up of $CO_2$ before having to pass it out of the system. (For example, it is high $CO_2$ in our lungs that makes us want to breathe out, not shortage of oxygen.)

Gases are exchanged with the environment, whether it is water or air, as body fluids are passed very close to the body surface. For example, blood flows close under the lung surface in our own breathing. As long as the environment has higher $O_2$ and lower $CO_2$ than the body, diffusion acts to pass $O_2$ in and $CO_2$ out. The rate depends on several factors: the surface area and the thickness of tissue through which the gases must diffuse; the rate at which the external and internal fluids pass across the surface; and the concentrations of gases in the internal fluid and in the external medium. In normal fishes, $CO_2$ and $O_2$ diffuse in opposite directions across the gill surface (Figure 7.10).

### Respiration in Early Fishes

The earliest fishes probably had a respiration system like that of living cephalochordates (Figure 7.2). Water is pumped into the basket-like pharynx, located behind the mouth, and food particles and oxygen are taken out of it. Oxygen is carried in a blood system pumped by a heart, and travels through the body tissues, delivering oxygen, until it reaches the heart again.

This system was inherited by later jawless fishes. They may have evolved sophisticated gills (Figure 7.7) but their system had a basic flaw: blood arrived at the heart depleted in oxygen, because it had flowed all around the body first (Figure 7.18). Unlike in birds and mammals,

(a)

(b)

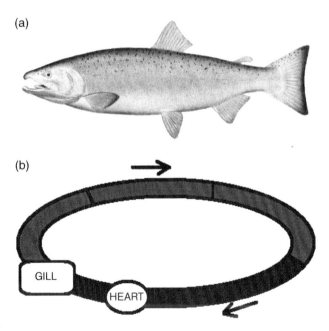

Figure 7.18 Respiration in a fish that uses gill breathing, such as an early fish, or the coho salmon shown here. The heart pumps blood to the gills, then oxygenated blood circulates round the body before it reaches the heart again. There is a considerable danger that the heart will become oxygen deprived. *Source:* Based on Farmer (1999).

fishes have a two-chambered heart, so blood flowing in and out mixes, and they do not separate one blood circuit to the lungs and one to the body, as we do using our four-chambered heart. The more active the fish, the more likely it was to suffer heart failure! This is not an ideal piece of engineering, but you inherit what your ancestors give you.

Relatively slow-moving and rather small, jawless fishes flourished for millions of years with this system. (When you think about it, their success on Cambrian sea floors confirms that global oxygen levels were reasonably high, at least in shallow water.)

With the evolution of jaws and the radiation of jawed fishes, more lineages must have become more active foragers and predators, at larger body sizes. There must have been strong selective pressure to modify the ancient system to cope with the extra oxygen demands of a more active life.

### Oxygen Intake

It is easier and cheaper to get oxygen from air rather than water. Water is hundreds of times denser and more viscous than air, and even at best it contains less oxygen. Many gill-breathing animals have to pump external water across their gill surfaces at 10 times the rate they pump their internal blood. Gills have to be designed to resist the leakage of dissolved body salts, and the tissues across which oxygen is exchanged cannot be as thin as they can in air, so gas exchange is rarely anywhere near 100% efficient.

Because oxygen diffuses 100 000 times more quickly in air than in water, oxygen-poor air is rare. But oxygen-poor *water* does occur quite often, especially in tropical regions, wherever warm freshwater or saltwater lakes, ponds, or lagoons are partly or completely isolated, especially in a hot season. Warm, rotting debris can quickly use up oxygen, especially if there is little or no natural water flow. Even if the effect is only seasonal, it may still be critical for fishes and other organisms living in the water. The water is stagnant, hot, and full of rotting debris, often teeming with bacteria that may also release toxic substances.

Even today, there are often natural **fish kills** in which massive mortality occurs among fishes (Figure 7.19). Many fish kills are related to a lowering of oxygen in the environment, for example in shallow pools, rivers, or lagoons that heat up too much. The immediate culprit is the environmental crisis, but it is made worse because the fishes were using a gill system that could not handle the oxygen shortage.

Why would fishes swim into oxygen-poor water, where gill breathing is difficult? The food supply may be rich for fishes that can tolerate it, and there are situations in which fishes might benefit from swimming into areas of warm, often oxygen-poor water near the surface.

Figure 7.19 Lack of oxygen in the waters of Greenwich Bay, Rhode Island, killed about a million fish in August 2003. *Source:* Photograph by Tom Ardito (Wikimedia).

Many carnivorous fishes today are bottom feeders, hunting for small prey that live on or in the surface of the sediment. In warm latitudes, the bottom waters are often much cooler than the surface waters, which are heated by the sun. Digestion can be very slow in cold-blooded animals, especially if they live in cold environments. That may be a critical factor holding back growth and development. In such cases, increasing the digestive rate by swimming into the warm surface water can produce faster growth, earlier maturity, and more successful reproduction.

### Air Gulping and Fishy Lungs

But what happens to a fish that swims into surface water because it is warm, only to find that it is also oxygen poor? Even if surface waters are generally low in oxygen, there is always a thin surface layer of water, about a millimeter thick, that gains oxygen from the air by diffusion. Many living fishes in tropical environments come to this surface layer to bathe their gills in the surface oxygen layer. They can breathe but they have to solve other problems too. If they break the surface, their bodies extend out into the air, losing some buoyancy.

Some living fishes in this situation bite off bubbles of air and hold them in their mouths for positive buoyancy, to remain at the surface without active swimming. Some living species of gobies use this action to breathe. Once they have an air bubble, they can extract oxygen from it in the back of their mouths much more efficiently than at the gills. When the oxygen level in the mouth bubble falls, reducing its size and its buoyant effect, the fish must then get rid of the bubble and bite off another. Rhythmic air breathing might have evolved this way, as fishes get rid of $CO_2$ from their mouth bubbles while they are still losing it at their gill surfaces as well.

Oxygen intake in the mouth enriches the blood supply there, and a fish can store oxygen in an air bubble. An air bubble that takes up only 5% of body volume can increase oxygen storage by 10 times compared with a fish without a bubble. Therefore, bubble breathing does not mean that a fish is completely tied to the surface; it can make extended dives to the bottom. This is true today of all air breathers with low metabolic rates, including crocodiles and turtles. Many water-dwelling insects use air bubbles too, and the wonderful diving-bell spider uses a silken dome to make a bubble-filled underwater home for itself.

In fishes that breathe air, freshly oxygenated blood flows directly to the heart (Figure 7.20). Colleen Farmer suggests that this made an immense difference to early bony fishes, enabling them to escape from the danger presented by an oxygen-starved heart (Figure 7.18). Bony fishes then radiated dramatically as they became able to live more active lives in surface waters. Thus, all rayfins (as far as we know) and all lobefins were capable of breathing air by Devonian time, because it made them more efficient in the water.

Most living rayfins have now lost the ability to breathe air, and the structure that was once their lung has evolved into an enclosed gas-filled organ called the **swim bladder**, which helps them to maintain buoyancy in the water. This means that they reverted to the older system of Figure 7.18, which seems counterproductive. Yet it can only have happened for good functional reasons. Colleen Farmer suggests that most rayfin lineages reverted to gill respiration (in Mesozoic times) after they became vulnerable to newly evolved aerial hunters at the surface, first pterosaurs and then seabirds. This is speculative, of course, but air breathers are vulnerable since they must come to the surface to breathe. Remember that nineteenth-century whalers relied on spotting whales "spouting" at the surface to locate and kill them.

## Actinopterygii (Rayfin Fishes)

Actinopterygians, or rayfins, have very thin fins that are simply webs of skin supported by numerous thin, radiating bones (called rays). Typically, rayfins are lightly built

(a)

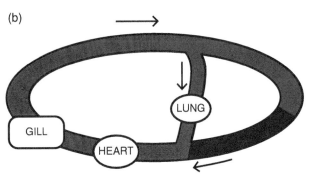

(b)

**Figure 7.20** Respiration in fishes that use air breathing as well as a gill. The living tarpon shown here has evolved to use its swim bladder as a lung. But in early bony fishes, some of the blood from the gill was circulated to the mouth, throat, or lung, where it was charged up with oxygen and delivered directly to the heart. This removed a major disadvantage of the earlier system shown in Figure 7.18. *Source:* Photograph from NOAA, diagram based on Farmer (1999).

fishes that swim fast or maneuver very well. They have dominated marine and freshwater environments of the world since the end of the Devonian, but really diversified in the Early Carboniferous after the Late Devonian mass extinctions (see Chapter 6). It is tempting to suggest that their evolutionary success reflects their mastery of swimming and feeding in open water.

In general, the evolution of the rayfin fishes resulted in a lightening of the bony skeleton and the scaly armor, both of which improved locomotion. Increasing sophistication and variation in the shape and arrangement of the paired fins led to patterns that were optimum for specialized sprinters, cruisers, or artful dodgers. In the most advanced rayfins, swimming has come to depend more and more on the tail fin rather than on body flexing, while the other fins are modified as steering devices and/or defensive spines. Even flying fishes had evolved by Triassic times.

The jaws and skull of rayfins were gradually modified for lightness and efficiency. In particular, intricate systems of levers and pulleys allow advanced fishes to strike at prey more effectively by extending the jaws forward as they close. The same system also allows more efficient ways of browsing, grazing, picking, grinding,

Figure 7.21 *Cheirolepis* from the Devonian of Scotland, a well-studied early actinopterygian. It has a bony head, a simple open-and-shut jaw, and thick bony scales. *Source:* Art by Smokeybjb (Wikimedia).

Figure 7.22 *Guiyu*, from the Silurian of southern China, is the best-known early osteichthyan (bony fish), but it is also clearly a sarcopterygian (lobefin). That means that the rayfin fishes had already diverged from the lobefins, and we can hope that good specimens of their ancestors will be found soon. *Source:* Artwork by Nobu Tamura (Wikimedia).

and nibbling, all encouraging the evolution of the tremendous variety of living fishes.

Devonian rayfins include *Cheirolepis* (Figure 7.21), which looked modern and probably swam well, but it had heavy bones over the skull and thickened, bony scales. Modern groups of actinopterygians, the neopterygians, appeared and diversified rapidly in the Triassic, especially the teleosts in the Jurassic and Cretaceous which today make up most of the 30 000 species of fishes in the oceans and fresh waters (see Chapter 11).

## Sarcopterygii (Lobefin Fishes)

Sarcopterygians, the lobefin fishes, are distinguished as a separate group because they evolved several pairs of fins that are stronger than any found in a rayfin fish. Other differences in scale and skull structure confirm the separate evolution of these groups. They separated as early as the Silurian; the recently discovered fish *Guiyu* (Figure 7.22) from the Silurian of southern China is the earliest known lobefin (Zhu et al. 2009). The major lobefin groups diverged during the Devonian in shape, structure, and ecology into three major clades: coelacanths, lungfishes, and a group named tetrapodomorphs which includes the ancestors of all tetrapods, including us (Figure 7.23).

The central part (the lobe) of a lobe fin is sturdy and contains a series of strong bones, while the edges have radiating rays as in ray fins (Figure 7.24). A lobe fin must beat more slowly than a ray fin of the same area because there is more mass to accelerate and decelerate, but the resultant stroke is more powerful. Furthermore, and fundamental to later vertebrate history, a lobe fin that

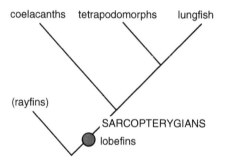

Figure 7.23 A cladogram of early lobefin fishes (sarcopterygians). They radiated in the Devonian from a Silurian ancestor like *Guiyu*. The Devonian forms were all powerful swimmers, unlike all their living descendants.

imparts a powerful stroke to the water has to have some kind of support at its base, just as an oar has to be stabilized in a rowlock. Therefore, lobefin fishes have internal systems of bones and muscles that help to tie one dorsal and two ventral pairs of lobe fins to the rest of the skeleton (Figure 7.24). These ventral linkages evolved to become the pectoral and pelvic girdles of land vertebrates but of course, that was not why they evolved: they evolved originally to allow early lobefin fishes to swim more effectively. All Devonian lobefins seem to have been effective swimmers and predators.

Lungfishes and coelacanths still survive but only as rare and unusual fishes. Two species of coelacanth survive as small populations in South African and Indonesian waters, and three species of lungfish survive, one living in each of the three southern continents: Australia, South America, and Africa. All these living lobefins have such unusual biology and ecology that they must be interpreted with caution. They are much evolved from their Devonian

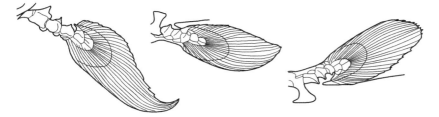

Figure 7.24 From left to right: the anterior ventral, the posterior ventral, and the posterior dorsal lobe fins of the living coelacanth *Latimeria*. The anterior (or pectoral) fin is significantly larger than the other two, but the posterior dorsal fin is just as large as its ventral counterpart and has just as strong an internal bony skeleton. *Source:* After Millot and Anthony.

Figure 7.25 *Latimeria*, the last surviving coelacanth. Coelacanths grow to about 5 ft long (close to 2 m) and may live to be 200 years old. *Source:* Artwork by Robbie Cada for Fishbase (www.fishbase.org).

Figure 7.26 *Dipterus*, a Devonian lungfish from Scotland. Even in this crushed specimen, the powerful lobe fins, like those of the living coelacanth (Figure 7.24), contrast with those of a rayfin fish. *Source:* Photograph by Haplochromis (Wikimedia).

ancestors in structure and in habits, so they may not be very good guides to the biology of those ancestors.

## Coelacanths

A living coelacanth, *Latimeria* (Figure 7.25), was unexpectedly discovered in 1938. Coelacanths had been known as fossils for decades, dating back to the Devonian, but it was thought that they had died out after the Cretaceous. We know now that they are very rare, very long-lived, and endangered. Living coelacanths are lazy swimmers, do not have lungs, and do not breathe air. The females bear live young, as many as 26 at a time, which develop internally from very yolky eggs. They can move by swimming using their fins or by stilting along the seabed, using those same powerful lobe fins like legs.

## Lungfishes

Living lungfishes are medium-sized, long-bodied fishes found in seasonal freshwater lakes and rivers in tropical areas. They seem best designed for rather slow swimming. Living lungfishes can breathe air, allowing them to survive periods of drought or low oxygen in seasonal lakes and rivers in tropical climates. Lungfishes probably survive today because they can tolerate environments that would kill most other fishes. The African lungfish can even tolerate a dry season in which its river dries up. It digs a burrow, seals itself inside, and estivates (turns its body metabolism to a

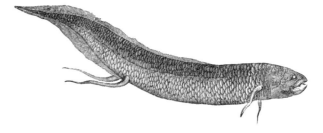

Figure 7.27 *Protopterus*, a living lungfish from Africa. Its pathetically weak fins can beat synchronously, especially the pelvic fins, to allow it to push effectively against the substrate. *Source:* From an old engraving.

very low level) until the rainy season sends water down the river and into the burrow, reviving it.

Lungfishes have evolved considerably since the Devonian. The first lungfishes were marine fishes and look as if they were much more active swimmers than their living descendants (Figures 7.26 and 7.27). Living lungfishes are descended from a clade of Devonian ancestors that evolved the ability to live in fresh water, where they evolved changes in teeth and jaws that mark a shift in feeding from other fishes to mollusks and crustaceans. (Living forms have flattened teeth shaped like plates, for crushing their prey.) But lungfishes had evolved burrowing for dealing with drought by Permian times, because many specimens have been found fossilized in their burrows!

## Tetrapodomorpha

Tetrapodomorphs are the sister group of lungfishes (Figure 7.23). The old name for them, rhipidistians, does not meet modern standards of cladistic precision. They include the ancestors of the land-going vertebrates and are discussed in more detail in Chapter 9.

While placoderms were the dominant fishes of the Devonian, at least in size, the lobefins were most successful in shallow waters around coasts and in inland waters but were hardly dominant. After the Devonian, the rayfin fishes came to be the most successful group, with their combination of lightness and maneuverability, while lobefins were gradually confined to unusual habits and habitats. Perhaps in the process of being squeezed, ecologically speaking, a lineage of Late Devonian lobefins evolved adaptations that allowed them to expand in an unexpected direction – toward life on land and in air. But there were huge changes in terrestrial habitats and landscapes during the Silurian and Devonian, before those first tetrapodomorphs ventured onto land, as we shall see in Chapter 8.

## References

Anderson, P.S.L. and Westneat, M.W. (2009). A biomechanical model of feeding kinematics for *Dunkleosteus terrelli* (Arthrodira, Placodermi). *Paleobiology* 35: 251–269.

Anderson, P.S.L., Friedman, M., Brazeau, M.D., and Rayfield, E.J. (2011). Initial radiation of jaws demonstrated stability despite faunal and environmental change. *Nature* 476: 206–209. [Devonian disparity among gnathostomes].

Farmer, C.G. (1999). Evolution of the vertebrate cardio-pulmonary system. *Annual Reviews of Physiology* 61: 573–592.

Gai, Z., Donoghue, P.C.J., Zhu, M. et al. (2011). Fossil jawless fish from China foreshadows early jawed vertebrate anatomy. *Nature* 476: 324–327. [The galeaspid *Shuyu*].

Long, J.A., Mark-Kurik, E., Johanson, Z. et al. (2015). Copulation in antiarch placoderms and the origin of gnathostome internal fertilization. *Nature* 517: 196–199.

Rücklin, M., Donoghue, P.C.J., Johanson, Z. et al. (2012). Development of teeth and jaws in the earliest jawed vertebrates. *Nature* 491: 748–751.

Zhu, M., Zhao, W., Jia, L. et al. (2009). The oldest articulated osteichthyan reveals mosaic gnathostome characters. *Nature* 458: 469–474. [And comment by M. I. Coates, pp. 413–414].

## Further Reading

Amaral, D.B. and Schneider, I. (2018). Fins into limbs: recent insights from sarcopterygian fish. *Genesis* 56: e23052.

Brazeau, M.D. and Friedman, M. (2015). The origin and early phylogenetic history of jawed vertebrates. *Nature* 520: 490–497.

Donoghue, P.C.J. and Keating, J.N. (2014). Early vertebrate evolution. *Palaeontology* 57: 879–893.

Friedman, M. and Sallan, L.C. (2012). Five hundred million years of extinction and recovery: a Phanerozoic survey of large-scale diversity patterns in fishes. *Palaeontology* 55: 707–742.

Janvier, P. (1996). *Early Vertebrates*. Oxford: Clarendon Press.

Jorgensen, J.M. and Joss, J. (2016). *The Biology of Lungfishes*. Enfield: CRC Press.

Lowe, C.J., Clarke, D.N., Medeiros, D.M. et al. (2015). The deuterostome context of chordate origins. *Nature* 520: 456–465.

Miyashita, T. (2016). Fishing for jaws in early vertebrate evolution: a new hypothesis of mandibular confinement. *Biological Reviews* 91: 611–657.

Sallan, L.C., Friedman, M., Sansom, R.S., Bird, S.M., and Sansom, I.J. (2018). The nearshore cradle of early vertebrate evolution. *Science* 362: 460–464.

Trinajstic, K., Boisvert, C., Long, J., Maksimenko, A., and Johanson, Z. (2015). Pelvic and reproductive structures in placoderms (stem gnathostomes). *Biological Reviews* 90: 467–501.

## Question for Thought, Study, and Discussion

Summarize Colleen Farmer's idea that bony fishes would operate more efficiently if they were air breathers. (You might need to draw diagrams to persuade yourself!) It seems that many Devonian fishes breathed air but now most of them do not. What happened to their lungs, and why did they evolve to become what seems to be less efficient?

# 8

## Leaving the Water

---

**In This Chapter**

Leaving the water is difficult for both plants and animals, for a variety of reasons associated with food or nutrients, exposure to extremes of heat and cold, respiration, and reproduction. We treat plants first because the earliest land plants were probably Cambrian-Early Ordovician in age. We do not simply rely on fossils to document this history but can also look at molecular phylogenetic evidence, and geological evidence for how the first land plants reengineered their landscapes in terms of soils, sediment flows, and oxygen in the atmosphere. Vascular plants had evolved by the Silurian and, by the end of the Devonian, the first forests. This process established a habitat for the first land animals which were arthropods, later joined by annelids and mollusks; vertebrates moved onto land much later. Some of the early terrestrial arthropods, including spiders and their relatives, millipedes, and insects show astonishing internal detail of their respiratory and digestive organs.

---

## Problems of Life in Air

Plants, invertebrates, and finally vertebrates evolved to live on land in the Middle Paleozoic. There were major problems in doing so, related not so much to the land surface as to exposure to air. Many marine animals and plants spend their lives crawling on the sea floor, burrowing in it, or attached to it. As a physical substrate, the land surface is not very different but land organisms are no longer bathed in water. There are predictable consequences for the evolutionary transitions involved, many of them based on the laws of physics and chemistry.

Organisms weigh more in air without the buoyant effect of water, so support is more of a problem. Air may be very humid but it is never continuously saturated, so organisms living in air must find a way to resist desiccation. Tiny organisms are particularly sensitive to drying out in air, because they have relatively large surface areas but cannot hold large reserves of fluid. Therefore, reproductive stages and young stages of plants and animals are very sensitive to drying. Temperature extremes are much greater in air than they are in water, exposing plants and animals to heat and cold. Oxygen and carbon dioxide behave differently as gases than they do when dissolved in water, so respiration

and gas exchange systems must change in air. The refractive index of light is lower in air than in water, and sound transmission differs too, so vision and hearing must be modified in land animals.

There are also ecological consequences. Seawater carries dissolved nutrients but air does not, so some organisms, especially small animals and plants, have a food supply problem in air. It's unlikely that the same food sources would be available to an animal that crossed such an important ecological barrier, so invasion of the land would often be associated with a change in feeding style.

Gravity also becomes important on land. Plants and animals in water are usually neutrally buoyant, meaning that their body fluids and tissues match the buoyancy of the water, so they float. Fishes and swimming invertebrates can pump fluid or air within their bodies so they match the density of water at their current depth. On land, plants have to defy gravity in order to grow upwards, so they need some means of support. Animals too require skeletal adaptations to support their internal organs and for locomotion. When a whale or dolphin is washed up onto the beach, it dies from suffocation because the weight of its internal organs on land crushes the lungs; under water, the whale effectively weighs nothing and it does not need special adaptations to keep its internal organs in good shape.

All the major adaptations for life in air had to be evolved first in the water, as adaptations for life in water. Only then would it have been possible for organisms to emerge into air for long periods. We must reconstruct a reasonable sequence of events during the transition, then test our ideas against evidence from fossil and living organisms.

There might be a problem with the quality of the fossil record, however. Being sure we have a rich enough fossil record of early land plants and animals is tricky because terrestrial fossils are generally less well preserved than marine fossils. But there are two smart ways to plug the gap – looking for geological evidence of the impacts of terrestrialization on the landscape and using genomic data to reconstruct the deep-time phylogenetic tree.

## How Plants Engineered the Landscape

Plants caused profound changes to Earth systems, most notably on the land. Although the first decent fossil plants are known from the Silurian, simple plants such as green algae almost certainly greened the land around the shores of lakes and seas by the end of the Precambrian and during the Cambrian. It is hard to identify such early

moves onto land because the plants involved were so simple and microscopic. Better perhaps to identify their impact on the environment.

Unexpected evidence comes from a detailed study of mud in ancient rivers. William McMahon and Neil Davies from the University of Cambridge, England, surveyed over 700 river deposits dating from the Archaean to the Carboniferous (3000–300 Ma), and measured the proportion of mudrock (mudstone and siltstone) to sandstone in the sedimentary logs of these formations. Levels of mudrock remained low, about 1%, through the Precambrian and Cambrian, and then moved rapidly to 25% in the Late Ordovician and, in fits and starts, eventually to 40% in the Carboniferous.

The uptick in mudrock percentage matches the progressive appearance of different groups of land plants (Figure 8.1), microbes forming mats, crusts, and biofilms from the Late Precambrian onwards, and then bryophytes from the Cambrian and Early Ordovician onwards, and finally vascular plants from the Silurian onwards. The mudrock evidence points to earlier dates than expected – the key rise was in the Ordovician, not the Silurian, from which the oldest macro remains of fossil plants have been recovered. This shows the biological switch had already occurred even though fossil evidence is scant.

Why would an increase in mud indicate a rise in land plants? It could be either that sand in rivers reduces or that mud increases. Most obvious, and long suggested, is that as plants spread onto the early Paleozoic land, they progressively stabilized the surface. With no plant

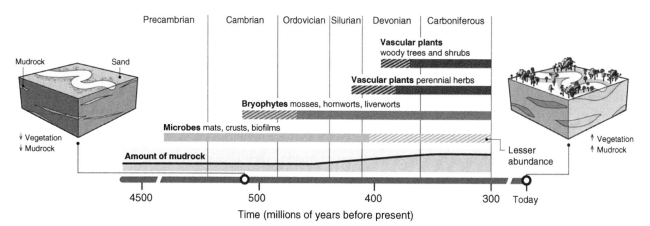

**Plants reshaped the sedimentary deposits left by a river**
Muddy floodplains were rare on prevegetated landscapes, compared with those developed after the evolution of plants and their colonization of the landscape. The rise of mudrock coincided with some of the earliest events in plant evolution.

Figure 8.1 Plants engineered the landscape by replacing sand in rivers with mud and silt. Muddy floodplains were rare in the Precambrian and early Paleozoic but from the Ordovician onwards, mudrock levels jumped up. The rise of mudrock coincided with some of the earliest events in plant evolution. *Source:* Image courtesy of G. Altounian, *Science* magazine.

cover, the surface of the land erodes fast as a result of chemical and physical processes of weathering. Plants send roots into the ground, which can break up rocks, but the rock fragments are incorporated into soil, which is a mixture of sand and organic matter. Today, when forests are cut down on mountainsides, the soil soon washes away and then catastrophic erosion can occur – this is a rerun of the preplant world. So, the spread of green plants around the water bodies, and then further and further away from permanent water, gradually stabilized the land through the 250 million years from Cambrian to Carboniferous. Stable land leads to a huge reduction in the wash-off of sand into rivers, and so the mudrock:sandstone ratio rises.

The other effect is that plants produce mud. By contributing to soil production, grain sizes of sand are reduced in the soil and then, when washed away, they are mud rather than sand. Further, the soil and mud associated with plants can form new habitats such as mudflats, and these contribute to slowing down rivers. Indeed, the stems of the plants themselves, crowding into rivers, also physically slow down the water flow and trap sediment particles.

The raging torrents on a bare early Paleozoic landscape, forming **braided streams** that head straight downhill, eroding and transporting masses of coarse sediment, slowed down to form lazy **meandering rivers**, associated with flatter landscapes. The transition from braided to meandering streams matches a reduction in the energy and speed of water flow, and also in the maximum size of sediment particles they can carry, from boulders to sand and down to mud-sized particles in meandering streams. So, as they evolved, plants engineered early Paleozoic landscapes, giving them the variety that we know, greatly reducing the **mass wasting** rate (the rate of weathering and erosion and of removal of rock to the ocean) and changing forever the great biogeochemical cycles of carbon, nitrogen, phosphorus, and sulfur, for example.

The other major way in which land plants engineered the Earth was in terms of oxygen production. It is well understood that the simple photosynthesizing cyanobacteria drove increases in atmospheric oxygen levels, rising in several steps through the Archaean and Neoproterozoic (see Chapters 2 and 3), but levels remained much lower than they are today until the Late Devonian. A variety of geochemical methods show that oxygen levels reached perhaps 10% in the Ediacaran, then fell through the Cambrian and Ordovician, when ocean anoxia persisted in many locations, and finally rose to modern levels of around 21% in the Late Devonian, coincident with a time when plants on land had clothed much of the landscape and the first trees and forests were emerging.

## Genomic Data and the Origin of Land Plants

Genomic data can be used to resolve large-scale evolutionary trees, as is currently being done around the origin of life and the origin of animals (see Chapters 1, 3, and 4). These methods necessarily depend on genomic evidence from living species, informing us about their evolutionary relationships and genetic distances from one another. However, the fossil record remains crucial, calibrating the genetic distances on evolutionary trees to geological time. This "molecular clock" approach works by modeling the process of genomic evolution within the age constraints provided by fossils on the minimum ages of living lineages (see Chapter 4).

Jenny Morris and colleagues at the University of Bristol have done this kind of molecular clock analysis for evolutionary emergence of land plants (Morris et al. 2018). This was a challenge because there is little current agreement on the fundamental evolutionary relationships among the living land plant lineages. They tested seven possible resolutions of the land plant phylogeny, and all of them gave the same inferred dates of origin. The key land plant group, which includes bryophytes (mosses, hornworts, liverworts) and vascular land plants, arose in the Middle Cambrian-earliest Ordovician (515–470 Ma), and the vascular land plants themselves arose at some point between the latest Ordovician to Late Silurian (472–419 Ma).

These dates are perfectly in accord with the implications of the mudstone record, as noted earlier, and they extend the origins of the key groups back to earlier points in time than the oldest fossils. In the case of vascular land plants, the oldest fossils come from the Middle Silurian, some 421 Ma.

Therefore, sedimentological and genomic data point to similar times for the greening of the land, and these match the fossils. But before looking at the fossils, it is important to consider the biological problems faced by plants in making the transition to life on land. Plants must have emerged gradually into air and onto land from water, and the first "land" plants must have been largely aquatic, living in swamps or marshes.

## How Plants Adapted to Life on Land

Almost all the major characters of land plants are solutions to the problems associated with life in air. Land plants grow against gravity, so they have evolved structural or hydrostatic pressure supports (hard **cuticles** or wood) to help them stay upright. They cannot afford evaporation from moist surfaces, so they have evolved some kind of **waterproofing**. **Roots** gather water and nutrients from soil and act as props and anchors. **Internal transport systems** distribute water, nutrients, and the products of photosynthesis around the plant. Even so, all these adaptations for adult plants are useless unless the **reproductive cycle** is also adapted to air. Cross-fertilization and dispersal require special adaptations in air. All these adaptations must have evolved in a rational and gradual sequence. But because the first stages would have been soft-bodied water plants, the fossil record of the transition is difficult to find.

A scenario for the evolution of land plants was presented by Edwards et al. (2015). Water-dwelling plants, probably green algae, were already multicellular. Green algae grow rapidly in shallow water, bathed in light and nutrients. One might think that cells in a large alga are comparatively independent of one another. In the water, each cell has access to light, water, nutrients, and a sink for waste products. But the fastest growing points of algal fronds need more energy than the photosynthesis of the cells there can supply, so some green algae have evolved a transport system between adjoining cells to move food quickly around the plant. They presumably do this because they can then grow more rapidly.

Their scenario begins with green algae living in habitats that were subject to temporary drying. The algae might already have evolved to disperse spores more effectively by releasing them into wind instead of water. Spores, even in algae, are reasonably watertight and could easily have been adapted for release into air from special spore containers (**sporangia**) growing high enough to extend out of the water on the uppermost tips of otherwise aquatic plants. As plant tissue extended into air, photosynthesis increased because light levels in air are higher than they are in water, especially at each end of the day, and are free from interference by muddy water. Furthermore, $CO_2$ is more easily extracted from air than it is from water.

As plants grew out into air, some tissues were no longer bathed in the water that had provided nutrients and a sink for waste products. Internal fluid transport systems between cells became specialized and extended. Photosynthesis was concentrated in the upper part of the plant that was exposed to more light. Photosynthesis fixes $CO_2$, so there had to be continual intake of $CO_2$ from the air. However, plant cells are saturated with water but air is usually not, so the same surfaces that take in $CO_2$ automatically lose water. Sunlight heats the plant, encouraging evaporation. The water loss had to be made up by transporting water up the stem to the photosynthesizing cells.

Water is transported much more effectively as liquid than as vapor. Early land plants evolved a simple piping system called a **conducting strand** of cells to carry water upward. The conducting strand, found in living mosses, is powerful enough to prevent water loss in small, low plants, if soil water is abundant. But mosses quickly dry up if soil water is in short supply.

Early land plants began to evolve a **cuticle** (a waxy layer) over much of their exposed upper surfaces. The cuticle helps the plant through alternating wet and dry conditions. In wet times, it acts as a waterproof coating. It prevents a film of water from standing on the plant that could cut off $CO_2$ intake. In dry times, it seals the plant surface from losing water by evaporation. A cuticle may also have added a little strength to the stem of early plants, and its wax probably helped to protect the plant from UV radiation and from chewing arthropods.

But the cuticle also cut down and then eliminated, from the top of the plant downward, the ability to absorb water-borne nutrients over the general plant surface. Nutrients were taken up more and more at lower levels of the plant, eventually taking place on specialized absorbing surfaces at the base (**roots**) which probably evolved from the runners that these plants often used to reproduce asexually. As roots grew larger and stronger, they helped to anchor and then to support the plant. Roots also extended deep into the soil to extract water, and, often with the help of fungi and bacteria, to extract nutrients.

As cuticle evolved, it sealed off $CO_2$ uptake over the general plant surface, so plants evolved special pores called **stomata** where $CO_2$ uptake could be concentrated. If it is too hot or too dry, stomata can be closed off by **guard cells** to control water loss. As $CO_2$ uptake was localized, plants evolved an **intercellular gas transport system** that led from the stomata into the spaces between cells, improving $CO_2$ flow to the photosynthesizing cells. The same system was also used to solve an increasingly important problem. As roots enlarged, more and more plant tissues were growing in dark areas where photosynthesis was impossible, yet those tissues needed food and oxygen. Soils are low or lacking in $O_2$, especially when they are waterlogged. The intercellular gas transport system feeds $O_2$ from the air down through the plant to the roots, sometimes through impressively large hollow spaces.

Later plants evolved **xylem**, an improved piping system for better upward flow of water from the roots. Xylem is made of very long dead cells arranged end to end to form long pipes up and down the stem. Even a narrow xylem can transport water much faster than can normal plant tissue and, once begun, the evolution of xylem was probably a rapid process that immediately

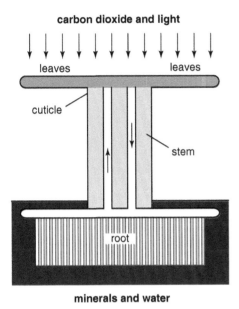

**carbon dioxide and light**

leaves          leaves

cuticle

stem

root

**minerals and water**

↑ Xylem transports water and minerals
  upward from the roots to the leaves.

↓ Phloem transports food around the
  plant from the leaves where it is made.

Figure 8.2 The basic land plant, in terms of transport of water and nutrients and capture of energy through photosynthesis.

gave plants greater tolerance of dry air. Plants that have xylem are called **vascular plants** (Figure 8.2).

Xylem cells are dead so xylem transport is passive, driven entirely by the suction – or negative pressure – of evaporation from the upper part of the plant, and it takes place at no cost to the plant. The forces generated can be very large, so the long narrow walls of xylem cells may tend to collapse inward. Xylem cell walls came to be strengthened by a structural molecule, **lignin**. Once lignin had evolved, it was used later to strengthen the roots and stem as plants grew taller and heavier. Later still, it also became a deterrent to animals trying to pierce and chew on plant tissues.

As plants became increasingly polarized, with nutrient and water being taken up at the roots and photosynthesis taking place in the upper parts, the xylem and gas transport systems improved but neither of them could transport liquid downward. This problem was solved by the evolution of another transport system called **phloem** that works like the cell-to-cell transport system of green algae. Phloem cells carry photosynthate from photosynthesizing cells to growing points such as reproductive organs and shoots, and to tissues such as roots that cannot make their own food.

Throughout the process, the advantage that encouraged plants to extend into air in spite of the difficulties involved was the tremendous increase in available light. Marine plants are restricted to the narrow zone along the shore where light has to penetrate sediment-laden, wave-churned water. Growth above water increases light availability. Furthermore, competition for available light tended to encourage even more growth of plant tissues above the water surface, and more effective adaptations to life in air (Figure 8.2). Once plants could grow above the layer of still air near the water surface, spores could be released into the breeze. Greater plant height and the evolution of sporangia on the tips of branches were both adaptations for effective dispersal.

## The Earliest Land Plants

Land plants are divided into two main groupings: the tracheophytes (vascular plants) and the bryophytes (mosses, hornworts, liverworts). Tracheophytes comprise the majority of plant diversity today, including the lycophytes (e.g., monkey puzzle trees), monilophytes (ferns and horsetails), gymnosperms (e.g., cycads, conifers, pines), and the angiosperms (flowering plants) which encompass more than 80% of living plant diversity. The other main grouping, the bryophytes, is made up of mosses, liverworts, and hornworts. Unlike most animals, plants have alternating haploid (gametophyte) and diploid (sporophyte) life cycles. Tracheophytes and bryophytes differ in that their life cycles are dominated by sporophyte and gametophyte stages, respectively. Tracheophytes have all the key adaptations that make life on land tolerable for plants, but bryophytes have chimeric subsets of these characters – perhaps representing the gradual assembly of land plant characters culminating in a tracheophyte bodyplan or loss of terrestrial adaptations in a more tracheophyte-like ancestor.

### The Bryophytes: Mosses and Relatives

The earliest spores that belonged to land plants come from Middle Ordovician rocks (Figure 8.3) (Wellman and Strother 2015). They look very much like the spores of living liverworts, and there are fragments of their parent plants preserved with them. All evidence from living plants suggests that the simplest ones fall into a group called **bryophytes** (liverworts, mosses, and hornworts).

Today, there are 17 000 species of bryophytes and they never grow very large. The dominant gametophyte stage of liverworts and hornworts is flattened, sometimes branching structures, while mosses often have tiny spirally arranged leaf-like structures. Hornworts and mosses also have an upright sporophyte stage that develops from the gametophyte and, you guessed it,

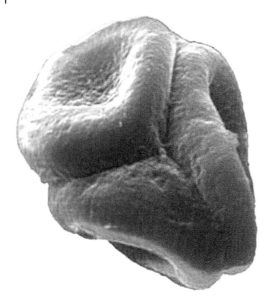

Figure 8.3 One of the earliest known spores from a land plant, from the Ordovician of Libya. *Source:* Courtesy of Charles Wellman, Sheffield University, UK.

Figure 8.4 Living liverworts. They prefer damp shady places, probably much like their Ordovician ancestors. This is *Conocephalum* from Scotland. *Source:* Photograph by Lairich Rig (Wikimedia).

yields spores from which the subsequent gametophyte develops. Mosses are so familiar that they are often overlooked but these small plants are perhaps our best living model for the ancestral land plant, living in damp environments on land.

Bryophytes show adaptations to life on land, such as a waterproof cuticle over their leaves and stems. Many hornworts and mosses have simple stomata, used for controlling water loss, but they are absent in liverworts. A few of the larger mosses and liverworts have a very simple water-conducting system. Some bryophytes have the unusual ability of being able to dry up completely, and they can then rehydrate when rain falls, and they carry on as if nothing had happened.

Bryophytes have a very poor fossil record principally because their life cycle is dominated by the gametophyte stage which has very little fossilization potential. Therefore, while there are claims of bryophyte remains dating back to the Ordovician, evidence is absent until the Mesozoic, hundreds of millions of years after the origin of bryophytes.

Until recently, many botanists regarded the bryophytes as an artificial group, of which the liverworts (Figure 8.4) were the most primitive (since they lack key land plant adaptations, like stomata), followed by mosses and then, perhaps, hornworts as closest relatives of the vascular plants. However, there is now growing evidence that that the bryophytes form a natural group (Figure 8.5), a sister clade to the vascular plants (Puttick et al. 2018). This suggests that the patchy distribution of land plant adaptations in bryophytes is a consequence of losses from a more complex and better adapted ancestral land plant.

## The Earliest Tracheophytes

The earliest records of tracheophytes are from the Late Ordovician, but they occur in the form of isolated spores with a characteristic Y-shaped "trilete" mark on their surface, reflecting their development as nonpermanent "tetrads" of spores squished together. It's not altogether certain whether these were produced by members of the ancestral lineage, before the divergence of living vascular plants, or whether they represent descendants of that diversification event. More convincing macroremains are not found in rocks older than Late Silurian. Although it is Devonian in age, *Aglaophyton* (Figure 8.6a) has a grade of structure that evolved in Silurian times. It grew to a height of less than 15 cm (about 6 in.). It had most of the adaptations needed in land plants (cuticle, stomata, and intercellular gas spaces) but it did not have xylem, only a simple conducting strand, so it was not a fully evolved vascular plant.

*Cooksonia* (Figure 8.6b), known from the Middle Silurian to Early Devonian, was only a couple of centimeters high and had a simple structure of thin, evenly branching stems with sporangia at the tips, and no leaves. But it also had central structures that were probably xylem, including tracheids, rather than simple conducting strands. Later species of *Cooksonia* from the earliest Devonian have definite strands of xylem preserved, and cuticles with stomata, so they probably had intercellular gas spaces and were better adapted for life in air. Although they are limited to the sporophyte stage in the plant life cycle, these remains are remarkably well preserved and all the details of their internal anatomy – even the spores packed into the sporangia – have been identified by Dianne Edwards from the University of Cardiff, using

**Figure 8.5** The early evolution of land plants has been debated, with many finding that the bryophytes are not a natural group. However, the most recent genomic analyses confirm that Bryophyta is a clade, sitting between the green algae and the vascular plants in the evolutionary tree. Blue icons represent the molecular clock estimates for the ages of groupings. *Source:* After Morris et al. (2018).

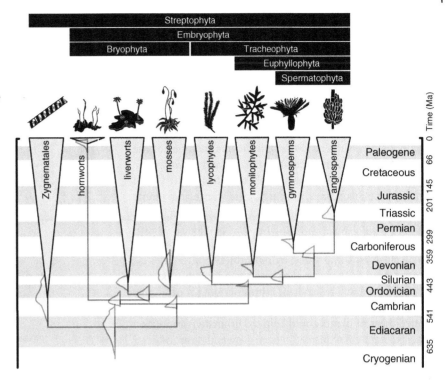

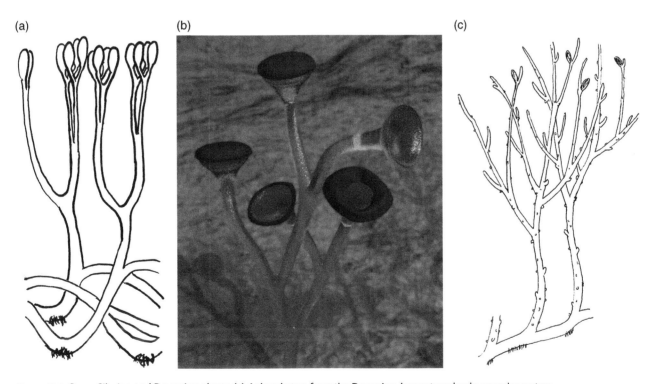

**Figure 8.6** Some Silurian and Devonian plants. (a) *Aglaophyton*, from the Devonian, has not evolved a vascular system. *Source:* Reconstruction by Grienstiedl (Wikimedia). (b) Reconstruction of *Cooksonia*, the earliest vascular plant, to show its funnel-shaped sporangia in various stages of development. *Source:* Image by Smith609 (Wikimedia). (c) Reconstruction of *Rhynia*, a vascular plant from the Devonian of Scotland. *Source:* Grienstiedl (Wikimedia).

acids to dissolve the rock entombing the tiny organic remains so that they can be dissected and analyzed with a scanning electron microscope. It is likely that some species of *Cooksonia* are members of the ancestral vascular plant lineages, while others are members of the living clade, including members of the lycophyte lineage.

Early Devonian land plants were dramatically more diverse, including definitive members of the living clade of tracheophytes, such as *Zosterophyllum*, a member of the lycophyte lineage, and *Eophyllophyton* and *Psilophyton*, which belong to the "euphyllophyte" lineage, to which ferns, horsetails, pines, conifers, and flowering plants belong. These early tracheophytes grew up to a meter high, although they were slender (1 cm diameter). For support, they must have grown either in standing water or in dense clusters, some aided by reproductive budding systems of rhizoids for asexual, clonal reproduction, as strawberries do today. This style of reproduction not only gave mutual support to individual stems but, by "turfing in," a cloned mass of plants could help to eliminate competitive species. Plants like this could have grown and reproduced very quickly, a way of escaping the consequences of relatively poor adaptations for living in air.

### The Rhynie Chert as a Window into Early Life on Land

The Rhynie Chert flora from the Early Devonian of northern Scotland provides a remarkable window into the early evolution of plants and animals on land. One notable form, *Rhynia* (Figure 8.6c), like *Cooksonia*, is a vascular plant. Its lateral branches could break off and develop into new plants, so it probably grew in dense thickets of identical plants.

The Rhynie Chert plants were occasionally flooded by silica-rich water from hot springs and preserved perfectly with all their tissues (Figure 8.7). The water likely left the hot spring at temperatures from 90 to 120 °C (194–248 °F) but may have cooled considerably before it flowed around the plants. The hot springs were only active from time to time, and their activity is preserved in 53 beds, each about 8 cm (3 in.) thick. Scientists can study modern hot springs, such as those at Yellowstone, to work out how the extraordinary preservation model at Rhynie might have worked. The chert probably formed in a marshy area, of which 55% was covered by vegetation, 30% with plant debris, and the remaining 15% of the ground being bare.

The Rhynie Chert fossils include seven species of vascular plants, including *Rhynia*, as well as fungi, one species of lichen, and bacteria. There are also six groups of terrestrial arthropods. If you were transported back to the Early Devonian in a time machine and walked among the Rhynie fossils, the taller plants

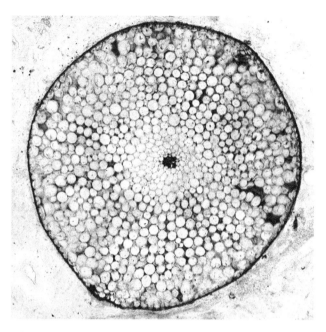

**Figure 8.7** A microscopic examination of the stem of *Rhynia* shows its cell structure preserved beautifully. The xylem is the small dark area in the center of the stem, and the phloem is the large zone of circular tubes surrounding it. *Source:* Image by Plantsurfer (Wikimedia).

such as *Asteroxylon* and *Zosterophyllum* would barely have grazed your knees. These two taller plants had stems covered with scale-like leaves, and they pumped water through their simple vascular canals, powered by stomata under the leaves. Between the stems of these plants, and even inside their stems, tiny spider-like and insect-like arthropods crept.

At some point in the Early Devonian, a lineage of small plants evolved thick-walled cells called **tracheids** within their xylem, first seen in species of *Cooksonia*. The thickened cell walls allowed the secondary xylem to transport liquid under greater pressure, increasing its efficiency, and made the tracheids relatively rigid. In some living plants, the tracheid system is very large and quite rigid, and we know it now as wood. Late Devonian plants were able to grow to greater heights, competing with one another for light and therefore for living space.

### Later Devonian Tracheophytes – Diversification of Lycophytes and Euphyllophytes

Structural advances are seen in later Devonian and Early Carboniferous plants. Successive floras all lived in lowland floodplains, which have a good fossil record. It looks as if we are seeing waves of ecological and evolutionary replacement on all levels, from individual plants to world floras, as structural innovations allowed each plant group to outcompete its predecessor.

For example, *Rhynia* had only 1% of its stem cross-section made of xylem (Figure 8.7). Other Devonian plants had 10%, and the whole stem was more strongly built. Plants could grow taller (up to 2 m high) and compete for light more efficiently than *Rhynia*. Other improvements in reproduction and light gathering, through the evolution of leaves rather like those of living ferns and through more complex branching, also aided plant efficiency.

Some Devonian and Carboniferous lycophytes grew to be very large trees, as we shall see in Chapter 9. Their sister clade, the euphyllophytes, are named after their characteristic leaves, which first evolved in the Devonian (Figure 8.8), as seen in plants such as *Elkinsia*. Mosses and lycophytes also have leaves but these evolved independently, converging on the same trick of increasing surface area to enhance photosynthesis and gas exchange. This convergence is reflected in their different structure, moss leaves having interesting vein structures while lycophyte leaves have very simple veins. In truth, leaves appear to have evolved multiple times within the euphyllophytes, too. Oddly, roots and wood have also evolved independently in lycophytes and euphyllophytes – evidently, these innovations were evolutionary opportunities too good to miss.

Trees are woody and large, so they stand out in the fossil record, sometimes literally because they may be preserved still upright in life position. Fossil tree trunks from the Middle Devonian of New York suggest plants over 10 m (30 ft) high, with woody tissue covered by bark. Once plants reached these heights, shading of one species by another would have led to fairly complex plant communities. We see the first large forests by the Late Devonian. *Archaeopteris* grew to be 30 m high (100 ft) and was the dominant tree in forests that were globally widespread (Figures 8.9 and 8.10).

It is commonly assumed that trees became tall in order to reach the light. After all, in modern forests, trees compete with each other by racing up to the light, spreading their branches over neighboring trees. The losers remain weedy and soon die. However, Kevin Boyce of Stanford University and colleagues have argued that Devonian plants would have had low photosynthetic abilities in comparison to modern plants, so other drivers might have been at work (Boyce et al. 2017). He suggested that early trees became larger so they could spread their seeds further, have larger leaves, and especially have longer roots so they could extract more water and nutrients from the soil. Therefore, as Boyce says, "They were growing down to the water, not up to the light."

Plants evolved **seeds** (rather than spores) in the Late Devonian, too. A Late Devonian seed-like structure called *Archaeosperma* (Figure 8.11) looks as if it belongs to a tree very much like *Archaeopteris*. This was a great advance; all previous plants had needed a film of water in which sperm could swim to fertilize the ovum, but seed plants can reproduce away from water.

Seed plants invaded drier habitats and seed dispersal by wind (rather than water) became important; we have winged seeds from Late Devonian rocks. Seed dispersal allowed some plants to specialize as invaders into new

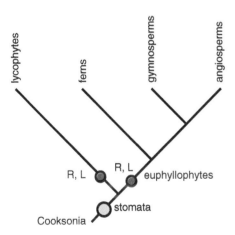

**Figure 8.8** Cladogram of early vascular plants. Stomata evolved after *Cooksonia* but true roots and leaves (R, L) evolved separately in lycophytes and euphyllophytes. Gymnosperms include conifers, gingkos, and cycads.

**Figure 8.9** The late Devonian tree *Archaeopteris*, based on the 1962 reconstruction by Charles Beck which connected the fossil foliage of *Archaeopteris* (see Figure 8.10) with the fossilized wood of its tree trunk.

Figure 8.10 A frond from the top of an *Archaeopteris* tree. *Source:* Image by Richard Cowen.

areas, avoiding the increasingly dense and competitive habitats in wetlands and along rivers.

Thus, by the end of the Devonian, all the major innovations of land plants except flowers and fruit had evolved. Forests of seed-bearing trees and lycopods had appeared, with understories of ferns and smaller plants. The evolution of seeds seems to have been the foundation for success in the Early Carboniferous (see Chapter 9).

### Early Plant Innovation

The increasing success of land plants, especially their growth to the size of trees, must have produced ever larger amounts of rotting plant material in swamps, rivers, and lakes, leading to very low $O_2$ levels in any slow-moving tropical water ($O_2$ is used up in decay processes). At the same time, the increasing photosynthesis by land plants drew down atmospheric $CO_2$ and increased atmospheric oxygen.

All this probably helped to encourage air breathing among contemporary freshwater arthropods and fishes, and it led to better preservation of any fossil material deposited in anoxic swamp water. Some coals are known from Devonian rocks but truly massive coal beds formed for the first time in Earth history in the Carboniferous Period, which was named for them (see Chapter 9).

The dominant process in Devonian plant evolution seems to have been selection based on efficiency – in size and stability, photosynthesis, internal transport, and reproductive systems. Plant groups replaced one another as innovations appeared. Perhaps the most interesting

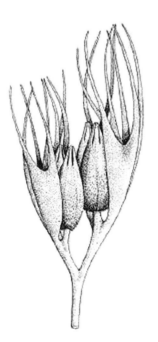

Figure 8.11 The Late Devonian seed-bearing structure *Archaeospermum*, reconstruction (*left*) and fossil (*right*). *Source:* From the work of John Pettit and Charles Beck.

part of this story of early plants is the slow rate at which innovations appeared. The innovations we have discussed should have given immediate success whenever they appeared but it took the length of the Devonian (about 50 m.y.) for seed plants to emerge.

## Comparing Plant and Animal Evolution

Whether you count spores or plant macrofossils, there was a striking increase in land plant diversity from the Silurian to Middle Devonian, when a diversity plateau was reached that extended into the Carboniferous (Figure 8.12). A second increase in Carboniferous land plant diversity was followed by a long period of stability. A third, Late Mesozoic expansion in land plants and animals raised diversity to current levels (see Chapter 15).

The pattern looks rather like the pattern of Sepkoski's three major faunas in the oceans (see Chapter 6). But the radiations among land plants and marine animals did not occur at the same time, so they were not linked. Extinctions among plants are different from those among animals, which suggests that plants and animals may respond to quite different extinction agents. Andrew Knoll suggested three major factors.

- Plants are more vulnerable to extinction by competition.
- Plants are more vulnerable to climatic change.
- Plants are less vulnerable to mass mortality events.

These differences reflect basic plant biology. All plants do much the same thing. They are all at the same primary trophic level, so they cannot partition up niches as easily as animals can. A new arrival in a flora may be competitively much more dangerous than a new arrival in a fauna.

For example, $CO_2$ uptake must be accompanied by the loss of water vapor, since the plant is open to gas exchange. Many plant adaptations are responses to the problem of water conservation. Because it is so basic a part of their biology, an innovation here could provide a new plant group with an overwhelming advantage. Other plant systems such as light gathering are equally likely to be improved by innovation.

Plant distributions are sensitive to climate. If climate changes, plants must adapt, migrate, or become extinct. In extreme circumstances, there may be no available refuges. Thus, the tree species of northwest Europe were trapped early in the ice ages between the advancing Scandinavian glaciers to the north and the Alpine glaciers to the south and were wiped out. In contrast, similar species in North America were able to move their range south along the Appalachians, then north again as the ice retreated.

On the other hand, plants are well adapted to deal with temporary stress, even if it is catastrophic to animals. Plants readily shed unwanted organs such as leaves or even branches in order to survive storms and extreme weather. Many weeds die, to overwinter as seeds or bulbs. Even when plants are removed, by fire or drought,

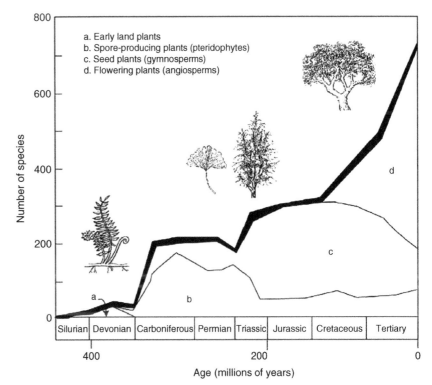

Figure 8.12 The diversification of land plants occurred in four phases, with early land plants in the Devonian (a), spore-producing plants in the Carboniferous to Triassic (b), seed plants in the Carboniferous to Cretaceous (c), and finally angiosperms, from the Cretaceous to the present day (d). *Source:* Modified from work by Karl Niklas.

a. Early land plants
b. Spore-producing plants (pteridophytes)
c. Seed plants (gymnosperms)
d. Flowering plants (angiosperms)

Number of species

Silurian | Devonian | Carboniferous | Permian | Triassic | Jurassic | Cretaceous | Tertiary

Age (millions of years)

the soil is always rich with seeds so that mass mortality of full-grown plants does not mean the end of the population. Plants are the dominant biomass in communities recovering after volcanic eruptions, tropical storms, or other catastrophes have devastated an area. In many land communities, the removal of dominant trees by storm, fire, or human agency is followed by the rapid growth of species that are very good at colonizing disturbed areas.

## The First Land Animals

As plants extended their habitats into swamps and on to riverbanks and floodplains, they would have provided a food base for animal life evolving from life in water to life in air. The marine animals best preadapted to life on land were arthropods. They already had an almost waterproof cover and were very strong for their size, moving on sturdy walking legs. The incentive to move out into air might have been the availability of organic debris washed ashore on beaches, or perhaps the debris left on land by the first land plants. (Foraging crabs are obvious members of many beach communities.) Plant debris, whether it's on a beach or on a forest floor, tends to be damp; it provides protection from solar radiation and is comparatively nutritious. Thus, it is not surprising that the earliest land animals were arthropods that ate organic debris, and other arthropods that ate them.

Different arthropods probably moved into air by different routes. The easiest transition would seem to have been by way of estuaries, deltas, and mudflats, where food is abundant and salinity gradients are gentle. This idea has been tested in a molecular clock study by Lozano-Fernandez et al. (2016). This team studied the three main independent shifts from water to land by the arthropods, namely millipedes, insects, and spiders. In all three cases, the most likely environmental transition turns out to have been from the sea to land, and all three transitions are dated to have occurred between the Cambrian and Silurian. The molecular dates match early fossil finds of spider relatives, insect relatives, and millipedes but also those for the emergence of land plants.

The oldest traces of arthropods on land are known from the Cambrian of Canada but they represent only periodic forays onto land for, perhaps, reproduction in otherwise marine organisms (MacNaughton et al. 2002). There are also burrows in paleosols and fragmentary remains, both attributed to annelids, from the Upper Ordovician of Pennsylvania and Michigan, respectively. However, the oldest certain remains of terrestrial arthropods are known from Late Silurian trace fossils of their footprints and coprolites, as well as body fossils. For example, the oldest fossil that provides evidence of air breathing is the millipede *Pneumodesmus* from the Late Silurian or Early Devonian of Scotland (Figure 8.13). The specimen is a short segment, 1 cm (0.5 in.) long, showing six body segments, each with some bars and knobs and a long slender leg below. The specimen is especially important because, under the microscope, the cuticle can be seen to bear numerous tiny openings, interpreted as spiracles that would have been used for gas exchange. Other Late Silurian fossils from England include spider-like critters and centipedes, extracted from among remains of some of the oldest land plants (Jeram et al. 1990).

Very small arthropods have been found at several Early Devonian localities. Most of them (mites and springtails) were eating living or dead plant material, and in turn were probably eaten by larger carnivorous arthropods such as early spiders. Larger arthropods are usually found in tiny fragments but it is clear that some were large by any standard. There are fossil pieces of a scorpion that was probably about 9 cm (over 3 in.) long, and a very large millipede-like creature, *Eoarthropleura*, which probably lived in and ate plant litter. At 15–20 cm (6–8 in.) long, this was the largest terrestrial animal of the Early Devonian.

The Rhynie Chert land animals include a nematode worm, some trigonotarbids, the earliest harvestman, a spider-like animal with very long thin legs, and the oldest mites. All these are arachnids, close relatives of spiders, and equipped with eight legs (instead of six, as in insects). There are also a centipede, a probable millipede, and among insects, the oldest springtail and another species.

**Figure 8.13** The world's oldest air-breathing animal, the first millipede, *Pneumodesmus* from the Late Silurian of Scotland. *Source:* Photo from Xenarachne (Wikimedia).

Figure 8.14 An array of Carboniferous terrestrial arthropods from classic sites in France and the UK, all reconstructed with X-ray microtomography. A, D, Harvestmen arachnids, identified as Opiliones, both members of living orders. B, Close relative of spiders. C, Harvestman arachnid, showing close-up of the mouthparts. E, H, Juvenile insects, of which H might be related to roaches. F, G, Trigonotarbid arachnids. *Source:* Original images and montage courtesy of Russell Garwood, taken from his many publications, including Garwood and Edgecombe (2011).

These fossils are so well preserved that Jason Dunlop and Russell Garwood were able to identify plant debris in the stomach of one of the insects, the book lungs of the trigonotarbid, and tracheae and genitals in the harvestman.

Tracheae (singular, trachea) are openings through the cuticle, the external hardened skin, of arthropods. In modern arthropods, the tracheae extend through layers of the cuticle and penetrate the soft tissues inside. Most arthropods breathe passively by allowing oxygen to diffuse in through the tracheae and supply the tissues. This marks a limit to arthropod size because passive diffusion of oxygen cannot fuel a large body. Spiders and trigonotarbids, however, had so-called book lungs, as they do today, which allow them to pump oxygen.

The Paleozoic arthropod specimens are often incomplete but by the Carboniferous, many more complete specimens have been found. Sometimes, the specimens are entirely or largely enclosed in sediment but

Russell Garwood of the University of Manchester, England, has used X-ray tomography to identify astonishing detail of these early land animals (Figure 8.14), even showing details of the hairs on their legs and the lenses of their eyes.

This early terrestrial ecosystem did not include any vertebrates as permanent residents but no doubt the entire food chain, including fishes in the rivers, lakes, and lagoons, benefited from the increased energy flow provided by plants and their photosynthesis.

Fishes were abundant in Devonian ecosystems, in fresh and marine waters, and some were evolving to take advantage of the new life on land. Early fishes were all air breathers to some extent, so they must have snatched insects and other animals from the surface of the water and from the shore. It was but a modest step for them to spend more time on land. We explore the origin of tetrapods in Chapter 9.

## References

Boyce, C.K., Fan, Y., and Zwieniecki, A. (2017). Did trees grow up to the light, up to the wind, or down to the water? How modern high productivity colors perception of early plant evolution. *New Phytologist* 215: 552–557.

Edwards, D., Cherns, L., and Raven, J.A. (2015). Could land-based early photosynthesizing ecosystems have bioengineered the planet in mid-Palaeozoic times. *Palaeontology* 58: 803–837.

Garwood, R.J. and Edgecombe, G.D. (2011). Early terrestrial animals, evolution, and uncertainty. *Evolution Education Outreach* 4: 489–501.

Jeram, A.J., Selden, P.A., and Edwards, D. (1990). Land animals in the Silurian – arachnids and myriapods from Shropshire, England. *Science* 250: 658–661.

Lozano-Fernandez, J., Carton, R., Tanner, A.R. et al. (2016). A molecular palaeobiological exploration of arthropod terrestrialization. *Philosophical Transactions of the Royal Society of London. Series B, Biological Sciences* 371: 20150133.

MacNaughton, R.B., Cole, J., Dalrymple, R. et al. (2002). First steps on land: arthropod trackways in Cambrian-Ordovician eolian sandstone, southeastern Ontario, Canada. *Geology* 30: 391–394.

Morris, J.L., Puttick, M.N., Clarke, J.W. et al. (2018). The time-scale of early land plant evolution. *Proceedings of the National Academy of Sciences of the United States of America* 115: E2274–E2283.

Puttick, M.N., Morris, J.L., Williams, T.A. et al. (2018). The interrelationships of land plants and the nature of the ancestral embryophyte. *Current Biology* 28: 733–745.

Wellman, C.H. and Strother, P.K. (2015). The terrestrial biota prior to the origin of land plants (embryophytes): a review of the evidence. *Palaeontology* 58: 601–627.

## Further Reading

### Plants

Beerling, D. (2017). *The Emerald Planet: How Plants Changed Earth's History, second edition.* Oxford: Oxford University Press.

McMahon, W.J. and Davies, N.S. (2018). Evolution of alluvial mudrock forced by early land plants. *Science* 359: 1022–1024.

Stein, W.E., Berry, C.M., VanAller Hernick, L. et al. (2012). Surprisingly complex community discovered in the mid-Devonian fossil forest at Gilboa. *Nature* 483: 78–81.

### Animals

Dunlop, J. and Garwood, R.J. (2017). Terrestrial invertebrates in the Rhynie chert ecosystem. *Philosophical Transactions of the Royal Society of London. Series B, Biological Sciences* 373: 20160493.

Wilson, H.M. and Anderson, L.I. (2004). Morphology and taxonomy of Paleozoic millipedes (Diplopoda: Chilognatha: Archipolypoda) from Scotland. *Journal of Paleontology* 78: 169–184.

## Question for Thought, Study, and Discussion

Imagine a world without life on land. If you remove all forests, grass, insects, and other animals, what would the land surface look like? How do land plants and animals shape the landscape today? Think about temperate lands like Canada or Germany, then tropical zones like Brazil or Indonesia.

# 9

# Early Tetrapods and Amniote Origins

---

### In This Chapter

Tetrapodomorph fishes showed many adaptations to life at the water's edge, being able to breathe in air and water and probably hunting prey among the new plant-lined pond margins. The transition from fish to tetrapod involved small changes to the bones in the pectoral and pelvic lobe fins so they could work as props or legs. The skull separated from the shoulder girdle to enable walking. Tetrapods likely moved onto land in search of new food supplies but the first tetrapods of the Late Devonian, such as *Acanthostega* and *Ichthyostega*, were more adapted for swimming than walking. These first tetrapods diversified in the Early Carboniferous, although their fossil record is patchy and confined to rather special habitats. The maximum size increased and the skeletons became stronger, although many of these creatures would have been clumsy on land. The ecological diversity is surprising, however, with some early tetrapods evolving to look like little water snakes or crocodile-like predators, or small and large lizards. Somewhere in this diverse array, some early tetrapods evolved a shelled egg with specialized compartments to foster the growth of an advanced hatchling. This is the amniotic egg, the basis for the dramatic radiation into truly land-going vertebrates (reptiles, birds, and mammals) that followed. An amniotic egg must be laid on land and it hatches into air, so the final link with a water existence was broken. This evolutionary innovation took place in giant swampy forests that covered much of the tropical Earth and formed many of the coalfields we are mining today. The forest ecosystem was rich in diversity, with plants, insects, reptiles, and amphibians all thriving.

---

## Vertebrates Conquer the Land

We saw in Chapter 8 that plants and invertebrates, mostly arthropods such as insects and spiders, had moved onto land in the early Paleozoic. By the beginning of the Devonian, there was abundant life on land, at least around the margins of rivers, lakes, and seas. We see the first signs that fishes were beginning to exploit the newly enriched habitats near the shore and near the surface. But we must not imagine that vertebrates adapted quickly to life in air or that they readily left the water.

We saw in Chapter 7 that lobefin fishes evolved towards different ways of life by the end of the Devonian. Most of the Middle and Late Devonian lobefins are called **tetrapodomorphs**, meaning "tetrapod-like" because, while they are close relatives of the ancestors of modern lungfish and coelacanths, these were closer in evolutionary terms to tetrapods. The Devonian tetrapodomorphs hunted fishes in shallow waters along sea coasts and into brackish shoreline lagoons and freshwater lakes and rivers. They were probably fast-sprinting ambush predators. These were the ancestors of one of the greatest transitions in the history of life, the origin of tetrapods, the group that includes all amphibians, reptiles, birds, and mammals, and we explore later stages in this remarkable story in later chapters.

## Tetrapodomorphs

Tetrapodomorphs had long, powerful, streamlined bodies with strong lobe fins and a tail with a fin (Figure 9.1), adapted for strong swimming. They had long snouts, especially the larger ones. Perhaps as a result, they evolved a skull joint that allowed them to raise the upper jaw as well as, or instead of, lowering the lower jaw as they gaped to take prey. This could have had two important effects, both related to life in shallow water. First, the snout movements would have changed mouth volume, perhaps allowing extra water to be pumped over the gills without moving the lower jaw. Second, tetrapodomorphs could have caught prey in shallow water by raising the snout without dropping the lower jaw. Some tetrapodomorphs may have been able to chase prey right up to or even beyond the water's edge. Their powerful lobe fins, set low on the body, may have allowed them to drive after prey on, over, or through shallow mudbanks, a first step, or rather belly flop, in occupying the land.

The main sprinting propulsion in tetrapodomorphs came from the tail. They also had numerous lobe fins, as seen in *Osteolepis* (Figure 9.1) and *Eusthenopteron* (Figures 9.2–9.4) – the paired **pectoral fins** at the front and **pelvic fins** at the back, as well as two unpaired fins in the midline on the back. The pectoral fins could be used against the bottom as supports, strengthening the posture of the trunk and acting as props in chasing; the pelvic fins acted to grip and push on the substrate, adding to the thrust. The fishy pectoral fins of the tetrapodomorphs eventually became the arms or forelimbs of tetrapods, and the pelvic fins became the legs. But these paired lobe fins evolved toward limbs not as an adaptation for walking, but to make the tetrapodomorph a more efficient fish.

Tetrapodomorphs evolved an adaptation for air breathing that we still have. If you breathe through your nose, air reaches your lungs through a passage called a **choana** that runs from your nostrils, through your sinuses, and through the back of your throat. That same passage evolved in the earliest and most basal of the tetrapodomorphs, *Kenichthys* from the Middle Devonian of China. This is clear evidence that all tetrapodomorphs are closer to tetrapod ancestry than any other lobefins.

*Osteolepis* and *Eusthenopteron* (Figures 9.2–9.4) have long been recognized as probably close to the tetrapod ancestors. The skull bones, the pattern of bones in the lobe fins, and the general size, shape, and geographic distribution of these fishes are close to those of the earliest

Figure 9.2 *Eusthenopteron foordi*, from the Devonian of Canada. *Source:* Photograph by Ghedoghedo (Wikimedia).

Figure 9.1 An early tetrapodomorph, *Osteolepis* from the Devonian of Scotland. *Source:* Artwork by Nobu Tamura (Wikimedia).

Figure 9.3 Reconstruction of *Eusthenopteron*, from the Devonian of Canada. Strong ventral fins could have been used to help swimming in very shallow water. *Source:* Artwork by Nobu Tamura (Wikimedia).

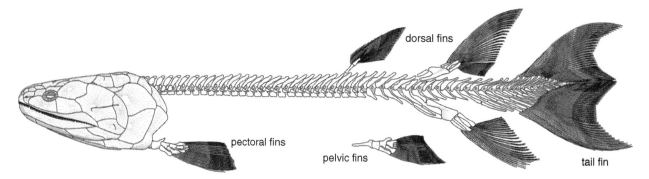

Figure 9.4 Skeleton of *Eusthenopteron*, showing the lobe fins, two centrally placed dorsal fins on the back and two pairs of fins, the pectoral and pelvic, below. In swimming, it is likely that all these lobe fins were deployed but when pulling itself along the bottom of a pond or over some wet mud, the tetrapodomorph used its pectoral and pelvic fins only. *Source:* After Erik Jarvik.

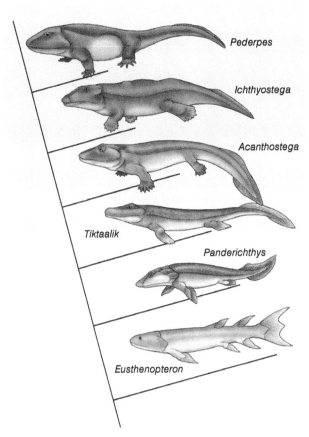

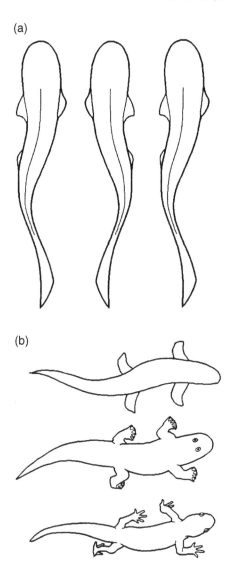

Figure 9.5 A series of cartoon tetrapodomorphs, from Middle Devonian to early Carboniferous, arranged along a cladogram to show the evolutionary transition from tetrapodomorph (first three) to tetrapod (last three). *Source:* Diagram by Maija Karala (Wikimedia).

tetrapods. The bones inside the lobe fins match those of tetrapods. For example, in its fore paddle, *Eusthenopteron* (Figure 9.4) has a single upper bone, equivalent to our humerus, then a pair of longish little bones, equivalent to the radius and ulna of our forearm, and then a number of shorter bones that match our wrist bones. It's the same in the pelvic fin (= leg), where both the tetrapodomorph fishes of the Devonian and we have the femur, tibia + fibula, and we add to this the ankle bones (tarsals), and foot bones. We and all other tetrapods share the same pattern, inherited from tetrapodomorphs.

We have a picture, then, of varied Devonian tetrapodomorphs, all hunters and most adapted to shallow-water habitats. None was adapted to be active out of water for any length of time. These creatures (Figure 9.5) show how their strong ventral fins begin to look like feet.

## From Tetrapodomorph to Tetrapod

Tetrapodomorph locomotion in shallow water and on shallow mudbanks would have been improved by stronger

Figure 9.6 Locomotion from fish to tetrapod. (a) Fishes swim by undulating the body significantly, while the head swings less. (b) The same basic body movements are used by a tetrapodomorph fish swimming or squirming over a mudbank, by an early tetrapod crawling, and by a salamander walking on dry land. No sudden or large shifts in locomotory mechanism were required for the transition, even though the fish has fins and the salamander has feet.

fins, especially stronger fin edges. Land locomotion consisted at first of the same undulatory twisting that salamanders still have, with the fins acting simply as passive pivots (Figure 9.6). The fins gradually exerted stronger traction on the substrate, which may have encouraged the multiple rays in the fins to become fewer and stronger until toed feet evolved. In the process, the pectoral fins came to support the thorax, while the pelvic fins came to be better suited to push the body forward. The pelvic fin evolved a hinge joint at a "knee" and a rotational joint at an "ankle," a pattern that

Figure 9.7 *Tiktaalik*, a tetrapodomorph from the Late Devonian of Arctic Canada. The skull is about 20 cm (8 in.) long.
*Source:* Photograph by Eduard Solà (Wikimedia).

Figure 9.8 *Tiktaalik*, reconstruction by Zina Deretsky, in conjunction with the scientific team that discovered *Tiktaalik*.
*Source:* Courtesy of the National Science Foundation.

persisted into tetrapods. This difference was inherited by all later vertebrates: elbows flex backward, knees flex forward. As the pectoral and pelvic girdles evolved better linkage with the fins, the fins evolved to become clearly defined limbs.

We use the word **tetrapod** to describe an animal that has feet rather than fins. The bones of tetrapod limbs, all the way to the toes, are coded by the same sets of genes that once coded for the bones in the fins of their fishy ancestors. This can be proved by studies of modern animals, where the same fundamental genes code for fins and for legs, and the sequence of expression of those genes determines the succession of bones in a modern fish from the root to tip of each fin, and in modern frogs or mammals from the shoulder to the tips of the fingers, and from the hip to the tips of the toes.

Other changes also took place as tetrapodomorphs evolved into tetrapods. A leathery skin evolved to resist water loss, and senses changed so they would work in the air. Ecologically, tetrapods and tetrapodomorphs divided up the habitat as they diverged. Derived tetrapodomorphs (evolving tetrapods) spent more and more time at and near the water's edge, sunning and basking, while basal tetrapodomorphs remained creatures of open water.

Since the year 2000, the number of links in the evolutionary chain of fishapods has increased (Figure 9.5) as new fossils have come to light. One of the most extraordinary is *Tiktaalik* from the Late Devonian of Arctic Canada (Figures 9.7 and 9.8). One of the key advances in *Tiktaalik* is that it has a neck, in other words a separation between the bones of the skull and shoulder girdle. It might seem odd to us but in fishes, the thin shoulder blade is attached to the bones of the skull. This was true of tetrapodomorphs such as *Osteolepis* and *Eusthenopteron* (Figure 9.4), but it worked well in stabilizing the shoulder girdle. This is fine in a swimmer which moves smoothly through the water but a land walker jolts along, and so the head must be freed from

the limbs or it would have shaken its (diminutive) brain too much. *Tiktaalik* then was a walker but also a swimmer. Perhaps it used its pectoral fins to push its head up to enable breathing in air. The ribs were broad and overlapping, as part of the same way of life. *Tiktaalik* was found in rocks that were laid down in broad coastal river systems.

## Limbs and Feet: Why Become Tetrapod?

Why would tetrapodomorphs, as fast-swimming predators in the water, have evolved lobe fins that increasingly looked and operated like the tetrapod limb? How would a tetrapodomorph have benefited from an ability to push on a resistant substrate, rather than using a swimming stroke in water? Why take excursions out into air rather than simply breathe air at the water surface? In other words, why would a tetrapodomorph become a tetrapod? There must have been strong evolutionary advantages.

One older view sounded smart and yet paradoxical: it was that tetrapodomorphs learned to walk so they could stay in the water. We know that climates were monsoonal in the Devonian of the Euramerican continent and so rivers and lakes were seasonal. As pools dried up, the idea was that the best walkers among the tetrapodomorphs would be most likely to find another pool. However, this seems unlikely. Modern animals in drying pools tend to stay put rather than striking off into parched country in the hope of finding more; it's simply a better bet for survival. So why did they move onto land?

### Fresh Food Supplies?

The obvious argument is that tetrapodomorphs went on land to secure more food. As we have seen in Chapter 8, land life had become diverse around the water margins by the Early and Middle Devonian. There were succulent plants with their rhizome systems in the water but extending further and further from the water. The tetrapodomorphs were not interested in eating plants but rather the little beasts that lurked in and on their stems. Fossils of the Early Devonian plants from Rhynie in Scotland have been found with ancestors of spiders, mites, crustaceans, and insects tucked in among the plant tissues. Other succulent and tasty animals were also on land in the damp areas among the plants, including worms, snails, and millipedes.

The evolution of strong lobe fins in tetrapodomorphs probably helped them to hunt small prey in shallow water by poling their bodies through and over mudbanks. The fins became powerful enough to support the weight of the fish, at least briefly, while it gasped and thrashed its way along. The brief exposures to air would not have been long enough to pose much danger of drying out, but they would have preadapted tetrapodomorphs for longer periods of exposure.

The tetrapodomorphs would gather in the murky, warm waters around these plants, snatching beasts from the water's surface and foraging ashore. The further they dared go, the more food they would get.

### Basking?

Another idea is that tetrapodomorphs went on land to improve their digestions. If they evolved the habit of sunning themselves on mudbanks to warm up their bodies, their digestion would have been faster than in the water. Other things being equal, they would have grown faster, matured earlier, and reproduced more successfully than their competitors did. Basking behavior would have been effective even if the fish exposed only its back at first, supported mainly by its own buoyancy. But such effectiveness would have encouraged longer and more complete exposure. Some fishes, and many living amphibians and reptiles, bask while they digest.

As a basking, air-breathing tetrapodomorph became more exposed, more of its weight would have rested on the ground, threatening to suffocate it by preventing the thorax from moving in respiration. The pectoral fins in particular would therefore have become stronger, to take more and more of the body weight during basking (Figure 9.8).

### Reproduction?

Maybe the key was to protect their spawn. The most vulnerable parts of the life cycle of a fish are its early days as an egg and hatchling. If some tetrapodomorphs could make very short journeys – even a meter or so to begin with – over land, or over very shallow water, they would have been able to find isolated pools or backwaters to spawn in. There would have been fewer predators in these side pools than in open water, and eggs and young would have survived better there. In much the same way and for the same reasons, salmon struggle to swim far upstream to spawn, and many freshwater fishes swim into seasonally flooded areas, such as rice paddies, to breed.

Isolated warm ponds would also have been ideal breeding grounds for small invertebrates such as crustaceans and insects, which would have formed a rich food supply for the young tetrapodomorphs. Then, reaching a size at which they could handle larger prey and that would give them some protection against being eaten themselves, the young tetrapodomorphs could make their way back to the main stream and take on their adult way of life as predators on fishes. Among young crocodiles today, the greatest cause of death (apart from human hunting) is being eaten by an adult crocodile. Crocodiles provide intensive parental care while their young are small. Tetrapodomorphs perhaps solved the same problem by arranging for their young to spend time away from other adults.

## The First Tetrapods: Devonian

The best-known and securely identified early tetrapods are *Ichthyostega* and *Acanthostega* from the Late Devonian of Greenland, both known from complete skeletons, and *Tulerpeton*, which is a pile of bones from Russia. Other less well-preserved "stem tetrapods" have been found within a 2 m.y. time span close to the Devonian/Carboniferous boundary (363 Ma). Small arthropods and plants were not a suitable food supply for these early tetrapods, which were all large (more than a meter long). These animals ate fishes in the water. After that, there are interesting variations.

The stem tetrapods had many digits. The number varied but it was not five. *Acanthostega* had eight toes and *Ichthyostega* had seven (Figure 9.9); *Tulerpeton* had six. The number of examples is limited but the loss of toes seems to be linked to the relative use of the foot in pushing on the bottom. *Tulerpeton* could have walked quite well on land, *Acanthostega* was much more adapted to life in water, and *Ichthyostega* was somewhere in between.

*Acanthostega* (Figures 9.10 and 9.11) had the most fish-like biology of the first tetrapods. Perhaps its ancestors had become ground-dwelling but it then went back to a more aquatic lifestyle. Its forelimbs were rather weak, its ribs did not curve round to support its weight well, and its eight-toed limbs were flipper-like. It may have been best adapted to eating fish in weed-choked shallows, and it may not have been able to support its weight for long (or at all!) out of water.

*Ichthyostega* had a massive skeleton (Figure 9.12) but was otherwise very much like Late Devonian

(a)

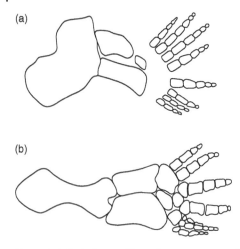

(b)

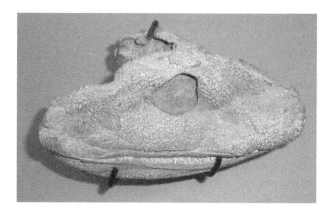

Figure 9.9 Early feet and toes, from two tetrapods from the Late Devonian of Greenland. (a) The front limb of *Acanthostega*. The limb clearly has (eight) toes but it looks more like a functional flipper rather than a walking foot. (b) The hind limb of *Ichthyostega*. There are seven toes and it is more like a foot than a flipper. *Source:* After Jenny Clack and Mike Coates.

Figure 9.11 A reconstruction of *Acanthostega*, showing it was very much a swimming tetrapod. *Source:* Artwork by Günter Bechly (Wikimedia).

Figure 9.12 The Late Devonian tetrapod *Ichthyostega*, a digital model based on CT scans, with transparent flesh to indicate the overall body shape. Note the broad, flat skull, short neck, broad ribs, and hands and feet with more than five digits.
*Source:* Artwork by Julia Molnar, courtesy of Stephanie Pierce, Jenny Clack, and John Hutchinson.

Figure 9.10 The heavy 12 cm long skull of *Acanthostega*. *Source:* Photograph by Ghedoghedo (Wikimedia).

tetrapodomorphs in spine, limb, tooth, jaw, palate, and skull structure, and probably in diet and locomotion. Like them, it had a tail fin but unlike them, it had a strong rib cage and limbs and feet rather than lobe fins. *Ichthyostega* solved the problem of supporting the chest for breathing on shore by having a massive set of ribs attached to the backbone.

As an adult, *Ichthyostega* was probably much like a living crocodile in ecology. In functional studies using digital models constructed from CT scans, Stephanie Pierce, Jenny Clack, and John Hutchinson showed that the limb girdles were not strong enough for walking. In fact, *Ichthyostega* could barely lift its body from the ground. So, they conclude that it was mostly a swimmer and when on land might have hauled itself about like an elephant seal (though at smaller size). Elephant seals

basically pull themselves around on the beach, using the strong front feet for propulsion and the hind feet as props and skids.

The aquatic hunting of *Ichthyostega* was aided by a unique ear structure. A large air-filled pocket in the skull probably amplified any underwater sound reaching it, then transmitted the signals through a long thin stapes bone to the inner ear.

## Tetrapods Diversify: Carboniferous

Once the first tetrapods evolved, they radiated quickly into a great variety of sizes, shapes, and ways of life. Current understanding of their phylogeny (Figure 9.13) is that the Devonian

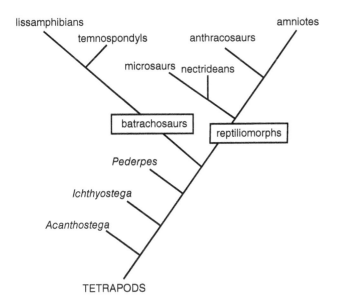

Figure 9.14 *Pederpes*, an early tetrapod from the Early Carboniferous of Scotland. About 1 m (3 ft) long. *Source:* Artwork by Dmitry Bogdanov (Wikimedia).

Figure 9.13 Outline evolutionary tree of the basal tetrapods, including the basal Devonian forms *Acanthostega* and *Ichthyostega*, followed by the Early Carboniferous *Pederpes*, and then the two great divisions, batrachomorphs, including extinct temnospondyls and living lissamphibians (frogs, salamanders), and the reptiliomorphs, including extinct microsaurs, nectrideans, and anthracosaurs and living amniotes.

forms like *Acanthostega* and *Ichthyostega* are at the base of the tree, followed by some Early Carboniferous forms and then a split into major groups, one leading eventually to modern amphibians such as frogs and salamanders and the other leading to amniotes, the reptiles, birds, and mammals. It might help first to understand a little about modern amphibians before exploring these early tetrapods.

Living amphibians are all small-bodied and soft-skinned, and in these respects are quite unlike early tetrapods. They are newts and salamanders, frogs and toads, and caecilians, which are burrowing legless amphibians. Living amphibians are usually classed together as "smooth amphibians" or Lissamphibia, comprising about 7200 species, mostly frogs. The oldest lissamphibian is a frog-like creature from the Early Triassic of Madagascar, and isolated fossils of frogs, salamanders, and caecilians are known from the Mesozoic and Cenozoic. Modern amphibians live both in water and on land, laying their eggs as spawn in ponds or rarely in hollows in trees and leaves. The young hatch as fish-like tadpoles with long tails, and eventually transition to the adult tail-less form that lives on land. In an odd way, they are still repeating today the water-to-land transition that their ancestors undertook back in the Devonian.

Tetrapods of the Early Carboniferous are known from few fossils, meaning there is a gap of some 25 m.y. between the Devonian tetrapods and abundant Carboniferous forms. Some of the best examples to fill

the gap have been found in Scotland, at sites such as East Kirkton, which dates to 335 Ma. East Kirkton was a complex tropical delta environment at the time, including shallow pools fed by hot springs. The tetrapods lived in the rivers and swamps near the pools, walked by them and sometimes fell into them or were washed into them. Additional finds included scorpions and millipedes, the earliest known harvestman, and several tetrapod groups.

An example of a Scottish Early Carboniferous tetrapod, from another locality, is *Pederpes*, a meter-long tetrapod (Figure 9.14). Its feet have only five toes, unlike earlier tetrapods. It still looked and probably behaved like a small crocodile, spending most of its time in the water. It is the first tetrapod with feet that are genuinely adapted for walking on land. Other tetrapods from these Scottish localities include examples of the two main groups: the **batrachosaurs** and the **reptiliomorphs**.

## Ancestors of Living Amphibians: Batrachosaurs

Batrachosaurs, meaning "frog-like" (Figure 9.13), include the **temnospondyls**, an entirely extinct group, and the lissamphibians, the living frogs and salamanders, as we have seen. The temnospondyls include 40 families and 160 genera from the Carboniferous. There are over 30 skeletons of temnospondyls in the East Kirkton collections.

Temnospondyls were large, with teeth like those of osteolepiforms and *Ichthyostega*. A typical example is *Capetus*, from the Late Carboniferous of the Czech Republic (Figure 9.15). It probably had habits much like a crocodile, feeding on fish with its broad jaws and low skull. The jaw was designed to slam shut on prey, and the skull was therefore strongly built. Temnospondyls might have hunted actively, pursuing their fishy prey, or they might have been lurkers, sitting on the bed of a lake or stream and waiting for a fish to swim by. On seeing the flash of the fish, the temnospondyl would have opened its jaw by lifting the skull rapidly; this would create a sudden suction force, drawing the fish in. No wonder the temnospondyls seemed to smile!

Figure 9.15 The Late Carboniferous temnospondyl *Capetus*, 1.5 m (5 ft) long. *Source:* Artwork by Dmitry Bogdanov (Wikimedia).

More terrestrially adapted temnospondyls, such as *Eryops* from the Early Permian of Texas, had very massive, strong skeletons capable of supporting them on land (Figure 9.16). Some temnospondyls even became more terrestrial during their lifetimes. For example, young *Trematops* had a jaw designed for eating small, soft food items but adults had a carnivorous jaw and a lightly built skeleton capable of rapid movement. As adults, they were probably land-going predators.

As part of their adaptation to life in air, temnospondyls had an ear structure that could transmit air-borne sound. The stapes bone is strong and seems to have conducted sound to an amplifying membrane that sat in a special notch in the skull.

Some Triassic temnospondyls were giants like *Metoposaurus* (Figure 9.17), at up to 3 m (10 ft) long. This group ranged globally, from Russia to Antarctica. Temnospondyls survived into the Cretaceous in parts of

Gondwana. The last big temnospondyl *Koolasuchus*, up to 5 m (16 ft) long, survived in rivers in Australia, no doubt hunting like a crocodile.

At some point in the Permian, a lineage of temnospondyls evolved into the ancestors of living amphibians. While most temnospondyls were large, the amphibian ancestors were small-bodied, and probably specialized in small-bodied prey, especially insects. Texas is usually the home of everything that is large but the Early Permian *Gerobatrachus*, at 11 cm (4.3 in.) long, is a small tetrapod from Texas that is of great importance. This little goggle-eyed animal (Figure 9.18) was dubbed the "frogamander" by its describer Jason Anderson, and it is the nearest temnospondyl relative of lissamphibians.

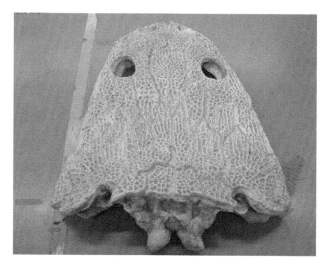

Figure 9.17 Skull of the Late Triassic temnospondyl *Metoposaurus*, viewed from above. Note the eyes set well forward and the heavily sculptured skull bones. These large tetrapods have been found in "mass graves," showing they lived in huge numbers, feeding on fish in warm lakes of Europe. *Source:* Photo by Jeyradan (Wikimedia).

Figure 9.16 Skeleton of the Early Permian temnospondyl *Eryops*, in the National Museum of Natural History, Washington. *Source:* Photo by Daderot (Wikimedia).

Figure 9.18 *Gerobatrachus*, the so-called frogamander, from the Permian of Texas. It is a possible temnospondyl ancestor of living amphibians. *Source:* Image by Nobu Tamura (Wikimedia).

Figure 9.19 *Microbrachis* is a microsaur, about 15 cm (6 in.) long, from the Late Carboniferous of the Czech Republic. It still had gills as an adult. *Source:* Image by Nobu Tamura (Wikimedia).

## Ancestors of Living Amniotes: Reptiliomorphs

The tetrapod branch that led eventually to amniotes, the reptiliomorphs, meaning "reptile-like," includes three or four groups of quite small Carboniferous and Permian tetrapods. Some of these small forms were aquatic and had reduced their limbs, sometimes losing them altogether, whereas others were fully terrestrial. For example, the **microsaurs** were small, with weakly calcified skeletons (Figure 9.19). They include many species of a variety of forms, some like small lizards and adapted to live largely in dry areas, and others much more aquatic and still retaining gills as adults (like grown-up tadpoles).

**Nectrideans** had a short body and a long, laterally flattened tail that made up two-thirds of their total length and was probably used for swimming. The vertebrae were linked in a way that allowed extremely flexible bending. Nectrideans probably swam like salamanders. The most famous nectridean is the horned *Diplocaulus* from the Early Permian of North America and Europe (Figure 9.20). The broad projections at the sides of its skull gave the head the shape of a swept-wing jet fighter, and this has been explained as an adaptation to bottom feeding. By raising its skull, the water current would lift the animal's head, which acted like an airfoil wing, perhaps important in maneuvering after speedy fish prey.

**Anthracosaurs** are the other large and diverse group of early tetrapods. There are a few anthracosaurs in the East Kirkton fauna. Most anthracosaurs were adapted for life primarily in water, as long-snouted and long-bodied predators, presumably crocodile-like fish eaters, with jaws designed for slamming shut on prey. Their limbs were not very sturdy but they may have been very good at squirming among dense vegetation in and around

Figure 9.20 *Diplocaulus* is a horned nectridean, about a meter (3 ft) long, from the Permian of Texas. It is difficult to imagine any other function for the horns than hydrodynamic control during swimming. *Source:* Image by Nobu Tamura (Wikimedia).

shallow waters. It's unlikely that they had the speed and power to compete with tetrapodomorphs in open water, even though some were quite large, up to 4 m (13 ft) long.

A few anthracosaurs were smaller, slender animals, adapted to terrestrial life. These tetrapods seem to be the closest relatives of early reptiles (Figure 9.13), even though their ears remained adapted for low-frequency water-borne sound. *Seymouria* (Figure 9.21) is a well-known example, known from many skeletons in the dry lands of the Early Permian in North America and Europe.

The consensus is that amniotes evolved somewhere near or in anthracosaurs, but no one hypothesis is strong. The problem is that amniotes began small, whereas most anthracosaurs were large. Since a major shift in habitat and ecology was probably involved here, convincing evidence is going to be difficult to find.

## Amniotes and the Amniotic Egg

**Amniotes**, the clade that includes reptiles, birds, and mammals, originated in the mid-Carboniferous from a small reptiliomorph. The name "amniote" refers to the

Figure 9.21 *Seymouria* was one of the few anthracosaurs well adapted for terrestrial life. It may be close to the origin of amniotes. *Source:* Photograph by Ryan Somma (Wikimedia).

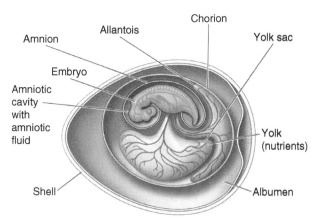

Figure 9.22 The amniotic egg protects the embryo in various ways, through the crystalline (or toughened) eggshell and its lining, the chorion. The yolk sac supplies food and the allantois collects waste. The embryo itself sits inside the amnion for further protection. *Source:* Image from Biology Pictures.

fact that they all have the amniotic egg, and that term is based on one of the internal membranes within the egg: the amnion.

The oldest fossil eggs of an amniote are known from much later, from the Early Jurassic, some 190 Ma, so how do we know that the Carboniferous amniotes, dating from 310 Ma, laid amniotic eggs? There are two reasons. First, amniotes are distinguished from other tetrapods not only by the amniotic egg but also by particular characters of the skull and ankle which can be seen in the fossils. Second, when we look at the eggs of all modern amniotes – including turtles, lizards, snakes, crocodiles, birds, and mammals – they all share exactly the same structure in detail. This means it is most likely that the amniotic egg arose once, and we track that back down the cladogram to the root, which is the first amniote of the mid-Carboniferous.

As we have seen, amphibians have successfully solved most of the problems associated with exposure to air but their reproductive system was and is linked to water, and it remains very fish-like. Almost all amphibians spawn in water and lay a great number of small eggs that hatch quickly into swimming larvae (tadpoles). The eggs do not need any complex protection against drying because if the environment dries, the larvae are doomed as well as the eggs. Thus, selection has acted to encourage the efficient choice of suitable sites for laying eggs, rather than devices to protect eggs. Both fishes and amphibians may migrate long distances for spawning, and favored sites are often disputed vigorously.

Living reptiles have a different system. Their juveniles hatch into air as competent terrestrial animals, often miniature adults. Yet the stages of embryological development are strikingly similar to those of amphibians. The difference is that reptiles develop for a longer time

inside the egg, which in turn means that the egg must be larger and must provide more food and other life-support systems. Reptiles typically lay far fewer eggs than amphibians of comparable body size, so they have evolved more complex adaptations to ensure greater chances of survival for each individual egg.

A reptile (**amniotic**) egg (Figure 9.22) is enclosed in a tough membrane, the **chorion**, covered by an outer shell made of leathery or calcareous material. The membrane and shell layers allow gas exchange with the environment (water vapor, $CO_2$, and oxygen) for the metabolism of the growing embryo, but also resist water loss. Reptiles lay eggs on land, so eggs are not supported against gravity by water. Instead, the shell gives the egg strength, protects it, holds it in a shape that will allow the embryo room to grow freely, and buffers it against temperature change and desiccation. Birds' eggs are much like reptile eggs, usually with harder shells, and in most mammals the whole egg (without a shell) is nurtured internally so that the embryo emerges from the amnion at the time it emerges from the mother ("live birth").

Inside the amniotic egg, the embryo is nourished by a large **yolk** and waste material is stored in the **allantois**, a bag that expands as the embryo grows. The most fundamental innovation, however, is the evolution of another internal fluid-filled sac, the **amnion**, in which the embryo floats. The amniotic egg acts toward the embryo like a spacecraft nurturing an astronaut in an alien environment: it has food storage, fuel supply, gas exchangers, and sanitary disposal systems.

Because the embryo inside an amniotic egg is encased in membranes, and often inside a shell, the female's eggs must be fertilized before they are finally packaged. Internal fertilization must have evolved along with the amniotic egg.

The evolution of the amniotic egg broke the final reproductive link with water and allowed tetrapods to take up truly terrestrial ways of life. Its evolution demanded changes in behavior patterns and in soft-part anatomy and physiology. The amniotic egg probably evolved in an early tetrapod that looked like a little reptile. The evolutionary model for the origin of amniotes is probably similar to the reasons that tetrapods first went onto the land – to find new food sources and to escape competition.

During the Devonian and Carboniferous, plants evolved to survive further and further from permanent water. Soils developed, and plants evolved roots to capture moisture. With the expansion of the amount of landscape covered in plants, animals of various kinds followed. The first amniotes then were exploiting new food sources that other early tetrapods could not. To make the move, the first amniotes had to evolve the protective cover over their eggs, together with internal fertilization and direct development of the young.

The major problem in laying unshelled eggs away from water is drying, at spawning time and during development. A crude sort of egg membrane would have been a partial solution to the developmental problem. Further refinement of the system is then fairly easy to imagine. Internal fertilization was probably a preliminary solution to the problem of desiccation while spawning. (Some living amphibians have independently evolved a crude kind of internal fertilization.)

Direct development, where the tadpole stage is cut out, might seem harder to evolve. However, modern frogs show all kinds of intermediate stages. In some species, the mother or father carries the tadpoles in special brood pouches, so they develop from egg to tadpole and to adult form, but not swimming freely in a pond. A few frogs today lay large eggs, 10 mm across, which hatch into miniature adults. These eggs show no sign of evolving toward amniotic eggs but they show that some living amphibians can lay large eggs that then develop without the larval stage. Presumably, as the amniotic egg evolved, the reproductive problems faced today by living frogs were avoided by simply allowing the embryo to develop longer and longer inside the egg. Longer development could, of course, have evolved gradually along with increased size and complexity of the egg.

In this chapter, we have followed the evolution from fish to tetrapod, and from basal tetrapod ("amphibian") to amniote. In both cases, evolutionary change is linked with successful reproduction, driven by exploitation of expanding food supplies on land.

## Why Were the First Amniotes Small?

All early amniotes were about the same size as living lizards, and much like them in body proportions, posture, and jaw mechanics (Figure 9.23). They were probably like them in ecology too. They had a notably small skull with a short jaw well suited to hold, bite, and crush small, wriggling prey, and to shift the grip for repeated bites. The small head was set on a neck joint that allowed very swift three-dimensional motion. But why small size?

Carboniferous forests were rich in worms, insects, and grubs. Young or small animals would have been best suited for foraging after this kind of food. Worms, insects, and grubs are small, though highly nutritious; they are easy to seize, process, and swallow; and they can be found among cracks and crevices in a maze of plant growth in a complex three-dimensional forest. Most of these potential prey items are slow-moving, and a successful predator need not have been quick and agile at first. But small and light-bodied animals could have quickly evolved greater agility as their repertoire of prey extended to the expanding number of large insects in Carboniferous forests.

Small body size may also have been favored by a thermoregulatory effect. Animals encounter greater temperature extremes in air than they do in water, and small animals can shelter more easily from chilling or overheating among vegetation, in cracks and crevices, or in hollow tree trunks than can animals the size of *Ichthyostega*. Small bodies are also quicker and easier to heat by basking in the sun. Again, this suggests that juvenile or small animals would have been adapted to survive away from water.

A scenario of amniote evolution, mostly due to Robert Carroll, is that they evolved on a forest floor covered with rotting material, leaf litter, fallen branches, and tree stumps, ideal places for prey to hide and amniotes to search (Figure 9.24).

The best candidates for the first amniotes are found in the Late Carboniferous. Many tree stumps and tree trunks

Figure 9.23 Life reconstruction of the early amniote *Hylonomus* as an analog of a living lizard. Its fossils were found inside upright fossil tree trunks in Late Carboniferous rocks of Nova Scotia, Canada. *Source:* Artwork by Nobu Tamura (Wikimedia).

Figure 9.24 Reconstruction of a Carboniferous swamp forest. Trees were tall but often had shallow roots and weak structure, so they were frequently felled by storm and flood. A rich fauna of insects, spiders, and other arthropods lived in this ecosystem, and it is likely that small early reptiles lived as much in the tree tops as under them. This scene has much growth on and near the forest floor because it is near natural open spaces around water bodies. *Source:* Artwork by Mary Parrish, under the direction of Tom Phillips and William DiMichele, and used by permission.

(a)  (b)

Figure 9.25 (a) *Sigillaria*, a fossil tree found still standing upright in rocks to the north of the town of Stanhope, County Durham, England. It was removed and remounted in 1862 in Stanhope, next to the Church of St Thomas. *Source:* Photograph © Andrew Curtis (Wikimedia). (b) A famous engraving from J. W. Dawson's 1868 account of the geology of Nova Scotia, Canada, the first scientific account of a tree fossilized upright.

have been fossilized upright in life position in rocks associated with coal forests (Figure 9.25), and amniotes have been found in Nova Scotia, Canada, preserved inside some of the hollow stumps. *Hylonomus* (Figure 9.23) is one example. This may not be a freak of preservation. The amniotes may have lived inside hollow tree trunks, as little insectivorous mammals do today in tropical rainforests, or perhaps they sheltered in the hollow stumps during storms or were washed into them in floods.

Whichever suggestion one prefers, amniotes were feeding in the canopy forest in the Late Carboniferous. Vertical climbing is easy with a small body size, so small Carboniferous vertebrates could have been tree dwellers, as many salamanders are today. Trees offer damp places in which to lay eggs, and rich insect life high in the canopy forest provides abundant food. Even today, salamanders (and spiders) are the top carnivores in parts of the Central and South American canopy forest. The rich fossil record of Late Carboniferous insects, scorpions, spiders, and amniotes may reflect the ecosystem of the canopy rather than the forest floor.

## Carboniferous Land Ecology

Little is known yet about the land ecology of Early Carboniferous times; the East Kirkton fossils are the best-known tetrapod fauna from this time, and they are preserved in an unusual setting. All the Devonian and Carboniferous tetrapods so far discovered lived close to the Equator.

The evidence is much better when we turn to the Late Carboniferous. Late Carboniferous coalfields have been intensely studied for economic reasons, yielding a lot of information that gives us a good picture of the flora and global paleoecology of the time. Swamp forests in tropical lowlands were dominated by lycopods, and the vegetation and organic debris deposited in oxygen-poor water formed thick accumulations of peat, now compressed and preserved as giant coal beds stretching from the American Midwest to the Black Sea.

By the time all this carbon was buried, there may have been high levels of oxygen in the seas and atmosphere of the Carboniferous. The evolution of flight in insects, a very fuel-intensive activity, may have been made possible by a richly oxygenated atmosphere but at the moment both the data and the inference are very speculative.

There were no herbivores among early land arthropods, possibly because of lignin (Chapter 8). This universal substance in vascular plants is formed through biochemical pathways that include toxic substances which are often stored in cell walls and dead plant tissue. From the Silurian to the Late Carboniferous, lignin and its associated biochemistry probably protected vascular plants from potential herbivores.

But eventually, of course, both invertebrates and vertebrates made the breakthroughs that allowed direct herbivory. Bacteria and fungi can break down the toxins in dead plants, and it's possible that symbiosis (see Chapter 3) with one or both allowed some animals to eat living plant material for internal enzyme-assisted digestion. Also, early land plants evolved larger sporangia and seeds (see Chapter 8) that were very nutritious and low in toxins, and therefore susceptible to attack by arthropods. Insects quickly evolved the anatomy to feed on the reproductive tissues of plants. Seeds may have evolved not only for better waterproofing of the embryo but also to deter insect predation.

The first insect is Devonian but the dominant fact of early insect evolution is the explosive radiation of winged insects in the Late Carboniferous, about 325 Ma. Some had mouthparts for tearing open primitive cones, and their

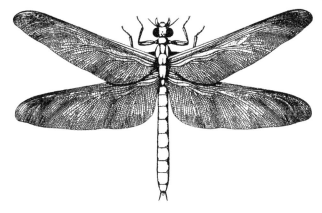

**Figure 9.26** Reconstruction of the giant Carboniferous dragonfly *Meganeura*, which had a wingspan up to 60 cm (2 ft).
*Source:* Diagram by Dodoni (Wikimedia).

guts were sometimes fossilized with masses of spores inside. Others had piercing and sucking mouthparts for obtaining plant juices. Overall, it seems that leaf eating was rare among early insects; instead, they ate plant reproductive parts, sucked plant juices, or ate other insects. Gigantic dragonflies were flying predators on smaller arthropods; Late Paleozoic dragonflies were the largest flying insects ever to evolve, with wingspans up to 60 cm (Figure 9.26).

Explosive evolution had occurred among land-going invertebrates by the Late Carboniferous, much of it linked with the evolution of herbivory among insects; 137 genera of terrestrial arthropods are recorded from the Mazon Creek beds of Illinois, including 99 insects and 21 spiders, with millipedes present also. Most of the living groups of spiders had evolved by the Late Carboniferous, with only the sophisticated orb-web spiders missing. Centipedes were important predators.

Millipedes are important forest recyclers today, feeding on decaying plant material. They include flattened forms that squirm into cracks in dead wood and literally split their way in, reaching new food and making space for shelter and brood chambers at the same time. Carboniferous millipedes reached half a meter in length and a giant relative, *Arthropleura*, reached 2.3 m (7 ft) long, and 50 cm (18 in.) across. The gut contents of *Arthropleura* suggest that it ate the woody central portion of tree ferns.

These lush forests across the Euramerican continent disappeared quite rapidly just before the end of the Carboniferous, about 305 Ma, and that marked the end of many of the plants and animals we associate with the classic Carboniferous coal forest. In particular, the damp-loving temnospondyls, nectrideans, and anthracosaurs had to become adapted to much more limited forests and more extensive dry lands. As Sarda Sahney suggested, this marks the changeover from the first phase in tetrapod evolution to a new phase dominated by amniotes in the Permian, as we shall explore in Chapter 10.

## Further Reading

Anderson, J.S., Reisz, R., Scott, D. et al. (2008). A stem batrachian from the Early Permian of Texas, and the origin of frogs and salamanders. *Nature* 453: 515–518. [*Gerobatrachus*].

Clack, J.A. (2004). From fins to fingers. *Science* 304: 57–58. [Who ever said fish don't have fingers?].

Clack, J.A. (2012). *Gaining Ground: The Origin and Evolution of Tetrapods*, 2e. Bloomington: University of Indiana Press. [A very well-written and complete picture of research by one of the leading investigators of the earliest tetrapods].

Clack, J.A., Bennett, C.E., Carpenter, D.K. et al. (2016). Phylogenetic and environmental context of a Tournaisian tetrapod fauna. *Nature Ecology and Evolution* 1: 0002. [An overview of the fossils and ancient environments of early tetrapods].

Daeschler, E.B., Shubin, N.H., and Jenkins, F.A. Jr. (2006). A Devonian tetrapod-like fish and the evolution of the tetrapod body plan. *Nature* 440: 757–763. [*Tiktaalik*].

Janis, C.M. and Farmer, C. (1999). Proposed habitats of early tetrapods: gills, kidneys, and the water–land transition. *Zoological Journal of the Linnean Society* 126: 117–126. [Ideas about physiology through the water to land transition].

Pierce, S.E., Clack, J.A., and Hutchinson, J.R. (2012). Three-dimensional limb joint mobility in the early tetrapod *Ichthyostega*. *Nature* 486: 523–526. [Application of CT scanning and computerized 3D reconstruction to find new details of anatomy and function].

Sahney, S., Benton, M.J., and Falcon-Lang, H.J. (2010). Rainforest collapse triggered Pennsylvanian tetrapod diversification in Euramerica. *Geology* 38: 1079–1082. [Evidence that the dramatic climatic switch from humid to dry near the end of the Carboniferous drove tetrapod evolution].

Schoch, R.R. (2014). *Fossil Amphibians: The Life of Early Land Vertebrates*. New York: Wiley. [An excellent account of everything about fossil amphibians].

## Question for Thought, Study, and Discussion

Summarize the biology of the early tetrapods on land, including the need for them to lay their eggs in water. How much of the land surface of the world could they survive in? Then think about the early amniotes, who laid their eggs on land. Clearly, they could succeed in different environments but could they expand over a lot of other land environments? Summarize the problems in geographic and climatic expansion of the early amniotes.

# 10

## Early Amniotes and Thermoregulation

### In This Chapter

The radiation of amniotes continued into the Permian and by the end of the Permian, land vertebrates existed on all the continents and from pole to pole. Permian climates showed dramatic changes, from being dominated by the great Gondwanan glaciation at the start of the Early Permian, through increasing warmth and dryness in the Middle and Late Permian. The dominant amniotes were the synapsids (mammal ancestors), which included two main diversifications in the Permian: the basal forms, or "pelycosaurs," and the more advanced therapsids. We discuss the pelycosaurs, an early group of synapsids that reached large body sizes as predators and as browsing herbivores. Any ver-tebrates that are fully adjusted to life on land must be adapted toward changing temperatures, and there has been a lot of discussion about thermoregulation in pelycosaurs. In particular, pelycosaurs like *Dimetrodon* probably used a great bone-supported sail on their backs to help regulate their body temperature. The therapsids diversified as small, medium, and larger forms, and there were some truly huge plant eaters among the dinocephalians and dicynodonts. These lived side by side with equally diverse pareiasaurs, nonsynapsid parareptiles. The Permian–Triassic mass extinction wiped out these diverse and complex terrestrial ecosystems.

## The Amniote Radiation

The first amniotes evolved in the Carboniferous in warm, humid, tropical regions along the southern shores of the great northern continent Euramerica. Life away from such swamps and forests demands adaptations for dealing with seasons, where temperature, rainfall, and food supply vary much more and are less predictable than in the tropics. In many ways, such challenges to early land vertebrates were simply extensions of the problems involved in leaving the water. In this chapter, we will follow the early history of amniotes and discuss the adaptations that allowed them their great terrestrial success.

Amniotes came to be dominant large animals in all terrestrial environments in the Permian. The radiation probably began in Euramerica, because hardly any land vertebrates are known from Siberia, from East Asia, or from the whole of Gondwana before the Middle Permian. Seas and mountain ranges may have blocked land migrations or problems of thermoregulation may have confined land vertebrates to the tropics of Euramerica until the Middle Permian. The invasion of other continents and/or climates was accompanied by a spectacular evolution of varied body types. Since amniotes rather than amphibians radiated so successfully in the Permian, perhaps it was their solution to thermoregulatory problems that allowed them to invade regions in higher latitudes.

Three major groups of amniotes diverged in the Late Carboniferous and Early Permian. The earliest amniotes had **anapsid** skulls (they had no openings behind the eye) (Figure 10.1a). This character was inherited from fishes and early tetrapods. The two major amniote clades have derived or advanced skull types, in which there are one or two large openings behind the eye socket. **Synapsids** (with one skull opening behind the eye socket) (Figure 10.1b) diverged first, from an anapsid ancestor that we have not yet identified. Synapsids dominated

*Cowen's History of Life*, Sixth Edition. Edited by Michael J. Benton.
© 2020 John Wiley & Sons Ltd. Published 2020 by John Wiley & Sons Ltd.

(a)

(b)

(c)

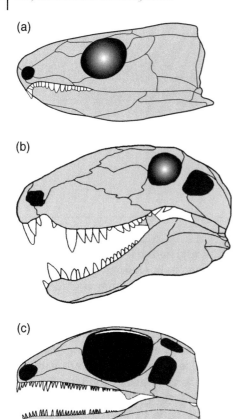

**Figure 10.1** Three different skull types among amniotes are defined by the number of holes in the skull behind the eye socket. (a) Anapsid (no holes), represented by *Captorhinus*. (b) Synapsid (one hole), represented by *Dimetrodon*. (c) Diapsid (two holes), represented by *Petrolacosaurus*. *Source:* Images (a) and (b) are © Dalton Harvey and used by permission.

**Figure 10.2** Reconstruction of the earliest well-known diapsid, *Petrolacosaurus* from the Late Carboniferous of Kansas. It was about 40 cm long (16 in.) and looked externally like a modern lizard. *Source:* Image by Nobu Tamura (Wikimedia).

The major radiation of diapsids took place much later, in the Triassic (see Chapter 12). The dominant Late Carboniferous and Permian reptiles were synapsids, including five of the six other amniotes found with *Petrolacosaurus*. We focus in this chapter on the synapsids, looking first at the basal forms, informally termed "pelycosaurs," and then the derived therapsids, which came on the scene in the Middle and Late Permian.

## Pelycosaurs

The early synapsids, or **pelycosaurs**, include famous forms such as the sail-backed Early Permian *Dimetrodon* (Figure 10.1b). They were the key land animals in the Late Carboniferous and Early Permian, comprising over 50% and over 70% of amniotes during those time intervals. Despite their variety, early pelycosaurs are rare. The earliest one is *Archaeothyris*, found in the fossil tree trunks of Nova Scotia. It was small and lizard-like but it had the characteristic synapsid skull. All early reptiles, in summary, were small insect eaters. The pelycosaurs were the first to evolve to larger size, and perhaps because of that evolved into groups that were more abundant as fossils, varied in diet, and more widespread geographically than the other reptiles – for a while at least.

## Pelycosaur Biology and Ecology

### Locomotion

Pelycosaurs are well enough known that we can reconstruct how they walked. The massive front part of the body was supported by a heavy, sprawling fore limb (Figure 10.3). The lighter hind limb had a greater range of movement, although it was also a sprawling limb. There was no well-defined ankle joint, and the toes were long and splayed out sideways as the animal walked. Thus, the feet provided no forward thrust but

Late Paleozoic land faunas. They include the Late Paleozoic pelycosaurs and their descendants, the therapsids and mammals. Synapsids never evolved the water-saving capacity to excrete uric acid rather than urea, a character that all other surviving amniote groups share.

**Diapsids** are amniotes with two skull openings behind the eye socket (Figure 10.1c). They include the dominant land-going groups of the Mesozoic (including dinosaurs and pterosaurs) and all living amniotes except mammals (that is, reptiles and birds). Turtles have no skull openings so are technically anapsid, but their ancestors were diapsids that lost the two skull openings. Turtles evolved in the Middle Triassic (see Chapter 11).

The earliest well-known diapsid is *Petrolacosaurus* (Figure 10.1c), which looked like and probably lived like a lizard (Figure 10.2). Compared with later diapsids, *Petrolacosaurus* had a heavy ear bone, the stapes, that could not conduct air-borne sound. As in most early tetrapods, the massive stapes probably transmitted ground vibrations through the limb bones to the skull.

Figure 10.3 Skeleton of the pelycosaur *Ophiacodon* showing the heavy front quarters and fore limbs, and the powerful, driving hind limbs. *Source:* Photo by Smokeybjb (Wikimedia).

simply supported the limbs on the ground. The fore limbs were entirely passive supports that prevented the animal from falling on its face, while the hind limbs provided all the forward thrust in walking with powerful muscles that rotated the femur in the hip joint.

Think of two children (or adults) playing "wheelbarrow." The propulsion and steering are both from the rear, and the wheelbarrow is pushed along, the arms simply supporting weight. The whole construction of two people is stable only as long as the leader stays stiff. Spinal flexibility is important to many swimming animals, particularly those that actively pursue fishes. But in pelycosaurs, a strong stiff backbone prevented the body from collapsing in the middle under its own weight, and allowed thrust from the hind limb to be converted directly into forward motion. Therefore, most pelycosaurs were predominantly terrestrial animals. If they swam, they were slow swimmers that hunted by stealth rather than speed.

## Carnivorous Pelycosaurs

Early pelycosaurs were all carnivorous: they all have the pointed teeth and long jaws of predators. Two groups remained completely predatory. **Ophiacodonts** became quite large. *Ophiacodon* itself was over 2 m (7 ft) long and probably weighed over 200 kg (450 pounds). Many ophiacodonts have long-snouted jaws with many teeth set in a narrow skull. The hind limbs tended to be longer than the fore limbs.

Ophiacodonts may have hunted fishes in streams and lakes of the swamps and deltas of the Late Carboniferous and Permian (Figure 10.4), although they were perfectly capable of walking on drier, higher ground and, like crocodiles, their prey may well have included terrestrial animals coming down to the water to drink. Their general

Figure 10.4 *Ophiacodon*, 2 m (7 ft) or more in length, from the Early Permian of Texas. *Source:* Reconstruction as a fish-eating pelycosaur by Dmitry Bogdanov (Wikimedia).

lack of spinal flexibility may suggest that they were slow swimmers, possibly eating more tetrapods than fishes.

**Sphenacodonts** were specialized carnivores on land. Many of their skull features betray the presence of very strong jaw muscles, and the teeth were powerful. They were unlike typical early amniote teeth in that they varied in shape and size and included long stabbing teeth that look like the canines of mammals. The sphenacodont body was narrow but deep, and the legs were comparatively long. Both of these characters suggest that sphenacodonts were reasonably mobile on land.

The earliest sphenacodont was *Haptodus* (Figure 10.5), a little less than a meter long and fairly lightly built. Similar forms existed throughout the Permian but later sphenacodonts were much larger. The group is best known from spectacular fossils of *Dimetrodon* which show the classic spines supported on its dorsal vertebrae (Figure 10.6). (We will discuss ideas that these

Figure 10.5 *Haptodus*, the earliest sphenacodont, from the Late Carboniferous of Kansas. *Source:* Reconstruction by Dmitry Bogdanov (Wikimedia).

Figure 10.6 *Dimetrodon*, the best-known sphenacodont, from the Permian of Texas. *Source:* Photograph by H. Zell (Wikimedia).

structures might have been used in controlling temperature below.)

Evolution within carnivorous pelycosaurs reflected their prey capture. The jaws slammed shut around the hinge, with no sideways or front-to-back motion for chewing. With this structure, a long jaw made it easier to take hold of prey but the force exerted far from the hinge was not very great. Small prey could perhaps be killed outright by slamming the jaw on them.

In ophiacodonts, which may have hunted in water for fish, the difficult part of feeding would have been seizing the prey; their teeth were subequal in length in a long, narrow jaw. Most fish eaters swallow their prey whole.

In sphenacodonts, which were terrestrial carnivores, the head was bigger and stronger. Long, stabbing teeth were set in the front of the jaw (Figure 10.1b). Struggling prey could be held between the tongue and some strong teeth set into the palate, and could be subdued by powerful crushing bites from the teeth at the back of the jaw. Robert Carroll suggested that the success of pelycosaurs in the Carboniferous and Permian, compared with diapsids, was due to their massive jaw muscles, which

were strong enough to hold the jaws steady against the struggles of large prey. Carnivorous pelycosaurs thus could become large predators, not simply small insectivores.

## Vegetarian Pelycosaurs

Carroll's suggestion cannot be the whole story, because there were also vegetarian pelycosaurs. Caseids and edaphosaurs were the first abundant large terrestrial animals, and were among the first terrestrial herbivores. They had similar body styles, presumably because they were similar ecologically. They had about the same range of body size as the carnivorous sphenacodonts but they had smaller, shorter heads that gave more crushing pressure at the teeth. There were no long canines and the teeth were short, blunt, and heavy. In addition, smoothing of the bones at the jaw joint allowed the lower jaw to move backwards and forwards slightly, grinding the food between upper and lower teeth. Caseids ground their food between tongue and palatal teeth, while edaphosaurs had additional tooth plates in their lower jaw that they used to grind food against palatal teeth. Vegetation is low-calorie food compared with meat, so herbivores need a large gut to contain a lot of food (Figure 10.7). As one would expect, the bodies of all these vegetarian pelycosaurs were wide to accommodate a large gut. The limb bones were short but heavy.

**Caseids** were more numerous than edaphosaurs. They included *Cotylorhynchus*, which was over 3 m long (10 ft), and weighed over 300 kg (650 pounds) (Figure 10.7).

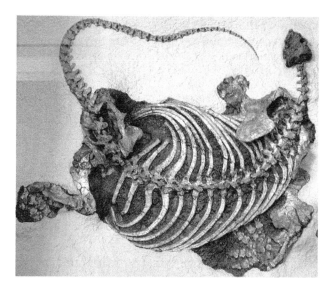

Figure 10.7 *Cotylorhynchus*, a big caseid from the Permian of Texas. The small head and capacious gut mark it as a vegetarian, even without looking at the teeth. *Source:* Photograph by Ryan Somma (Wikimedia).

Figure 10.8 *Ianthasaurus* is the earliest edaphosaur, from the Late Carboniferous of Kansas. It was a small insect eater, not a vegetarian, about 15 cm long (6 in.). *Source:* Image by Dmitry Bogdanov (Wikimedia). (He forgot to draw the rock that *Ianthasaurus* was sunning itself on!)

Figure 10.9 *Edaphosaurus* from the Permian of Texas was a big vegetarian pelycosaur, about 3 m (10 ft) long, and weighing 300 kg (over 600 pounds). *Source:* Photograph © Ken Angielczyk/Field Museum, used by permission.

Caseids had small heads for their size, which perhaps implies that they did not chew very much, and perhaps had powerful digestive enzymes or gut bacteria to help break down plant cellulose.

The earliest **edaphosaur**, *Ianthasaurus*, was not a vegetarian but a small insect eater (Figure 10.8). It had a small sail on its back and sharp, pointed predatory teeth. Vegetarian edaphosaurs evolved in the Early Permian and are best known from *Edaphosaurus* itself (Figure 10.9), which carried a very large sail made of vertebrae extended into spines, rather like *Dimetrodon*. Unlike *Dimetrodon*, the spines carry small cross-bars, whose function is unknown, but this is evidence that the two sails probably evolved independently.

## How Does Herbivory Evolve in Tetrapods?

Most plant material is difficult to digest. Vertebrates can break down cellulose only if they chew it well and have some way of enlisting fermenting bacteria as symbionts (see Chapter 3) to aid digestion. Living vegetarians do this; for example, cattle, horses, and rabbits have gut bacteria in the digestive tract. Any vertebrate that begins to eat comparatively low-protein plant material must process large volumes of it, and so must have a rather large food intake at a rather large body size. Some plant material is high in protein or sugar, especially the reproductive parts, but only a small animal can selectively feed on plant parts.

In other words, there are only two possible evolutionary pathways toward herbivory. One of them begins with animals that are small, active, and selective in their food gathering, eating high-calorie foods such as juices, nectar, pollen, fruits, or seeds from plants. Examples today are small mammals, hummingbirds, and insects. If an animal then enlists gut bacteria as symbionts, however, the diet can contain more and more cellulose, and a larger vegetarian can evolve, as in many mammal groups, including leaf-eating monkeys and gorillas. Large birds can also be herbivores: the extinct moas of New Zealand are good examples. *Ianthasaurus* suggests that edaphosaurs evolved herbivory this way.

The other pathway begins at rather large body size with rapid and rather indiscriminate feeding, possibly omnivory, so that a large volume of low-calorie food can be processed. Bear-like mammals are examples of a group in which some members have evolved away from a carnivorous way of life toward omnivory and then to a completely vegetarian diet, as in pandas.

Because vegetarianism depends so much on body size, diets must change with growth. Most living reptiles and amphibians change their diet as they grow. Food requirements and opportunities change as they reach greater size and can catch a different set of prey. Among living reptiles, small, and young iguanas are carnivorous or omnivorous, while large iguanas are largely vegetarian but take meat occasionally. Living amphibians today are almost all small and carnivorous.

What plant food was available for the first herbivores? As we have seen in Chapter 9, the giant Carboniferous coalfields of the Euramerican continent provide evidence of rich plant communities, and these were first exploited only by insects and other arthropods. Large terrestrial herbivores appeared in the Late Carboniferous of Euramerica. The anthracosaur *Diadectes*, for example, ranged up to as much as 4 m (13 ft) long as an adult, and synapsids of this size were also common. Large herbivores appeared at the same time as a major change in plants, when upland plants replaced the coal swamp forests.

Why were tetrapods relatively slow to evolve herbivory? First, because the wet tropical forest in which they evolved is a poor habitat for ground-dwelling herbivores. As in today's tropical forests, most leaves were in the canopy and leaf litter was broken down quickly by fungi and arthropods. The first vertebrate herbivores could not have found much green material on the floor of the coal forests (see Chapter 9, Figure 9.15). Vertebrates could not have evolved herbivory until they could survive well on the forest margin, away from the watery habitats most likely to be preserved.

Second, any large-bodied vegetarian eats large volumes of low-calorie plant material and needs gut bacteria to help digest the cellulose. Gut bacteria work well only in a fairly narrow range of temperature, so an additional requirement for the first successful large-bodied vegetarians was some kind of thermoregulation.

## Thermoregulation in Living Reptiles

Body functions are run by enzymes, which are sensitive to temperature. Other things being equal, each enzyme works best at an optimum temperature; at any other body temperature, they are less efficient – in digestion, in locomotion, in reaction time, and so on. Birds and mammals have a sharp peak of efficiency that drops off radically with a small rise or fall in body temperature. Reptiles are called cold-blooded but in fact they take on the temperature of their surroundings, so can be warm or cold. Their bodies can function over quite a range of

internal temperature but they also have an optimal temperature, and reptiles try to control it at that level by **behavioral thermoregulation**.

The basics of behavioral thermoregulation are simple (Figure 10.10): the reptile creeps out in the cool morning and suns itself on a rock, picking up heat directly from the Sun above and reflected from the rock below. Toward noon, as it gets hot, the lizard might scuttle into the shade so it can radiate heat to the air and rock.

Generally, reptiles try to maintain their body temperatures at the highest level that is consistent with safety and cost. Although it takes energy to stay warm, the higher activity levels that are possible at higher temperatures give greater hunting or foraging efficiency, greater food intake, faster digestion, and faster growth; remember the section on basking in Chapter 9. As long as the climate is warm and food supply is abundant enough to fuel a reptile, thermoregulation that produces or maintains warm body temperature gives a net gain in reproductive rate and so is selectively advantageous. The same principles should apply to all cold-blooded vertebrates.

Body size is a vital factor in thermoregulation. Small bodies have a low mass with a relatively large surface area. Small reptiles bask in the sun, sit in the shade, hide in burrows or in leaf litter, or exercise violently (often with push-ups) to change their body temperatures. Their small mass allows them to respond quickly to temperature changes by behavioral means, giving them sensitive control over their body processes. Large reptiles have a natural resistance to temperature change because of

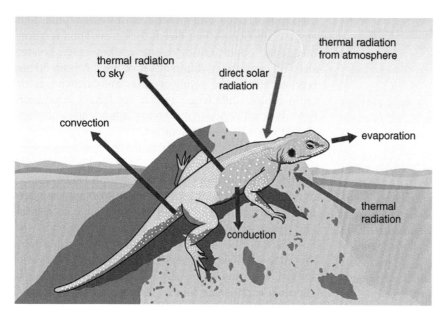

Figure 10.10 **Behavioral thermoregulation.** A basking lizard showing how heat is absorbed from the Sun above and the ground below when it is cold, and how heat is transmitted when it becomes overheated. *Source:* Artwork from OpenLearn.

their mass: it takes a long time to heat them up or cool them down (just as it takes a long time to boil a full kettle of water).

Behavioral thermoregulation is more energy consuming and much less responsive for larger reptiles than for smaller ones. So large reptiles today live in naturally mild tropical climates with even temperatures day and night and season to season (like the large monitor lizards of Indonesia, Australia, and Africa), or they live near or in water, which buffers any changes in air temperature (like crocodiles and alligators, which even so are never found far outside the tropics). There are no large lizards at high latitudes.

## Thermoregulation in Pelycosaurs

The spectacular pelycosaur *Dimetrodon* was large, over 3 m long. It evolved very long spines on some of its vertebrae, forming a row of long vertical spines along its back (Figure 10.6). In life, the bones were covered with tissue to form a huge vertical sail (Figure 10.11). Most people think that the sail was used for thermoregulation.

Here is the simplest version of the story. *Dimetrodon* was too large to hide from temperature fluctuations (in a crevice or tree stump or burrow, for example). It probably used its sail to bask in the early morning and the late afternoon, turning its body so that the sail intercepted the sunshine. By pumping blood through the sail, it could collect solar heat and transfer it quickly and efficiently to the central body mass (solar panels work this way to heat water). Once warm and active, *Dimetrodon* would face no further problem unless it overheated. It could shed heat from the sail by the reverse process, turning the sail end-on to the Sun. At night, heat would be conserved inside the body by shutting off the blood supply to the sail.

The sail, as an add-on piece of solar equipment, allowed rapid and sensitive control over body temperature.

Figure 10.11 Reconstruction of *Dimetrodon* to show the spines covered with tissue to form a "sail." *Source:* Artwork by Dmitry Bogdanov (Wikimedia).

Enzyme systems could have been fine-tuned to work at high biochemical efficiency within narrow temperature limits, and the animal could have foraged even in environments where air temperatures fluctuated widely. The activity levels, locomotion, and digestive systems of *Dimetrodon* were all improved. Smaller reptiles that lived alongside them would have been able to heat up quickly in the morning, simply because they were smaller, and it would have been important for *Dimetrodon* to be equally active at that time, for effective hunting.

Some pelycosaurs did not have a sail at all, and the small pelycosaur *Ianthasaurus* had only a small sail (Figure 10.8). Young *Dimetrodon* had a small sail too. The area of the sail was related to body size, which makes sense if it was used for behavioral thermoregulation like living lizards. A large body warms and cools slowly. Birds and mammals burn large amounts of food in a built-in, high, internal metabolism that allows them to be continuously "warm-blooded." The sail, as add-on solar technology, may have made *Dimetrodon* into a super-pelycosaur, but it did not make it a mammal.

But *Edaphosaurus* has a sail too (Figure 10.9), which was also thought to be a thermoregulatory device. Studies of the bone structure of *Edaphosaurus* spines by Adam Huttenlocker and colleagues suggest that there was no system of canals through the bone for the blood vessels that would be needed for efficient thermoregulation, either dominantly heating or dominantly cooling. Most probably, then, *Edaphosaurus* had a sail for some other reason. The most likely alternative is for display.

How can we resolve the debate between thermoregulation and display functions for the pelycosaur sail? The simple answer is that the sails were probably used for both functions. The sails were distinctive enough to signal the species they belonged to, perhaps the gender and/or age, but there was also enough thermoregulation going on to add to the evolutionary advantage of the sails and to help to defray the cost of building them. There are many examples in modern animals where structures have multiple functions and yet may not be perfectly adapted for either. For example, African elephants flap their great ears to lose heat but also use the ears to signal their mood and scare predators. Elephant ears are not perfect radiators but they do the job, and they have other functions too. Perhaps this was true of *Dimetrodon* and *Edaphosaurus* in the Early Permian.

If pelycosaurs with sails thermoregulated, then other pelycosaurs (*Cotylorhynchus*, for example; Figure 10.7) probably thermoregulated too, in behavioral ways that left no traces on the skeleton. After the Permian, we see little sign of thermoregulatory devices as advanced pelycosaurs evolved into therapsids. There is indirect evidence, however, that therapsids had limited thermoregulation but that evidence is presented in another chapter.

## Permian Changes

Shifting continental geography resulted in major biogeographic changes in the Permian (see Chapter 6). The large southern continent Gondwana moved north to collide with Euramerica, and by the Middle Permian these blocks formed a continuous land mass. A little later, Asia crashed into Euramerica from the east, buckling up the Ural Mountains to complete the assembly of the continents into the global supercontinent Pangea.

These tectonic events put an end to the wet climates that had fostered the system of large lakes, swampy deltas, and shorelines along the south coast of Euramerica, where the Carboniferous coal forests had flourished. Permian floras, in contrast, were dominated by gymnosperms, mostly ginkgos, conifers, and cycads. Compared with other Late Paleozoic plants, conifers were better adapted for drought resistance and they probably evolved in much drier uplands, because they are rare on lowland floodplains. Tree-sized lycopods disappeared from coal swamps in the Late Carboniferous as climates became drier and Permian conifers extended into lowlands and replaced them.

## The Invasion of Gondwana

Geological evidence from Gondwana shows that a huge ice sheet was centered on the South Pole (Figure 10.12) in Late Carboniferous and Early Permian times. Ice sheets moving northward scoured rock surfaces and deposited stretches of glacial debris on a continental scale.

The continental collisions that formed Pangea allowed land animals to walk into Gondwana but pelycosaurs, which had always had a narrow tropical distribution, remained in the tropical areas, much reduced in diversity. Late Permian pelycosaurs are found only in North America and Russia but they did not last long. Instead, Gondwana was invaded by their descendants, the synapsid reptiles called **therapsids**.

There has always been a gap in geological time between the pelycosaurs of the Early Permian, especially from North America, and then the therapsids, first known from the Middle Permian of Russia, China, and South America. The debate is about whether this is an artificial gap, a failure in sampling, because the Middle and Late Permian in North America was largely marine and the Russian Early Permian was also marine, so we cannot find synapsid fossils. The balance of evidence, as presented by Neil Brocklehurst and colleagues, suggests that this was an extinction event between the Early and Middle Permian, sometimes called "Olson's Extinction" in honor of the paleontologist Everett Olson who was one of the first to compare Permian synapsids from North America and Russia in the 1950s.

## Thermoregulation in Therapsids

Therapsids lived mainly at middle or high latitudes rather than in tropical regions; almost all therapsid clades evolved in Gondwana and spread outward from there. The restriction, or adaptation, of therapsids to drier and more seasonal habitats may have encouraged their success in southern Gondwana, away from the tropics and toward higher latitudes. Thousands of

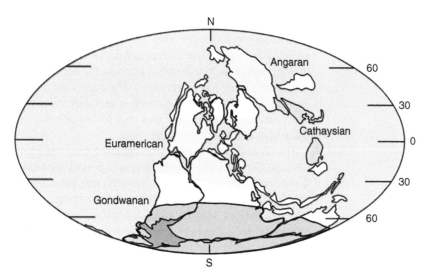

Figure 10.12 During the Late Carboniferous and Early Permian, a great ice sheet covered much of Gondwana around the South Pole (shown gray). Evidence is seen in the form of glacial deposits, and scratches from glaciers moving out over the southern lands of South America, South Africa, Australia, and India. The Gondwanan, Euramerican, Angaran, and Cathaysian are floral-climatic belts in the nonglaciated world.
*Source:* Artwork by GeoPotinga (Wikimedia).

specimens of therapsids have been collected from Late Permian rocks in South Africa, for example, and someone with time on their hands estimated that these beds contain about 800 billion fossil therapsids altogether! There are literally dozens of species, and we have a good deal of evidence about their environment. The glaciations were over and vegetation was abundant, with mosses, tree ferns, horsetails, true ferns, conifers, and a famous leaf fossil, *Glossopteris*. The climate may well have been mild considering that South Africa was at 60°S latitude (Figure 10.12). But it must have been seasonal, so the supply of plant food would have been seasonal too.

When we find large extinct synapsids at such high latitudes, we can be reasonably sure that they were unlike living reptiles in their metabolism. We know that mammals evolved from therapsids in the Late Triassic but did Permian therapsids already have a mammalian style of thermoregulation, with automatic internal control, a furry skin, and a high metabolic rate? Their anatomy could suggest both conclusions. On the one hand, they had sprawling fore limbs and did not move very efficiently compared with later reptiles and mammals, so might have been cold-blooded. But, on the other hand, unlike other reptiles, many therapsids had short, compact, stocky bodies, with short tails: good adaptations for conserving body heat.

Two studies published in 2017 may have resolved the question definitively. It seems that Late Permian therapsids were already warm-blooded like mammals, and this change arose in a stepwise fashion through the Middle and Late Permian. The new studies use smart methods – bone histology and oxygen isotopes. **Bone histology** is the study of the cellular structure of bone, which can show differences in growth rate and overall metabolism. Oxygen isotopes in fossil bones can tell geochemists something about ancient environmental temperatures but also the temperatures of the animals themselves. In the bone histology study, Chloe Olivier and her colleagues show that some latest Permian therapsids already had mammalian-type thermoregulation. Second, Kévin Rey and his colleagues show that dicynodonts and cynodonts evolved warm-bloodedness in the Late Permian, based on comparisons of oxygen isotopes in their bones with similar measurements from contemporary cold-blooded reptiles.

## Therapsid Evolution

Therapsids radiated very rapidly into several groups, and by Late Permian time they had spread globally all over Pangea. Therapsids as a group had larger skull openings than pelycosaurs did, indicating that they had more powerful jaws. The whole skull was strengthened and thickened. Therapsids also had much better locomotion

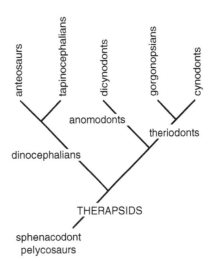

Figure 10.13 The pattern of evolution of the major groups of therapsids. The only surviving clade of therapsids is the cynodonts, a group of advanced therapsids that included mammals in the Late Triassic.

than pelycosaurs. There is little doubt that therapsids evolved from a lineage of sphenacodont pelycosaurs, as relatively small to medium-sized carnivores. The key groups in the Middle and Late Permian were the dinocephalians, gorgonopsians, and dicynodonts. Figure 10.13 shows the likely pattern of therapsid evolution.

## Dinocephalians

Dinocephalians were the first abundant therapsids, occurring from Russia to South Africa, and dominating Middle Permian ecosystems. They moved much better than pelycosaurs. Their spine was quite stiff, and limb length and stride length were longer than in pelycosaurs. The fore limbs were still sprawling but the hind limbs were set somewhat closer to the vertical, accentuating the wheelbarrow mode of walking we have already described for pelycosaurs.

Dinocephalians became very large. They had large skulls and, like all therapsids, they had strong canines. They also had well-developed incisors that seem to have been both efficient and important in feeding. Dinocephalians had unusual front teeth: their upper and lower incisors, and sometimes the canines too, interlocked along a line when the mouth closed, forming a formidable zigzagged array that would bite off pieces of food (animal or plant) as well as piercing and tearing (Figure 10.14).

The earliest dinocephalians, the anteosaurs, were carnivores with skulls up to a meter long. Like sphenacodonts, they killed prey mainly by slamming their long, sharp front teeth into them, then tearing and piercing.

(a)

(b)

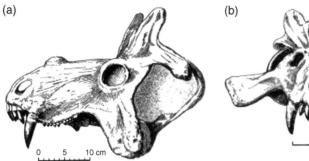

0   5   10 cm

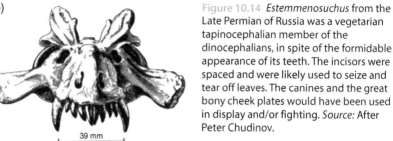

39 mm

Figure 10.14 *Estemmenosuchus* from the Late Permian of Russia was a vegetarian tapinocephalian member of the dinocephalians, in spite of the formidable appearance of its teeth. The incisors were spaced and were likely used to seize and tear off leaves. The canines and the great bony cheek plates would have been used in display and/or fighting. *Source:* After Peter Chudinov.

Apparently, the back teeth were not used very much; they were fewer and smaller than in sphenacodonts.

Most other dinocephalians, the tapinocephalians, look carnivorous at first sight because they have large canines and incisors in the front of the jaw (Figure 10.14). But they had a broad, hippo-like muzzle, a large array of flattened back teeth, and massive bodies with a barrel-like rib cage that must have contained a capacious gut (Figure 10.15). These animals may have been omnivorous but more likely the incisors were cropping, cutting teeth used on vegetation, and the canines were fighting tusks, not carnivorous weapons. Look inside the mouth of a hippo sometime! The jaw exerted most pressure when closed, for efficient food processing rather than slamming.

Some tapinocephalians were particularly bizarre, with horns; some of them, probably males, had great bony flanges on the cheeks (Figure 10.14). All tapinocephalians had thick skull bones, sometimes up to 11 mm (0.5 in.) thick. Herbert Barghusen suggested that individuals butted heads, presumably to establish dominance within a group. Large vegetarians today tend to fight by head butting or pushing, while carnivores today are quick and agile and tend to use claws and teeth as they fight. Early therapsids, even the carnivores, were heavy and clumsy; they had sprawling limbs that were so committed to supporting the body that they could not have used claws as weapons.

## Advanced Therapsids

The rapid evolution of therapsids brought a new wave of advanced forms across the world in the Late Permian. These are the theriodonts and dicynodonts. Theriodonts were all carnivores, with low flat snouts and very effective jaws, and the best known are the **gorgonopsians**, so called for their ferocious appearance. They were the dominant large carnivores of the Late Permian, measuring up to 3.6 m (11.5 ft) long. They look as if they specialized on large prey because of their large canine teeth, up to 12 cm (4.7 in.) long. They were like sabertooth cats, using the canines as killing weapons (Figure 10.16). The feeding action involved a very wide gape of the jaw and a

Figure 10.15 Therapsids had compact stocky bodies and short tails. *Keratocephalus* is a tapinocephalian from the Middle Permian of South Africa. It was about 3 m (10 ft) long and weighed perhaps 500 kg (1000 pounds). *Source:* Richard Cowen.

slamming action that drove the huge canines deep into the prey (Figure 10.17). The incisors were strong but the back teeth were small and must have been practically useless. The snout was rather short, but the jaws were deep to hold the long roots of the canines. The limbs were long and slender, and gorgonopsians may have been comparatively agile. Limb joints were lightly built, and the hind limb could be swung into an erect position, extending the stride length, and altogether indicating some speed of movement. They did not go in for much chewing but simply tore large chunks from prey that was too large to eat in one bite (sharks, crocodiles, and Komodo dragons do that too). Their front teeth had serrations on them to slice through muscle and tendon.

## Dicynodonts

Dicynodonts evolved in the Late Permian and very quickly radiated to become the most important herbivores of the Late Permian. They were the first truly abundant worldwide herbivores, with a great variety of sizes and specializations. They make up 90% of therapsid

Figure 10.16 Skeleton of *Sauroctonus*, a 2 m (7 ft) long gorgonopsian from the Late Permian of Tanzania, Africa. *Source:* Photo by H. Zell (Wikimedia).

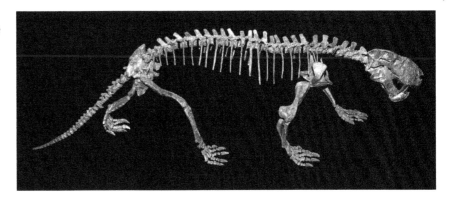

Figure 10.17 The gorgonopsian *Inostrancevia* menaces the herbivorous pareiasaur *Scutosaurus*, in this scene from the Late Permian of Russia. *Source:* Artwork by Dmitry Bogdanov (Wikimedia).

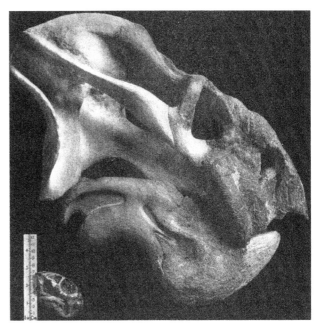

Figure 10.18 The box-like skull of two dicynodonts from the Permian of South Africa. The large skull is *Odontocyclops* and the small one is *Diictodon*. *Source:* American Museum of Natural History digital library, in the public domain.

specimens and much of the therapsid diversity preserved in Late Permian rocks. At their peak in the Late Permian, dozens of species of dicynodonts were living in Gondwana. However, their history shows bursts of high diversity in the Late Permian and the Middle Triassic, with a strong pinching across the P–Tr boundary, when the mass extinction reduced them to only one or two species and nearly finished them off.

Early dicynodonts already had a secondary palate, separating the mouth from the nostrils so that they could breathe and feed at the same time. They differed from other therapsids in having very short snouts, and they had lost practically all their teeth except for the tusk-like upper canines, which were probably used for display and fighting rather than for eating. Because there were no chewing teeth, the jaws must have had some sort of horny beak (like that of a turtle) for shearing off pieces of vegetation at the front and grinding them on a horny secondary palate while the mouth was closed.

The jaw joint was weak, and the lower jaw moved forward and back in a shearing action instead of sideways or up and down. As part of this system, the jaw muscles were unusual, set far forward on the jaw, and they took up a good deal of space on the top and back of the skull. These unusual jaw characters had their effect on the whole shape of the skull, which was short but high and broad, almost box-like. The extensive muscle attachments resulted in the eyes being set relatively far forward on a short face (Figure 10.18). Dicynodonts look as if they cropped relatively tough vegetation with their beaks, and then ground it up in a rolling motion in the mouth.

Dicynodont jaws varied a lot, presumably because of their diet. There were dicynodonts with cropping jaws and with crushing jaws (perhaps for large seeds), and

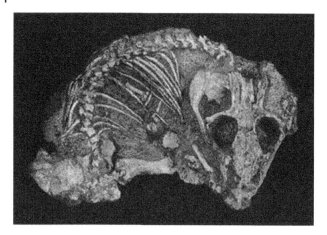

Figure 10.19 One of several specimens of the little Permian dicynodont *Diictodon* that have been found fossilized inside their burrows. *Source:* Courtesy of Dr. R. M. H. Smith of the South African Museum.

Figure 10.20 The Middle Permian parareptile *Mesosaurus*, 1 m (3.3 ft) long, and armed with needle-like teeth for catching small fish. *Source:* Artwork by Nobu Tamura (Wikimedia).

many browsers and grazers. Some dicynodonts were specialized for grubbing up roots. As in other herbivores, the body was usually bulky, with short, strong limbs.

The success of dicynodonts is astonishing. Most dicynodonts were rather small, though they ranged from rat-sized to cow-sized. Presumably, the fact that the horny feeding structures of dicynodonts were replaced continuously throughout life had a great deal to do with their success. Reptiles with teeth replace them throughout life, but intermittently, so it is difficult for them to achieve continuously effective tooth rows. Other therapsids evolved effective cutting and grinding teeth, but teeth do wear out with severe and prolonged use.

Climates in the Late Permian were hot and steamy, and small dicynodonts could use their digging claws to excavate burrows. In a spectacular discovery in South Africa, skeletons of a little dicynodont were found at the bottom of sophisticated spiral burrows (Figure 10.19). It is suggested that they constructed these burrows so they could escape the heat of the midday sun, and keep cool and damp until later in the day. Sometimes, sadly, the burrow was filled by a rock slide or a rising flooding river, and we find the skeletons preserved perfectly.

Dicynodonts declined abruptly at the end of the Permian but a few lineages persisted and then recovered in the earliest Triassic. The best-known example was the dicynodont *Lystrosaurus* (see Chapter 12, Figure 12.6), a famous survivor of the P–Tr mass extinction that diversified worldwide.

## Parareptiles

Throughout the Permian, there were some anapsid reptiles that never achieved great ecological dominance. One example is the little aquatic animal *Mesosaurus*

Figure 10.21 The pareiasaur *Scutosaurus* from the Late Permian of Russia, showing its massive body, short limbs, and massive, armored skull. Note the relatively small teeth, adapted for plant eating. *Source:* Photo from anonymous source (Wikimedia).

(Figure 10.20), famous for having lived in marine basins on both the east coast of South America and the west coast of Africa. It was used as early evidence that the South Atlantic did not exist in the Permian, and in favor of continental drift. *Mesosaurus* was about 1 m (3.3 ft) long and swam by beating its deep-sided tail from side to side.

The other important parareptiles were the **pareiasaurs**, important herbivores of the Middle and Late Permian. Some, such as *Scutosaurus* from the Late Permian of Russia (Figures 10.17 and 10.21), were the size of hippos, 3 m (10 ft) long, with tiny heads and jaws lined with leaf-like teeth. The body was massive and supported by rather short legs. Pareiasaurs were armored, with bony plates set in their skin and thickened skulls carrying bony horns, to protect them from the saber-toothed gorgonopsians. Pareiasaurs lived in hot, tropical climates, and there is evidence they wallowed in mud baths by the rivers to cool off. Pareiasaurs were dominant in many Late Permian faunas but disappeared during the P–Tr mass extinction

## Synapsids and Their Diapsid Replacements

With their generally large bodies, their radiation as herbivores and carnivores of varying sizes, and their experimentation with horns, fangs, and fighting, a Late Permian therapsid community viewed from a long distance would not seem totally strange to a modern ecologist, especially one familiar with the large mammals of the African savanna. However, the comparison would not stand close examination. All therapsids moved in a slow clumsy fashion, especially the larger ones. Slow

motion is fine as a way of life if it applies to prey and predators too, otherwise both would eventually die out (think about it).

This seemingly peaceful scene, with dicynodonts and pareiasaurs lumbering around, constantly chomping at their plant food and tracked by slow-moving predatory gorgonopsians, was to come to an abrupt end with the P–Tr mass extinction (see Chapter 6). The pareiasaurs and gorgonopsians were wiped out, and the dicynodonts barely survived. In the new, empty, and ravaged world of the earliest Triassic, new groups, primarily diapsids, came to the fore and there was a long span of change among both synapsids and diapsids; this was the "Triassic Takeover," during which reptiles all became faster-moving and dinosaurs exerted their dominance step by step. We will explore these next stages for vertebrate life on land in Chapter 12, after looking first at the impact of the P–Tr mass extinction on life in the sea, in Chapter 11.

## Further Reading

Benson, R.B.S. (2012). Interrelationships of basal synapsids: cranial and postcranial morphological partitions suggest different topologies. *Journal of Systematic Palaeontology* 10: 601–624. [An overview of the basal synapsids].

Brocklehurst, N., Day, M.O., Rubidge, B.S. et al. (2017). Olson's extinction and the latitudinal biodiversity gradient of tetrapods in the Permian. *Proceedings of the Royal Society B: Biological Sciences* 284: 20170231. [Extinction and paleobiogeography].

Huttenlocker, A.K., Mazierski, D., and Reisz, R.R. (2011). Comparative osteohistology of hyperelongate neural spines in the Edaphosauridae (Amniota: Synapsida). *Palaeontology* 54: 573–590. [Explanation of the pelycosaur sail].

Olivier, C., Houssaye, A., Jalil, N.-E. et al. (2017). First palaeohistological inference of resting metabolic rate in an extinct synapsid, *Moghreberia nmachouensis* (Therapsida: Anomodontia). *Biological Journal of the Linnean Society* 121: 409–419. [Evidence from bone histology that some Permian synapsids were warm-blooded].

Rey, K., Amiot, R., Fourel, F. et al. (2017). Oxygen isotopes suggest elevated thermometabolism within multiple Permo-Triassic therapsid clades. *elife* 6: e28589. [Evidence from oxygen isotopes that some Permian synapsids were warm-blooded].

Sues, H.-D. and Reisz, R.R. (1998). Origins and early evolution of herbivory in tetrapods. *Trends in Ecology & Evolution* 13: 141–145.

## Question for Thought, Study, and Discussion

Explain carefully the history of ideas about the sail on the back of *Dimetrodon*. Now think about baby *Dimetrodons*. They were small and they did not have a sail. Did they have any thermoregulation? And if not, why not? And then think about all the other pelycosaurs in the Permian world, and ask the same question.

11

## The Mesozoic Marine Revolution

### In This Chapter

The Mesozoic Marine Revolution marks a series of stages in continuing arms races between predators and prey. Crustaceans, gastropods, and fishes evolved fiendish new predatory modes, boring, snipping, and crushing the shells of their prey. Much of the new behavior tracks back to the recovery of life after the devastating Permian–Triassic mass extinction. New reef environments and new groups of fishes and reptiles were part of the revolution. Several diapsid lineages invaded the sea to become powerful air-breathing carnivores, apparently competing on equal terms with the sharks and other fishes that filled the same ecological niches. For example, ichthyosaurs were streamlined powerful swimmers with long jaws that carried fish-eating teeth. The largest specimens, from the Triassic, were up to 50 ft long. Placodonts seem to have crushed mollusks between big tooth plates. The sauropterygians began at fairly small size but eventually evolved into plesiosaurs, which had large strong paddles that probably moved in an up-and-down motion, generating an underwater "flight" pattern. Again, some plesiosaurs were over 50 ft long. Mosasaurs are simply water-going lizards but they evolved long, powerful, streamlined bodies and limbs that made hydrofoils for efficient swimming. Amazingly, rare finds of embryos inside fossils show that mosasaurs, ichthyosaurs, and plesiosaurs all had lineages that had live birth at sea, just like living whales and dolphins. Turtles were in the sea by Late Triassic times, and they include the 10 ft long giant turtle *Archelon* in the Late Cretaceous. Crocodiles are mostly freshwater and estuarine today but there were marine crocodiles in the Jurassic.

## Mesozoic Ocean Ecosystems

Life was nearly destroyed at the end of the Permian. The Mesozoic was a time of rebuilding of ecosystems, when the world's fauna and flora started to take on many of the characteristics of the modern world. A scuba diver in Permian seas would not have seen many familiar creatures. However, by the Triassic, the life of the oceans had changed to more modern forms, such as bivalves and gastropods, lobsters, and modern-style fishes. The P–Tr mass extinction punctuated the history of life, marking the replacement of the so-called Paleozoic fauna of Sepkoksi by the "modern fauna" in the oceans (see Chapter 6).

After the P–Tr mass extinction, fishes became the mid-sized predators of the ocean. Ammonites were still abundant but the major additions to the global oceans in terms of large-bodied predators were not fishes but fish-eating marine reptiles. Most of these reptile groups evolved in Triassic times but reached their greatest abundance in the Jurassic and Cretaceous. Several different clades of reptiles evolved spectacular adaptations to life at sea. All these changes point to higher productivity of the oceans (= more life in general), and everything was probably moving faster (the Mesozoic fishes and reptiles had adaptations to swim faster than their armored Paleozoic cousins).

Geerat Vermeij noted in a masterful overview in 1977 that the modern oceans, which originated in the Mesozoic, had undergone a revolution, which he called the Mesozoic Marine Revolution (MMR). This was characterized not only by higher food productivity and greater speeds of all the animals but also by the fact that Mesozoic predators were meaner and nastier than Paleozoic predators.

Vermeij dated the MMR as beginning in the Jurassic and speeding up in the Cretaceous. However, new evidence takes the MMR back into the Triassic, and shows

*Cowen's History of Life*, Sixth Edition. Edited by Michael J. Benton.
© 2020 John Wiley & Sons Ltd. Published 2020 by John Wiley & Sons Ltd.

that it was triggered by the profound ecological effects of the P–Tr mass extinction. Therefore, in this chapter, we review the whole Mesozoic history of the life of the oceans, mentioning some key groups that are commonly found as fossils, such as ammonites and belemnites, bivalves, and gastropods. We look first at how life recovered during the Triassic, then explore some key marine animal groups and their predator–prey interactions, and then look more closely at the amazing Mesozoic marine reptiles.

## Recovering Seas

At one time, it was thought that life would simply bounce back after a mass extinction. It was assumed that the grim conditions of the P–Tr mass extinction, involving flash heating, acid rain, wash-off of soil into the oceans, ocean heating and acidification, and seabed anoxia (see Chapter 6), would simply have stopped and life could recover in peace. However, it wasn't like that. In fact, flash heating and other environmental crises continued through the Early Triassic, and of course only 5% of species had survived and so recovery was patchy and slow.

Life took many millions of years to recover. Some groups such as ammonites recovered fast, but then they were hit hard several times during the Early Triassic. Overall, species numbers in ecosystems did not recover until some 6 m.y. after the event, in the Middle Triassic. Complex ecosystems took even longer to recover.

The evidence for continuing harsh environmental conditions through the first 6 m.y. of the Triassic comes from oxygen and carbon isotopes, which show at least a further three episodes of flash heating, presumably associated with acid rain, ocean anoxia, and ocean acidification. These pulses in the isotope signals (Figure 11.1) are the same size as, or even larger than, the isotope spike at the P–Tr event, so they could have had just as devastating effects. The isotope spikes might have been driven by continuing, massive eruptions of the Siberian Traps.

The devastation of the P–Tr mass extinction and the continuing harsh environmental conditions are borne out by three important "gaps," each of which lasted for about 10 m.y. These are the chert gap in the deep oceans, the coral gap in shallow seas, and the coal gap on land. The **chert gap** reflects the absence of silica-based skeletons falling to the seabed, especially planktonic radiolarians and glass sponges. The **coral gap** is evidence that the Paleozoic coral groups had disappeared in the Late Permian, and it took until the Late Triassic before new coral groups became established. The **coal gap** is marked by the absence of coal deposits in terrestrial sedimentary sequences, evidence that plants had been devastated and forests wiped out.

## Crisis on the Seabed

Seabed organisms such as brachiopods and mollusks, and reef builders such as corals, sea lilies (crinoids), and bryozoans were largely wiped out by the P–Tr mass extinction. Comparison of seabed scenes from South China (Figure 11.2) show how a rich community with dozens of species in the latest Permian was reduced to only four or five species in the earliest Triassic – the burrowing brachiopod *Lingula* and some thin-shelled clams such as *Claraia* sitting on the sea floor and flapping their shells to move. Some very rare ammonites and sharks also survived.

The loss of life is evident when you collect fossils through sections that span the Permian–Triassic boundary. Below the extinction level (Figure 11.2a), the rock is often solid with fossils; in fact, it's what is called a bioclastic limestone, made up of shells of brachiopods and coral skeletons. After the boundary (Figure 11.2b), the rock is black, reflecting rich organic material that is not being eaten by anything. In the black mud, iron pyrites crystals (pure iron sulfide) also confirm the absence of oxygen. Life is so rare that many rocks contain no fossils at all. It took until the Middle Triassic for seabed life to diversify and recover to something like normal (Figure 11.2c): there are mollusks and lobsters on the seabed, diverse fishes above, and a variety of burrowing animals going deep into the sediment and feeding on organic material.

## Reefs

The reef gap of the Early and Middle Triassic is famous. This meant that one of the most diverse settings for life in shallow seas was absent; think of how tropical reefs in shallow seas today teem with hundreds of species – not just the corals and other reef-structural builders but also the shrimps and lobsters creeping among the coral twigs, and dozens of fishes large and small prowling on top.

New coral groups emerged in the Middle and Late Triassic, including the **scleractinian corals** that dominate reefs today. Some were solitary but most were colonial, forming structures composed of numerous tiny individuals that communicate and organize feeding, protection, and reproduction together. Many also were photosynthetic, harboring algae that require sunlight to survive, and so the majority of reefs, both today and in the Mesozoic, were in sunny, shallow waters.

Other reef builders in the Mesozoic included some unusual oyster-like clams, bryozoans (colonial organisms with tiny filter-feeding individuals making up their sheet-like

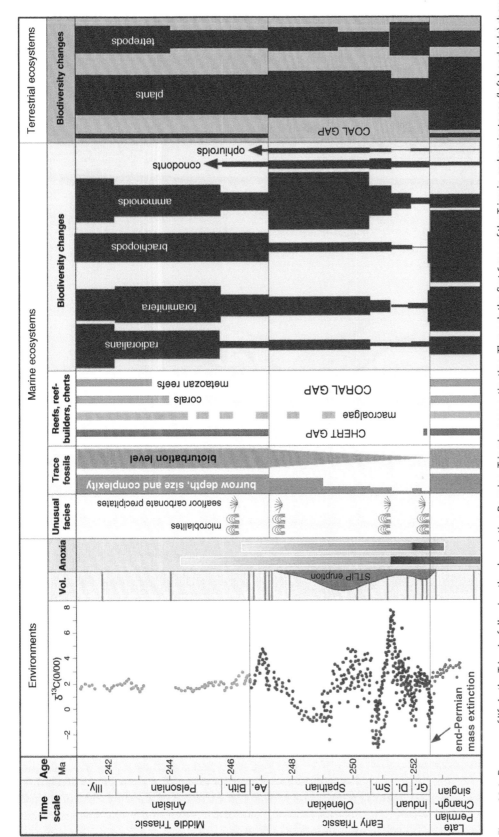

Figure 11.1 Recovery of life in the Triassic, following the devastating Permian–Triassic mass extinction. Through the first 6 m.y. of the Triassic, carbon isotopes (left-hand side) show repeated spikes, indicating episodes of flash warming, possibly caused by continuing volcanic eruptions of the Siberian basalt volcanoes. In the sea and on land, there were major gaps in key habitats and fossil groups, and key groups of animals and plants took up to 10 m.y. to recover to preextinction levels of diversity. *Source:* Diagram from Chen and Benton (2012).

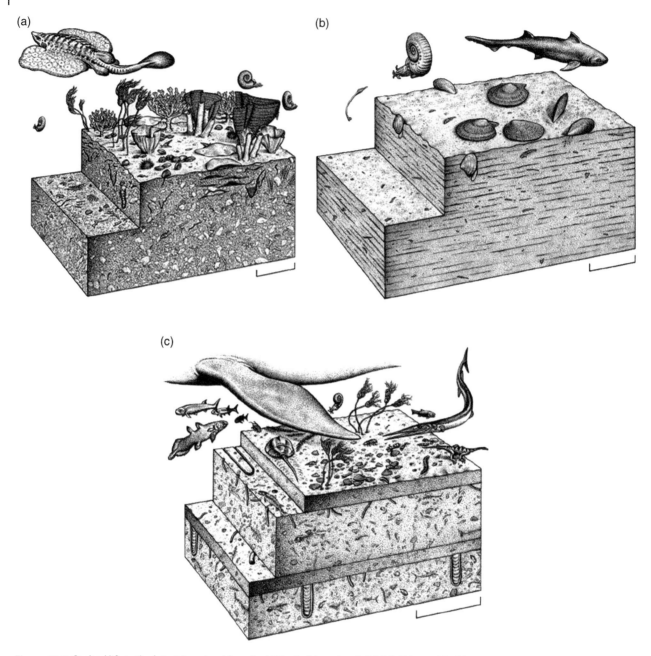

Figure 11.2 Seabed life in the latest Permian (a), earliest Triassic (b), and early Middle Triassic (c) of South China. *Source:* Drawings by John Sibbick.

structures), and sponges. The reefs grew and shrank, depending on the extent of the tropics and how sea level was rising and falling. The reefs provided feeding zones and hiding places for many mollusks, crustaceans, and fishes.

## Ammonites and Belemnites

Swimming over the sea floor, and over the reefs, were ammonites and belemnites. Indeed, Mesozoic rocks around the world are often stuffed full of **ammonites**,

and their remarkably beautiful, coiled shells (Figure 11.3) have long been used as index fossils for dating the rocks. This is because ammonite species evolved fast, often changing into new forms within as little as 200 000 years, so they allow geologists to identify several stratigraphic zones in every million years. It helps also that many ammonite species existed worldwide, so can help correlate rock units from continent to continent. The fossils vary a great deal and the different species can be easy to identify.

(a)  (b)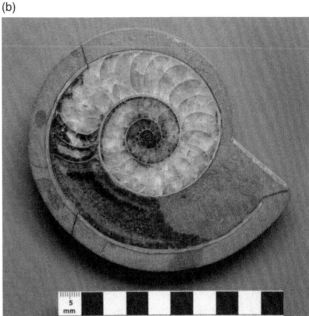

Figure 11.3 Jurassic ammonite, in external (a) and internal (b) views. The outer face of the shell shows intricate suture lines (the wiggly lines), and these correspond to chambers within the shell. The chambers are seen in the sliced specimen (b), as curving structures, and the animal lived in the gray mud-filled part of the shell. *Source:* From the Jurassic Coast website.

Ammonites were **cephalopods**, relatives of modern octopus and squid. Whereas octopus and squid have internal shells, ammonites wore their coiled shells on the outside, and they lived inside the shell, goggling at the world with their huge eyes and snatching for food with their tentacles. Ammonite shells provided some protection from predatory lobsters, fishes, and reptiles, but not completely – there are famous examples of fossil ammonite shells that show neat rows of tooth marks where they were bitten by a marine reptile. But they could confuse their predators, as squid and octopus do today, by squirting black ink and swimming away from the ink cloud.

**Belemnite** fossils are like large bullets, about the size of your index finger, cylindrical and coming to a neat point at one end (Figure 11.4). Like ammonites, belemnites are common throughout the Mesozoic, and are used for stratigraphic purposes in the Cretaceous. These are internal shells and in life, the belemnite animal would have looked like a squid, strengthened internally by the shell.

Ammonites and belemnites were fast-evolving but they also responded dramatically to environmental changes. So, they were hit hard by the end-Triassic mass extinction (see Chapter 6), then recovered fast and became hugely abundant through the Jurassic, had another wobble during sea-level changes in the Cretaceous and plummeted to extinction in the end-Cretaceous mass extinction (see Chapter 17).

Figure 11.4 A pile of the guards (= internal shells) of belemnites from the Jurassic of Wyoming. *Source:* Photo by Mark A. Wilson (Wikimedia).

## Fishes

Change in the major groups of fishes happened piecemeal through the Mesozoic. Two new groups emerged in the Triassic: the neoselachians and the neopterygians. The **neoselachians** are the modern types of sharks, generally shaped like torpedoes for fast swimming, and with mouths that can be projected to cause maximum damage with their sharp teeth. They lived side by side in the

Mesozoic with other shark groups that had arisen in the Carboniferous (see Chapter 7), but neoselachians prevailed and became the dominant sharks in the Cretaceous.

Just as sharks show advances in speed and feeding through the Mesozoic, so too did the bony fishes (see Chapter 7). New, faster-swimming, and unarmored fishes called **neopterygians** became important in the Triassic, and some of these are especially well represented in sites from China. In a reconstructed scene from a location called Luoping in Yunnan Province (Figure 11.5), where 20 000 exquisite fossils of lobsters, fishes, and marine reptiles have been excavated, we get a glimpse of the new ocean scene, teaming with modern-style neopterygians. They lived in vast shoals, some long and slender, predators that lurked among the seaweeds, and others with remarkable humps and color patterns.

Later in the Mesozoic, a major new group of neopterygians, the **teleosts**, took over. These are the dominant fishes today, comprising 30 000 species, ranging from cod and herring to sea horses and catfish, eels, and flying fish. Teleosts are so hugely successful because they are much faster than the other neopterygians, with their thin, flexible scales, and because their mouth parts are very adaptable, allowing them to project the jaws to suck in food (Figure 11.6) and modify the jaws to specialized diets such as coral snipping or scraping algae from rocks.

## Arms Races and New Predatory Modes

The insight that alerted Geerat Vermeij to the idea of the MMR was that he saw a huge increase in predatory action when he compared Cretaceous and Jurassic fossil assemblages. The Cretaceous shells were more often bored, chewed, and crushed than those of the Jurassic. He noticed a rise of particular groups of gastropods, lobsters, and fishes that were especially equipped to deploy fiendish new modes of attack.

For example, many gastropods feed on other shelled animals. Gastropods include marine whelks and terrestrial slugs and snails – they often have coiled shells. But their hunting equipment is focused around the mouth. They have a rasping tongue-like structure called a **radula** which limpets, for example, use to scrape algae off rocks and snails use to chew up your garden cabbages and lettuces. But the hunting marine gastropods have a variety of advanced radular systems to maximize their success rate. Some project the radula on a long mouth structure and they drill neat holes, up to 1 mm across, through the shells of clams and sea urchins and sometimes other gastropods (Figure 11.7). Many predatory gastropods also use acid to help create the hole quickly. Once they are through the shell, they suck out the prey animal.

The other hunting mode that Vermeij studied was shell snipping. This is a specialty of crabs and lobsters today – with their great claws they can snatch a clam and cut off lumps of the shell. Sometimes the prey can escape by leaping away or burrowing, and there are many fossils showing where a clam was chewed but recovered and continued growing. Very little shell snipping happened before the Mesozoic, and then it became common.

Finally, there were the shell crushers, mainly fishes and reptiles. As mollusks thickened their shells and learned to burrow, the shell eaters grew smarter. Several groups of sharks, bony fishes, and marine reptiles independently evolved shell-crushing dentition. These specialists on shell crushing, called **durophages** (= hard eaters), all had rounded and thickened teeth, often arranged in great pavements inside their mouths (see Figure 11.14), so they could grab an oyster and crush it. They would spit out the shell and swallow the flesh.

Figure 11.5 Lobster lunch of Luoping. Over a muddy seabed during the Middle Triassic, an early rock lobster attracts the unwanted attention of a coelacanth and an armored placodont *Sinosaurosphargis*, allowing a pair of horseshoe crabs to scuttle away unnoticed. Behind them cruise other aquatic reptiles: the ichthyosaur *Mixosaurus* and the little pachypleurosaur *Dianopachysaurus*. Swarming around is a diverse community of six neopterygians, the newly evolved, fast-moving fishes of the early Mesozoic. *Source:* Artwork by Brian Choo, with permission.

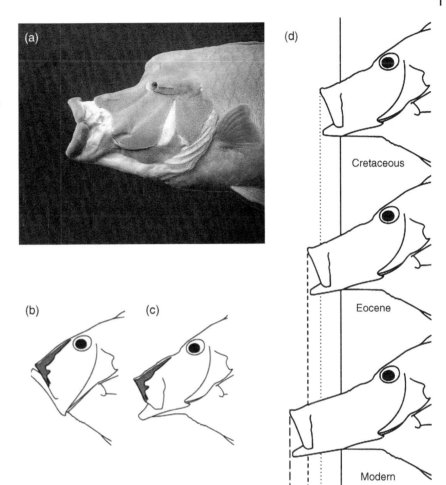

**Figure 11.6** Jaw protrusion in teleosts. (a) The modern teleost *Cheilinus* extending its jaws to feed. (b,c) The jaws of a John Dory closed (b) and open (c), showing how the premaxilla bone (in blue) projects and extends the jaws forward, forming a tube. (d) Differences in maximum jaw protrusion through time, rising from 8% in the Cretaceous teleost through 13% in an Eocene form, to 21% in the modern form. *Source:* Photo by J. P. Krajewski; artwork courtesy of David Bellwood.

The prey did not simply sit passively and let all these predators experiment with their new feeding modes. These were arms races and as the predators got smarter and nastier, the prey found new ways to protect themselves. These protection mechanisms included burrowing (get out of sight as fast as you can when threatened) as well as thickened shells and spines all over the shells (make the predator give up and try someone else).

The gastropods that bored into shells and slurped animals out alive, the lobsters and crabs that snipped and chopped their prey, and shell-crushing sharks and bony fishes were small terrors of Mesozoic seas. But more impressive and astonishing were the marine reptiles that acted as top predators throughout the Mesozoic.

## Ichthyosaurs

Ichthyosaurs were highly derived diapsids (see Chapter 12, Figure 12.1) that arose in the Early Triassic. They were shaped much like dolphins, except that the tail flukes are horizontal in dolphins and vertical in ichthyosaurs. Advanced ichthyosaurs had a continuation of the spine running into the lower tail fin (Figure 11.8). The main propulsion was by a side-to-side body motion, like a fish rather than a dolphin. The limbs were modified into small, stiff fins for steering and attitude control (Figure 11.9), again like dolphins, so that ichthyosaurs would have been very maneuverable up and down in the water as well as sideways. The tail fin was usually very deep, which is characteristic of swimmers that use fast acceleration in hunting prey. Ichthyosaurs were beautifully streamlined but would have been unable to move on land.

Beautiful ichthyosaur fossils have been known for 200 years, and they figured in many early discussions of evolutionary theory because everyone could recognize their exquisite adaptations for life in water. Ichthyosaurs all had good vision, with large eyes sighting right along the line of the jaw (Figures 11.8 and 11.10). In advanced ichthyosaurs, the jaw was long and thin, with many piercing conical teeth that were well designed for catching fish. Preserved stomach contents include fish scales and hooklets from the arms of cephalopods, possibly soft-bodied squids. One spectacular Jurassic ichthyosaur,

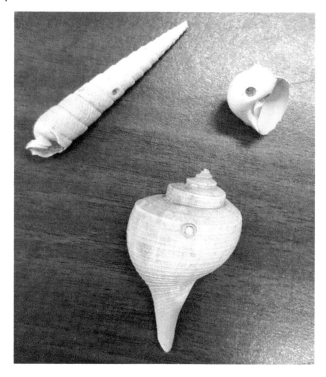

Figure 11.7 Three different species of gastropods with boreholes by a predatory gastropod, all from the Pliocene of Florida. Note that the drilling attempt at the bottom did not penetrate right through. *Source:* Photos courtesy of Kristina Barclay.

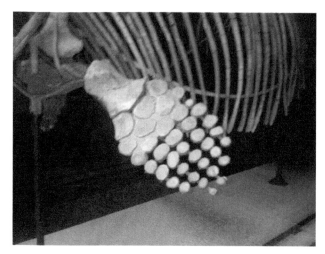

Figure 11.9 A front limb of the Late Jurassic ichthyosaur *Ophthalmosaurus*, with the arm, wrist, and hand bones all modified to form the paddle. The shoulder girdle and rib cage are seen in the background. *Source:* Photo by Emöke Dénes (Wikimedia).

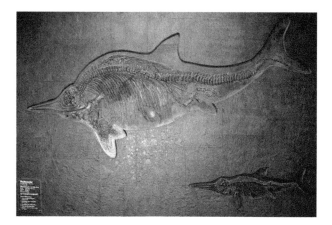

Figure 11.8 An adult and juvenile ichthyosaur, *Stenopterygius*, from the Lower Jurassic of Germany. Note how the fossil preparators have removed the rock to show the body outline, sometimes marked by black traces of skin tissues. *Source:* Photo by Haplochromis (Wikimedia).

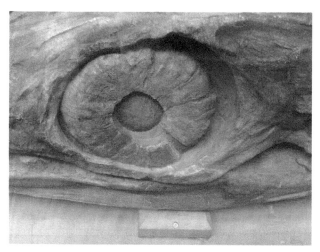

Figure 11.10 Ichthyosaurs all had huge eyes to allow them to see in murky water. This eye socket of *Temnodontosaurus* from the Early Jurassic of England has a bony sclerotic ring, which held the giant eyeball in shape. *Source:* Photo by Emöke Dénes (Wikimedia).

*Eurhinosaurus*, had a sword-like upper jaw projecting far beyond the lower, with teeth all along its length. It was probably an ecological equivalent of the swordfish, using its upper jaw to slash its way through a school of fish, then spinning around to catch its crippled victims.

Most early ichthyosaurs had blunt, shell-crushing teeth and may have hunted and crushed ammonites and other shelled cephalopods in a way of life that did not demand high levels of hydrodynamic performance. The first ichthyosaurs, from the Early Triassic, were small, about 1 m (3 ft) long, but they were already specialized for marine life. *Mixosaurus* is a typical small, early ichthyosaur, from Middle Triassic rocks ranging from the Arctic to Nevada to Indonesia, but the best-preserved specimens come from the Alps. The spine had not yet turned down to form the lower tail fin, but almost all the other features show excellent adaptation to swimming, with the limbs totally modified into effective fins.

Figure 11.11 The huge Triassic ichthyosaur *Shonisaurus*, with huge paddles and a deep body shape. *Source:* Artwork by Dmitry Bogdanov (Wikimedia).

*Shonisaurus* from the Late Triassic, at 15 m (50 ft) long, is one of the largest ichthyosaurs known (Figure 11.11). It was robust too, with a huge, strong, deep body and long, powerful fins, and nearly 200 vertebrae in the spine and tail. It also had a very unusual mode of life. It had no teeth and apparently fed by suction feeding, pulling in large quantities of water along with its prey of fishes and soft-bodied cephalopods. Some living whales (the beaked whales) feed in this way, using movements of a very large tongue to create the suction. The suction-feeding ichthyosaurs did not survive the end of the Triassic, but they formed an interesting evolutionary radiation and they were the first vertebrates to evolve this unusual way of life.

The huge diversity of Triassic ichthyosaurs was substantially reduced by the end-Triassic mass extinction, and only a few species survived. In the Jurassic, ichthyosaurs became common again in seas worldwide (Figures 11.8–11.10), but they never again reached the huge size of *Shonisaurus*. Ichthyosaurs are not so well known from the Cretaceous but one genus, *Platypterygius*, became widespread and survived into the Late Cretaceous (Figure 11.12). Ichthyosaurs survived until 30 m.y. before the end of the Cretaceous, and it is debated whether they died out because of some environmental changes or through competition with new marine reptiles such as mosasaurs (see below).

## Sauropterygians

Sauropterygians (Figure 11.13) are a large clade of reptiles, classified among the diapsids (see Chapter 12). Sauropterygians had unusually large limbs for land animals that had evolved toward life in water (compare crocodiles, seals, and whales). Most had small heads and comparatively long necks for their body size, so their prey (presumably fishes) must have been relatively small.

Placodonts are early but very specialized sauropterygians known only from the Triassic; in other words, they were an early offshoot from the basal part of the sauropterygian clade (Figure 11.13). They had their own set of

Figure 11.12 One of the last ichthyosaurs, *Platypterygius* from the Late Cretaceous of Russia. *Source:* Artwork by Dmitry Bogdanov (Wikimedia).

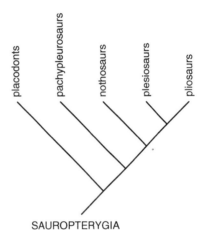

SAUROPTERYGIA

Figure 11.13 A cladogram of the sauropterygians, which included some of the more spectacular marine reptiles.

adaptations that may reflect a specialized ecology like the living walrus, which dives down to shallow sea floors to dig and crush clams. *Placodus* itself had unusual teeth that suited it for this way of life. The large comb-like teeth at the front of the jaw (Figure 11.14a,b) were probably used to dig into the sea floor to scoop up clams, and sediment could be washed off them by shaking the head with the mouth open. The clean clams were then crushed between flat molar teeth in the lower jaw and flat plates on the roof of the mouth (Figure 11.14b). Placodonts did not need great maneuverability or speed, and many had heavy plated carapaces (Figure 11.5) that covered them dorsally and ventrally, rather like a turtle.

Pachypleurosaurs were small marine sauropterygians (Figure 11.13) that are well known from Middle Triassic rocks of the Alps and China. They were small animals, typically 30 cm (1 ft) long. Pachypleurosaurs had thick

(a)

(b)

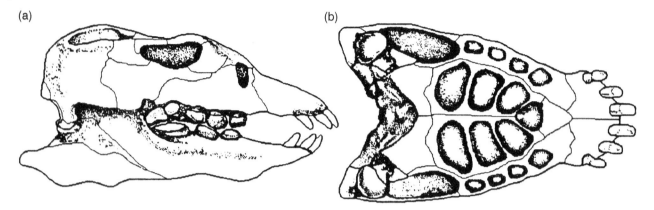

Figure 11.14 (a) The skull of *Placodus*, a marine reptile from the Triassic of Germany that may have had an ecology much like the living walrus. (b) The lower jaw of *Placodus*, showing the clam-crushing teeth of the palate. (What did it do with its tongue during this process?) *Source:* After Broili.

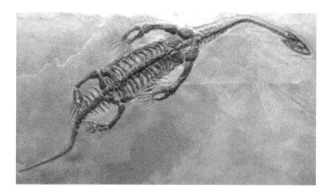

Figure 11.15 The pachypleurosaur *Keichousaurus* from the Middle Triassic of China. Note the powerful pectoral structure and front limbs. *Source:* Photograph by Daderot (Wikimedia).

ribs that presumably made the thorax quite stiff (Figure 11.15). They probably swam like living monitor lizards, with propulsion from the tail, with the front limbs tucked away against the rib cage and the hind limbs used as rudders. *Keichousaurus*, however, had distinctly powerful fore limbs, set on a strong pectoral girdle (Figure 11.15). Its hands probably made powerful paddles for swimming, but they might also have been useful for dragging the animal out onto land for breeding and for egg laying (or even live birth).

Nothosaurs were larger Triassic sauropterygians (Figure 11.13), averaging 3 m (10 ft) in length. Nothosaurs extended the rigid thorax of pachypleurosaurs by evolving ribs far back along the body (Figure 11.16). With their bodies stiffened in this way and a short tail, nothosaurs probably used their strong fore limbs for swimming power, and employed a rowing action for propulsion. The hind limbs were not very well adapted for a swimming stroke either. One fossil trackway from China shows that nothosaurs could prop themselves along the sea floor while hunting for buried prey, using the fore

limbs in pairs to push back, and with the hindquarters and feet floating behind.

The placodonts, pachypleurosaurs, and nothosaurs all went extinct in the Late Triassic, and the only surviving sauropterygians were the clade **Plesiosauria**. The Plesiosauria had large bodies and limbs that were very strong, equally well developed front and back, and highly modified for swimming. They swam with all four limbs that used the stiffened body as a solid mechanical base, in a further extension of the swimming style of nothosaurs. The limbs were strengthened and further modified for efficient swimming strokes (Figure 11.17). The jaws have modifications that look very well evolved for fish eating.

The Plesiosauria flourished worldwide in marine ecosystems from the Early Jurassic until the end of the Cretaceous. They came in two versions: **plesiosaurs** and **pliosaurs** (Figure 11.13). Plesiosaurs had low, muscular bodies and four very long paddles, and a very long neck and small head (Figure 11.17). An average adult was about 3 m (10 ft) long, with a neck that had 40 vertebrae. Some plesiosaurs were very large too. *Elasmosaurus* from the Cretaceous of Kansas was 12 m (40 ft) long, with 76 neck vertebrae. Plesiosaurs fed on belemnites and fishes.

Pliosaurs had short necks and long, large heads (Figure 11.18). They swam mainly with their strong limbs, all four of which were large, paddle-shaped structures, shaped into effective hydrofoils. Some pliosaurs were huge: *Liopleurodon* reached 10 m (33 ft) long. It is likely that pliosaurs preyed on plesiosaurs, ichthyosaurs, and marine crocodiles.

The plesiosaur paddles were jointed to massive pectoral and pelvic girdles, presumably by very strong muscles and ligaments. Jane Robinson suggested that these structures could be explained if all four limbs were used in an up-and-down power stroke, in underwater "flying" like that of penguins except, of course, that four limbs were

Figure 11.16 Skeleton of the nothosaur *Nothosaurus* from the Middle Triassic Muschelkalk of Germany, on show in the Berlin Museum für Naturkunde. *Source:* Photo by Elke Wetzig (Wikimedia).

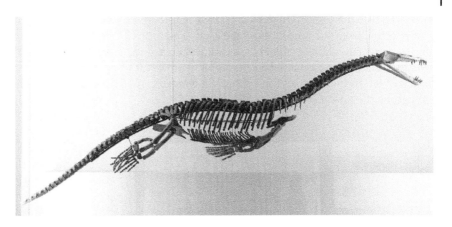

However, it had always been mysterious why plesiosaurs had four equally large paddles, when most other swimmers with paddles, such as marine turtles, penguins, and ichthyosaurs, used only their fore limbs. Did all four plesiosaur limbs work in synchrony? Did the power stroke of both front limbs alternate with the power stroke of both back limbs? Or did right front and left back limb strokes coincide with left front and right back? In recent studies of biomechanics using model paddles in water tanks, paleontology student Luke Muscutt showed that the hind flippers of plesiosaurs generated up to 60% more thrust when they operated in phase with their front paddles, compared with a model where front and back paddles operated independently.

Figure 11.17 The plesiosaur *Seeleyosaurus* from the Early Jurassic of Germany, about 3.5 m (11.5 ft) long. *Source:* Artwork by Dmitry Bogdanov (Wikimedia).

## Mosasaurs

Mosasaurs were essentially very large Late Cretaceous monitor lizards, up to 10 m (30 ft) in length, the largest lizards that have ever evolved. Their evolution of aquatic adaptations in parallel with ichthyosaurs and plesiosaurs is astonishing.

Mosasaur bodies were long and powerful, with tails and limbs adapted for swimming. In early mosasaurs, the main propulsion came from flexing the body and sculling with the tail, which was flat and deep, as it is in living crocodiles. But, in addition, the limbs were modified into beautiful hydrofoils (Figure 11.19). The elbow joint was rigid and the shoulder joint was designed for up-and-down movement. Although the fore limbs could have given some lift, most mosasaurs probably used them as steering surfaces, as dolphins do.

Some forms like *Platecarpus*, however, had well-developed fore limbs (Figure 11.20), and may have used them in a kind of underwater flying, like penguins. The hind limbs were like the fore limbs, although smaller, with the major muscle attachments also giving up-and-down movement. Because the pelvis was not strongly fixed to

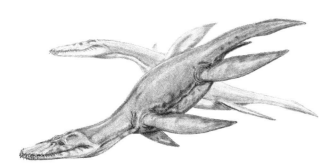

Figure 11.18 The pliosaur *Simolestes* from the Middle and Late Jurassic of England, about 4.6 m (15 ft) long. *Source:* Artwork by Dmitry Bogdanov (Wikimedia).

involved instead of two. But plesiosaur limb joints did not allow the paddles to be lifted above the horizontal. Steven Godfrey suggested instead that the propulsion stroke was downward and backward in a combination of "flying" and rowing (living sea lions swim this way).

the backbone, the hind limb strokes cannot have delivered much power. The hind limbs could rotate and probably worked like aircraft elevators to adjust pitch and roll. But the tail must have been large and powerful, because it had the backbone bent downward to strengthen the power stroke (Figure 11.20).

Mosasaurs had long heads set on a flexible but powerful neck. The large jaws often had a hinge halfway along the lower jaw, which may have served as a shock absorber as the mosasaur hit a large fish at speed. This hinge and the powerful stabbing teeth (Figure 11.21) suggest that most mosasaurs ate large fishes. Other mosasaurs had large, rounded, blunt teeth like those of *Placodus* (Figure 11.14), and they probably crushed mollusk shells to reach the flesh inside.

It is rare to find fossilized skin but the skin of the Cretaceous mosasaur *Ectenosaurus* (Figure 11.22) gives special insight into its biology. The skin has diamond-shaped scales, tightly bound to one another by layers of muscle fibers. Thus *Ectenosaurus* (and possibly other mosasaurs) had a stiffened coat over the body, which would have allowed efficient water flow at speed. This is consistent with the teeth, jaws, and size of late mosasaurs,

Figure 11.19 The fore limb of a mosasaur evolved into an excellent swimming structure. *Source:* Photograph by Dr Mark A. Wilson (Wikimedia).

Figure 11.21 The skull of *Platecarpus*, showing its fish-stabbing teeth. *Source:* Photograph by Yaakov (Wikimedia).

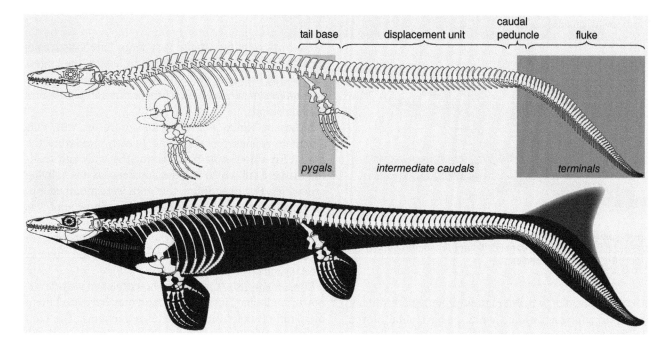

Figure 11.20 (a) Skeleton and (b) reconstruction of the body outline in the Cretaceous mosasaur *Platecarpus*. Source: From Lindgren et al. 2010, Figure 8 in PLoS One 5(8): e11998. Publication in PLoS One places the images in the public domain.

(a) (b)

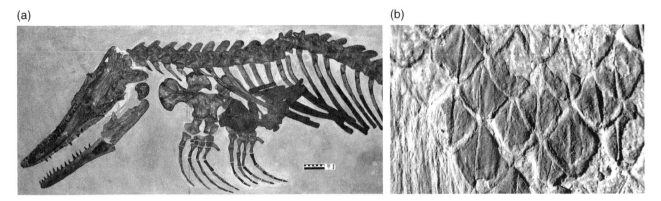

Figure 11.22 (a) The Cretaceous mosasaur *Ectenosaurus* (scale in cm). (b) A rare find of fossilized skin on this animal. It has diamond-shaped scales that were tightly bound to one another by strands of muscle fibers under them, helping to form a very stiff skin that probably helped fast swimming. The scales are about 2 × 2 mm in size. *Source:* From Lindgren et al. 2011, Figure 1 and Figure 2 in PLoS One 6(11): e27343. Publication in PLoS One places the images in the public domain.

Figure 11.23 Reconstruction of the world's oldest turtle, *Pappochelys*, from the Middle Triassic of Germany, showing the broadened ribs (*orange*) and belly ribs (*red*), which are already expanded, and would become the carapace and plastron respectively in later turtles. *Source:* Artwork by Rainer Schoch, with permission.

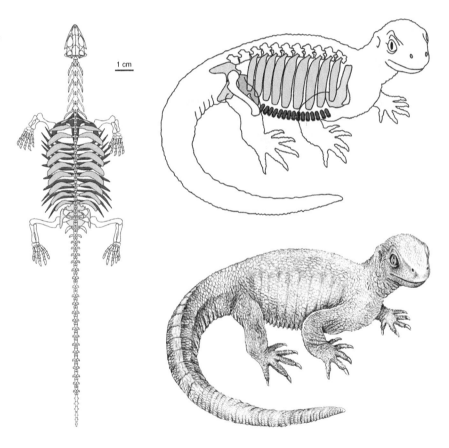

and makes them even more convincingly the ecological equivalent of the huge fish-chasing ichthyosaurs and plesiosaurs that preceded them in Cretaceous seas.

## Turtles

Turtles are diapsids (see Chapter 12). The oldest turtle is *Pappochelys*, described by Rainer Schoch and Hans-Dieter Sues from lake sediments of the Middle Triassic of Germany (Figure 11.23). At first, *Pappochelys* does not look much like a turtle, as it has teeth in its jaws and it lacks a shell. However, it shares features of the limbs and shoulder girdle with later turtles, which confirms it is a so-called "stem-turtle," meaning "nearly, but not quite" a turtle.

*Pappochelys* solved a long-running dispute among paleontologists about the origin of turtles. Researchers used to argue that turtles were parareptiles, related to Permian animals like the pareiasaurs (see Chapter 10), yet all molecular evidence said that turtles were diapsids. Now, with its diapsid skull, *Pappochelys* shows the molecular results were right. In addition, the skeleton of this Middle Triassic pond dweller shows that the turtle shell, comprising a **carapace** over the top and a **plastron** underneath, had formed from broad ribs and belly ribs respectively.

In the Late Triassic, there were marine turtles such as *Odontochelys* from China and terrestrial forms such as *Proganochelys* from the Late Triassic of Europe. This turtle already had bony plates on its surface, though it had not yet accomplished the turtle trick of having the shoulder blades inside the ribs.

Turtles were widespread and successful in Jurassic and Cretaceous seas and estuaries. Perhaps the most famous is the giant Cretaceous turtle *Archelon*, which was 3 m (10 ft) long and nearly 4 m (13 ft) in flipper span. It was so large that it could not carry the weight of a complete solid carapace, so it had only a bony framework (Figure 11.24). Large marine turtles have hydrofoil-like paddles and they "fly" underwater. Marine turtles can navigate precisely over thousands of kilometers and are warm-blooded, maintaining their body temperatures at levels significantly higher than the water around them.

## Crocodiles

Crocodiles are archosaurs and their ancestry is clearly terrestrial (see Chapter 12). All crocodiles were terrestrial predators in the Late Triassic. There were large, powerful crocodile-like aquatic phytosaurs in the Triassic, and true crocodiles did not become aquatic until these others became extinct. *Saltoposuchus*, a crocodile from the Triassic of Britain, had long, slim, erect limbs and probably ran quite fast on land (see Chapter 12, Figure 12.14).

Ever since the Early Jurassic, most crocodiles have been amphibious. Indeed, during the Jurassic and Cretaceous, several lines became entirely marine, with their feet modified into paddles and bearing a tail fin to improve their swimming ability. These marine crocodiles fed on fishes, which they snatched in their often slender, long jaws, lined with long thin teeth that formed a kind of cage as the jaws closed around some wriggling fish. Larger marine crocodiles like *Dakosaurus* (Figure 11.25) had massive skulls and broad-based teeth, evidently adapted to snapping down on other marine reptiles and puncturing their heads.

Some fossil crocodiles evolved to large size and were reasonably common in Mesozoic seas and rivers. There are several candidates for the largest crocodile that ever lived: four separate lineages had species over 10 m (33 ft) long and perhaps 5 tons in weight (Figure 11.26). *Deinosuchus* from the Late Cretaceous of Texas may have taken duckbilled dinosaurs as prey (they are found in the same rock formations) in the same way that living Nile crocodiles take hippos. All of these crocodiles were much larger than the largest living crocodiles, which reach 6 m at most.

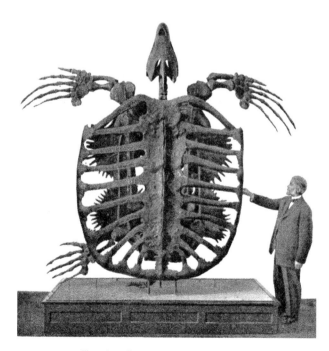

Figure 11.24 The giant Cretaceous marine turtle *Archelon* evolved a carapace that was lightened so that the turtle could maintain buoyancy in the water. This 1902 image is in the public domain.

Figure 11.25 The marine crocodile *Dakosaurus* vied with pliosaurs and ichthyosaurs to be top predator in Late Jurassic seas. This 5 m (16.5 ft) crocodile leaps from the water, startling two pterosaurs. *Source:* Artwork by Dmitry Bogdanov (Wikimedia).

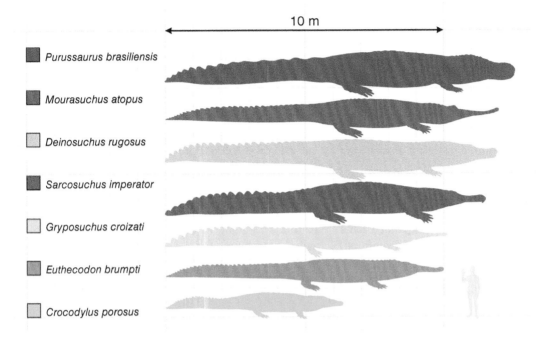

10 m

Purussaurus brasiliensis

Mourasuchus atopus

Deinosuchus rugosus

Sarcosuchus imperator

Gryposuchus croizati

Euthecodon brumpti

Crocodylus porosus

Figure 11.26 Huge fossil crocodiles were much larger than those living today. The silhouettes are placed on a meter grid. Four different genera reached over 10 m long. From top to bottom, the crocodiles are *Purussaurus* and *Mourasuchus* from the Miocene of South America; *Deinosuchus* from the Late Cretaceous of Texas; *Sarcosuchus* from the Early Cretaceous of Africa; *Gryposuchus* from the Miocene of South America; *Euthecodon* from the Miocene of East Africa; and the largest living species, the salt-water crocodile of Australia, *Crocodylus porosus*. *Source:* Diagram by Smokeybjb (Wikimedia).

Crocodiles today are not equipped to kill large prey quickly. They can use two hunting modes: twist-pulling and drowning. In twist-pulling, the crocodile lunges at the prey animal, say a wildebeest, and grabs a limb or fleshy part in its jaws and then flings its body into a twist or death roll. Underwater, it can readily do this, and this then means it is using all the muscles of its body, rather than just the jaw muscles, to rip off the meaty mouthful. In drowning, the crocodile simply holds its prey under water until it can breathe no more. It's likely that the giant Cretaceous dinosaurs used these hunting modes when they preyed on dinosaurs.

## Air Breathers at Sea

All these Mesozoic reptiles were air breathers and therefore faced special problems for life in the sea. Precisely the same problems are faced today by marine mammals. The major one, of course, is the fact that air breathers must visit the surface for air, but there are also problems in introducing air-breathing young to a complex and dangerous world where they must be ready to use sophisticated skills immediately after birth.

Many marine reptiles, mammals, and birds return to the shore for reproduction. Turtles simply lay large clutches of eggs and leave them buried in the sand, a method that results in horrific mortality but has obviously worked successfully for 200 million years. Seals, sea lions, and penguins have their young on shore in safe nurseries, so that they can breathe air, be fed, and grow for a while before they take to a swimming and foraging life at sea.

But dolphins and whales never come ashore. They have special adaptations for air breathing, breeding, giving birth and caring for the young at sea. The young are born tail first, and mothers and other related adults will push them to the surface until they learn to breathe properly. The young must be able to dive immediately to suckle, and whales feed their babies milk under high pressure. There is increasing evidence that all the major groups of Mesozoic marine reptiles solved the same kinds of problems in spectacular fashion.

Several fossils of ichthyosaurs have been found with young preserved inside the rib cage of adults, evidence that ichthyosaurs had evolved live birth. The preserved fetuses have long, pointed jaws, showing that they would have been able to feed for themselves immediately after birth, and they were born tail first as whales are. Some specimens even show a tiny embryo in the sediment close behind the birth canal of the mother, but it is likely these were expelled by decay gases.

The early sauropterygians had limbs that would have allowed them to haul themselves out onto a beach to lay eggs or to give birth, rather like sea lions. However, it is likely that all sauropterygians had evolved live birth from the Middle Triassic onwards, as embryos have been found inside the body of a pachypleurosaur. Nothosaurs and plesiosaurs are usually much larger and would have had to work much harder to drag themselves up a beach, so they were likely to have been totally sea-going, with live birth at sea. This has been confirmed for one plesiosaur at least; a *Polycotylus* from the Late Cretaceous was fossilized with a large fetus inside it, certain evidence of live birth at sea.

Mosasaurs do not look as if they would readily have come ashore to lay eggs, and no fetuses or even juveniles have been found associated with adults. However, the mosasaur pelvis is very unusual in being expanded. This may have resulted simply from adaptation to swimming, but perhaps the normal pelvis was expanded to give birth to live offspring much bigger than any normal egg. Embryos have been found inside an aigialosaur, which is a mosasauroid, a basal relative of mosasaurs proper.

Therefore, all mosasauroids may have had live birth, though it would be reassuring to have more evidence.

Many large and powerful swimming creatures today are warm-blooded to some extent, including some sharks, tuna, and several turtles, as well as dolphins. The metabolic effort of swimming contributes to a warm body. There is direct evidence from the microscopic structure of the bone that ichthyosaurs and plesiosaurs were warm-blooded. This is confirmed by evidence from oxygen isotope measurements of their bones which show that they maintained constant body temperatures around 35–39 °C, similar to body temperatures of mammals today.

Almost all these magnificent marine reptiles became extinct at the end of the Cretaceous, along with dinosaurs, pterosaurs, and a significant number of marine invertebrates. Only the egg-laying crocodiles and turtles have survived to give us some clues about the mode of life of large reptiles.

## References

Chen, Z.Q. and Benton, M.J. (2012). The timing and pattern of biotic recovery following the end-Permian mass extinction. *Nature Geoscience* 5: 375–383.

Lindgren, J., Caldwell, M.W., Konishi, T. et al. (2010). Convergent evolution in aquatic tetrapods: insights from an exceptional fossil mosasaur. *PLoS One* 5 (8): e11998.

Lindgren, J., Everhart, M.J., Caldwell, M.W. et al. (2011). Three-dimensionally preserved integument reveals hydrodynamic adaptations in the extinct marine lizard *Ectenosaurus* (Reptilia, Mosasauridae). *PLoS One* 6 (11): e27343.

## Further Reading

Bellwood, D.R., Goatley, C.H.R., Bellwood, O. et al. (2015). The rise of jaw protrusion in spiny-rayed fishes closes the gap on elusive prey. *Current Biology* 25: 2696–2700. [Reasons for success of the teleost fishes].

Benton, M.J., Zhang, Q.-Y., Hu, S.-X. et al. (2013). Exceptional vertebrate biotas from the Triassic of China, and the expansion of marine ecosystems after the Permo-Triassic mass extinction. *Earth-Science Reviews* 123: 199–243. [New evidence from China about Triassic seas, and the early start of the MMR].

Blackburn, D.G. and Sidor, C.A. (2015). Evolution of viviparous reproduction in Paleozoic and Mesozoic reptiles. *International Journal of Developmental Biology* 58: 935–948. [Live birth in marine reptiles].

Buchholtz, E.A. (2001). Swimming styles in Jurassic ichthyosaurs. *Journal of Vertebrate Paleontology* 21: 61–73.

Fischer, V., Bardet, N., Benson, R.B.J. et al. (2016). Extinction of fish-shaped marine reptiles associated with reduced evolutionary rates and global environmental volatility. *Nature Communications* 7: 10825. [The last ichthyosaurs].

Moon, B.J. (2019). A new phylogeny of ichthyosaurs (Reptilia: Diapsida). *Journal of Systematic Palaeontology* 2: 129–155.

Motani, R. (2009). The evolution of marine reptiles. *Evolution: Education and Outreach* 2: 224–235.

Muscutt, L.E., Dyke, G., Weymouth, G.D. et al. (2017). The four-flipper swimming method of plesiosaurs enabled efficient and effective locomotion. *Proceedings of the Royal Society B* 284: 20170951. [Physical experiments in plesiosaur flipper function].

Schoch, R.R. and Sues, H.-D. (2015). A Middle Triassic stem-turtle and the evolution of the turtle body plan. *Nature* 523: 584–587.

Stubbs, T.L. and Benton, M.J. (2016). Ecomorphological diversifications of Mesozoic marine reptiles: the roles of ecological opportunity and extinction. *Paleobiology* 42:

547–573. [Changing shape and structure of Mesozoic marine reptiles].

Vermeij, G.J. (1977). The Mesozoic Marine Revolution: evidence from snails, predators and grazers. *Palaeobiology* 3: 245–258. [The paper that set it all off].

Watson, T. (2017). How giant marine reptiles terrorized the ancient seas. *Nature* 543: 603–607.

## Questions for Thought, Study, and Discussion

1   Make a list of the typical Paleozoic and Mesozoic marine animal groups, such as brachiopods, crinoids, trilobites (Paleozoic) and bivalves, sea urchins, and neopterygian fishes (Mesozoic), and find out some key facts about them, such as their maximum body size, their position in food chains (e.g., herbivore, predator, top predator), and even some culinary facts (do we eat them?). Compare the Paleozoic and Mesozoic faunas.

2   Choose one of the large marine reptiles that lived in Mesozoic seas, and describe the similarities and differences between that group and living whales and dolphins, in terms of anatomy, locomotion, and ecology. If you see some general principles at work, describe them

# 12

# The Triassic Takeover

**In This Chapter**

As we saw in Chapter 6, there was a huge extinction at the end of the Permian. In the following Triassic period, synapsid amniotes were replaced by an impressive array of diapsid reptiles, the ancestors of lizards and snakes, dinosaurs and birds, and crocodiles. Other diapsids returned to living in fresh water or in the ocean, and a few gliding diapsids are known. The reason for the diapsid takeover was a mix of advances in physiology and some extinction events. The physiological advances in both synapsids and diapsids included upright, rather than sprawling, posture and a faster metabolism that was enabled by their ability to run and breathe at the same time. New evidence shows that dinosaurs originated in the Early Triassic and then expanded by steps, generally following mass extinctions dated at 232 and 201 Ma. These perturbed the environments, drove other groups to extinction, and gave dinosaurs their opportunity to take over.

## Diapsids

During the long history of tetrapods, there has been a to-and-fro between the two great clades, the synapsids and the diapsids. First, as we saw in Chapter 10, synapsids were the key animals in ecosystems on land in the latest Carboniferous and Permian. Then, they were hit hard by the Permian–Triassic mass extinction, and diapsids, especially dinosaurs, took over during the Mesozoic. Then they in turn were killed by the end-Cretaceous mass extinction, and synapsids in the form of mammals took over the land in the Cenozoic. The replacement of synapsids by diapsids in the Triassic was so dramatic that it is a debating ground for the general question of replacement of one vertebrate group by another. The variety of Triassic diapsids leads to a mass of unfamiliar names, and we have tried to keep the list as simple as we can. Questions and some possible answers relating to the recovery of life and the diapsid takeover form the major themes of this chapter.

## Basal Diapsids

The first diapsids were animals like *Petrolacosaurus* (see Chapter 10, Figure 10.2), and most basal diapsids were basically lizard like in size, structure, and behavior. But basal diapsids (Figure 12.1) also evolved into some ecologically diverse roles in the Late Permian, perhaps because they were unable to compete with the synapsids. For example, *Coelurosauravus* was a glider in the forest and *Hovasaurus* was aquatic.

Fossils of *Coelurosauravus* have been found in the Late Permian of Germany, Britain, and Madagascar, so it was widespread across Pangea. All these areas were near tropical shores at the time. *Coelurosauravus* is an ordinary, small diapsid reptile in the structure of its skull and body, about the size of a small squirrel. But the trunk is dominated by 20 or more long, curving, lightly constructed rod-shaped bones that extended outward and sideways from the body. They supported a skin membrane that was close to an ideal airfoil in shape, 30 cm

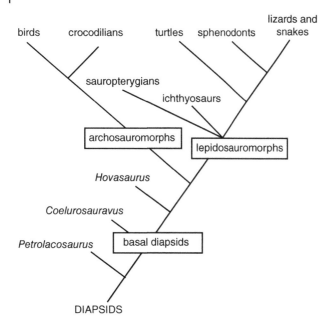

Figure 12.1 Cladogram of major groups of diapsids. The split into archosauromorphs, lepisodosauromorphs, and basal diapsids is clear. More controversial has been the assignment of turtles and the marine reptiles (sauropterygians, ichthyosaurs) within lepidosauromorphs.

Figure 12.2 The Late Permian gliding reptile *Coelurosauravus*. The supports for its airfoil are clearly not ribs: they do not begin at the vertebrae. *Source:* Artwork by Nobu Tamura (Wikimedia).

Figure 12.3 *Hovasaurus*, a swimming diapsid from the Permian of Madagascar. *Source:* Artwork by Nobu Tamura (Wikimedia).

(1 ft) across, and could only have been used for gliding (Figure 12.2). More impressive still, the bones are jointed so that the airfoil could have been folded back along the body when it was not in use. Extra-long vertebrae allowed space for this folding along the spinal column. These bones are not ribs but must have evolved specifically under the skin as a gliding structure.

*Hovasaurus* is known from the Late Permian of Madagascar. Overall, it was lizard-like, perhaps only 30 cm (1 ft) long from snout to vent. But the tail was exceptionally long, strong, and deep (Figure 12.3), so the whole animal was close to a meter (3 ft) in length. The tail had at least 70 vertebrae with long spines above and below, making it a perfect flat-sided swimming aid. The abdominal cavity usually contains a mass of small quartz pebbles. Presumably they were swallowed by the animal during life. They are too small to be food-grinding pebbles and too far back in the abdomen to have occupied the stomach in life. *Hovasaurus* almost certainly swallowed the stones as ballast for diving. Living Nile crocodiles do the same thing, and the extinct plesiosaurs may have done so too. This means that a very early diapsid had evolved a relatively sophisticated adaptation to an aquatic way of life.

The radiation into major diapsid clades on land began in the latest Permian but was truly spectacular in the Triassic. The key driver was the Permian–Triassic mass

extinction which, as we saw in Chapter 10, hit the synapsids hard. Synapsids never really recovered and through the Triassic, they got smaller and smaller, ending in the mammals (see Chapter 16), whereas diapsids got larger and larger, ending in the dinosaurs.

There are two living clades of diapsids: the **lepidosaurs** (lizards, snakes, and the tuatara of New Zealand) and the **archosaurs** (crocodiles and birds) (Figure 12.1). We use a crown-group definition of the clades Lepidosauria and Archosauria: all those diapsids that are more closely related to the living survivors than to anything else. There were extinct clades that branched off below the base of these crown groups, and these are placed in larger clades called Lepidosauromorpha and Archosauromorpha – these names mean lepidosaur-like and archosaur-like.

## Lepidosauromorphs

On land, the crown-group lepidosaurs have been the dominant group of small-bodied reptiles since the Mesozoic. They consist of four major clades (Figure 12.1). **Squamata** are the lizards and snakes, comprising over 10000 species today, and ranging in size from the tiny dwarf gecko, only 16 mm (0.63 in.) long, to the green anaconda snake, at up to 5.21 m (17.1 ft) long. Among fossil forms, the mosasaurs were especially impressive marine lizards of the Late Cretaceous (see Chapter 13).

Close relatives of Squamata are **Sphenodontia,** including only one living species, the tuatara, *Sphenodon* (Figure 12.4), a seemingly lizard-like animal that survives today only on a few islands off the coast of New Zealand (see Chapters 17 and 21). Sphenodonts are known as far back as the Triassic, when many species lived worldwide, and the living form is sometimes called a **living fossil**. This is because it looks very like its early ancestors from the Triassic and Jurassic, and also because there is only the single living species. In the evolutionary tree, Squamata and Sphenodontia branched apart sometime in the Triassic — one branch led to 10000 species, the other to one. This is a classic comparative example in evolution: they started equal, so why did one clade become so species-rich and the other not?

Slightly unexpected among lepidosauromorphs are the turtles and marine reptiles. **Turtles** are certainly diapsids and may be basal lepidosauromorphs. Also branching here in the earliest Triassic were the extinct marine reptiles, the fish-like **Ichthyosauria** and the more diverse **Sauropterygia**, which includes placodonts and plesiosaurs. All these were hugely important marine predators throughout the Mesozoic, as we saw in Chapter 11.

Some of the uncertainties about relationships have led to squabbles among biologists and paleontologists. Turtles in particular have proved very hard to pin down. This is partly because all living and fossil turtles share a unique structure of the shell and much modified skeleton, and this in a sense "overprints" a lot of possible clues about their ancestry. But the molecular evidence is clear – they are lizard relatives. What was going on was probably an explosive burst of diapsid evolution in the first million years or so of the Triassic – sometimes called a **star phylogeny**, meaning the evolutionary branches exploded on all sides. Everything was changing fast in the empty postextinction world, and there was no time for shared novel anatomical features to become fixed.

## Archosauromorphs

Archosauromorphs include dinosaurs, the largest land animals that ever lived, and they rose to dominate land ecosystems by the Late Triassic. This is the big story of Triassic terrestrial ecosystems, how the archosauromorphs exploded onto the scene, taking advantage, ecologically speaking, of the devastatation wrought by the Permo-Triassic mass extinction. Archosauromorphs arose in the Permian, including their core clade, Archosauria, but they came to dominance only in the Triassic.

Archosauromorpha includes several groups (Figure 12.5), of which we consider two here. **Rhynchosauria** are basal archosauromorphs that were the dominant large herbivores for a brief period during the Triassic.

Figure 12.4 The tuatara *Sphenodon* in its natural habitat. *Source:* Photograph by PhillipC (Wikimedia).

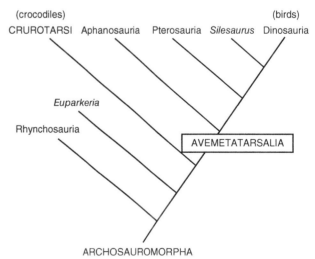

Figure 12.5 Cladogram of archosauromorphs. Crurotarsi is the clade that includes living crocodilians and extinct forms that are closely related to them, and Avemetatarsalia is the 'bird line', leading to dinosaurs and birds.

**Archosauria** includes two great clades. One is the crocodile-like Crurotarsi (crocodiles, alligators, and related extinct forms), and the other is the bird-like Avemetatarsalia (pterosaurs, plus dinosaurs and their subgroup the birds). The pterosaurs evolved true flapping flight much earlier than birds did, and they dominated the skies throughout the Mesozoic (see Chapter 13). The Dinosauria are discussed in Chapter 13 and birds, a side branch of the dinosaurs, are discussed in Chapter 14.

## The Triassic Diapsid Takeover: The Pattern

As we saw in Chapter 10, synapsids of all shapes and sizes were the key land animals in the Middle and Late Permian. Key herbivores were the dinocephalians in the Middle Permian and dicynodonts in the Late Permian, and these animals could walk all over Pangea – paleontologists have reported fossils of the same genera of dicynodonts and other therapsid synapsids from as far apart as modern South Africa, China, and Russia.

The Permo-Triassic mass extinction wiped out 95% of synapsid species. For example, the dominant Late Permian dicynodonts were reduced from 40 or 50 species before the crisis to only two or three survivors. Most famous was *Lystrosaurus* (Figure 12.6), known in the latest Permian from South Africa, but which survived and then spread worldwide in the earliest Triassic. It is a famous **disaster species**, an organism that either by good adaptation or more likely good luck survived into the devastated postextinction world.

In the Early and Middle Triassic, dicynodonts were extraordinarily abundant at larger sizes, and cynodonts included medium-sized herbivores. There were few therapsid predators; most of them were small- and medium-sized cynodonts such as *Cynognathus* (see Chapter 15). Some of the early archosauromorphs were

small and carnivorous, although therapsids outnumbered them 65 to 1 at first. But by the end of the Early Triassic, some archosauromorphs were 5 m (16 ft) long, with massive skulls a meter long. In South Africa, *Euparkeria* was a fast, lightly built carnivore that is very close to the ancestry of the Archosauria (Figure 12.7).

Therapsids, especially dicynodonts, were the dominant herbivores well into the Late Triassic but by the end of the Middle Triassic, rhynchosaurs became abundant vegetarians alongside the dicynodonts. There were even greater changes among the carnivores. Diapsid carnivores of various sizes became abundant from the start of the Triassic. Among therapsids, cynodont carnivores were at most medium-sized but were still abundant and diverse.

Therapsids and rhynchosaurs declined in ecological importance in the Late Triassic. The vegetarians of the latest Triassic were almost all prosauropod dinosaurs; diapsid carnivores were larger, more diverse, and more mobile than before, and they were joined by the first theropod and ornithischian dinosaurs. The first mammals were few and small.

Finally, at the end of the Triassic, dinosaurs quickly overwhelmed terrestrial ecosystems throughout the world, replacing crurotarsans and therapsids alike in every medium- and large-bodied way of life, to form a land fauna dominated by dinosaurs that lasted throughout the Jurassic and Cretaceous Periods.

The replacement of therapsids by archosaurs was worldwide. There have been two kinds of explanations for the takeover – competition and catastrophe. The competition idea is that the diapsids, especially the archosaurs, were superior to the synapsids and they overwhelmed them step by step through the Triassic. The other view is that dinosaurs benefited from one or more extinction events that wiped out the dicynodonts and rhynchosaurs and gave them their opportunity. In fact, it seems to have been a bit of both, based on recent discoveries. But let us look first at the rhynchosaurs and the Triassic archosaurs in a little more detail.

Figure 12.6 **The great survivor.** *Lystrosaurus*, a dicynodont synapsid that survived the Permian–Triassic mass extinction and spread all over the world. This is a disaster taxon, meaning it recovered fast after the crisis but then died out. *Source:* Artwork by Nobu Tamura (Wikimedia).

Figure 12.7 *Euparkeria* from the Late Permian of South Africa seems to be close to the ancestor of all later archosaurs. *Source:* Image by Taenadoman.

## Rhynchosaurs

Rhynchosaurs evolved in the Middle and Late Triassic. They were all herbivores, pig-sized animals with hooked snouts bearing a powerful cutting beak and hind limbs that look as if they might have been used for digging. Strong jaws bore batteries of slicing teeth, which are unusual among reptiles in that they were fused to bone at the base, not set into normal sockets. The teeth were ever-growing and were not replaced during life. As rhynchosaurs grew, they simply added more bone and more teeth at the back of the growing jaw as the teeth at the front became worn out. This style of tooth addition allowed rhynchosaurs great precision in tooth emplacement, so their bite was very effective for slicing vegetation with a scissor-like action (Figure 12.8).

Rhynchosaurs are basal archosauromorphs (Figure 12.1). They were abundant and widespread in the Middle and Late Triassic. When they occur, they are the dominant herbivores, sometimes making up as many as 80% of all animals in a fauna. Rhynchosaurs apparently died out rather rapidly about 232 Ma, and we will see below why this might have been important in understanding how dinosaurs took over the Earth.

## Triassic Archosaurs: Crurotarsi

The line to archosaurs began in the latest Permian, with quadrupedal carnivores like *Archosaurus* from Russia (Figure 12.9). These animals then diversified rapidly in the Early Triassic, all initially carnivores, and some remaining at about 1 m (3 ft) long whereas others were five times that size. Their success seems to connect to a major change in their posture, from sprawling to upright. Whereas the Permian diapsids mostly held their limbs out to the side, like modern lizards, the Triassic archosaurs held their limbs tucked under the body, in an upright or erect posture.

The new upright posture in Triassic archosaurs meant that their ankle joints had to change. The limb movement had gone from a broad sweep out to the side, as seen in sprawlers, to a simple back-and-forwards motion. There are many bones in the ankle region and possibly because of this skeletal legacy, there are alternative joint layouts. Two lineages of archosaurs can be distinguished on this basis (and other characters): **Crurotarsi**, which are represented today only by crocodilians, and **Avemetatarsalia**, which evolved into pterosaurs and dinosaurs, and are represented today by birds. Each lineage exploited the ankle to achieve more erect gait, culminating not only in erect-limbed dinosaurs but erect-limbed crurotarsans too.

For most of the Middle and Late Triassic, the largest carnivorous groups were crurotarsan archosaurs such as rauisuchians and ornithosuchians, two groups of basal crurotarsans typified by *Ornithosuchus* (Figure 12.10). *Postosuchus*, from the Late Triassic of Texas, was about 4 m long including the tail, and stood 2 m high. It was lightly built and could have walked on four feet (Figure 12.11), but it surely ran bipedally. It was a hunter, with a heavy killing head, impressive wide-opening jaws,

Figure 12.9 One of the first archosauromorphs, *Archosaurus* from the Late Permian of Russia. It's one of the wonderful legal quirks of cladistics that *Archosaurus* is NOT an archosaurian, but merely an archosauromorph! *Source:* Artwork by Dmitry Bogdanov (Wikimedia).

Figure 12.10 A typical large powerful carnivorous Triassic archosaur, *Ornithosuchus* from the Late Triassic of Scotland. About 4 m (13 ft) long. *Source:* Artwork by Nobu Tamura (Wikimedia).

Figure 12.8 The Triassic rhynchosaur *Hyperodapedon*. About 1.3 m (4 ft) long. *Source:* Artwork by Nobu Tamura (Wikimedia).

Figure 12.11 *Postosuchus*, a 4 m long rauisuchian from the Late Triassic of Texas. *Source:* Photograph by Dallas Krentzel (Wikimedia).

Figure 12.12 *Stagonolepis*, a 3 m (10 ft) long aetosaur from the Late Triassic of Scotland. *Source:* Artwork by Arthur Weasley, based on original by Alick Walker (Wikimedia).

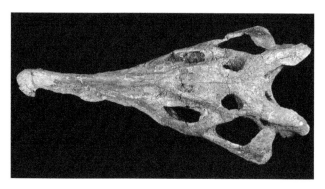

Figure 12.13 A phytosaur skull from the Triassic of Petrified Forest National Park, Arizona. *Source:* National Park Service photograph.

Figure 12.14 The terrestrial crocodilian *Saltoposuchus* from the Late Triassic of England, presumably a small fast agile predator. *Source:* Artwork by Nobu Tamura (Wikimedia).

and serrated stabbing and cutting teeth. The eyes were large, set for forward stereoscopic vision, with bony eyebrows to shade them. *Postosuchus* is uncannily like a small version of the much larger and later carnivorous dinosaurs *Allosaurus* and *Tyrannosaurus* in overall body plan and presumably in ecology.

Quite different in appearance were the **aetosaurs**, which were crurotarsan archosaurs of the Late Triassic that specialized in feeding on tough vegetation, probably grubbed out of the ground. Aetosaurs such as *Stagonolepis* (Figure 12.12) are characterized by triangular-shaped skulls, rather small teeth, and a cut-off snub nose. It is assumed they used their toughened snout to sniff for and dig up roots and tubers. Aetosaurs carried an impressive set of armor plates, generally square or rectangular, and arranged in neat rows all over their backs and bellies – presumably some form of protection against the hungry rauisuchian predators.

One group of crurotarsans explored a way of life that we now associate with living crocodiles: ambush hunting at the water's edge. **Phytosaurs** were large, long-snouted carnivores from the tropical belt of the Late Triassic which evolved toward a crocodilian appearance and ecology (Figure 12.13). Two phytosaurs from India more than 2 m (7 ft) long had stomach contents that included small bipedal archosaurs, and one had eaten a rhynchosaur!

Living crocodiles may well be some guide to the physiology, locomotion, and ecology of phytosaurs. Crocodiles have a good circulatory system, with more advanced heart and lung modifications than other living reptiles. Although they normally walk slowly on land, in a sprawling stance, they are also capable of a faster run in which the limbs are nearly vertical. The little freshwater crocodile of Australia can gallop (briefly) at 16 kph (10 mph).

Rauisuchians, aetosaurs, and phytosaurs disappeared at the end of the Triassic, with many other archosaur groups. They were replaced in that ecological niche by true crocodilians. Earlier crocodilians had been small,

long-legged, terrestrial predators, like *Saltoposuchus*, from the Late Triassic of western Europe (Figure 12.14). Jurassic crocodilians adapted to water, replacing the phytosaurs, and only then did they become much larger. They evolved a secondary palate so that they could bite and chew under water without flooding their nostrils, and they also lost some of the earlier changes of their terrestrial gait, becoming secondarily sprawling.

## Dinosaur Ancestors: Avemetatarsalia

Until the year 2000, it was understood that dinosaurs first appeared about 233 Ma, in the Late Triassic. However, several lines of evidence have pushed their origin back into the Early Triassic, about 250 Ma. This is geologically a short shift in time, of only 17 m.y., but it changes our understanding of dinosaur origins substantially.

The new evidence consists of footprints and skeletons. First, some small three-toed footprints were reported from the Early and Middle Triassic of Poland, and these

Figure 12.15 *Silesaurus*, a dinosauromorph, meaning a very close relative of dinosaurs, measuring 2.3 m (7.5 ft) long, from the Late Triassic of Poland. *Source:* Artwork by Nobu Tamura (Wikimedia).

could have been made by dinosaurs or close relatives. Then, a possible early dinosaur and a silesaurid were reported from the Middle Triassic of Tanzania. The silesaurids were the closest relatives of dinosaurs – their so-called sister group – and they were all small, agile, and bipedal at first. A typical silesaurid is *Silesaurus* (Figure 12.15), from the Late Triassic of Poland, which looks very like an early dinosaur but lacks only a couple of skull features; it was a small herbivore with a beak for cropping vegetation, like many later dinosaurs.

Even more impressive was the discovery of a whole new clade of avemetatarsalians. Reported in 2017, the Aphanosauria are all Middle Triassic in age. They are based around *Teleocrater*, a 1.5 m (5 ft) long, slender quadruped from Tanzania, which shows some dinosaur-like characters in the vertebrae and hind legs.

If dinosaurs were present in the latest part of the Early Triassic and throughout the Middle Triassic, they remained rare and elusive. Then, dinosaurs emerged in the early part of the Late Triassic, especially in South America. For example, *Herrerasaurus* and *Eoraptor* were 6 m and 1 m long, bipedal carnivores living in Argentina alongside a fauna dominated by rhynchosaurs, with synapsids present too.

Then, through the remaining 32 m.y. of the Late Triassic, dinosaurs became more and more diverse and some became truly large, such as the 10 m (33 ft) long *Plateosaurus* from Germany and *Riojasaurus* from Argentina. These were plant eaters and because of their size, they could switch between being bipedal when running and quadrupedal when feeding or resting. Importantly, dinosaurs became more and more important components of their ecosystems through these years of the Late Triassic, as they took over the world, which we shall explore further in Chapter 13.

## Respiration, Metabolism, and Locomotion

The story of how the dinosaurs took over the world in the Triassic is a mix of biological innovations and profound climate change driven by a mass extinction. The biological innovations centered around their upright posture linked with warm-bloodedness and the diaphragm. David Carrier was the first to point out the necessary links between respiration, locomotion, and physiology (Carrier 1987).

The key is the difference between sprawlers and upright walkers such as dinosaurs and mammals. Most modern amphibians and reptiles, such as salamanders and lizards, look awkward when they move. As the animal steps forward with its left front foot, the right side of the chest and the lung inside it are compressed while the left side expands (Figure 12.16). Then the cycle reverses with the next step. This distortion of the chest interferes with and essentially prevents normal breathing, in which the chest cavity and both lungs expand uniformly and then contract. If the animal is walking, it may be able to breathe between steps, but sprawling vertebrates cannot run and breathe at the same time. In the first edition of this book, Richard Cowen called this **Carrier's Constraint**, and the name seems to have stuck.

Animals can run for a while without breathing; for example, Olympic sprinters usually do not breathe during a 100 m race. Animals can generate temporary energy by *anaerobic glycolysis*, breaking down food molecules in the blood supply without using oxygen. But this process soon builds up an oxygen debt and a dangerously high level of lactic acid in the blood. Mammalian runners (cheetahs and humans, for example) often use anaerobic glycolysis even though they can breathe while they run;

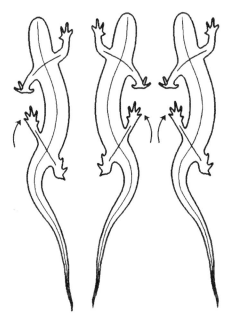

Figure 12.16 Lizard locomotion. David Carrier pointed out that the sprawling locomotion of a lizard or salamander forces it to compress each lung alternately as it moves (see text).

it's a useful but essentially short-term emergency boost, like an afterburner in a jet fighter.

Living amphibians and reptiles can hop or run fast only for a short time, first using up the oxygen stored in their lungs and blood, then switching to anaerobic glycolysis. They cannot sprint for long, however. If lizards want to breathe, they have to stand still (Figure 12.17). Lizards run in short rushes, with frequent stops. Modern amphibian and reptilian carnivores use ambush tactics to capture agile prey: chameleons and toads flip their tongues at passing insects, for example.

Early tetrapods all had sprawling gaits and faced a great problem. Their respiration and locomotion used much the same sets of muscles, and both systems could not operate at the same time. Imagine the laborious journey of *Ichthyostega* from the water to its breeding pools, with a few steps and a few gasps repeated for the whole journey. This may be why so many early tetrapods through the Devonian and Carboniferous remained adapted to life spent largely in water.

Mammals and birds have evolved a beautiful answer to Carrier's Constraint. They have freed the mechanics of respiration from the mechanics of locomotion by evolving an **upright** stance. With upright limbs, the thorax does not twist much as the animal walks or runs, allowing it to make its breathing movements with hardly any distortion of the lungs.

Mammals went one step further than birds and evolved the **diaphragm**, a set of muscles to pump air in and out of the chest cavity. Air is sucked in as the diaphragm contracts and forced out by the reaction of the elastic tissues of the lung. At the same time, the locomotion in most mammals has evolved to encourage breathing on the run. The backbone flexes and straightens in an up-and-down direction with each stride, alternately expanding and compressing the rib cage evenly (Figure 12.18). This rhythmic pumping of the chest cavity in the running action can be synchronized with the action of the diaphragm to move air in and out of the lungs with little effort. Thus, quadrupeds running at full speed – gerbils, jackrabbits, dogs, horses, and rhinoceroses – take one breath per stride, and wallabies take one breath per hop.

These changes all happened in the Triassic in both diapsids and synapsids, the ancestors of the modern erect-postured birds and mammals. There are two lines of evidence – footprints and bone histology. When Tai Kubo, a Japanese paleontologist, looked at a large database on tetrapod trackways, he noted a step change (if one may make that pun) at the Permian–Triassic boundary (Figure 12.19). All Permian tracks were made by sprawlers; all Triassic tracks were made by animals with erect posture. He ignored small tracks made by lizard-sized animals, which continued as sprawlers, but focused on tracks made by animal over 1 m long. So, synapsids and diapsids alike switched

(a)

(b)

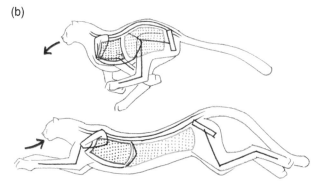

Figure 12.18 (a) A running cheetah. Photography by Malene Thyssen (Wikimedia). (b) A cheetah breathing while it runs, aided by the diaphragm. *Source:* Diagram by Coluberssymbol (Wikimedia).

Figure 12.17 This lizard has stopped to take a breath. *Source:* Photograph in Zion National Park by Thomas Schoch (Wikimedia).

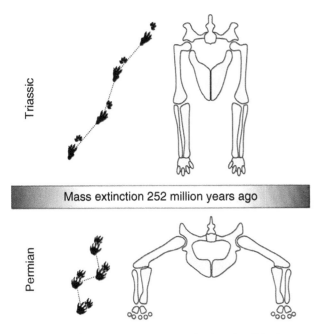

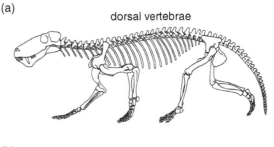

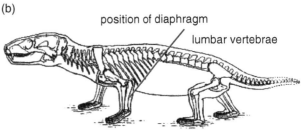

Figure 12.19 Change in posture across the Permian–Triassic boundary, from sprawling before to upright after. Note how, in the sprawling track, the footprints are wide apart but in the upright track, they are closer to the midline. *Source:* Image by Simon Powell, University of Bristol.

Figure 12.20 The transition to fast metabolism in synapsids, from the Permian *Lycaenops* (a) with ribs back to the pelvis, and probably no diaphragm, to Triassic *Thrinaxodon* (b) with ribs pushed forward, distinct lumbar vertebrae, and almost certainly a diaphragm. The diaphragm helped mammals, and Triassic synapsids, to pump oxygen as they ran.

from sprawling to erect posture, and this marks a switch in physiology, as predicted by Carrier, and evidence for a so-called **arms race**, where predators and prey were speeding up.

Second, Triassic tetrapods all show some evidence of warm-bloodedness in their **bone histology**, that is, the cellular structure of their bone under the microscope. Once, it was argued that only dinosaurs were possibly warm-blooded (see Chapter 13), but this evidence goes deeper, to their ancestors. Further, some of these Triassic forms might have possessed simple feathers. Triassic synapsids too, on the way to being mammals, already had diaphragms, as shown by the forward move of their rib cage (Figure 12.20), probably had hair, and were likely warm-blooded too. This is all part of the acceleration of tetrapod life in the Triassic.

## The Carnian Pluvial Episode (CPE) and the Rise of the Dinosaurs

As we saw earlier, the first dinosaurs emerged at the end of the Early Triassic, probably as part of the recovery of life from the Permo-Triassic mass extinction. Lots of new plants and animals emerged and began to build new ecosystems. But dinosaurs such as *Herrerasaurus* and *Eoraptor* (Fig. 13.2) only became reasonably abundant about 233 Ma, in the mid to late Carnian. The Carnian

was a stratigraphic stage that lasted from 237 to 227 Ma, and something was going on during that time.

In fact, this was a mass extinction that had remained quite obscure until about 2012, when geologists noted the coincidence of the explosion of dinosaurs, some profound changes in the climate, and the eruption of the Wrangellia basalts around British Columbia and the west coast of Canada and Alaska. All the evidence shows that the Wrangellia eruptions caused the same sequence of catastrophic events as during the Permo-Triassic mass extinction (see Chapter 6): global warming, acid rain and soils stripping on land, and warming and acidification of the ocean, and loss of oxygen from bottom waters.

Climatic consequences of the eruptions had been noted much earlier – a series of up to five flip-flops between humid and dry climates over a span of about 1.5 m.y. These drove sudden changes in the plants, and in insects and tetrapods. One consequence was that in the new, dry world that emerged at the end of the CPE, floras of seed ferns had given way to conifers, ancestors of firs and monkey puzzle trees that could survive with minimal water. The changing plants, however, were apparently not to the taste of the rhynchosaurs, which went extinct at this point. Perturbing the plants knocked out the herbivores, and this profoundly changed the nature of ecosystems.

The new view of the rise of dinosaurs (Figure 12.21) is that it was perhaps a three-step process, with their origin in the Early Triassic, about 250 Ma, and survival for 18 m.y. as very rare animals. Then, 233–232 Ma, during

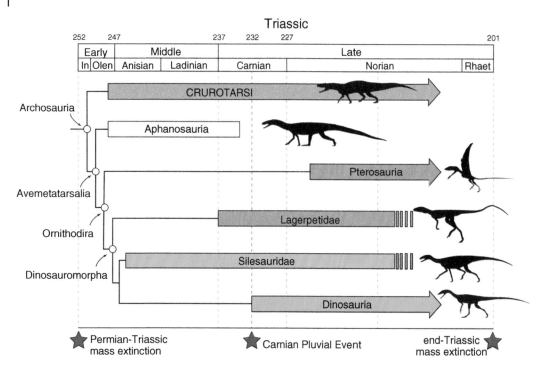

**Figure 12.21** Timing of key steps in the origin of dinosaurs. New evidence points to a very early origin of dinosaurs and pterosaurs, and their closest fossil relatives. In particular, Aphanosauria and Silesauridae are known from the beginning of the Middle Triassic, 247 Ma, and so dinosaurs originated then, in the late Early Triassic. They remained rare until the Carnian Pluvial Episode (CPE) (233–232 Ma) caused major changes in climates and floras, and rhynchosaurs died out. This gave dinosaurs a chance to expand. The end-Triassic mass extinction saw the loss of many crurotarsans and a further boost in dinosaurian dominance. Image modified from data provided by Sterling Nesbitt.

and after the CPE, predatory dinosaurs such as *Eoraptor* and *Herrerasaurus* preyed on rhynchosaurs first and lived side by side with some small plant-eating dinosaurs. These then diversified after the CPE, eventually becoming much larger. The third and final step was at the end of the Triassic, when a further mass extinction killed off most of the crurotarsans and cleared the way for the further diversification of dinosaurs (see Chapter 13).

## Reference

Carrier, D.R. (1987). The evolution of locomotor stamina in tetrapods: circumventing a mechanical constraint. *Paleobiology* 13: 326–341. [A breakthrough paper].

## Further Reading

Benton, M.J., Bernardi, M., and Kinsella, C. (2018). The Carnian Pluvial Episode and the origin of dinosaurs. *Journal of the Geological Society of London* 175: 1019–1026. [Evidence that the CPE drove a step-change in the composition of tetrapod faunas].

Benton, M.J., Forth, J., and Langer, M.C. (2014). Models for the rise of the dinosaurs. *Current Biology* 24: R87–R95. [New fossils and new evolutionary models].

Dzik, J. (2003). A beaked herbivorous archosaur with dinosaur affinities from the early Late Triassic of Poland. *Journal of Vertebrate Paleontology* 23: 556–574. [*Silesaurus*].

Kubo, T. and Benton, M.J. (2009). Tetrapod postural shift estimated from Permian and Triassic trackways. *Palaeontology* 52: 1029–1037. [Evidence that all medium-sized and large tetrapods switched from sprawling to erect gait across the Permian–Triassic boundary].

Nesbitt, S.J., Butler, R.J., Ezcurra, M.D. et al. (2017). The earliest bird-line archosaurs and the assembly of the dinosaur body plan. *Nature* 544: 484–487. [*Teleocrater* and the new picture of avemetatarsalian and dinosaur origins in the Early Triassic].

## Question for Thought, Study, and Discussion

Explain the advantages that the upright locomotion of archosaurs seems to have given them. Then summarize the evidence that the origin of dinosaurs was in the Early Triassic. Then think of reasons why the archosaurs did not take over the terrestrial world as soon as they evolved.

# 13

# Dinosaurs

**In This Chapter**

The whole chapter is devoted to dinosaurs. There were four main groups: theropods, prosauropods, sauropods, and ornithischians, and we briefly describe their probable biology and ecology. All dinosaurs laid eggs, and probably all of them had some sort of parental care for the nests and the hatchlings. Ironically, some of the best finds are of nests and hatchlings that did not survive but died and were buried as fossils for us to study. Dinosaur bones often have growth lines, so we now know how they grew and how long it took. (Dinosaurs had a "teenage" growth spurt just as we humans do!) The old arguments about dinosaur body temperatures are over: dinosaurs all had good temperature regulation and were warm-blooded, though we do not know exact temperatures. Vegetarian dinosaurs include many 5–7-ton animals that are comparable ecologically with rhinos, but nothing living is like the giant sauropods that weighed several tens of tons. Sauropods owed their huge size to a combination of many small offspring but limited parental care, fast food uptake without chewing or stomach stones, a bird-like lung allowing efficient breathing, and a steady high body temperature. Increasingly, we are discovering dinosaurs with feathers, usually small or young animals. The number of nonflying dinosaurs with feathers means that feathers originally had nothing to do with flight. They may have been for thermoregulation, although their bright color patterns suggest that they also played a role in display. Dinosaur stampedes are known from their footprints, and some bizarre offshoots of the nostril system suggest that they made sounds for communicating. The amazing array of dinosaur adaptations makes their extinction at the end of the Cretaceous all the more puzzling.

## Dinosaurs

We are familiar with dinosaurs in many ways; they have been with us since kindergarten or before, in comic strips, toys, stories, movies, nature books, TV cartoons, and advertising. Yet it's still not easy to understand them as animals. The largest dinosaurs were more than 10 times the weight of elephants, the largest land animals alive today. Dinosaurs dominated land communities for 150 million years, and it was only after they disappeared that mammals became dominant. It's difficult to avoid the suspicion that in the Mesozoic, dinosaurs were in some way competitively superior to mammals and confined them to small body size and ecological insignificance.

We are in a golden age of dinosaur paleontology. Dinosaurs are being discovered, described, and analyzed faster than ever, and new techniques are giving us better insight into their biology and ecology. Fortunately, the basic outline of dinosaur history has been stable for decades, and the iconic dinosaurs we all know and love remain iconic. We know that dinosaurs were all warm-blooded, with high metabolic rates. They lived from pole to Equator and on all continents.

The earliest dinosaurs were small bipedal carnivores, which appeared in the Late Triassic of Gondwana at about the same time as the first mammals. All the spectacular variations on the dinosaur theme came later but all four major dinosaur groups (Figure 13.1) had evolved by the end of the Triassic. We have seen (Chapter 12) how dinosaurs probably emerged in the Early Triassic, continued as small and obscure members of their faunas for 20 m.y. and finally burst forth in ecosystems worldwide after the Carnian Pluvial Episode, 232 Ma.

*Cowen's History of Life*, Sixth Edition. Edited by Michael J. Benton.
© 2020 John Wiley & Sons Ltd. Published 2020 by John Wiley & Sons Ltd.

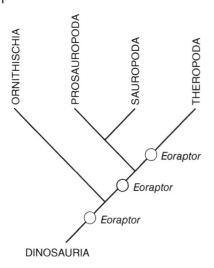

**Figure 13.1** A cladogram showing the major groups of dinosaurs. The earliest well-known dinosaur is *Eoraptor*. It is debated whether it is a basal dinosaur, an early theropod, or even an early sauropodomorph, so we show it in several places. Prosauropods are known from the Late Triassic, so the cladogram shows that all four groups of dinosaurs had diverged by that time.

Most people know now that birds are highly evolved dinosaurs. This means that in cladistic terms, the clade Dinosauria includes birds as well as those dinosaurs that are not birds (the "nonavian dinosaurs"). That is a clumsy phrase. In this chapter, we will use dinosaurs with a small "d" to refer to nonavian dinosaurs and Dinosauria to mean dinosaurs plus birds.

## Theropods

The earliest known dinosaurs, immediately after the Carnian Pluvial Episode, were theropod-like, being either basal dinosaurs, basal saurischians, or basal theropods – it can be hard to resolve. These were small flesh eaters and almost all later theropods retained their body plan as bipedal runners and their ecological character of a carnivorous way of life. *Eoraptor* and *Herrerasaurus*,

both from the Late Triassic of Argentina, are right at the point where theropods diverge from other saurischians. *Eoraptor* was a very small animal with a skull only about 8 cm (3 in.) long (Figure 13.2). Even so, it was a fast-running carnivore, with sharp teeth and grasping claws on its fore limbs. *Herrerasaurus* was very like *Eoraptor* but much larger, between 3 and 6 m (10–20 ft) long.

Early theropods included *Coelophysis* from the Late Triassic of North America (Figure 13.3a). At 2.5 m (8 ft) long and lightly built (perhaps only 20 kg or 45 pounds), it was clearly adapted for fast running. Famously, the Ghost Ranch site in New Mexico yielded dozens of *Coelophysis* skeletons that had all died and been washed together in a flood. *Coelophysis* is a member of the first major branch of theropods, the **Ceratosauria**, which includes also the Late Jurassic *Ceratosaurus* (Figure 13.3b).

All other theropods fall into the clade **Tetanurae**, basal members of which include *Megalosaurus* from the Middle Jurassic of England, the first ever dinosaur to be named, in 1824, as well as the remarkable spinosaurids. Spinosaurids are named for the long spines along their backs, which might have carried a sail. Their other unusual features are a long, slender-snouted skull like a crocodile (Figure 13.4), and the fact that the fore and hind limbs were of similar lengths. The long snout was certainly adapted for eating fish but did spinosaurids spend most of their time swimming, as has been suggested (Ibrahim et al. 2014), or did they mix bipedal walking, like other theropods, with rare swimming?

The spinosaurids could be huge, and the remaining tetanurans include at least three lineages that evolved to giant size: the Jurassic allosaurs and the Cretaceous tyrannosaurs and carcharodontosaurids (*Giganotosaurus* from Argentina and *Carcharodontosaurus* from North Africa). These varied theropods were the largest land carnivores of all time, each weighing 6 or 7 tons or more (more than an elephant) and standing about 6 m (20 ft) high, with a total length around 12 m (40 ft). Weight and mass estimates change all the time, but the very large

**Figure 13.2** The small early dinosaur *Eoraptor* from Argentina, the earliest theropod. *Source:* Photograph by Lord of the Allosaurs (Wikimedia).

Figure 13.3 (a) *Coelophysis*, a small Late Triassic theropod from Arizona. *Source:* Artwork by John Conway (Wikimedia). (b) The skull of *Ceratosaurus*, a large ceratosaur theropod from the Late Jurassic of North America. *Source:* Photograph by Tremaster (Wikimedia).

(a)

(b)

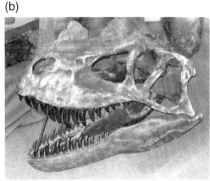

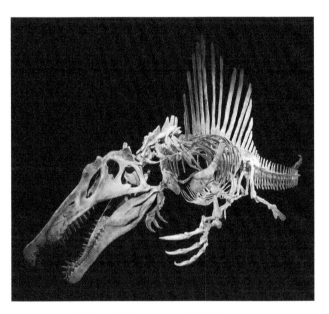

Figure 13.4 Skeleton of the spinosaurid *Suchomimus*, in swimming pose; this is a controversial suggestion. *Source:* Photo by Mike Bowler (Wikimedia).

Figure 13.5 The skull of *Tyrannosaurus rex. Source:* Photograph by Quadell (Wikimedia).

*Tyrannosaurus rex* called "Sue" is possibly the largest of them all, with an estimated mass of 9 tons (Hutchinson et al. 2011). All these giant theropods must have relied on massive impact from the head for killing, aided by huge stabbing teeth that would have caused severe bleeding, usually lethal, in a prey animal (Figure 13.5). *Tyrannosaurus* had the most powerful bite of any land carnivore that has ever lived (Bates and Falkingham 2012). **Tyrannosaurs** are characterized by their enormous heads and tiny arms, a combination of characters that is still not understood. It is still a debate whether tyrannosaurs were giant predators or giant scavengers (the easiest and most likely answer is both, based on analogy with modern hyenas).

Tyrannosaurs belong to the **Coelurosauria**, a subdivision of Tetanurae, as do some smaller theropods. *Compsognathus* is a basal coelurosaur that was small but an active predator with long arms and clawed fingers (Figure 13.6). As we shall see, the Chinese compsognathid *Sinosauropteryx* had feathers. **Ornithomimids** are the so-called *ostrich dinosaurs*. Their body plan is much like that of a living ostrich except that they had long arms and slim, dexterous fingers instead of wings. Ornithomimids had long legs and necks, large eyes and rather large brains, but no teeth. They could have been formidable carnivores, of course, but perhaps they specialized on smaller prey animals. Many of them lacked large claws and had long fingers that could have been used to manipulate objects.

Figure 13.6 *Compsognathus*, a small basal coelurosaur from the Late Jurassic of Germany, cast of the original fossil specimen. *Source:* Photograph by MatthiasKabel (Wikimedia).

Figure 13.7 The therizinosaur *Therizinosaurus*, showing its tiny head, huge arms and claws, and short hindlimbs and stubby tail. *Source:* Unknown artist, from Dinosaur World.

Climbing the evolutionary tree toward birds, the next major clade is **Maniraptora**, meaning "hand hunters"; whereas other theropods showed a long-term reduction in their arms (think of *T. rex* and its dinky little arms), the maniraptorans specialized in long arms and powerful hands. Among the maniraptorans are the **therizinosaurs**, which are a very strange group, including smaller, feathered forms and some real monsters (Figure 13.7). They look like a nightmare combination of ill-fitting bits and pieces – a tiny head with feeble teeth, massive gorilla-like arms with 1 m long scythe-like claws, tiny hind legs, and a silly little tail. When their fossils were first found in Mongolia in the 1950s, they were assigned variously to turtles, sauropods, ornithischians, and theropods. Only complete skeletons found in China in the 1990s showed how all the bits fitted together to make a strange plant-eating theropod that sat on its fat bottom and grabbed tree branches with its long claws. Relatives are the **oviraptors**, which were nest-building dinosaurs from Mongolia, to be discussed later.

More derived maniraptorans include the **dromaeosaurids**, which were mostly small, fast, and agile. Dromaeosaurs include *Velociraptor*, supposed star of the movie *Jurassic Park*, and *Deinonychus*, from the Early Cretaceous, which actually was the dinosaur on which the movie image was based. *Deinonychus* was one of the most impressive carnivores that ever evolved (Figure 13.8). It was about 3.5 m long, it was clearly fast

(a)

(b)

Figure 13.8 (a) The skull of *Deinonychus. Source:* Photograph by Didier Descouens (Wikimedia). (b) Most of the skeleton of *Deinonychus*, posed in active running. *Source:* Photograph by dinoguy2, modified by Conty (Wikimedia).

Figure 13.9 The large prosauropod *Plateosaurus*, one of dozens of complete skeletons excavated in Germany in the early twentieth century. *Plateosaurus* could walk biopedally or on all fours, and it could use its powerful hands for grasping vegetation. *Source:* Photo by FunkMonk (Wikimedia).

and agile, had murderous slashing claws on both hands and feet, and a most impressive set of teeth. Close relatives were the slender, brainy **troodontids** mainly from the Cretaceous of North America, and of course *Archaeopteryx*. But *Archaeopteryx* is generally identified as the first bird and so this part of theropod evolution, the origin of birds, is included in Chapter 14.

## Sauropodomorphs

Some early dinosaurs evolved to become very large, heavy quadrupedal vegetarians with broad feet and strong pillar-like limbs. The sauropodomorphs had an early radiation as prosauropods and a later radiation as the famous sauropods with which we are all familiar.

The term "prosauropods" is informal because it includes a series of basal members of Sauropodomorpha, not a distinct group. Prosauropods were abundant, medium to large dinosaurs of the Late Triassic and earliest Jurassic. They were typically about 6 m (20 ft) long, but *Riojasaurus* was unusually large at 10 m (over 30 ft). Prosauropods lived on all continents except Antarctica, with rich faunas known from Europe, Africa, South America, and Asia. They ranged into the Early Jurassic, when they were replaced by sauropods.

Prosauropods were all browsing herbivores. The teeth were generally good for cutting vegetation but not for pulping it. Opposing teeth did not contact one another and all the grinding must have been done in a gizzard. (Masses of small stones have been found inside the skeletons of several prosauropods.) Prosauropods have particularly long, lightly built necks and heads, and light forequarters. They were clearly adapted to browse high in vegetation, perhaps reaching up from the tripod formed by the hind limbs and heavy tail. Only *Riojasaurus*,

Figure 13.10 The huge sauropod *Brontosaurus*, a family on the move. *Source:* Life reconstruction by Nobu Tamura (Wikimedia).

the largest, was always quadrupedal because of its weight but it had a very long neck to compensate. Prosauropods were the first animals to browse on vegetation high above the ground, and they represent a completely new ecological group of herbivores exploiting an important new resource in the zone up to perhaps 4 m (13 ft) above ground. The same adaptation was reevolved later in sauropods, and again in mammals such as the giraffe.

Prosauropods began small, like other dinosaur groups, but were soon the largest and heaviest members of their communities, and they were abundant. *Plateosaurus* (Figure 13.9) accounts for 75% of the total individuals in a well-collected site in Germany, and probably over 90% of the animal mass in its community.

Sauropods (Figure 13.10) include the largest land animals that ever lived. Remember that there is a natural human ambition to discover the largest or the oldest of anything; we must be careful to assess the evidence for some of the claimed huge sizes we read in the press. But even a cautious person must admit that well-documented sauropod body weights are at least 50 tons; *Argentinosaurus* (from Argentina, of course) has been estimated at between 60 and 88 tons, with 73 tons as a best guess. This is the best documented most massive dinosaur. Some impressive names have been invented, such as *"Ultrasaurus," "Supersaurus"* and *"Seismosaurus,"* but initial size estimates have often had to be reduced! *Brachiosaurus* and *Giraffatitan* had long fore limbs carrying it over 12 m (40 ft) high, as tall as a four-story building, and with a weight estimated at 50 tons – the specimen on show in Berlin is the largest display skeleton of any dinosaur (Figure 13.11).

Sauropods were all herbivores, of course; no land animals that size could have been carnivorous. They had curiously small heads and very long necks that allowed them to browse on anything within 10 m (33 ft) of ground level. The tails were long also but the body was massive, with powerful load-bearing limb bones and pelvis. All sauropods were quadrupedal. The major body mass was centered close to the pelvis, which was accordingly more massive than the shoulder girdle.

Sauropods showed two main feeding strategies characterizing their main lineages, and this is reflected in the shapes of their snouts (Figure 13.12). In a detailed biomechanical study, David Button and colleagues (Button et al. 2014) showed that the long slender snout of *Diplodocus* was used for snatching bunches of leaves and pulling back. It had teeth only at the front of the jaws, so it could hang on tightly, but the skull was not constructed for twisting or turning. *Camarasaurus*, on the other hand, with its high, short snout, could grab and wrestle with tough branches and twigs, chomping on each side of its jaws. This is how these two sauropods, which lived side by side in Morrison Formation times of North America, could divide up the feeding niches between them. This study used new methods in computational biomechanics to combine estimated jaw muscle forces with calculated physical properties of the skull as a means to actually test (not guess about) how the dinosaurs functioned.

## Ornithischians

The ornithischians were all herbivores. Judging by their teeth, most ornithischians ate rather coarse, low-calorie vegetation, so many of them tended to be at least medium-sized. They were the most varied and successful herbivorous animals of the Mesozoic, and were abundant in terrestrial ecosystems right to the end of the Cretaceous. Small bipedal basal ornithischians gave rise to several derived groups that were much heavier, with some animals weighing 5 tons or more (Figure 13.13). The armored dinosaurs (**Thyreophora**) form one clade that includes **stegosaurs** and **ankylosaurs**, and another major clade is the **Marginocephalia**, which includes the horned dinosaurs or **ceratopsians** and the heavy-skulled **pachycephalosaurs**. However, most of the larger ornithischians were **ornithopods,** which include **iguanodonts** and the so-called "duckbill" dinosaurs, the **hadrosaurs**.

The best-known early ornithischians are small bipedal dinosaurs from the Early Jurassic of Gondwana. *Lesothosaurus* was small, agile, and fast-running, but it clearly had vegetarian teeth. *Heterodontosaurus* had teeth that were even more specialized for a vegetarian diet (Figure 13.14). Small teeth at the front of the upper jaw bit off vegetation against a horny pad on the lower jaw. The back teeth evolved into shearing blades for cutting vegetation. (The sharp incisors were for display or

**Figure 13.11** The formidable skeleton of *Giraffatitan* from Tanzania, in the Humboldt Museum, Berlin. *Source:* Photo by Shadowgate (Wikimedia).

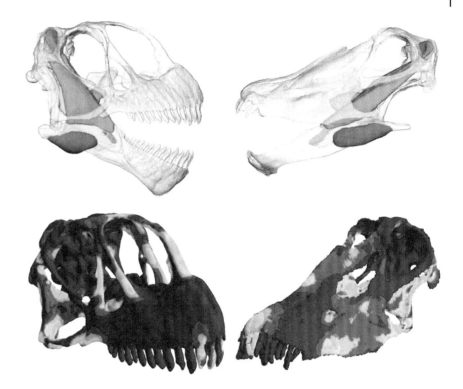

Figure 13.12 Functional comparison of the skulls of *Camarasaurus* (*left*) and *Diplodocus* (*right*), showing the restored jaw muscles (*top row*), and the stress diagrams (*lower row*), with high stresses shown by warm colors (*red, yellow*). *Source:* Image courtesy of David Button and Emily Rayfield.

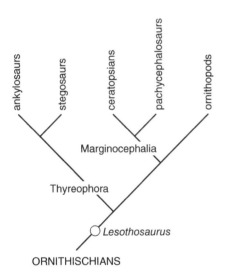

Figure 13.13 Cladogram of ornithischian dinosaurs.

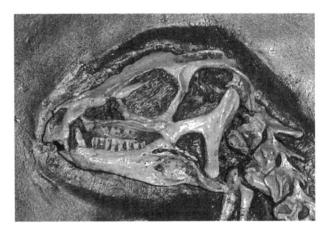

Figure 13.14 *Heterodontosaurus* had teeth that varied greatly along the jaw in size, shape, and presumed function. Skull about 10 cm (4 in.) long. *Source:* Photograph by Sheep81 (Wikimedia).

fighting.) The cheek teeth were set far inward, with large pouches outside them to hold half-chewed food for efficient processing.

The earliest ornithopods were less than a meter long but they soon increased significantly in size. A general theme of ornithopod evolution was the successive appearance of groups that in different ways evolved toward the 5–6-ton size that seems to have been a weight limit for most terrestrial herbivores. Even at this size, many ornithopods remained bipedal. Others probably walked most of the time on all fours but raised them-

selves up on two limbs for running or browsing on high vegetation.

Ornithopods evolved large batteries of teeth, and newly evolved modifications of the jaws and jaw supports allowed complex chewing motions. Iguanodonts were particularly abundant in the Early Cretaceous; they reached 9 m (30 ft) in length and stood perhaps 5 m (16 ft) high. They cropped off vegetation with powerful beaks before grinding it. Most iguanodonts were replaced ecologically by a variety of hadrosaurs (duckbilled dinosaurs) in the Middle and Late Cretaceous (Figure 13.15). Hadrosaurs were about the same in size and body plan as

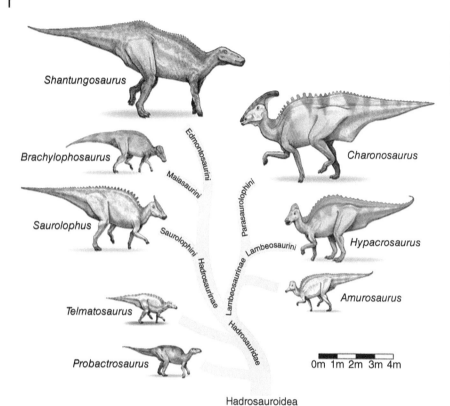

Figure 13.15 A variety of hadrosaurs. Body size varied but the basic body plan did not. The main differences are in the skull, where some hadrosaurs had elaborate head crests. *Source:* Image by Debivort (Wikimedia).

Figure 13.16 *Stegosaurus. Source:* Artwork by Nobu Tamura (Wikimedia).

Figure 13.17 The ankylosaur dinosaur *Euoplocephalus. Source:* Artwork by Nobu Tamura (Wikimedia).

iguanodonts but had tremendous tooth batteries, with several hundred teeth in use at any time.

The other ornithischians were dominantly quadrupeds, but they betrayed their bipedal ornithopod ancestry with hind limbs that were usually longer and stronger than the fore limbs. Stegosaurs (Figure 13.16), with their characteristic plates set along the spine, were the major quadrupedal ornithischians in the Jurassic but were replaced in the Early and Middle Cretaceous by the armored ankylosaurs (Figures 13.17 and 13.18). Later in the Cretaceous, the ornithischians were particularly abundant and varied in their body styles. Many quadrupedal forms lived alongside the hadrosaurs, including the ceratopsians, or horned dinosaurs (Figure 13.19), which used their face horns and neck frill in defense against predators such as *T. rex*.

## Dinosaur Paleobiology

Now that we have surveyed the dinosaurs, it's time to try to reconstruct their biology. Fortunately, new discoveries over the past 50 years have given us much more detail about their daily lives.

Figure 13.18 The ankylosaur dinosaur *Minotaurosaurus*. *Source:* Artwork by Nobu Tamura (Wikimedia).

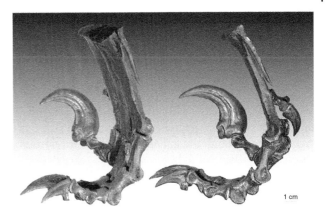

Figure 13.20 Left foot of *Deinonychus*, showing the way the second claw is reflexed upward and backward for a powerful strike. *Source:* Photograph by Didier Descouens (Wikimedia).

Figure 13.19 The ceratopsian dinosaur *Nedoceratops*. *Source:* Artwork by Nobu Tamura (Wikimedia).

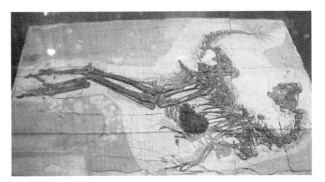

Figure 13.21 Gastroliths preserved within the rib cage of the little theropod dinosaur *Caudipteryx* from China. Theropods, like birds, probably swallowed their food without chewing and so the stones helped break up the lumps. *Source:* Photograph by Kabacchi (Wikimedia).

## Feeding

From the time of their first discovery, paleontologists have been fascinated by dinosaur feeding. They quickly recognized that dinosaurs with big, sharp teeth (Figures 13.2 and 13.3b) ate flesh and those with other kinds of teeth ate plants. The big theropods in particular (*Tyrannosaurus*, for example) have teeth with serrations that are beautifully shaped to break through membranes and muscle fibers in meat. While tyrannosaurs must have killed prey with their skulls, the more lightly built maniraptorans had long arms and vicious claws, and many deinonychosaurs had huge foot claws that would have made fearsome ripping weapons (Figure 13.20). Some theropods swallowed grit, as chickens do today, to assist in digestion (Figure 13.21).

There is more diversity in the herbivorous dinosaurs, in terms of tooth shapes, jaw shapes and operation, and overall body size. This reflects the fact that there are always more herbivores than carnivores in an ecological community, and the broad range of plant types they fed on. Most herbivorous dinosaurs had teeth but some theropods that had secondarily adopted plant diets, such as oviraptorids and ornithomimids, lacked teeth and chopped food with their horn-edged jaws.

The largest dinosaurs, the sauropods, had very small heads for their size. This has sometimes been thought to indicate a soft diet that did not need much chewing. However, sauropods probably had a small head for the same mechanical reason that giraffes do: the head sits on the end of a very long neck. In both sets of animals, the food is gathered by the mouth and teeth, then swallowed and macerated later. It used to be thought that sauropods had abundant stomach stones, or **gastroliths**, to help them break down their rough, unchewed plant food. This is not the case, however, for sauropods, and so-called gastroliths have been shown by Wings and Sander (2007) to be chance finds of stones inside dinosaur rib cages.

Large animals on purely vegetarian diets almost always have bacteria in their guts to help them break down cellulose. Large animals have slower metabolic rates than small ones and for vegetarians, this means a slower passage of food through the gut and more time for fermentation. Alternatively, a large vegetarian can digest a smaller percentage of its food and live on much poorer quality forage. Vegetarians usually grind their food well so that it can be digested faster, but sauropods must have relied on chemical and bacterial fermentation.

Ornithischians, on the other hand, generally had impressive batteries of teeth, especially in advanced hadrosaurs and ceratopsians, and they would have chewed up their food thoroughly, as living mammals do.

## Locomotion

Dinosaur locomotion has been much debated. The first dinosaurs such as *Eoraptor* (Figure 13.2) were doubtless fast little creatures, whereas the majority of dinosaurs, being large, presumably moved rather slowly. There has been a fashion for envisaging dinosaurs like *Triceratops* and *T. rex* charging around at high speed, like modern deer or ostriches, but 10 times larger. However, fossilized tracks and calculations from the skeleton and inferred musculature show that the larger dinosaurs walked or trotted at best. They certainly did not gallop.

Dinosaur running mode and speed can be studied both by looking at ancient trackways and by digital modeling. Tracks of footprints show the speed a dinosaur was moving because the greater the speed, the greater the stride length. If you know the dinosaur who made a track, you can estimate its size, and then the stride length is proportional to speed. These results can be checked by dynamic digital modeling, in which a 3D skeleton is studied in the computer. The different leg joints (hip, knee, ankle) can be estimated and possible movements established. Then the major leg muscles are reconstructed from marks on the bones and their relative strengths estimated. The whole contraption is set moving (Figure 13.22) and the experiments show that, for example, *T. rex* managed a gentle trot but could not run at high speed.

## How Could Dinosaurs Be So Huge?

There are small dinosaurs: *Compsognathus* was only the size of a chicken, for example. But the dominant feature of dinosaurs, and the dominant aspect of their paleobiology, is the enormous size of the largest ones. Ornithischian dinosaurs are easier to understand than the others because they were vegetarians in the 5-ton range, comparable with living elephants or rhinos, perhaps. On the other hand, there are no 5-ton carnivores alive today on land that we can compare with dinosaurs such as *Tyrannosaurus*, and there are no 50-ton vegetarians that we can compare with the sauropods. Despite this, we can make some reasonable inferences about dinosaur biology.

A few dinosaur species are known from enough specimens to get an idea of their life history. But we need a clock to find out how fast they grew. Fortunately, the limb bones (the femur, for example) of dinosaurs show growth rings, and they are spaced regularly enough to be annual growth rings, recording seasonal fluctuations of the environment during growth (Figure 13.23). The growth rings in dinosaur bone (and also in the bone of modern lizards and turtles) are laid down just like the growth rings in a tree – fast growth in summer, slow growth in winter. This has revolutionized our understanding of dinosaur life history (Erickson 2005).

Figure 13.22 **The dynamic model of** *Tyrannosaurus rex* developed by William Sellers and colleagues, and set in motion (see the animation here: www.manchester.ac.uk/discover/news/tyrannosaurus-rex-couldnt-run-says-new-research).

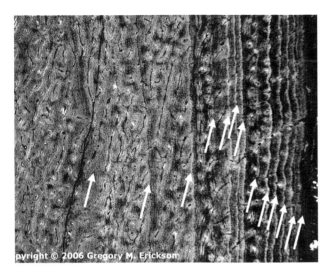

Figure 13.23 LAGs (Lines of Annual Growth) in a limb bone of *Tyrannosaurus rex*. You can see rapid growth (well-spaced growth lines) over much of this specimen, during its growth spurt, followed by slow growth as it reached its full adult size. *Source:* © Gregory M. Erickson of Florida State University, used by permission.

Juvenile dinosaurs grew moderately slowly but adolescents went through a growth spurt that is uncannily like that of humans. At this stage in their lives, they were large enough to be a lot safer from predators, and the growth spurt takes them quickly to sexual maturity. After that, growth slows down as more energy is put into reproduction.

What do these new studies of growth rate tell us about how the sauropods could commonly reach a size of 50 tons? In the past, some scientists speculated that they could achieve these huge weights only because gravity in the Jurassic was lower than it is today, or because they lived in water and so achieved neutral buoyancy. However, there is no evidence for either of these (rather wild) speculations. We have to stick to facts!

Martin Sander at the University of Bonn, together with a large team of collaborators, has presented a plausible model. It combines a bunch of things dinosaurs could do but which neither reptiles nor mammals today can do: lots of small babies and no parental care (this saves huge amounts of energy and risk to the mother), fast feeding (but no chewing), a bird-like lung (air goes one way, more efficient than the mammal "tidal" lung; air goes in and out), and warm-bloodedness. This model (Figure 13.24) looks complex, but follow the arrows and boxes and it all works together. Among the largest land animals today, elephants cannot get larger because they break some of these rules (they have large babies, they care for their babies, they chew their food, they have a tidal lung). As we shall see, the warm-bloodedness of dinosaurs was "smart" warm-bloodedness, which saved food.

It should be noted that this model is speculative, because we cannot test a living sauropod. However, the components are all true (long neck and small head, tiny babies, high growth rate, warm-bloodedness). We know the last two from study of the bone under the microscope, and the other connections come from studies of the physiology of modern animals.

## Metabolic Rate and Temperature in Dinosaurs

The debate about whether dinosaurs were warm-blooded or not has rumbled on since 1970. At first, the debate was poorly informed because of lack of good fossil data, and even lack of understanding of the great spectrum of physiology in living animals – in other words, animals need not be simply cold-blooded like a modern lizard or warm-blooded like a human. These two are the ends of a spectrum; we devote nine-tenths of our food to maintaining our body temperature at a very precise level, whereas the lizard can function over a range of body temperatures, and it spends very little energy on doing so. If a lizard is cold, it curls up and sleeps or lies on a rock to soak up warmth from the sun. We now know that living animals show all kinds of intermediate conditions between these two extremes.

New work on dinosaur bone shows that, apart from the growth lines, the overall microscopic structure shows evidence of rapid growth and constant remodeling (Figure 13.23), proof that they were warm-blooded or, as the physiologists would say, they had a high basal metabolic rate (BMR), just as modern birds and mammals do, but unlike modern lizards and turtles.

In addition, paleontologists have long known that dinosaurs had a bird-like breathing system (Figure 13.25). Modern birds and crocodiles both use bone and muscle systems, and a system of air spaces set into the body, to aid breathing by moving the pelvis and guts to generate a pumping action that in turn affects the lungs. Air is moved through the lungs in a one-way system (Cieri and Farmer 2016), rather than the two-way system we have, increasing its efficiency. Furthermore, the pumping system is closely linked with locomotion. All these observations allow this same pumping system to be reconstructed in dinosaurs, linking their respiration with their active motion in an analogous way to that seen in mammals and birds (see Chapter 10).

Dinosaurs all show extensive air cavities in their skeletons, including, as in birds, cavities throughout the vertebral column, and even in some other bones such as the humerus in the arm (Wedel 2006). In life, great air sacs were located down the flanks on either side of the backbone and when the dinosaur breathed in, it took a great gulp that filled the lungs and the air sacs, and as it breathed out, these structures drained directly to the

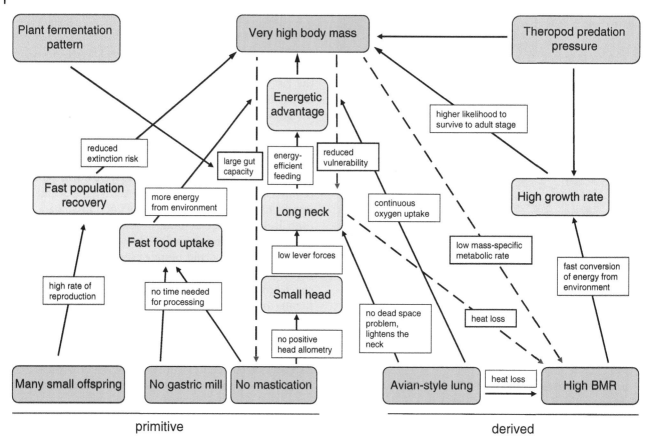

Figure 13.24 How to be huge. The flow chart shows the key adaptations of dinosaurs, and sauropods in particular, that allowed them to grow to 10 times the size of the largest modern elephant. *Source:* Courtesy of Martin Sander, Bonn, redrafted.

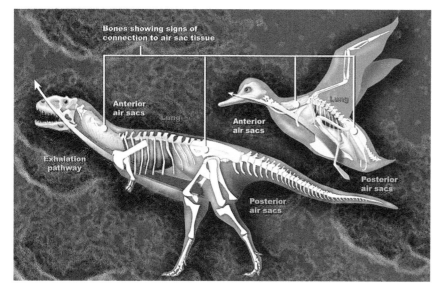

Figure 13.25 A comparison of the respiration system that is known in living birds and reconstructed for dinosaurs. *Source:* Drawn by Zina Deretsky of the National Science Foundation, in the public domain.

mouth (O'Connor and Claessens 2005). When mammals breathe out, we leave some unoxygenated air in the dead space of the lungs and the newly inhaled air, rich in oxygen, mixes. Birds (and dinosaurs) kept the oxygen flowing in with no such mixing.

## Dinosaur Eggs and Nests

As we saw, dinosaur eggs and babies were unusually small, relative to the size of their parents. In modern birds, egg size scales roughly with body size but not very

Figure 13.26 (a) A *Troodon* nest. This specimen came from "Central Asia," probably under dubious circumstances, and was to be auctioned off. (b) Part of a Cretaceous dinosaur nest in Indroda Fossil Park in India (the rest was broken off and fell down the ravine). The eggs are carefully spaced. *Source:* Photo by SBallal (Wikimedia).

precisely. In fact, some large birds lay small eggs, while the flightless kiwi of New Zealand is famous for laying an egg that is about one-fifth of the mother's size. Ouch.

Nonetheless, dinosaur eggs were small, usually not much bigger than a football, and so the hatchlings were tiny. As we saw (Figure 13.25), tiny babies and little or no parental care is a good strategy for huge dinosaurs because it saves them energy. As humans, we are used to the idea of producing very few babies, usually one at a time, and devoting great efforts to caring for them. Most animals today do the opposite – a codfish produces a million eggs at a time and abandons them. So long as two or three survive to adulthood, she has done her job, in evolutionary terms.

All dinosaurs laid eggs and as far as we know, they laid them in carefully constructed nests, usually scooped out of the ground. Major finds of fossilized dinosaur eggs and nests have been made in Jurassic and Cretaceous rocks worldwide, and we have eggs from all major dinosaur groups. Dinosaur nests were rounded hollows scooped in the soil. In each nest, the eggs were laid or arranged (by the mother) in a single layer and in a neat pattern so they would not roll around (Figure 13.26a,b). Sometimes many nests are clustered together at regular close intervals, suggesting that they were in communal breeding grounds: nesting colonies, if you like.

Late-stage embryos are preserved inside some dinosaur eggs. Once they hatched, very young dinosaurs seem to have stayed in or around the nest, sometimes until they grew to twice their hatching size. One nest from the Cretaceous of Montana contained 15 baby duckbilled dinosaurs. They were not new hatchlings because they were about twice as large as the eggshells found nearby, and because their teeth had been used for long enough to have wear marks. But they were together in the nest when they died and were buried and fossilized, with an adult close by – named *Maiasaura*, the "good mother." The degree of parental care has been

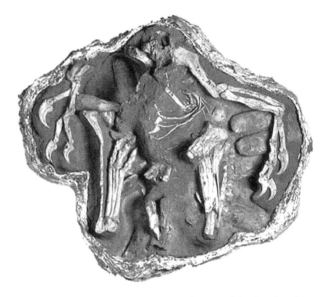

Figure 13.27 A fossil nest with an incubating adult, showing the foot and leg bones in the middle, and the elongated eggs located on either side, and the long arms round the sides. *Source:* AMNH press image.

debated – did *Maiasaura* parents hang around the nest, feeding their babies, or did they abandon them?

Dinosaur eggs and nests were found in Mongolia in the 1920s, associated with the most abundant dinosaur in the area, *Protoceratops*. To everyone's surprise, when an embryo was finally discovered inside one of the eggs, it was well enough preserved to be identified not as *Protoceratops* but as *Oviraptor*. The irony here is that the genus *Oviraptor* was named as "egg thief" because it was supposedly harassing the *Protoceratops* parents and threatening to eat their eggs. Since then, an amazing fossil showing a mother *Oviraptor* incubating her nest of eggs has been found (Figure 13.27): she put her feet carefully in the middle of the nest, shuffled her bottom around over the eggs to split them into two rows,

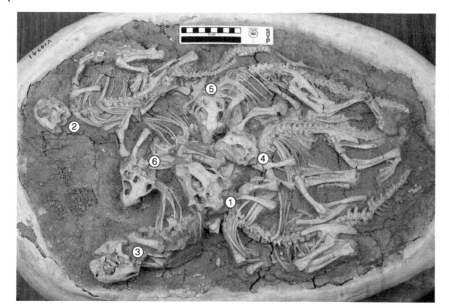

Figure 13.28 A cluster of juveniles of the little ornithischian *Psittacosaurus*, from the Early Cretaceous of north China, showing six individuals, all facing the same way, as if caught while moving; perhaps they were overwhelmed by falling ash from a volcanic eruption. They were all two years old, except number six which was three years old when it died. *Source:* Photo by Qi Zhao, in the public domain.

and then sank down, with her legs on the ground and great feathered arms resting lightly over the eggs on either side.

Since this discovery, hundreds of clusters of juvenile examples of the early ceratopsian *Psittacosaurus* have been found. In one case, Qi Zhao and colleagues were able to work out the age of each individual from growth rings in the bone; in the cluster of juveniles (Figure 13.28), they all turned out to be two years old, except baby number six which was three at the time it died. These remarkable juvenile clusters are preserved in volcanic sediments in North China and may have been overwhelmed by ash falling after an eruption.

These discoveries all confirm that dinosaurs laid eggs, that the eggs were relatively small, and that some, such as *Oviraptor*, incubated their eggs. Modern crocodiles also care for their young to some extent, staying near the nest and helping the hatchlings. Birds, of course, show much more parental care, incubating the eggs, feeding the hatchlings, and teaching them how to fly (or at least pushing them off the branch or cliff when the time comes). Were dinosaurs crocodile-like or bird-like in their parental behavior? We do not know yet for sure.

## Dinosaur Feathers

Feathers have always been regarded as structures unique to birds; in fact, for 200 years they were used as one of the most important characters that define birds. But fossils found since 1995 have shown that dinosaurs had feathers too.

Figure 13.29 The feathered dinosaur *Caudipteryx* from the Cretaceous of China. Compare with Figure 13.21. *Source:* Artwork by M. Martyniuk (Wikimedia).

Theropods from Early Cretaceous beds in China are of great interest because they are so well preserved. *Sinosauropteryx* and *Beipiaosaurus* have a halo of very fine structures on the body surface that look like down. *Protarchaeopteryx* has down feathers on its body, tail, and legs, and a fan-shaped bunch of long feathers, several inches long, at the very end of its tail. *Caudipteryx* also has down and strong tail feathers, but it has feathers on its arms as well. They are shorter toward the fingers and longest toward the elbow, in contrast to the feathers on the wings of flying birds (Figures 13.21 and 13.29).

Figure 13.30 The feathered dinosaur *Microraptor* from the Cretaceous of China. Feathers are indicated by white arrows. *Source:* David Hone (Wikimedia).

Figure 13.31 The early ceratopsian *Psittacosaurus* showing detailed color patterns and the amazing quill-like feathers on its tail. Some exceptionally preserved fossils show both the feathers and scales, but also the distribution of light and dark color, evidence for countershading. *Source:* Illustration is from a model by Bob Nicholls (Palaeocreations, Bristol) (Wikimedia).

Finally, and most important, *Microraptor* (Figure 13.30) and *Sinornithosaurus* have true branching feathers on all four limbs and on the tail.

These two dinosaurs used their sets of two or four wings in flight – probably not flapping flight, as in birds, but gliding flight, where the limbs were held out to the side. We will explore more in Chapter 14 about the different modes of flight, but it is interesting to see how the small theropods from China show us intermediate stages from no flight at all in short-armed forms like *Sinosauropteryx* to quite effective and sophisticated gliding in *Microraptor*, but still not flapping or powered flight, as we see in birds.

There is now a question of how deep in the evolutionary tree feathers go. It's no surprise that some of the closest relatives of birds among theropod dinosaurs might have had feathers but feathers have also been reported in some ornithischian dinosaurs. The most amazing is *Psittacosaurus*, a small ceratopsian from the Early Cretaceous of China, whose nests and clusters of babies have been noted (Figure 13.28). This small herbivore is well known from some exceptional fossils that preserve all the scales over the body and the palisade of long quill-like feathers down the crest of the tail (Figure 13.31). The scales carry light and dark colors, and Vinther et al. (2016) were able to show that this was a form of countershading, a kind of camouflage that is commonly seen in modern mammals. Deer and antelope, for example, have pale-colored bellies and disruptive darker patterns over their sides and backs. This helps them escape detection as they move through forests or the edges of grasslands, and it seems *Psittacosaurus* used the same adaptation to avoid being spotted by large theropods.

Now, the bristles or quolls of *Psittacosaurus* are very odd-looking feathers but another ornithischian, *Kulindadromeus* from the Late Jurassic of Russia, also has exceptionally preserved feather-like structures and Godefroit et al. (2014) interpreted these as definitely feathers, many of them very similar in detail to theropod feathers. It is still controversial but it now seems that feathers might have arisen right at the origin of Dinosauria in the Triassic. They might go even deeper in the phylogeny to encompass pterosaurs (see Chapter 14), the flying reptiles, which also show simple feather-like structures over their bodies.

Critics of this idea argue that most dinosaurs did not have feathers, and indeed some such as the giant sauropods were too huge for feathers, and others such as ankylosaurs were covered in bony armor plates. However, hair is fundamental to mammals today but hair is lost in whales and adult elephants and rhinos. Baby pigs are actually like puppies, covered in hair, and it is largely lost as the animals get older and their large size helps regulate their body temperature. Perhaps dinosaurs fundamentally all had the possibility of producing feathers, and they were lost during evolution in some groups and during growth in some forms. We can imagine a baby *T. rex*, maybe the size of a pheasant, richly covered in feathers, and these would then be lost so the 5-ton adult had no feathers at all, or perhaps a quiff on the head for showing off.

This is perhaps a good moment to reflect on how major groups originate. People once argued that birds had evolved suddenly from reptiles – some even imagined a sparrow hatching directly from a crocodile egg! Well, up to 1990, there were far fewer fossils and so the transition from reptile to bird was not well documented. Now, thanks to all the remarkable new fossils from China, we have 50 or more "missing links" along the way, from typical large theropods such as *Megalosaurus* and *Allosaurus*, through steps of body size reduction, elongation of the arms, and redeployment of feathers to form gliding surfaces. At one time, birds were distinguished from reptiles by 30 or more major characters; now 29 of these characters have been found in theropods, and the only unique characteristic of *Archaeopteryx* and other birds (see Chapter 14) is that they show powered flapping flight.

## The Origin of Feathers

Feathers in birds have three main functions: insulation, flight, and display. But which came first? Feathers are related in many ways to mammalian hairs, being made of the protein keratin and carrying colors in the same way through melanosomes that hold the pigment melanin embedded in the keratin structure. Hairs and feathers emerge from follicles in the skin. Indeed, modern studies of development and genomics show there are shared patterns of developmental regulation for reptilian scales, mammalian hairs, and bird feathers. In fact, the production of hairs or feathers is fundamental, and it has to be switched off by genomic regulation so we do not have hairs on the palms of our hands, for example.

Feathers may have evolved to aid thermoregulation. The feathered Chinese theropods all have down, probably as insulation to keep their bodies at an even temperature. It does not matter whether they used their feathers to conserve heat in cold periods or to keep heat out in hot periods, or both. Insulation would have been useful in either case.

Even before the Chinese feathered dinosaurs were discovered, everyone assumed that thermoregulation came first, flight second. After all, you need only simple down feathers to provide a warm coat but flight requires more complex feathers. This is confirmed by the fact that the earliest feathered theropods had short arms and certainly could not glide or fly. However, some of the early theropods had some long feathers, sometimes arranged as a crest on top of the head, and we now know that others sported remarkable color patterns.

The discovery of color in dinosaurs, by two teams at the same time, independently (Li et al. 2010; Zhang et al. 2010), confirmed that there must have been another function. Zhang et al. (2010) showed that *Sinosauropteryx* had regular ginger and white stripes down its tail, while Li et al. (2010) showed that *Anchiornis* had an amazing array of sharp color patterns – black and white stripes and spots on its wings, ginger spots on its cheeks and a ginger crest on top of its head (Figure 13.32).

The secret to identifying color in ancient feathers was the discovery that they contained very well-preserved **melanosomes**. Melanosomes are found in the hair of modern mammals and feathers of modern birds, and they contain small quantities of the pigment melanin. Melanin is well known for making our hair and skin brown or black, and in small quantities is responsible for gray or blonde hair. However, another kind of melanin

Figure 13.32 How to tell the color of a dinosaur. (a,b) The pigment-bearing melanosomes, sausage-shaped eumelanosomes (a) that contain emelanin, giving black, brown, and gray colors, and spherical phaemelanosomes (b) that contain phaeomelanin, giving ginger colors. (c,d) Dinosaurs with their true original colors and patterns, *Sinosauropteryx* from the Early Cretaceous of China (c) and *Anchiornis* from the Middle Jurassic of China (d). Microphotographs (a and b) from Stuart Kearns, with permission. (c) *Source:* Artwork by Jim Robins, by permission. (d) *Source:* Artwork by M. DiGiorgio, by permission.

gives a ginger color. Each type of melanin resides in a differently shaped melanosome – sausage-shaped for the black-brown melanin, ball-shaped for the ginger melanin. This is how the two teams could make such accurate color reconstructions.

But why have stripes, spots, and crests? Everybody now reads this as evidence for a display function. If it was camouflage, the patterns would be disruptive and they would cover the whole body. Stripes on the tail or wings, or head crests, say "display." These small dinosaurs must have strutted and hopped about just as impressively as any modern male pheasant or peacock. This brings us to the final topic for this chapter, which is dinosaur behavior.

Figure 13.33 A dinosaur stampede: trackways at the Lark Quarry Conservation Park, near Winton, Queensland, Australia (see text). The track site is now enclosed in a protective building, and the photograph shows hundreds of prints of dinosaurs racing around. *Source:* Photograph by me_whynot (Wikimedia).

## Dinosaur Behavior

Knowing now that some dinosaurs at least engaged in premating displays, just like certain modern birds, is a remarkable conclusion. It's part of a major shift in what we know about dinosaurs, from speculation to evidence, because at one time paleontologists just had to guess that dinosaurs might have had feathers. Then the fossils were found which proved that many, if not all, of them did have feathers. Then, dinosaur color was purely speculative: how could we ever know the color of a dinosaur? Now we know the colors of quite a few dinosaurs. The color patterns then let us step into a more speculative area, which is their use in premating display.

Dinosaur behavior can also be judged by footprints; for example, a dinosaur stampede has been discovered (Figure 13.33). Rocks laid down about 90 Ma as sediments near a Cretaceous lakeshore in Queensland, Australia, bear the track of a large dinosaur heading down toward the lake with a 2 m (6 ft) stride. Superimposed on this track are thousands of small footprints made by small, bipedal, lightly built dinosaurs, running back up the creek bed away from the water. More than 3000 footprints have now been uncovered, showing all the signs of a panicked stampede. At least 200 animals belonging to two species were stampeding. One of the species, probably a coelurosaur, ranged up to about 40 kg (90 pounds), and the other, probably an ornithopod, ranged up to twice as large. Juveniles and adults of both species were digging in their toes as they tried to accelerate: 99% of the footprints lack heel marks. The footprints show slipping, scrabbling, and sliding,

and the smaller species usually avoided the tracks of the larger one. They may have felt hemmed against the lakeshore, breaking away in a terrified group.

The stampede sheds light on ecology as well as behavior, telling us that at least some dinosaurs gathered in herds and behaved just as African plains animals do today at waterholes on the savanna, responding immediately and instinctively to the approach of larger animals. The stampede interpretation has been questioned because there is evidence that the different tracks might have been made over the course of some days or weeks, and so perhaps not all these events happened at the same time. Nonetheless, it's an example of dinosaurs going about their business and leaving a detailed record.

As a third example of dinosaur behavior, we can look at the hadrosaurs. Some hadrosaurs had huge crests on the head (Figure 13.34). The crests were not solid but contained tubes running upward from the nostrils and back down into the roof of the mouth. Only large males had large crests; females had smaller ones and juveniles had none at all. The tubes are unlikely to have evolved for additional respiration or thermoregulation. (If so, adults would have needed large tubes whether they were female or male.) In 1981, David Weishampel suggested that the tubes were evolved for sound production. In *Parasaurolophus*, for example, they look uncannily like medieval pipes (Figure 13.35). (Reconstructed tubes can be blown to give a note.) The varying sizes of crests allow us to infer differences between the sounds produced by young, by adult females, and by adult males, to go with the different visual signals provided by the crests (Figure 13.36). These hadrosaurs may have had a

Figure 13.34 The skull of *Saurolophus*, showing the crest that extends upward and backward on the skull. *Source:* Photograph by Didier Descouens (Wikimedia).

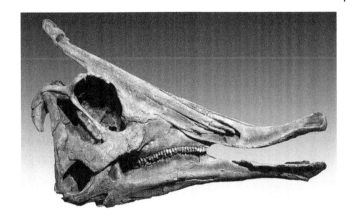

Figure 13.35 (a) Two skulls of the hadrosaur *Parasaurolophus*, from two different fossil localities. These skulls show the very long bone tubes that lead from the nostrils in a recurved tube to the back of the mouth. They could be adult and juvenile, or they could be male and female, or they could be two different species. More collecting and better age dating could resolve the issue. Whatever the answer, it is obvious that the two dinosaurs would have made very different sounds by blowing through their tubes! *Source:* Illustration © Dr Paul E. Olsen, used by permission. (b) A collection of Renaissance wind instruments called crumhorns. David Weishampel (1981) made this striking analogy with *Parasaurolophus*. *Source:* Drawings from the book *Syntagma musicum*, published in 1620.

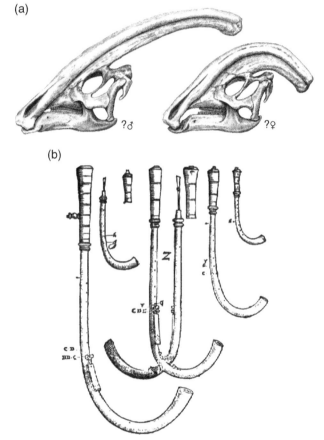

sophisticated social system, as complex as those we take for granted in mammals and birds.

With advances in technology, CT scans have become powerful ways to look inside solid objects such as dinosaur skulls. It turns out that the same sound-producing resonant tubes occur in the skulls of other hadrosaurs (Figure 13.37a), and once we have seen that, then one can look for (and find) anomalous (and analogous) nostril pathways in other dinosaurs such as ankylosaurs (Figure 13.37b).

In all aspects of their biology, therefore, dinosaurs seem as modern as mammals and birds. That makes their extinction at the end of the Cretaceous even more puzzling (except for the survival of the smallest dinosaurs, which we call birds). We will discuss that extinction in Chapter 16.

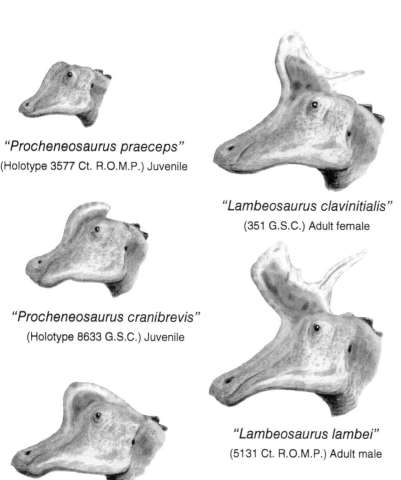

*"Procheneosaurus praeceps"*
(Holotype 3577 Ct. R.O.M.P.) Juvenile

*"Lambeosaurus clavinitialis"*
(351 G.S.C.) Adult female

*"Procheneosaurus cranibrevis"*
(Holotype 8633 G.S.C.) Juvenile

*"Corythosaurus frontalis"*
(Holotype 5853 Ct. R.O.M.P.) Juvenile

300 mm

*"Lambeosaurus lambei"*
(5131 Ct. R.O.M.P.) Adult male

*"Lambeosaurus clavinitialis"*
(Holotype 8703 G.S.C.) Adult female

Figure 13.36 Reconstructed heads of several "species" of lambeosaurine dinosaurs, drawn to imply that all of them could be different age and gender members of only one species, *Lambeosaurus lambei*. *Source:* Artwork by Nobu Tamura (Wikimedia).

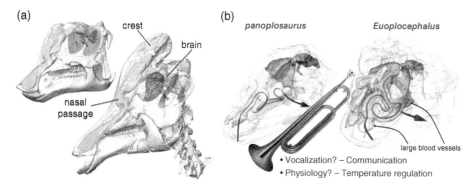

(a)

crest

brain

nasal
passage

(b)

*panoplosaurus*

*Euoplocephalus*

large blood vessels

• Vocalization? – Communication
• Physiology? – Temperature regulation

Figure 13.37 (a) Images from CT scans of the skulls of a young and a subadult hadrosaur *Corythosaurus*. The older individual has a nostril pathway that was more complex and was divided into resonating lengths. Research details in Evans et al. (2009). (b) Images from CT scans of two ankylosaurs, with trumpet for visual effect. The nasal passages could have been used to emit sounds. Of course, nostrils can also be used in thermoregulation (cooling), but both functions would have worked here. Look back at Figure 13.17, which shows the ankylosaur *Euoplocephalus*, and compose a signature trill for it. *Source:* Both images courtesy of the Witmer Lab at Ohio University, https://people.ohio.edu/witmerl/lab.htm.

## References

Bates, K.T. and Falkingham, P.L. (2012). Estimating maximum bite performance in *Tyrannosaurus rex* using multi-body dynamics. *Biology Letters* 8: 660–664.

Button, D.J., Rayfield, E.J., and Barrett, P.M. (2014). Cranial biomechanics underpins high sauropod diversity in resource-poor environments. *Proceedings of the Royal Society B: Biological Sciences* 281: 20142114.

Cieri, R.L. and Farmer, C.G. (2016). Unidirectional pulmonary airflow in vertebrates: a review of structure, function, and evolution. *Journal of Comparative Physiology. B, Biochemical, Systemic, and Environmental Physiology* 186: 541–552.

Erickson, G.M. (2005). Assessing dinosaur growth patterns: a microscopic revolution. *Trends in Ecology and Evolution* 20: 677–684.

Evans, D., Ridgely, R., and Witmer, L. (2009). Endocranial anatomy of lambeosaurine dinosaurs: a sensorineural perspective on cranial crest function. *Anatomical Record* 292: 1315–1337.

Godefroit, P., Sinitsa, S.M., Dhouailly, D. et al. (2014). A Jurassic ornithischian dinosaur from Siberia with both feathers and scales. *Science* 345: 451–455.

Hutchinson, J.R., Bates, K., Molnar, J. et al. (2011). A computational analysis of limb and body dimensions in *Tyrannosaurus rex* with implications for locomotion, ontogeny, and growth. *PLoS One* 6: e26037.

Ibrahim, N., Sereno, P.C., Dal Sasso, C. et al. (2014). Semiaquatic adaptations in a giant predatory dinosaur. *Science* 345: 1613–1616.

Li, Q., Gao, K.Q., Vinther, J. et al. (2010). Plumage color patterns of an extinct dinosaur. *Science* 327: 1369–1372.

O'Connor, P. and Claessens, L. (2005). Basic avian pulmonary design and flow-through ventilation in non-avian theropod dinosaurs. *Nature* 436: 253–256.

Vinther, J.A., Nicholls, R., Lautenschlager, S. et al. (2016). 3D camouflage in an ornithischian dinosaur. *Current Biology* 26: 2456–2462.

Wedel, M.J. (2006). Origin of postcranial skeletal pneumaticity in dinosaurs. *Integrative Zoology* 1: 80–85.

Weishampel, D.B. (1981). Acoustic analyses of potential vocalization in lambeosaurine dinosaurs (Reptilia: Ornithischia). *Paleobiology* 7: 252–261.

Wings, O. and Sander, P.M. (2007). No gastric mill in sauropod dinosaurs: new evidence from analysis of gastrolith mass and function in ostriches. *Proceedings of the Royal Society B: Biological Sciences* 274: 635–640.

Zhang, F., Kearns, S.L., Orr, P.J. et al. (2010). Fossilized melanosomes and the colour of Cretaceous dinosaurs and birds. *Nature* 463: 1075–1078.

## Further Reading

Barrett, P.M. and Rayfield, E.J. (2006). Ecological and evolutionary implications of dinosaur feeding behaviour. *Trends in Ecology and Evolution* 21: 217–224.

Benson, R.B.J. (2018). Dinosaur macroevolution and macroecology. *Annual Review of Ecology, Evolution, and Systematics* 49: 379–408.

Benton, M.J. (2019). *Dinosaurs Rediscovered: How a Scientific Revolution is Rewriting Their Story*. New York: Thames & Hudson. [The application of scientific method to understanding dinosaurian paleobiology].

Brett-Surman, M.K.,.T.H.J.,.J.F.,.B.W. (ed.) (2012). *The Complete Dinosaur*, 2e. Bloomington: Indiana University Press. [Many fine essays on all aspects].

Brusatte, S.L. (2012). *Dinosaur Paleobiology*. New York: Wiley. [The best textbook on dinosaurs].

Brusatte, S.L. (2018). *The Rise and Fall of the Dinosaurs*. New York: Simon & Schuster. [The best account of what it is like to excavate dinosaurs around the world].

Chiappe, L.M. and Meng, Q.J. (2016). *Birds of Stone: Chinese Avian Fossils from the Age of Dinosaurs*. Pittsburgh: Johns Hopkins University Press.

Fastovsky, D.E. and Weishampel, D.B. (2012). *Dinosaurs: A Concise Natural History*. New York: Cambridge University Press.

Hutchinson, J.R. and Gatesy, S.M. (2006). Dinosaur locomotion: beyond the bones. *Nature* 440: 292–294.

Long, J. and Shouten, P. (2009). *Feathered Dinosaurs: The Origin of Birds*. Oxford: Oxford University Press.

Norell, M.A. and Xu, X. (2005). Feathered dinosaurs. *Annual Reviews of Earth and Planetary Sciences* 33: 277–299.

Rayfield, E.J. (2007). Finite element analysis and understanding the biomechanics and evolution of living and fossil organisms. *Annual Review of Earth and Planetary Sciences* 35: 541–576.

Reisz, R.R., Evans, D.C., Roberts, E.M. et al. (2012). Oldest known dinosaurian nesting site and reproductive biology of the early Jurassic sauropodomorph *Massospondylus*. *Proceedings of the National Academy of Sciences of the United States of America* 109: 2428–2433.

Sander, P.M., Christian, A., Clauss, M. et al. (2010). Biology of the sauropod dinosaurs: the evolution of gigantism. *Biological Reviews* 86: 117–155.

Sellers, W.I., Pond, S.B., Brassey, C.A. et al. (2017). Investigating the running abilities of *Tyrannosaurus rex* using stress-constrained multibody dynamic analysis. *PeerJ* 5: e3420. https://doi.org/10.7717/peerj.3420.

Weishampel, D.B. (1997). Dinosaur cacophony. *Bioscience* 47: 150–159.

Weishampel, D.B., Dodson, P., and Osmólska, H. (eds.) (2004). *The Dinosauria*, 2e. Berkeley: University of California Press.

Zhao, Q., Benton, M.J., Sullivan, C. et al. (2013). Histology and postural change during the growth of the ceratopsian dinosaur *Psittacosaurus lujiatunensis*. *Nature Communications* 4: 2079.

## Questions for Thought, Study, and Discussion

1  We know that dinosaurs laid large clutches of eggs, and clearly a female dinosaur would lay a LOT of eggs during her lifetime. Yet on average, only two of her eggs would survive to be full adults (or the world would have been overrun with dinosaurs). So, do the best you can to describe the hazards of being a dinosaur egg (or hatchling, or adolescent), citing evidence where you can.

2  Think about how sauropods could have reached such huge sizes. Look for all the hypotheses you can find online (e.g., they lived underwater all the time, gravity was lower, they had unique physiologies; Figure 13.21). What is the evidence for and against each idea? Then work through the elements of the illustrated behavioral-functional model and write down key evidence for each observation, including data from living animals. Which modern animals could perhaps some day evolve this array of adaptations?

14

Birds and the Evolution of Flight

**In This Chapter**

Flight in animals began with insects in the Carboniferous coal forests, but the earliest animals with powered flapping flight were the pterosaurs. They were mostly fish eating but came in a variety of sizes and shapes, consistent with living in different habitats and catching different prey. Some were filter feeders with many very fine teeth, rather like a flamingo. Pterosaurs were undoubtedly warm-blooded, and they include the largest flying animals at a wingspan of over 10 m. The earliest bird is Late Jurassic in age, and birds evolved from small theropod dinosaurs that already had feathers. Flight evolved from maniraptoran theropods that had small body size and elongated arms lined with feathers. Early forms of flight included gliding, and it is debated whether these theropods were adapted to leaping from the ground to glide, or down from trees, where they exploited new food sources. The first bird, *Archaeopteryx*, had the skeleton of a dinosaur but with strong feathers that probably gave it the ability to fly. More modern-looking birds were well evolved in the Cretaceous, and there is no question that many of them were strong fliers with ecologies that we would easily recognize today. Birds survived the extinction at the end of the Cretaceous, and had a dramatic radiation in the Cenozoic that has given us birds ranging in size and biology from a hummingbird to an ostrich.

## Flying Animals

Plants and insects took to the air long before vertebrates attempted to leave the ground. As we saw in Chapter 8, flightless insects took off in the Devonian and Carboniferous, some of them reaching truly huge sizes. One of the first reptile fliers was *Coelurosauravus* from the Late Permian (see Chapter 12, Figure 12.2), and there were other gliding reptiles in the Triassic. Most of these had extra-long ribs that stuck out at the side and in life were covered by a membrane, like the living "flying lizard" *Draco* (Figure 14.1). The Cretaceous gliding lizard *Xianglong* (Figure 14.2) used the same kind of structure, and yet is not closely related to *Draco*. In both of these lizards, ligaments and muscles between the ribs give precise control of the gliding surface, while all four limbs remain free for walking, grasping, and climbing. All of them can fold up the airfoil when it is not in use.

These animals were parachuting or gliding. **Gliding** is any kind of flight that does not involve flapping the wings or other airfoils. **Parachuting** is a controlled descent, where the flight structures allow the animal (or plant seed) to fall more slowly than otherwise. A great array of vertebrates today parachute, from frogs with membranes between their toes, the lizard *Draco* with its rib-supported membranes, and even snakes that can flatten their bodies sideways. Several groups of mammals have also evolved parachuting adaptations, including flying squirrels and three lineages of Australian gliding marsupials (greater gliders, squirrel gliders, and feathertail gliders). This suggests that parachuting adaptations evolve in animals of the forest canopy that habitually jump from branch to branch, from tree to tree, or from trees to the ground. Any method of breaking the landing impact or of leaping longer distances would be advantageous and might evolve rapidly.

None of these parachuting animals has powered flight, however. The energy for gliding flight is gravitational, generated as the animal climbs in the tree and released as it parachutes off the branch. Parachuting can evolve in

Figure 14.1 *Draco volans*, a species of flying lizard, with its wing membranes stretched out to show the ribs built into them. The wings can also be used in display. *Source:* Photograph by Zhuxiaoyufish, Flickr.

Figure 14.2 The Cretaceous gliding lizard *Xianglong* from China. *Source:* Artwork by Nobu Tamura (Wikimedia).

Most phylogenetic analyses place pterosaurs within archosaurs, among the bird-like archosaurs, the Avemetatarsalia, as closest relatives of dinosaurs. If so, then you remember from Chapter 12 that pterosaurs must have evolved in the Early Triassic, with a very long ghost lineage that we have not yet found, probably consisting mostly of their terrestrial ancestors.

Pterosaurs have very lightly built skeletons, with air spaces in many of the bones. Their fore limbs were extended into long struts that supported a wing, as in birds and bats. Pterosaurs were unique, however, in that most of the wing membrane was supported on one extraordinarily long finger, while three other fingers were normal and bore claws (Figure 14.3). The fourth finger was about 3 m (10 ft) long in the largest pterosaurs. In contrast, birds support the wing with the whole arm, and bats use all their fingers as bony supports through their wing membranes. Pterosaurs thus have a unique wing anatomy but as the largest flying creatures ever to evolve

animals with rather low metabolic rates. It does not require the high metabolic rate of birds and bats, and extinct pterosaurs, which have (or had) powered flight. Among these groups, we look at bats in Chapter 19 and birds in this chapter. But first, we should consider the pterosaurs, which were the first truly successful powered flying vertebrates.

## Pterosaurs

Pterosaurs are the most famous flying reptiles. The earliest pterosaurs known are Late Triassic, when they were already well evolved for flight. The earliest well-preserved pterosaur, *Austriadactylus*, already had a bony crest on its skull and a very long tail.

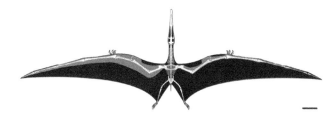

Figure 14.3 Body plan of a pterosaur. The leading edge of the wing membrane is supported for most of its length by the fourth finger, and its trailing edge is fixed to the hind limb. This pterosaur is *Anhanguera*, from the Cretaceous of Brazil. Scale bar, 10 cm. *Source:* Image by Leon Claessens, Patrick O'Connor, and David Unwin, Creative Commons.

and as a group that flourished for more than 160 m.y., they cannot be dismissed as primitive or poorly adapted.

## Pterosaur Feeding and Fossils

Most pterosaurs had large eyes sighting right along the length of long, narrow, lightly built jaws. The teeth were usually thin and pointed, often projecting slightly outward and forward, as in *Rhamphorhynchus* (Figure 14.4). This is most likely an adaptation for catching fish. Almost all pterosaur fossils are preserved in sediments laid down on shallow seafloors, and where stomach contents have been preserved with pterosaur skeletons, they always contain fish remains such as spines and scales. Some pterosaurs may have fished on the wing, like living birds such as gadfly petrels or skimmers, which fly along just above the water surface and dip in their beaks to scoop up fish or crustaceans, although this requires a strong beak. One can imagine *Anhanguera* doing this (Figure 14.5a). Other pterosaurs may have fed like terns, which dive slowly so that only the head, neck, and front of the thorax reach under the water, while the wings remain above the surface. Some pterosaurs with long sharp beaks may have fished standing in the water or slowly patrolling, like herons, or sitting on the water. It seems unlikely that pterosaurs crash-dived into water like pelicans or gannets or swam underwater like penguins: pterosaur wings were too long and too fragile. At least one pterosaur, *Pterodaustro* from

Argentina, had teeth that were so fine, long, and numerous that it must have been a filter feeder, perhaps like a flamingo (Figure 14.5b), and *Ctenochasma* looks like a filter feeder too (Figure 14.5c). Some short-jawed pterosaurs may have eaten shore crustaceans or insects.

(a)

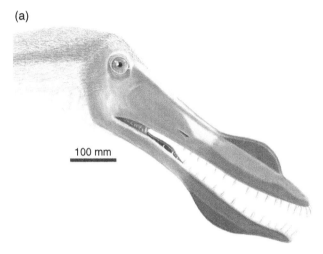

100 mm

(b)

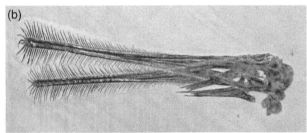

(c)

Figure 14.5 (a) *Anhanguera*, from the Cretaceous of Brazil, reconstructed by Larry Witmer. *Source:* Artwork by C. McQuilkin, courtesy of the Witmer Lab at Ohio University: https://people. ohio.edu/witmerl/lab.htm). (b) *Pterodaustro* may have been a filter feeder, perhaps like a flamingo. *Source:* Artwork by Nobu Tamura, Wikimedia. (c) *Ctenochasma* may also have been a filter feeder. *Source:* Image by Ghedoghedo (Wikimedia).

Figure 14.4 *Rhamphorhynchus*, a pterosaur from the Jurassic of Germany, model of the skull showing its fish-eating teeth. *Source:* Photograph by Amy Martiny, courtesy of the Witmer Lab at Ohio University: https://people.ohio.edu/witmerl/lab.htm.

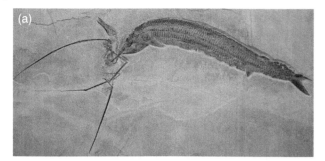

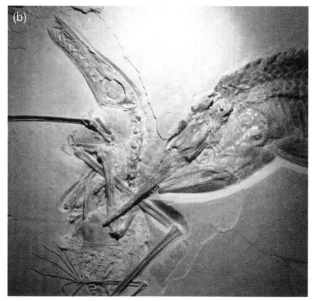

Figure 14.7 Some *Nyctosaurus* specimens have been preserved with a strange crest on the head. It is difficult to see this as anything other than a display structure. *Source:* Artwork by Matt Martyniuk (Wikimedia).

Figure 14.6 A bad day in the Solnhofen lagoon. A pterosaur had just caught a small fish right at the water surface when a larger fish struck at its wing. The pterosaur was too large for the fish to swallow, but its teeth were stuck so far into the elastic membranes of the wing that the fish could not get free and both animals died. This tragic accident was beautifully preserved for paleontological detectives to explain (Frey and Tischlinger 2012), as explained further in the text. *Source:* Photographs by Dino Frey and Helmut Tischlinger, Creative Commons.

A slab from the Late Jurassic Solnhofen Limestone of Germany preserves dramatic evidence of a very bad day in a tropical lagoon (Figure 14.6a,b). A *Rhamphorhynchus* had caught a little fish, presumably by dipping its beak into the water. The fish was still in the throat pouch of the pterosaur as it flapped strongly to regain height. The pterosaur may have touched the water with its wingtip or came so close to the water that it caused a strong shadow, because a large fish struck at it and seized the wing. The pterosaur was pulled into the water, but it was too big for the fish to swallow. The teeth of the fish were firmly fixed in the elastic membranes of the pterosaur wing, and although the fish struggled enough to severely damage the pterosaur

wing, it could not get free and all three animals died (Frey and Tischlinger 2012).

There are two main groups of pterosaurs. **Rhamphorhynchoids** (Late Triassic to Late Jurassic) are the stem group of early pterosaurs, rather than a clade. Most of them had wingspans under 2 m (6 ft), and some were as small as sparrows. Many, like *Rhamphorhynchus* itself (Figures 14.4 and 14.6), had long tails, sometimes with a vane at the end.

**Pterodactyloids** are a clade of advanced pterosaurs that replaced rhamphorhynchoids in the Late Jurassic and flourished until the end of the Cretaceous. Pterodactyloids had short tails and many were much larger than rhamphorhynchoids. At least some of them had extravagant crests on the head, for example *Nyctosaurus* with its astonishing headgear that dwarfed its body (Figure 14.7). The large forms were adapted for soaring rather than continuous flapping flight, although they all flapped for take-off. *Pterodactylus* itself was sparrow-sized but *Pteranodon*, from the Cretaceous of North America, had a wingspan of about 7 m (22 ft), and the gigantic pterosaur from Texas, *Quetzalcoatlus*, was 10–11 m (35–35 ft) in wingspan, the largest flying creature ever to evolve. (An incomplete set of pterosaur fossils from Romania may be pieces of an even larger form: guesses vary around 12 m.)

Although pterosaur bones were light and fragile, several examples of outstanding preservation have shown us many details of their structure. Black shales in Lower Jurassic rocks of Germany have shown details of rhamphorhynchoids; Late Jurassic members of both pterosaur groups have been found exquisitely preserved in the Solnhofen Limestone of Germany and in lake deposits in Kazakhstan in Central Asia. From the Lower Cretaceous of Brazil, we have partial skeletons preserved without crushing, and the Upper Cretaceous chalk beds of Kansas have yielded huge specimens of *Pteranodon*. Discoveries of skin, wing membranes, and stomach contents allow biological interpretations of these exciting animals.

### Pterosaur Flight

Interpreting the flight of pterosaurs is tricky because they had evolved their own way of doing it. However, it seems clear that all pterosaurs, including the giant forms, were capable of powered, flapping flight (Figure 14.8).

The pterosaur wing was attached low on the hind limb, in a "broad-wing" reconstruction (Figure 14.3). The wing itself was not simply a giant skin membrane; that would have been too weak to power flapping flight. Furthermore, with bones, joints, and ligaments only on the leading edge of the wing, a pterosaur needed a way to control the aerodynamic surface of the wing. Beautifully preserved specimens show that the wing had special adaptations. It was stiffened by many small, cylindrical fibers, which were probably tied together by small muscles. The combination of structural stiffeners and muscles allowed fine control

over the surface, and at the same time made the wing reasonably strong, not easily damaged or warped, and not likely to billow in flight like the fabric of a hang glider.

A research team led by Larry Witmer of Ohio University made CT scans of two uncrushed pterosaur skulls. The scans revealed the size and shape of the pterosaur brain. In both brains, the lobes associated with balance were very large, and this allowed the researchers to reconstruct the head to be arranged in the usual, or preferred, attitude it had in life. While the little Early Jurassic pterosaur *Rhamphorhynchus* apparently held its head horizontally (as birds do in normal flight), the later and larger Cretaceous pterosaur *Anhanguera* seems to have held its head angled downward, perhaps in fishing position (Figure 14.5a). This is not unreasonable. Herons spend hours in this kind of attitude as they stand waiting for fish, even though they fly with their heads horizontal. In completely different ways of life, kites and pelicans (at least the white-tailed kite and the white pelican of California) hold their heads "normally" as they fly from place to place, but kites hover over potential prey sites, and pelicans go into slow searching flight mode, both with their heads tilted dramatically downward (Figure 14.9).

All the small rhamphorhynchoids and many of the pterodactyloids had active, flapping flight. Naturally, the gigantic pterosaurs could not have flapped for long, and they probably spent most of their time soaring, as does the living albatross. **Soaring** is a mode of flying used by long-winged pterosaurs and birds that exploit rising air currents to stay aloft without much wing flapping. Aerodynamic analysis shows that pterosaurs were the best slow-speed soaring fliers ever to evolve.

Flapping flight involves very high energy expenditure. Birds are warm-blooded, as are bats and many large insects when they are in flight: dragonflies, moths, and bees are examples. Thus one might guess that pterosaurs too were warm-blooded. Several Jurassic pterosaurs have fur preserved on the skin; if pterosaurs had fur, they were probably warm-blooded. Pterosaur bones had air spaces

Figure 14.8 Lyrical reconstruction of two ornithocheirid pterosaurs in the Cretaceous skies above England. However, the artist has allowed some brutal reality: you will notice that a fish is being stolen. (This happens often among seabirds today.) *Source:* Artwork by Dmitri Bogdanov (Wikimedia).

Figure 14.9 A white-tailed kite hovering, with its head held still, looking downward to fix on a prey animal. *Source:* Photograph by Zoipe, Creative Commons.

running through them in the same way that living bird bones do. In birds, this system reduces weight and helps to provide air cooling, and it is reasonable to interpret pterosaur bone structure in the same way.

But the air spaces are more than that; we now realize that pterosaurs had much the same respiration system as dinosaurs and birds, including the one-way flow through the lungs that is much more efficient than our mammalian in-and-out system (see Chapter 13). This could suggest that not only do pterosaurs and dinosaurs share many aspects of anatomy and physiology but that these adaptations, including the feathers-hair, evolved from their common ancestor very early in the Triassic.

The social behavior of pterosaurs may have been complex. Many pterosaurs were dimorphic. Males were larger, with long crests on the back of the head and with relatively narrow pelvic openings. Females were smaller, with smaller crests but larger pelvic openings. New discoveries show that the soft tissues associated with some crests were extravagantly large and were much more likely to have been display structures than aids to flight (Figure 14.7).

The largest pterosaurs, *Quetzalcoatlus* and related forms (together called azhdarchids), lived right at the end of the Cretaceous. The fossils of *Quetzalcoatlus* were found in nonmarine beds in Texas, deposited perhaps 400 km (250 miles) inland from the Cretaceous shoreline. Perhaps it was the ecological equivalent of a vulture, soaring above the Cretaceous plains and scavenging on carcasses of dinosaurs. *Quetzalcoatlus* did have a strangely long, strong neck but its beak seems too

lightly built for this method of feeding. Witton and Naish (2008) have argued that it was more like a gigantic heron (Figure 14.10), standing and fishing in inland lakes and swamps, or picking up frogs, turtles, baby dinosaurs, or arthropods such as crayfish from shallow water.

We do not know why pterosaurs became extinct. As we have seen, they were most likely active, warm-blooded animals with flapping flight much like that of birds. Yet pterosaurs became extinct at the end of the Cretaceous, at the same time as the dinosaurs disappeared, while birds did not. We shall return to that question in Chapter 20.

## Getting Off the Ground

There are four kinds of flight: passive flight, parachuting, soaring, and powered flight. Passive flight can be used only by very tiny organisms light enough to be lifted and carried by natural winds and air currents, and light enough to suffer no damage on landing. Tiny insects, baby spiders, frogs' eggs, and many kinds of pollen, spores, and seeds can be transported this way. But their "flight" duration, direction, and destination are entirely at the mercy of chance events.

We have seen examples of parachuting and gliding flight in the earliest reptile flyers, as well as in many modern gliding frogs, lizards, snakes, and mammals. Some large pterosaurs, as we have seen, might have been soarers, taking advantage of thermal currents in the air to keep aloft. Parachuting organisms seek short-range

**Figure 14.10** Dawn patrol. A group of *Quetzalcoatlus* forages across a Late Cretaceous wetland in Texas. Note the baby sauropod! There is no specific evidence that *Quetzalcoatlus* ate them, though it certainly could have done so. *Source:* Image by Mark Witton and Darren Naish (Wikimedia).

travel from one point to another, and their landing point is reasonably predictable; they aim and leap. Parachuting is used in habitats where external air currents are minimal, especially in forests. Wind gusts and air currents are potentially disastrous to animal parachutists, just as they are to human paratroops.

Powered flight is usually accomplished by some sort of flapping motion with special structures (wings). It needs a lot of energy but gives independence from variations in air currents, and it is usually accompanied by a high level of control over flight movements. Because powered flight is achieved by controllable appendages, almost all powered fliers can glide to some extent, some very poorly (no better than parachutists) and some very well indeed. Raptors and soaring seabirds are examples of powered fliers that glide well.

Soaring is used by flying organisms that range widely over a broad habitat (Figure 14.9). It is a low-energy flight style because the lift comes from external air currents rather than muscular expenditure by the flier. Energy costs are mainly related to the maintenance and adjustment of gliding surfaces in the air flow. Soarers may need occasional bursts of flapping flight if there are no upcurrents, or in transferring from one upcurrent cell to another. Flapping is sometimes needed for take-off, until airspeed exceeds stalling speed, or for final adjustments of attitude and speed in landing. Because flapping flight is needed occasionally by all soarers today (especially in emergencies), soaring probably cannot evolve from parachuting but only from powered flight.

Flight of all kinds demands a light, strong body. Soaring especially emphasizes lightness in muscle mass as well as overall structure. Powered flight has more requirements, including a significant output of energy and strength. Even the best soarers among living birds, albatrosses and condors, cannot flap for long before they are exhausted, because their flight muscles are small relative to their size and total weight. It might be difficult for a specialized soarer to reevolve the ability to sustain flapping flight.

It might seem obvious that powered flight could evolve easily from gliding. Any evolving wing should be a fail-safe device, allowing a gliding fall during flight training. But an animal that has already evolved efficient gliding would not easily improve its flight by flapping in mid-glide, because that would disturb the smooth airflow over the gliding surfaces. Aerodynamic analysis shows that an evolutionary transition is possible from gliding to flapping, but only in very special circumstances. The glider must add fairly large, rapid wing beats, not little flutters, and because wing beats require considerable expenditure of energy, there must be a corresponding pay-off in energy saved (for example, the animal must save some walking or climbing, or must reach a larger food supply).

One of the issues to be overcome in a gliding-to-flapping transition relates to wing size and risk. Further, it could only happen where the glider uses limbs to support its flight structures – gliding lizards, for example, with their rib-mounted membranes could not start to flap those structures because they lack the muscles. No tetrapod has evolved additional flying limbs (if they had, they would become hexapods, like angels, with arms, legs, and wings), so the gliding limb, adapted to be a substantial area in proportion to the body size, has to flap up and down. But flapping brings risks to a large membrane or area of feathers, because the wing has to become resistant to breakage during flapping beats. Maybe small gliders with small wings were the first to become flapping fliers.

## Birds

Living birds are warm-blooded, with efficient thermoregulation that maintains body temperatures higher than our own. Birds breathe more efficiently than mammals, pumping air through their lungs rather than in and out. They have better vision than any other animals. Birds build extraordinarily sophisticated nests: bowerbirds are second only to humans in their ability to create art objects. New Caledonian crows learn to make tools faster than chimpanzees do. And above all, birds can fly better, farther, and faster than any other animals, an ability that demands complex energy supply systems, sensing devices, and control systems.

Birds include ostriches and penguins, which cannot fly, and hummingbirds, which can hardly walk. But birds share enough characters for us to be sure that they form a single clade that originated from among the theropod dinosaurs, as we saw in Chapter 13. The skull, pelvis, feet, and eggs of birds are so clearly archosaurian that Darwin's friend T. H. Huxley called birds "glorified reptiles."

### Archaeopteryx

*Archaeopteryx*, from Upper Jurassic rocks in Germany, is perhaps the most famous fossil in the world. It is a feathered dinosaur that looks remarkably like a bird until it is examined carefully (Figure 14.11). Only 11 specimens have been found, plus a single feather. The first complete *Archaeopteryx* was immediately seen as a fossil bird, because it had feathers on its wings and tail. But without feathers, it looks very much like a small theropod dinosaur. In fact, two of the specimens lay unrecognized for a long time, labeled as small theropods.

*Archaeopteryx* has a theropod pelvis, not the tight, box-like structure of living birds. It has a long, bony tail, clawed fingers, and a jaw full of savage little teeth. These

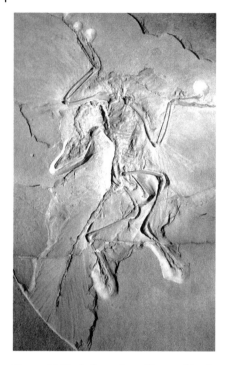

Figure 14.11 *Archaeopteryx lithographica* from the Late Jurassic of Germany, about the size of a large crow. Look carefully for feather impressions. *Source:* Photograph by H. Raab (Wikimedia).

are all theropod features. *Archaeopteryx* lacks many features of living birds. The only bird-like features on the entire bony skeleton of *Archaeopteryx* are a few characters of the skull, but a CT scan of its braincase shows that the brain was very bird-like (Alonso et al. 2004).

*Archaeopteryx* is always preserved in an unusual body attitude, with the neck severely ricked back over the body (Figure 14.11). We know why this happens. If an animal dies today on or near the beach or on a desert salt pan, it may be mummified by wind and salt spray before it rots or is eaten by predators. The muscles slacken and the tendons dry out. The long tendons that support the head contract severely, dragging the skull backwards over the spine. At the same time, any body feathers on a bird usually drop off, but the stronger wing and tail feathers stay fixed in position.

Occasionally, birds mummified on a beach may be washed out to sea on a high tide or blown into the sea by a gale. They may float for several weeks before becoming waterlogged, and even when they finally sink, they retain their peculiar body attitude.

But could *Archaeopteryx* fly? Over the years, paleontologists and ornithologists have argued both ways. Some would say it could not have flown by flapping its wings because the sternum, or breast bone, is too small, and so it could not have had large enough muscles to power its flight. But, on the other hand, *Archaeopteryx* has wings that are large enough to support its body

weight, and it had already amassed all the other special features of birds through its descent from small theropods such as the deinonychosaurs and troodontids (see Chapter 13). Those maniraptorans were very bird-like in being small, having long arms and hands, having hollow bones, and in being covered with feathers. It was just that their arms and wings were not large enough to bear their body mass for long enough; what about *Archaeopteryx*?

In fact, when you compare the key wing measurements among modern birds and airplanes, there is a standard relationship between wing shape and wing area. First, the wing area has to be large enough to carry the weight of the body, and the relationship between area and weight is the **relative wing loading**. In the classic diagram (Figure 14.12), the albatross has a huge wing in proportion to body size, so it can soar for long periods of time without flapping; at the other end of the wing loading axis are ducks and hummingbirds, which have relatively small wings compared to body mass, but they flap like mad. The **aspect ratio**, plotted on the y-axis, is the ratio of wing length to wing depth (front to back distance): wings with high aspect ratios are long and thin, like those of gulls and albatrosses. A low aspect ratio is seen in short, deep wings, as seen in birds like owls that fly in and out of trees with deep, powerful beats, but might suffer damage if their wings were too long and thin.

Note that various airplanes have their place in the diagram because they are subject to the same laws of the physics of flight. And *Archaeopteryx* is there, right in the middle of the diagram, with wings that are completely capable of flight. In a study of the wing bones of *Archaeopteryx*, Voeten et al. (2018) confirm this, finding that the bones share features of their cross-section with modern flying birds, and especially those that employ short-distance flapping.

In further detail, ornithologists had doubted the flying ability of *Archaeopteryx* because there is no hole through the shoulder joint through which to pass the large tendon that gives the rapid, powerful, twisting wing upstroke in living birds. This tendon passes *through* the shoulder joint and as well as raising the wing, it twists it. On the upstroke, the twist arranges the wing and feathers so that they slip easily through the air, with little drag. At the top of the upstroke, the wing is in exactly the right position to give a powerful downbeat. This is the **supracoracoideus system** (Figure 14.13) seen in modern birds and in most later fossil birds. Without it, *Archaeopteryx* swung its wing in a different way from modern birds, following a descending diagonal from in front of the body to behind the body, but without raising the wing much above the level of its back. Modern birds raise the wing high above the body before swinging the wing back and down forcefully to create the flight power stroke.

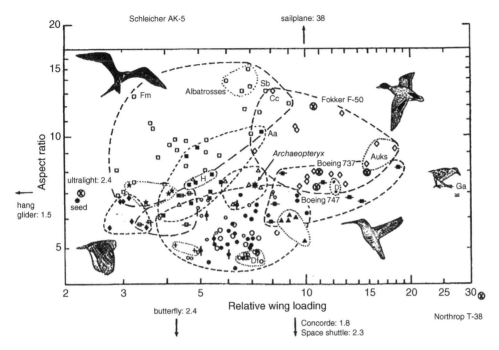

Figure 14.12 Much of the anatomy of birds is dictated by two fundamentals of aerodynamics: their relative wing loading (x-axis), which is body mass divided by wing area, and their aspect ratio, which is the ratio of length to width of the wing. *Source:* From Norberg (2002).

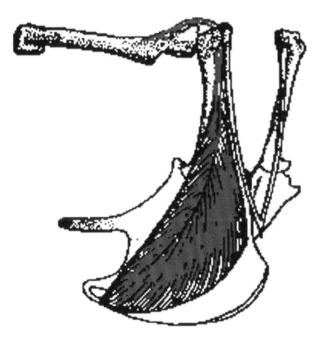

Figure 14.13 The supracoracoideus muscle in living birds attaches to the breastbone, then passes through the shoulder joint to insert on the upper side of the humerus. This is the muscle system that raises the wing. *Archaeopteryx* does not have it.

In small flying birds today, the wishbone or **furcula** is flexible and acts as a spring that repositions the shoulder joints after the stresses of each wing stroke. It is needed to give the rapid flaps necessary for flight (a starling flies with 14 complete wing beats per second). The wishbone

also helps to pump air in respiration, and to recover some of the muscular energy put into the downstroke. But the wishbone in *Archaeopteryx* and the wishbones of theropods are U-shaped and strong and solid; they could not have acted as effective springs, so *Archaeopteryx* flapped its wings relatively slowly and without the rapid recovery possible in modern birds. Further, the long bony tail, inherited from its theropod ancestors, would have slowed its flight.

Nevertheless, we can follow the common-sense view that if it looks like a bird and has wings like a bird, it could fly like a bird. *Archaeopteryx* was the first theropod to have wings that were big enough to provide lift, and even if it flew quite slowly and just for short distances, nonetheless it flew. It also scurried on the ground, using its powerful legs, just like a modern pheasant or road-runner, and clutched onto branches and used its toes, which pointed back and forwards, to hang on to tree trunks and climb about. Evidence from claw shape is equivocal, showing that *Archaeopteryx* could likely both run around and climb but did not specialize in either (Birn-Jeffery et al. 2012).

*Archaeopteryx* is reconstructed here flying after a small theropod, *Compsognathus*, found in the same rocks in southern Germany (Figure 14.14). Both are liberally covered in feathers, the *Compsognathus* shown ginger and stripy, following work on its Chinese relative *Sinosauropteryx* (see Chapter 13, Figure 13.32), and *Archaeopteryx* is shown as black and shiny based on an analysis of the melanosomes in one of its feathers (Carney et al. 2012).

Figure 14.14 Restoration of *Archaeopteryx* chasing a juvenile *Compsognathus*. *Source:* Artwork by durbed (Wikimedia).

### The Origin of Flight in Birds

The phylogenetic evidence for bird origins is clear: they are dinosaurs, they are theropods, and they are maniraptorans, whose closest relatives are deinonychosaurs and troodontids (see Chapter 13). Further, we know that the first birds such as *Archaeopteryx* had acquired many avian characters through their dinosaurian ancestry. For example, hollow bones and the furcula appeared in the first theropods. Feathers also appeared very early, possibly among the first dinosaurs and pterosaurs, or at least halfway through theropod evolution. Further, flight-type feathers appear first in maniraptoran theropods. Maniraptorans show the elongated, powerful arm, and many of them used the wing for gliding flight before the origin of birds. Importantly, a whole group of maniraptorans, including deinonychosaurs, troodontids, and birds, underwent a sharp phase of miniaturization at some point in the Early or Middle Jurassic – they got small when other theropods were increasing in size.

The big debate has always been whether birds evolved flight "from the ground up or from the trees down." In the 1980s, with the discovery of new deinonychosaurs, many paleontologists preferred a "ground-up" scenario, where the theropods charged about after their small prey on the ground and undertook leaps, perhaps aided by flapping of the arms, to overtake their prey. The idea was that achieving flight would replace the thrust of the feet on the ground by aerodynamic forward thrust from the wing. In the take-off run, energy expended by the fore limbs would replace energy expended by the hind limbs, after a transition period in which all limbs would be contributing to forward thrust.

Lift is not important at first. The first stages of this fast, low-level flight would be aided by the phenomenon of **ground effect**. Essentially, eddies generated by the wings interact with the ground immediately under the wings, providing enough lift at very low altitude to achieve take-off. Thus the wing stroke would not have to produce much lift as long as there was no advantage in acquiring height.

In this scenario, the bird is now capable of fast-flapping low-level flight, but its advantage ends if it ascends out of the shallow zone of ground effect. All the wing action is energetically expensive, especially in the early stages of lift-off. Rapid flapping is essential throughout the scenario. And finally, none of this scenario begins to work until (unless) wing thrust is powerful enough to replace the (powerful) leg thrust of a running theropod. (The earliest feathered wings would not have been very effective as thrust devices.)

The argument for the "trees-down" model for the origin of flight is that the ancestors of birds lived in trees, perhaps achieving flight by hopping from branch to branch, gliding on two or four wings, and gradually evolving larger and larger wings. As we have seen, there is much debate about whether *Archaeopteryx* was a tree climber, and the same is true for its feathered theropod relatives. But if they could climb trees, then this provides a neat explanation of why this one evolutionary line of theropods bucked the trend of all other theropods by becoming really tiny and elongating their arms at the same time.

The explanation could be that they were exploiting a new ecological opportunity. While the huge theropods like *Megalosaurus* and *Allosaurus* were charging after their large dinosaurian prey, the maniraptorans focused on smaller prey, including insects, and became smaller and evolved grabby hands so they could manipulate lively prey animals. With these strong hands, they could hoist themselves into the trees, so escaping the attentions of larger theropod predators and discovering a new world of small prey in the trees. While the insects and other tree dwellers could escape by flying, the bird-like theropods leapt after them, becoming efficient gliders. And so flight evolved.

This is all controversial and gives rise to numerous theropod and early bird adaptations to be tested. But a common-sense view might say that if you are going to evolve flight you might as well make gravity work for you (trees down) rather than against you (ground up).

### Cretaceous Birds

The radiation of birds was very rapid. Early Cretaceous rocks have yielded bird remains in all the northern continents and in Australia. Some of these birds became really abundant and diverse. For example, in China, paleontologists have excavated thousands of specimens of *Confuciusornis* (Figure 14.15). This bird was primitive in some ways, still having a rather reptilian skull and the

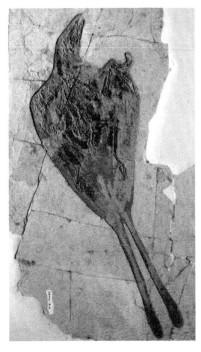

Figure 14.15 Specimens of the Cretaceous basal bird *Confuciusornis* often had long tail feathers that look as if they were for display. *Source:* Photograph by Laikayui (Wikimedia).

Figure 14.16 The Cretaceous basal bird *Sinornis* had more advanced characters (see text). *Source:* Artwork by Pavel Riha (Wikimedia).

long fingers with claws on its wings, but it had lost the teeth retained by *Archaeopteryx* and, importantly, the pelvis was more bird-like, and the bony tail had already shrunk to a short stump, as we see in modern birds whose tail is made from feathers, not bones.

*Sinornis*, a sparrow-sized bird from the Early Cretaceous of China, had many features directly related to much better flight and perching than was possible in *Archaeopteryx*. The body and tail were shorter, and the tail had fused vertebrae at its end that provided a firm but light base for strong tail feathers. The center of mass of the body was much farther forward, closer to the wings. *Sinornis* had a breastbone, a shoulder joint that allowed it to raise its wings well above the horizontal, and fingers that were adapted to support feathers rather than grasping and tearing claws. The wrist could fold much more tightly forward against the arm than the 90°

seen in *Archaeopteryx*, so the wing could be folded away cleanly in the upstroke or on the ground, reducing drag. The foot was much better adapted for perching. Even so, *Sinornis* still had some very primitive features: the skull and pelvis were much like those of *Archaeopteryx*, and it had teeth (Figure 14.16).

Paleontologists have a remarkable sample of forest-dwelling birds from the mid-Cretaceous of China, all dating from around 125–120 million years ago. Then, the record of birds in the Late Cretaceous is quite poor. Best known are diving birds such as *Hesperornis* (Figure 14.17a) and the tern-like *Ichthyornis* (Figure 14.17b), both found since the 1880s along the shores of the great Western Interior Seaway that divided North America in half at the time. *Hesperornis* was a flightless diving bird, up to 6 ft (1.8 m) long. Another Late Cretaceous flightless bird was *Patagopteryx* from Argentina, the size of a chicken and terrestrial. It seems amazing that, having evolved flight, birds began to become flightless already in the Cretaceous!

Much of bird evolution in the Cretaceous was associated with changing habitats on land, especially the changes brought about by the radiation of flowering plants and insects, the Cretaceous Terrestrial Revolution (see Chapter 15). There was then a substantial extinction of birds at the end of the Cretaceous (see Chapter 16), when nearly all the specialist Cretaceous groups disappeared, and just a few chicken-like and diver-like birds survived and gave rise to all modern birds.

Figure 14.17 Late Cretaceous marine birds from North America. (a) Skeleton of the Cretaceous diving bird *Hesperornis*. *Source:* Photograph by Quadell, Wikimedia. (b) The Cretaceous bird *Ichthyornis* seems to have been tern-like. *Source:* Artwork by Nobu Tamura (Wikimedia).

## Cenozoic Birds

When the dinosaurs died out at the end of the Cretaceous, there must have been a very interesting opportunity for surviving creatures to invade the ecological niches associated with larger body size on the ground. The two leading contenders were birds and mammals, and although mammals quickly became large herbivores, it was birds that became the dominant land predators in some regions in the Paleocene. These birds evolved to become flightless terrestrial bipeds once more.

Large, flightless birds called **diatrymas** lived across the northern hemisphere in the Paleocene and Eocene. They were close to 2 m (6 ft) tall and had massive legs with vicious claws and huge, powerful beaks (Figure 14.18). Diatrymas became extinct at the end of the Eocene. Their diet has long been debated – did they snatch early, tiny horses in their great jaws and chomp them, or did they use their vast beak to crack nuts? A detailed study of its inferred jaw musculature showed most similarity to modern herbivorous birds, but the clincher for a herbivorous diet came from carbon isotopes in the bones of fossil diatrymas – they showed clear evidence that the birds fed on plants (Angst et al. 2014).

Truly carnivorous birds with very similar appearance, the **phorusrhacids**, dominated the plains ecosystem of South America from the Paleocene to the Pleistocene (Figure 14.19). The skull and beak of phorusrhacids were much more rigid than they are in most birds, and they were particularly strong in resisting the stresses involved in a downward strike (Degrange et al. 2010). Some phorusrhacids were 2.5 m (8 ft) tall, and a spectacular late phorusrhacid, *Titanis*, crossed to Florida from South

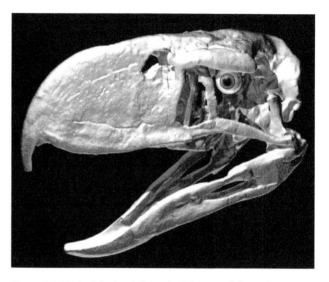

Figure 14.19 *Andalgalornis* from the Miocene of Argentina. (The eyeball is added for dramatic effect.) *Source:* Courtesy of the Witmer Lab at Ohio University: https://people.ohio.edu/witmerl/lab.htm

America less than 3 m.y. ago. It was larger than an ostrich and no doubt caused at least temporary consternation among the Floridian mammals of the time.

The southern continents have a number of large flightless birds. Living forms such as the ostrich, cassowary, rhea, and emu are familiar enough, but even more interesting forms are now extinct. The moas of New Zealand reached well over 3 m (10 ft) in height. *Aepyornis*, the "elephant bird" of Madagascar (Figure 14.20), was

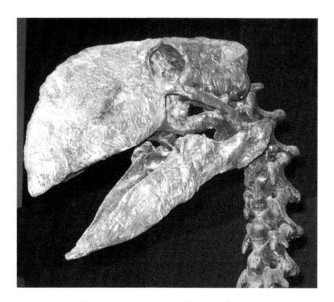

Figure 14.18 *Gastornis giganteus*, a diatryma from the Eocene of North America. The skull and beak of diatrymas were thought to indicate carnivory, but they may have been for cracking nuts! *Source:* Photograph by Mitternacht90 (Wikimedia).

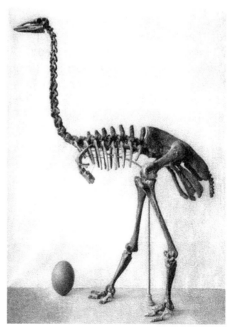

Figure 14.20 The recently extinct elephant bird *Aepyornis* from Madagascar. *Source:* Image from Monnier 1913, in the public domain.

Figure 14.21 The roc, Smizurgh, a fearsome legendary bird that captured elephants three at a time and gave Sinbad the Sailor a ride on its back. *Source:* Artwork © Fred Lu 2011, based on a nineteenth-century engraving, used by permission.

living so recently that its eggshells are still found lying loose on the ground. The eggs are unmistakable because they had a volume of 11 l (2 gal). Early Muslim traders along the African coast certainly saw these eggs, and they may even have seen living elephant birds in Madagascar, giving rise to folktales about the fearsome roc that preyed on elephants and carried Sinbad the Sailor on its back (Figure 14.21). *Aepyornis* and *Dromornis*, a giant extinct Australian bird related to ducks, are close competitors for Heaviest Bird Ever to Evolve. The *Guinness Book of World Records* currently favors *Dromornis*, which was powerfully built and weighed perhaps 500 kg (1100 pounds).

## The Largest Flying Birds

The largest flying birds so far discovered are **teratorns**, immense birds from South America, now extinct, which reached North America during the Pleistocene. Hundreds of specimens have been found in the tar pits of La Brea in Los Angeles, California, and from Florida and Mexico. But the largest teratorn was *Argentavis* from the Late Miocene of Argentina which had a wingspan of 7.5 m (24 ft). By contrast, the largest living bird is the royal albatross, just over 3 m (10 ft) in wingspan.

The beak of *Argentavis* suggests that it was a predator, not a scavenger. It probably stalked prey on the ground. With a skull 55 cm (2 ft) long and 15 cm (6 in.) wide, it could have swallowed prey animals 15 cm across. Its bones are associated with other vertebrate fossils but 64% of those are from *Paedotherium*, a little mammal about the size of a jackrabbit (in other words, an easy swallow for *Argentavis*).

In the same size range as teratorns were **pelagornithids**, gigantic marine birds that must have spent most of their time soaring over water (Figure 14.22). They ranged worldwide from the Eocene to the Late Miocene. They were lightly built but the wingspan was close to 6 m (nearly 20 ft) in the largest specimens. Their beaks were very long, with tooth-like projections built into their edges, presumably to help them hold squirming prey. More than any other living birds, pelagornithids were the ecological equivalents of pterosaurs, and it will be fascinating when further research allows us to reconstruct their mode of life accurately.

Figure 14.22 The huge marine hunter *Pelagornis* from the coast of Chile. Note the sharp points along the jaws – these are not teeth but bony projections that operated like teeth for grasping slippery fish. *Source:* Photograph by Ghedoghedo (Wikimedia).

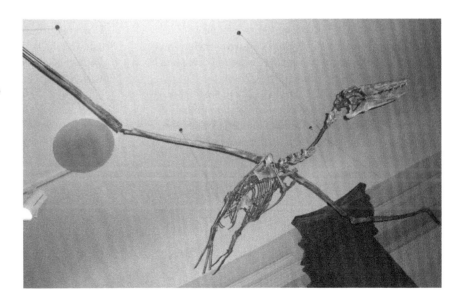

## References

Alonso, P.D., Milner, A.C., Ketcham, R.A. et al. (2004). The avian nature of the brain and inner ear of *Archaeopteryx*. *Nature* 430: 666–669.

Angst, D., Lécuyer, C., Amiot, R. et al. (2014). Isotopic and anatomical evidence of an herbivorous diet in the early Tertiary giant bird *Gastornis*: implications for the structure of Paleocene terrestrial ecosystems. *Naturwissenschaften* 101: 313–322.

Birn-Jeffery, A.V., Miller, C.E., Naish, D. et al. (2012). Pedal claw curvature in birds, lizards and Mesozoic dinosaurs – complicated categories and compensating for mass-specific and phylogenetic control. *PLoS One* 7 (12): e50555.

Carney, R.M., Vinther, J., Shawkey, M.D. et al. (2012). New evidence on the colour and nature of the isolated archaeopteryx feather. *Nature Communications* 3: 637.

Degrange, F.J., Tambussi, C.P., Moreno, K. et al. (2010). Mechanical analysis of feeding behavior in the extinct "terror bird" *Andalgalornis steulleti* (Gruiformes: Phorusrhacidae). *PLoS One* 5 (8): e11856.

Frey, E. and Tischlinger, H. (2012). The Late Jurassic pterosaur *Rhamphorhynchus*, a frequent victim of the ganoid fish *Aspidorhynchus*? *PLoS One* 7 (3): e31945.

Norberg, U.M.L. (2002). Structure, form, and function of flight in engineering and the living world. *Journal of Morphology* 252: 52–81.

Voeten, D.F.A.E., Cubo, J., Margerie, E. et al. (2018). Wing bone geometry reveals active flight in *Archaeopteryx*. *Nature Communications* 9: 923.

Witton, M.P. and Naish, D. (2008). A reappraisal of azhdarchid pterosaur functional morphology and paleoecology. *PLoS One* 3 (5): e2271.

## Further Reading

### Pterosaurs

Bestwick, J., Unwin, D.M., Butler, R.J. et al. (2018). Pterosaurs dietary hypotheses: a review of ideas and approaches. *Biological Reviews* 93: 2021–2048.

Claessens, L.P.A.M., O'Connor, P.M., and Unwin, D.M. (2009). Respiratory evolution facilitated the origin of pterosaur fight and aerial gigantism. *PLoS One* 4 (2): e4497.

Li, P.-P., Gao, K.Q., Hou, L.H. et al. (2007). A gliding lizard from the Early Cretaceous of China. *Proceedings of the National Academy of Sciences of the United States of America* 104: 5507–5509.

McGuire, J.A. and Dudley, R. (2011). The biology of gliding in flying lizards (genus *Draco*) and their fossil and extant analogs. *Integrative and Comparative Biology* 51: 983–990.

Witmer, L.M., Chatterjee, S., Franzosa, J. et al. (2003). Neuroanatomy of flying reptiles and implications for flight, posture and behaviour. *Nature* 425: 950–953.

Witton, M.P. (2013). *Pterosaurs: Natural History, Evolution, Anatomy*. Princeton: Princeton University Press.

### Birds

Chiappe, L.M. and Witmer, L.M. (eds.) (2002). *Mesozoic Birds: Above the Heads of Dinosaurs*. Berkeley: University of California Press. [Massive overview in 20 chapters, with full references. See especially chapters by Witmer, Clark et al., Chiappe, and Gatesy].

Lee, M.S.Y., Cau, A., Naish, D. et al. (2014). Sustained miniaturization and anatomical innovation in the dinosaurian ancestors of birds. *Science* 345: 562–566.

Murray, P.F. and Vickers-Rich, P. (2004). *Magnificent Mihirungs: The Colossal Flightless Birds of the Australian Dreamtime*. Bloomington: Indiana University Press.

Naish, D. (2014). The fossil record of bird behaviour. *Journal of Zoology* 292: 268–280.

Puttick, M.N., Thomas, G.H., and Benton, M.J. (2014). High rates of evolution preceded the origin of birds. *Evolution* 68: 1497–1510.

Zhou, Z., Barrett, P.M., and Hilton, J. (2003). An exceptionally preserved Lower Cretaceous ecosystem. *Nature* 421: 807–814.

## Questions for Thought, Study, and Discussion

1   Pterosaurs seem to have shared many adaptations for flying, feeding, nesting, and so on with birds. Try to find a reasonable suggestion for the fact that birds survived the Cretaceous extinction but pterosaurs did not. (I do not know of one, but there must have been some reason.)

2   Paleontologists debate heatedly whether birds evolved from the ground up or trees down. Read through the recent papers and make a list of phylogenetic, body size, ecological, and functional evidence for both viewpoints. Which do you prefer? What crucial observations or tests might resolve this debate once and for all?

3   Why would a bird lose the ability to fly? We know from their bone structure that ostriches, penguins, and many island birds lost the ability they once had to fly. This loss has to make evolutionary sense by giving an advantage to flightless birds over their flying relatives. Think of some reasons.

15

## The Cretaceous Terrestrial Revolution

### In This Chapter

Terrestrial ecosystems and landscapes changed forever about the middle of the Cretaceous, some 100 million years ago. The evidence is that life on land today represents 80–95% of all species on Earth, even though the oceans cover 70% of the surface of the Earth. Up to the Early Cretaceous, life in the sea had been more diverse than life on land. The driver for this changeover was the rise of the angiosperms (flowering plants) and their ability to dominate landscapes, capture more energy, and support highly biodiverse forests full of millions of species of insects. This changeover, termed the Cretaceous Terrestrial Revolution (KTR), happened well before the extinction of the dinosaurs but it set the shape of the postdinosaur terrestrial world. Angiosperms must have arisen long before the Cretaceous but the first fossils are known from the Early Cretaceous, including spectacular flowers in amber. Angiosperms evolved dramatically in the Cretaceous, until they were worldwide and successful in the Late Cretaceous. Much of their success seems to be linked to three factors: (i) their ability to attract insects or other visitors to them for pollination, using flowers that give visual or scent clues; (ii) using insects and other animals to transport their seeds; and (iii) being fast growers, able to outcompete gymnosperms in most circumstances.

## Life on Land in the Ascendant

Since the days of Charles Darwin, scientists have realized that the flowering plants, the **angiosperms**, changed everything on land. Darwin knew that angiosperms formed the framework of most ecosystems on land – they are all the trees, shrubs, and smaller plants in a typical tropical or temperate forest. They provide shelter for the multitudes of bugs and creeping animals, as well as the birds, reptiles, and mammals that feed on them. It's only in cold temperate and polar areas that conifers, such as spruce, fir, and pine, dominate the forests.

Angiosperms are also everything we eat. All grain crops such as wheat, oats, rye, barley, and maize (corn) are grasses, and they are angiosperms. All vegetables, such as potatoes, carrots, cabbage, and peas, are angiosperms. Farm animals bred for food, such as chickens, pigs, and cattle, all feed on angiosperms. All flowers, such as roses, daffodils, lilies, and lilacs, are angiosperms. Insects today are hugely diverse, with millions of species of beetles, ants, bees, bugs, and butterflies, and most of them depend on angiosperms. Insect eaters such as spiders, lizards, mammals, and birds depend on those insects. So, angiosperms are at the heart of modern terrestrial ecosystems. When did it all start?

It started in the Cretaceous – or perhaps earlier, in the Triassic – but the fossils are there from the Cretaceous onwards. We will see the amazing evidence of the first fossil flowers but also the ways in which these early flowers depended on insects for pollination. Darwin called the origin of the angiosperms an "abominable mystery" because they seem so different from other plants. It has been claimed that the angiosperm reproductive system was the key to their success – but was that all?

This was a major switching point in the history of the Earth and of life. We have seen how life colonized the land back in the early Paleozoic (see Chapter 8) and how the first fishes crept ashore soon after (see Chapter 9).

*Cowen's History of Life*, Sixth Edition. Edited by Michael J. Benton.
© 2020 John Wiley & Sons Ltd. Published 2020 by John Wiley & Sons Ltd.

Land life continued expanding and diversifying ever since that time, with new plant and animal groups in the Mesozoic, including of course the first lissamphibians, turtles, lizards, crocodiles, mammals, and indeed, dinosaurs and pterosaurs (see Chapters 12 – 14). This though represented a slow and steady increase of diversity and biomass. Life in the Mesozoic seas was also hugely successful, diversifying and conquering new living zones through the Mesozoic Marine Revolution (see Chapter 11).

But Geerat Vermeij and Richard Grosberg noted in 2010 that some time in the Early Cretaceous, about 100 Ma, life on land became more diverse than life in the sea for the first time. At one level, this is merely an interesting side note but it's more than just numbers. This was when the whole nature of the biosphere flipped over and today, life on land is as much as 20 times more diverse than life in the sea. The impact of angiosperms was not just that it gave us a million new species of ants, bees, and bugs, but the whole balance of carbon cycles, energy cycles, and climates changed as a result.

This event was called "The Great Divergence" by Vermeij and Grosberg (2010). Two years earlier, it had been named the "Cretaceous Terrestrial Revolution," abbreviated as KTR. (Why KTR and not CTR? Well, geologists use the letter C for the Carboniferous and K for the Cretaceous, based on the Greek *kreta*, which means chalk, a key Cretaceous rock.) The KTR was named by Graeme Lloyd and colleagues based on their study of the timing of bursts of activity in dinosaur evolution; they noted that dinosaurs had done a great deal of their explosive evolving back in the Triassic and Jurassic, and by Cretaceous times their evolution had largely slowed down. Even though, as we saw in Chapter 13, hadrosaurs and ceratopsians evolved and diversified fast in the Late Cretaceous, all the evidence suggests they were still mainly eating conifers. It seems that while the rest of life on land was benefiting from the angiosperms, dinosaurs plodded past the pretty flowers with their lovely scents, looking for ferns and conifer branches.

To put the KTR in context, it is worth reviewing the major groups of land plants and the origin of angiosperms. In Chapter 8, we explored the origins of land plants, and saw in Chapter 9 how they grew into trees and formed the great Carboniferous coal forests. Currently, there is contradictory evidence about when the angiosperms actually originated.

## Dating the Origin of Angiosperms

Paleontologists have long debated the timing of the origin of angiosperms. Fossil pollen of possible angiosperms is known from the Early Cretaceous, about 130 Ma, and fossil leaf and flower fragments from 125 Ma. Much earlier records of angiosperms have been reported repeatedly, but generally dismissed by most paleobotanists. There are examples of pollen from the Middle Triassic, 240 Ma, that might have come from an angiosperm ancestor, but these fossils are much debated. Can we solve the question of the age of the oldest angiosperms from a study of modern plant relationships?

The green land plants fall into several major groups, including the Devonian rhyniopsids, the lycophytes (= club mosses), ferns, seed ferns, gymnosperms, and angiosperms. Seed ferns were a successful component of Late Paleozoic floras (see Chapter 9), including the coal forests; they flourished into the Triassic. But Mesozoic gymnosperms perfected the seed system, making up 60% of Triassic and 80% of Jurassic species. Gymnosperms include conifers, cycads, and ginkgos. Mesozoic forests had trees up to 60 m (200 ft) high, forming famous fossil beds such as the Petrified Forest of Arizona. Conifers were the dominant land plants during the Jurassic and Early Cretaceous, and they are still by far the most successful of the gymnosperms.

In terms of solving the phylogeny, paleobotanists had long split up the gymnosperms into their subgroups. However, molecular phylogenies, based on DNA sequences of the living plants, all seem to agree that gymnosperms form a monophyletic group, and this was confirmed by Jose Barba-Montoya and colleagues in a definitive study in 2018.

The cladogram (Figure 15.1) shows a succession of land plants, roughly matching the sequence of their appearance in the fossil record. Then, we get the split into gymnosperms and angiosperms. This creates a real conundrum for understanding the timing of flowering plant evolution – the oldest fossil gymnosperms come from the Late Devonian or Early Carboniferous, some 360 Ma, much older than the oldest fossils of angiosperms. Therefore, if the cladogram (Figure 15.1) is correct, the oldest angiosperms must have occurred in the Carboniferous – this is a scary prospect for paleobotanists who have to understand why the fossils have never been found, but also what they might have looked like.

## The Oldest Fossil Angiosperms

Among all the reports of the world's first angiosperms, it is hard to interpret the cases of pollen. However, there have been reports of flowers from the Middle to Late Jurassic of China, named as *Euanthus*, *Juraherba*, and *Aegianthus*. Flower-like as some of these look, Patrick

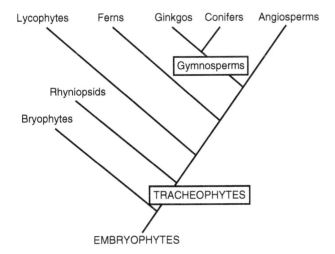

Figure 15.1 Phylogeny of the major plant groups, showing that conifers are a clade, sister group of angiosperms. This indicates both groups should have originated at the same time.

Figure 15.2 *Archaefructus*, a very early angiosperm from the Early Cretaceous of China. *Source:* Courtesy of Professor David A. Dilcher of the University of Indiana.

Herendeen and colleagues reinterpreted them all as cones and other structures of gymnosperms. This means that the oldest angiosperm fossils are samples of fossil pollen from the Early Cretaceous.

The earliest well-preserved angiosperm plants are from the famous sediments in northern China that have also yielded feathery dinosaurs and early birds. *Archaefructus* (Figure 15.2) is preserved almost completely and seems to be a water-dwelling weed. There are no petals but the plant has closed carpels with seeds inside, a classic feature of angiosperms. Cladistic analyses of *Archaefructus* place it as the most primitive as well as the earliest angiosperm.

A further Early Cretaceous angiosperm from China, *Archaeanthus* is reconstructed (Figure 15.3) with a large reproductive axis containing many pollen-bearing and seed-bearing organs. It is placed in a family of living angiosperms, the Liriodendraceae, the tulip trees. Today, the two species of tulip trees occur in eastern North America and China, although they are seen worldwide in gardens (Figure 15.4). If *Archaeanthus* is an ancient tulip tree, then this confirms that angiosperms were already quite diverse at this early point, which could be said to support the idea of a longer, but hidden, history.

Some of the most amazing Cretaceous flower specimens have been reported in amber from Myanmar (Burma) in southeast Asia. The amber from Burma was first reported over 100 years ago, but renewed interest in the past decade has brought to light thousands of astonishing specimens of plants and animals. The amber is fossilized tree resin but the conifers of those ancient Burmese forests must have been leaking huge amounts of resin, which has captured bugs, plants, and even

Figure 15.3 *Archaeanthus*, a very early angiosperm from the Early Cretaceous of China, is reconstructed with large flowers. *Source:* Courtesy of Professor David A. Dilcher of the University of Indiana.

Figure 15.4 The modern American tulip tree, *Liliodendron tulipifera*, showing the characteristic divided leaves and large, simple flower. *Source:* Photograph by Bruce Marlin (Wikimedia).

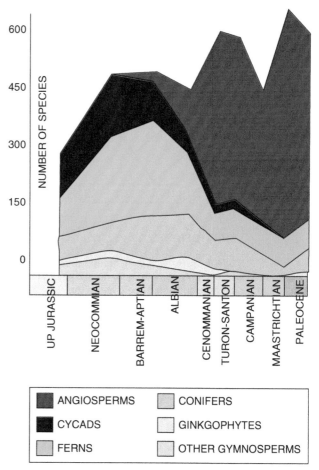

Figure 15.6 A classic diagram showing how angiosperms rose from nothing to something over 80% of species of plants in typical floras. Note that the ferns, conifers, ginkgos, and cycads, which dominated floras at the start of the Cretaceous, were all still present at the end (and indeed, survive today), but in much reduced roles. *Source:* Diagram based on classic work by Karl J. Niklas and colleagues.

tree frogs, lizards, birds, and parts of dinosaurs. One specimen described in 2017 is a perfect flower called *Tropidogyne*, which is nearly identical to modern *Ceratopetalum*, a flower from Australia (Figure 15.5).

Angiosperms were diverse by the mid-Cretaceous, especially in disturbed environments such as riverbanks. By the end of the Cretaceous, angiosperms had diversified even further, and as many as 40 families were present on all continents. In terms of impact (Figure 15.6), angiosperms had risen from being very rare (less than 5% of species) to being very common (more than 80% of species). So, angiosperms apparently did not cause any major groups of land plants to go extinct, but they had changed the landscape. How did the rise of angiosperms fit into the larger picture of Mesozoic terrestrial ecology?

Figure 15.5 A Cretaceous flower of *Tropidogyne* (*left*) compared with the modern New South Wales Christmas Bush, *Ceratopetalum*, a member of the Family Cunoniaceae, a group of flowers, bushes, and trees mainly occurring today in Australia and the Pacific. *Source:* Photographs by Oregon State University (*left*) and John Tann, Wikicommons (*right*).

## Angiosperms and Changing Mesozoic Ecology

At the end of the Jurassic, we saw a reduction of the sauropod dinosaurs that probably had been high browsers, and the rise of low-browsing ornithischians (see Chapter 13). More seedlings would now have been cropped off before reaching maturity, and any plant that could reproduce and grow quickly would have been favored.

Conifers reproduce slowly. It takes two years from fertilization until the seed is released from the cone, and wind dispersal typically does not take the seed very far. The whole reproductive system of conifers depends on wind and works best in a group situation such as a forest.

On the other hand, most angiosperms are adapted for pollination by animals, especially insects – this ensures rapid germination and growth and rapid release of seeds (within the year). An angiosperm is much more likely to succeed as a weed, rapidly colonizing any open space, and is more likely to be widely distributed because of its dispersal method. The earliest angiosperms were small, weedy shrubs, exactly the kind of plant that could survive heavy dinosaur browsing. A conifer forest, once broken up by dinosaur browsing or natural accident, would most likely have been recolonized by shrubs and weeds that could invade and grow rapidly (look at the results of clear-cutting in a conifer forest today). The weeds themselves would have reproduced quickly, so would have been more resistant to browsing than were young conifer seedlings.

Even without dinosaur browsing, angiosperms would have found habitats where they would have been very successful. In mid-Cretaceous rocks, for example, angiosperm leaves dominate sediments laid down in river levees and channels. Shifting and changing riverbank areas favor weeds because large trees are felled by storms and frequent floods. Most mid-Cretaceous pollen, on the other hand, comes from sediments laid down in lakes and near-shore marine environments. This is the wind-blown pollen from stable forests on the shores and on lowland plains away from violent floods, and it is dominantly conifer pollen.

Although angiosperm species numbers rose enormously through the Late Cretaceous, they did not take over the entire world. They were very slow to colonize high latitudes. This was probably for two reasons: that conifers were better adapted to the thin, dry soils of mountain sides, and that angiosperm insect pollinators drop off in both number and diversity in cooler climates.

It is important to be aware of the differences between diversity and ecological dominance. For example, Late Cretaceous fossil floras preserved in place under a volcanic ash fall in Wyoming show that even if angiosperms dominate a local flora in diversity of species, they may make up only a small percentage of the biomass. In the Big Cedar Ridge flora, angiosperms made up 61% of the species but covered only 12% of the ground. These are excellent study examples because they allow paleobotanists to see the shape of the flora as it was distributed across the landscape, rather than working through broken fragments that were transported and dumped in lakes or river beds.

## Seeds and Pollination

As plants invaded dry habitats from Devonian times onward, they evolved ways to retain water and protect their reproductive stages from drying out. The major advance was the perfection of seeds, which are fertilized embryos packed in a reasonably watertight container filled with food. The embryo can survive in suspended animation within the seed until the parent plant arranges for its dispersal. Germination can be delayed until after successful transport to a favorable location. The seedling then bursts its seed coat and grows, using the nutrition in the seed until its roots and leaves have grown large and strong enough to support and maintain the growing plant.

Seed plant reproduction has two phases: fertilization and seed dispersal. The plant must be pollinated, and after the seed has formed, it must be transported to a favorable site for germination. A major factor in the evolution of angiosperms is their manipulation of animals to do these two jobs for them.

Conifers and many other plants are pollinated by wind. They produce enormous numbers of pollen grains, which are released to blow in the wind in the hope that a grain will reach the pollen receptor of a female plant of the same species. Wind pollination works, just as scattering sperm and eggs into the ocean works for many marine invertebrates, but the process looks very expensive. The pollen receptor in conifers is only about one square millimeter in area, so to achieve a reasonable probability of fertilization, the female cone must be saturated with pollen grains at a density close to 1 million grains per square meter.

Parent plants do some things to cut the costs of wind pollination. Male cones release pollen in dry weather in just the right wind conditions, for example (Figure 15.7), and female cones are aerodynamically shaped to act as efficient pollen collectors. But for practical purposes, wind pollination is consistently successful only if many individuals of the same species live in closely packed groups: conifers in temperate

Figure 15.7 Pine pollen, released in clouds when wind conditions are right. *Source:* Photo by Dr Beatriz Moisset (Wikimedia).

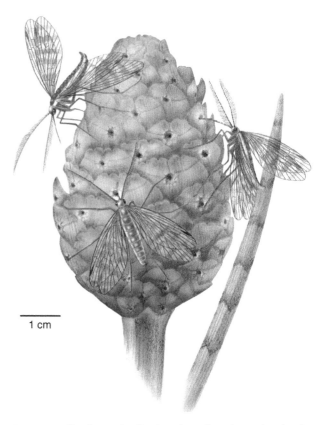

1 cm

Figure 15.8 Fossil scorpionflies have been found associated with gymnosperm cones in Jurassic sediments, suggesting that they were aiding pollination. Reconstruction by Mary Parrish under the direction of Conrad Labandeira. *Source:* Courtesy of the Smithsonian Institution, and used with the permission of Conrad Labandeira and Mary Parrish.

forests or grasses in prairies and savannas. An ecological setting like a tropical rainforest, where many species have well-scattered individuals, is not the place for wind pollination.

We can imagine Jurassic floras that depended on wind pollination, with plants that produced large supplies of pollen. But insects then, as now, probably foraged for the food offered by plentiful pollen and soft, unripe female organs waiting for fertilization. We know that there were large clumsy beetles and scorpionflies in the Jurassic thanks to the work of Dong Ren and others, and they probably visited plants looking for food. However, as they moved from plant to plant, they may have visited the same species frequently, collecting and transferring pollen by accident. Insects can help even by visiting one plant or one sex. In some living cycad gymnosperms, wind can only carry pollen to the surface of the female cone, but insects clustering around the cone carry it into the pollen receptors. It is therefore very likely that scorponflies and other insects were aiding in the pollination of gymnosperm plants before angiosperms even evolved (Figure 15.8).

Over time, the plant structure may have evolved toward cooperation with insects in certain ways. Perhaps delicate structures were protected but pollen was made easier to gather, and female pollen collectors were moved closer to the male pollen emitters. Such changes would have made pollen transfer by insects more likely and less costly to the plant. Devices to attract insects – strong scents at first, then brightly colored flowers – perhaps evolved side by side with rewards such as nectar. Those plants that successfully attracted insects would have benefited by increasing their chances of fertilizing and being fertilized. Insects deliver pollen much more efficiently than wind.

An ideal pollinator should be able to exist largely on pollen and nectar, so that it can gather all its food requirements by visiting plants. It should visit as many (similar) plants as possible, so it should be small, fast-moving, and agile. A nocturnal pollinator should have a good sense of smell, and a daytime pollinator should have good vision or a good sense of smell, or both.

The only Jurassic candidates to fit this job description are insects. Birds and bats had not yet evolved, and small mammals were probably too sluggish and/or nocturnal. Insect pollinators had an increasing incentive to learn and remember certain smells and sights, and those that evolved rapid, error-free recognition of pollen sources, and clever search patterns to find them, would have become superior food gatherers and probably superior reproducers. Today, insects discriminate strongly between plant species, even between color varieties of particular species. Some insects congregate for mating around certain plant species.

## Reasons for Angiosperm Success

Angiosperms were clearly a big hit in the Cretaceous, and they stepped up their dominance of terrestrial ecosystems in several stages through the Cenozoic. There are three reasons for this. First, they evolved to attract insects and other animals to act as pollinators, and we shall explore the role of seeds and pollination, especially by insects, in some detail. Second, many angiosperms also use insects and other animals to disperse their seeds, so achieving rapid spread away from the parent plant. Third, angiosperms grow fast, and we shall look at this idea also.

### Pollination and Insect Pollinators

The earliest known flowers, though small, had relatively large petals (Figure 15.3), and the flowers could have produced many small seeds. The angiosperm flower (Figure 15.9) is a marvel of construction and adaptation. The flower consists of **petals** and **sepals**, the green covers that form the bud just before the flower opens. In the center of the flower is the **carpel**, the female structure of the plant, consisting of an **ovary** in which the seeds develop and a tall structure called the **style** tipped by the **stigma**. This is the landing spot for pollen, the male reproductive material. Pollen is produced within the **stamens**, the male structures in the flower. Once the pollen grains are ready, the **anther** bursts open and pollen is either blown away on the wind or sticks to insects and other pollinating animals. Pollen lands on the stigma of the carpel and the grains burrow down through the style to reach the ovary where they fertilize the **ovule**, which then becomes a seed, ripens and is dispersed.

Waterlilies evolved early, and they have an intriguing pollination system. The plants bloom all summer but there are never more than a few flowers open at once (usually only one per plant). Each flower lasts for three days in the giant water lily *Victoria amazonica*. On the first day, it displays its carpels, which are white, and gives off an odor that attracts beetles (Figure 15.10). The odor is enhanced because the flower generates considerable

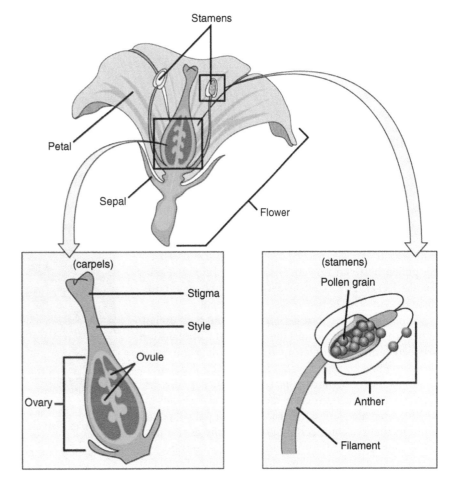

**Figure 15.9** The angiosperm flower (*top*) consists of petals and sepals, with the male and female reproductive organs (stamens, carpel) in the center. The stamens produce pollen which fertilizes the ovule in the carpel, which then matures into a seed. There are two dispersal phases – dispersal of pollen and dispersal of seeds. *Source:* Artwork by Mariana Ruiz Villareal, LumenLearning.

Figure 15.10 *Victoria amazonica,* the giant waterlily, with beetles feeding on nectar. *Source:* Photograph by Bilby (Wikimedia).

heat. Beetles are trapped overnight as the flower closes, and usually the flower is pollinated. On the second morning, it extends its stamens, which are pink, and any surviving beetles leave with a load of pollen. On the third day, the flower closes and its stalk bends to place it underwater. The flower develops its seeds there, and they are released when they are ripe into the water, where they float away or are spread by creatures that eat them.

No-one would seriously argue that this complex system was already present in the first waterlilies. But the style of pollination (tricking or trapping unwary beetles to achieve pollination) may be very ancient indeed, and perhaps a clue to the success of early angiosperms. Many living magnolias, also primitive flowering plants, also have large, fragrant flowers where insects congregate to feed and mate (and pollinate) (Figure 15.11). Some orchids have flowers that mimic female insects, so males visit and mate with the flower, tricked into picking up and delivering pollen between flowers of the single orchid species (Figure 15.12). This is a near-perfect example of a mutualism or symbiosis between plant and animal!

Animal pollination can deliver a large mass of pollen, rather than a few wind-blown grains. Competition between individual pollen grains to fertilize the ovule allows the female angiosperm more mate choice than in other plants (remember Chapter 3). Pollen grains are haploid, so they cannot carry hidden recessive genes (as we do). A female plant with an abundant supply of pollen could in theory select certain pollen grains over others by placing chemical or physical barriers around the ovule; pollen grains that can cross the barrier are selected

over others for fertilization. Experimentally, plants that can exercise pollen choice in this way have stronger offspring than others. Pollen choice may have been one of the most important factors in angiosperm success.

Of course, pollination encouraged tremendous diversity among the pollinators as they came increasingly to specialize on particular plants. The astounding rise in diversity of beetles and bees began in Cretaceous times, and there are now tens of thousands of species of each. The bees and beetles associated with angiosperms are

Figure 15.11 Beetles congregating on a magnolia flower. *Source:* Photograph by Dr Beatriz Moisset (Wikimedia).

Figure 15.12 The fly orchid *Ophrys insectifera* has a flower that resembles a female digger wasp. Undiscriminating and optimistic male digger wasps may be fooled into trying to mate with the flower, and in the process pollinate one flower after another. *Source:* Photograph by Ian Capper (Wikimedia).

many times more diverse than those associated with gymnosperms.

Pollination cannot be the whole story, however. Insects help to pollinate cycads too, yet angiosperms are enormously successful while cycads have always been a relatively small group of plants. Several other Mesozoic plants experimented with ways of persuading organisms to transport pollen, and flower-like structures evolved more than once.

Furthermore, if pollination were the key to angiosperm success, flowers could have evolved as soon as flying insects became abundant in the Late Carboniferous. There are some signs that insect pollination began then as something of a rarity. But angiosperms appeared much later and rather suddenly in the Early Cretaceous. Therefore, angiosperm success is not related simply to their evolution of flowers.

## Seed Dispersal

If seeds fall close under the parent plant, they may be shaded so that they cannot grow, or they may be eaten by animals or birds that have learned that tasty seeds are often found under trees. Many plants rely on wind to disperse their seeds. Sometimes seeds are provided with little parachutes or airfoils to help them travel far away from the parent; winged seeds evolved almost as soon as seeds themselves, in the Late Devonian.

But seeds dispersed by wind will often fall into places that are disastrous for them. Although wind dispersal works, it seems very wasteful; it can work only in plants that produce great numbers of seeds. Wind-dispersed seeds must be light, so cannot carry much energy for seedling growth. They have to germinate in relatively well-lit areas, where the seedling can photosynthesize soon after emerging above ground.

Alternatively, a plant could have its seeds carried away by an animal and dropped into a good place for growth. Many animals can carry larger seeds than wind can, and larger seeds can successfully germinate in darker places. As with pollination, animals must be persuaded, tricked, or bribed to help in seed dispersal.

Some animals visit plants to feed on pollen or nectar, and others browse on parts of the plants. Others simply walk by the plant, brushing it as they pass. Small seeds may be picked up accidentally during such visits, especially if the seed has special hooks, burrs, or glues to help to attach it to a hairy or feathery visitor. Such seeds may be carried some distance before they fall off. Small seeds may be eaten by a visiting herbivore, but some may pass unharmed through the battery of gnawing or grinding teeth, through the gut and its digestive juices, to be automatically deposited in a pile of fertilizer.

Plants face two different problems in persuading animals to disperse seeds and in persuading them to pollinate. In pollination, there is often a payment on delivery: the pollinator collects nectar or another reward as it picks up, and again as it delivers the pollen. There is no such payment on delivery of a seed. Any payment is made by the plant in advance, so that seed dispersers receive no payment for delivery of the seed. It would be better for them to cheat and to eat every seed. Thus, plants often rely on tricks (burrs, for example) to fix seeds to dispersers. Velcro was evolved by plants long before the idea was copied by an astute human. Alternatively, plants may pack many small seeds into a fruit so that the disperser will concentrate on the fruit and swallow the seeds without crushing them (in strawberries, for example).

Many plants actually invite seed swallowing. They have evolved a tasty covering around the seed (a berry or fruit), and if the animal or bird eats the seed along with the fruit, every surviving seed is automatically planted in fertilizer. Tiny seeds are likely to be swallowed without being chewed, but can carry little food for the developing embryo. Large seeds loaded with nutrition are often protected by a strong seed coat or packed inside a nut.

Seed dispersal by animals is not cost free. Many dispersers eat the seeds, passing only a few unscathed through their gut. So, there is a significant wastage of seeds, depending on a delicate balance between the seed coat and the teeth and stomach of the disperser. Too strong a seed coat, and the disperser will turn to easier food or germination will be too difficult; too weak a seed coat, and too many seeds will be destroyed. Some plants are so delicately adjusted to a particular disperser that the seeds germinate well only if they are eaten by that disperser.

Angiosperms evolved carpels as a new and unique protection for their ovules, and eventually for the developing seeds (Figure 15.9). Carpels probably evolved to protect against large, hungry insects. Soon, however, the angiosperm seed coat began to protect seeds as they passed through vertebrate guts. A seed with a strong coat was proof against many possible predators, but perhaps at the same time came to be desired food for one or a few animals that could break the seed coat. A plant could evolve to a stable relationship with a few such seed predators; the predators would receive enough food from the seeds to keep them visiting the plant regularly, but would pass enough seeds unscathed through the gut that the plant benefited too.

Seed dispersal by animals surely evolved after insect pollination. Jurassic insects may have become good pollinators but they were too small to have been large-scale seed transporters. Jurassic dinosaurs were large enough but any seeds they swallowed might have been exposed to digestive juices for a long time. Generally, dinosaurs did not have fur like mammals in which seeds could become entangled. But what about the feathered dinosaurs? Some day soon, a new fossil might be recovered showing seeds trapped in the feathers of a Jurassic or Cretaceous dinosaur or bird!

Seeds were undoubtedly dispersed by dinosaurs to some extent, since the huge vegetarian ornithischians and sauropods ate great quantities of vegetation. But despite the size of the deposit of fertilizer that must have surrounded seeds passing through a dinosaur, browsing dinosaurs probably damaged and trampled plants more than they helped them. It's unlikely that any Mesozoic plant would have encouraged dinosaur browsing.

Effective transport over a long distance can take a seed beyond the range of its normal predators and diseases and can allow a plant to become very widespread provided that there are pollinators in its new habitat. As angiosperms adapted to seed dispersal by animals, they probably dispersed into new habitats much faster than other plants. Other things being equal, we might expect a dramatic increase in the angiosperm fossil record as they adapted toward seed dispersal by animals rather than wind. (Some living angiosperms are pollinated by wind but have their seeds dispersed by animals. These include grasses, which evolved in the Cretaceous but did not become widespread until well into the Cenozoic.)

It is hard to establish the importance of seed dispersal in the early years of the angiosperms. In the Jurassic and Cretaceous, mammals, birds, and feathered dinosaurs might have carried seeds by chance in their hair and feathers. They undoubtedly fed on the nuts and fruits of early angiosperms, and so swallowed and passed their seeds. As active creatures, they then spread the seeds far and wide. But mammals and birds were still quite rare at these times, and there may be a third reason why angiosperms took over – neither their highly species-specific modes of insect-mediated pollination, nor the engagement with animals to spread their seeds, but their fundamental ability to grow fast.

### Fast-Growing Seedlings

For all the intricacies and advantages of their specialized reproductive system, and the need to enlist close cooperation with insects and other animals, the early success of angiosperms may be explained better by the "fast seedling" hypothesis. This idea is based on the fact that angiosperm seeds germinate sooner, and the seedlings grow faster and photosynthesize better than those of gymnosperms. Angiosperms may simply have outcompeted gymnosperms in the race for open spaces.

Angiosperms grow faster than gymnosperms because they have a greater capacity to photosynthesize, namely to capture energy from the Sun, and their water vascular system is more efficient at pulling in water and nutrients from the soil. In experiments with modern plants, both gymnosperm and angiosperm seedlings seem to grow at the same rate when resources are restricted, but when there is plenty of water and nutrients in the soil, angiosperms grow faster.

Broadly, then, if seedlings of pines and hazels are growing side by side, the hazels will generally capture more nutrient from the soil and quickly overshadow their conifer competitors. Also, angiosperms shed their leaves and dead branches more abundantly than do conifers, and this leads to the build-up of richer soil under the trees. This creates a so-called feedback system, where more angiosperms produce more leaf litter, which in turn feeds more plants, and because angiosperms grow faster, they capture more of the nutrients from the soils. Therefore, more angiosperms → more and richer soil → more angiosperms … and so it goes on.

Perhaps dinosaurs had a role. Even though dinosaurs did not generally eat angiosperms, or at least not according to the evidence of coprolites, they may have barged around knocking down trees and creating spaces, just as elephants do today. In these open spaces, angiosperms may have had the advantage over gymnosperms because of their ability to grow fast, so dinosaurian damage to forests could have given the early angiosperms an ecological boost.

## Explosion in Insects

Insects were crucial to the KTR, enabling pollination and seed dispersal. In fact, we can trace many of the most successful insect groups back to the time of the KTR, but insects had been around for a long time before angiosperms evolved. Primitive insects are known from Devonian rocks, but flying insects are not found until the Late Carboniferous, as insects radiated in the Carboniferous forest canopy. Many insects of all sizes are known from the coal beds of the Carboniferous, and half of all known Paleozoic insects had piercing and sucking mouthparts for eating plant juices. In turn, these smaller insects were a food source for giant predatory dragonflies and for early amniotes (see Chapter 9).

The early and continuing success of insects is down to their small size and their ability to fly. In living insects (except mayflies), only the last molt stage, the adult, has wings and there is a drastic metamorphosis between the last juvenile stage (the nymph) and the flying adult. Wings have to be as light and strong as possible. Most of the wing is simply a light mass of dead tissue that cannot be repaired. This gives great flying efficiency, though it usually means a short adult lifespan. The automatically short life expectancy of flying insects has played a strong part in the evolution of social behavior among some insects, in which the genes of a comparatively few breeding but nonflying adults are passed on with the aid of a great number of cheap, throwaway, sterile flying individuals (worker bees, for example). Some insects shed their wings. In many ants, for example, the wings are functional only for a brief but vital period during the mating flight. Insects do not have a long enough life expectancy to have the luxury of learning, so they operate simply by instinct – behaviors that are hard-wired in through their genes.

The first insects did not fly, and some primitive living forms such as silverfish also do not fly. But flight emerged widely in the Carboniferous coal forests, when some insects such as dragonflies were huge (see Chapter 9). There has long been debate about the origin of insect wings. One thing is for sure – unlike in birds and pterosaurs (see Chapter 14), insects did not evolve their wings by modifying limbs. It has been noted that insects and angels share one thing in common – they evolved flapping flight without sacrificing limbs to form the wings! Insects have thus lost little of their ability to move on the ground.

It was once argued that insect wings evolved either from modified gills or from entirely new structures, perhaps from the side of the body. Study of fossil nymphs of Paleozoic insects suggests that some of them had lobes at the side of the thorax, used in warming up, and some of these lobes could flap. Thus, the idea is that these forms went from basking to hopping and gliding, and eventually to flight.

Many groups of modern insects emerged from the Carboniferous to the Jurassic, but all evidence points to a great burst of diversification around the middle of the Cretaceous. A review of the fossil evidence by Fabien Condamine and colleagues shows no burst of evolution at that time, but fairly steady values for the diversity of families through the Cretaceous and Cenozoic. However, their analysis is at the level of families, and insect families are notoriously variable in species numbers – for example, the Family Reduviidae (the assassin bugs) contains 7000 species. Therefore, a steady number of families could mask an explosion of species diversity. (Equally it might not.)

But phylogenomic evidence keeps pointing to the KTR as marking an explosion in insect diversity, especially for the key clades Hemiptera, Hymenoptera, Coleoptera, and Lepidoptera. Hemiptera (bugs, 100 000 species) mostly show evidence for bursts of diversity in the Cretaceous and later, and they include forms mostly closely adapted to feeding on angiosperms. Hymenoptera (bees, wasps, sawflies, ants, 153 000 species) show explosions in diversity among all the key groups during the

Cretaceous, and especially among bees and wasps which have intimate relationships with angiosperms for nectar collection and pollination, and ants which feed among the leaf litter of the angiosperm forests. Coleoptera (beetles, 400 000 species) are the most diverse insects today (Figure 15.13), and they did most of their massive diversification in the Cretaceous, and many species today show close relationships with angiosperms in feeding and breeding. Indeed, as David Peris and colleagues showed, some beetles were caught in the act, carrying pollen from a gymnosperm flower (Figure 15.14), but now their living relatives have flipped to pollinating angiosperms. Finally, Lepidoptera (butterflies and moths, 180 000 species) also shows bursts of diversification in the Cretaceous and early Cenozoic, and of course most species have intimate relationships with angiosperms in feeding their caterpillars and nectar collecting as adults.

This new phylogenomic evidence has all emerged in the past 10 years, and the analyses are based purely on

Figure 15.13 The amazing diversity of beetles is merely sampled here. The image shows about 120 out of the 400 000 species of beetles alive today. Their bright colors are a feature of many tropical angiosperm forests, and most of the modern groups arose during the KTR. *Source:* Photograph by H. Zell (Wikimedia).

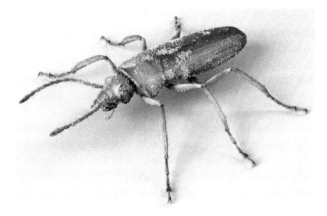

Figure 15.14 One of the most astonishing fossils ever – a 3D digital image from scans of the beetle *Darwinylus* preserved in amber from the mid-Cretaceous of Spain. On its head and back it carries pollen, suggesting it had been visiting flowers just before it was engulfed in tree resin. The pollen is identified as *Monosulcites*, indicating that in fact this beetle was pollinating a gymnosperm rather than an angiosperm. *Source:* Digital imaging by J. A. Peñas, courtesy of David Peris.

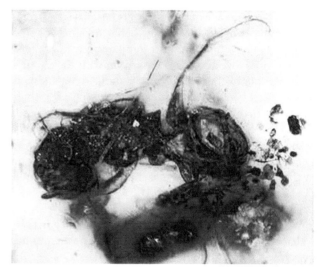

Figure 15.15 The world's oldest bee, *Melittosphex* from mid-Cretaceous amber from Myanmar. *Source:* Photograph by Hectarea (Wikimedia).

modern insects. The four groups just described account for 88% of modern insect diversity (833 000 out of 950 000 named species), and independent evidence tracks much of their explosion in diversity back to the KTR. This is a change of view in recent years, because formerly the diversifications of these groups were seen as either more drawn-out, say since the Triassic and Jurassic, or more focused on a bounce-back following the Cretaceous-Paleogene mass extinction 66 Ma (see Chapter 16). Perhaps the KTR really did kick off the explosive evolution of modern insects, our most diverse animal group.

## Ants and Termites

The success of angiosperms benefited pollinators and seed dispersers, and vice versa, and the later evolution of angiosperms was related to the ecology of large animal browsers. But today, some of the most effective tropical herbivores are leaf-cutting ants, and most terrestrial vegetation litter is broken down by termites. One-third of the animal biomass in Amazonia is made up of ants and termites. In the savannas of West Africa there are more than 2000 ants per square meter! There may be 20 million individuals in a single colony of driver ants, but the world record is held by a supercolony of ants in northern Japan, which has 300 million individuals, including a million queens, in 45 000 interconnected nests spread over 2.7 km$^2$ (one square mile).

The higher social insects (bees, ants, termites, and wasps) began a major evolutionary radiation in the Late Cretaceous, as angiosperms became dominant in terres-trial ecosystems. The earliest known bee, found in Cretaceous amber from Myanmar (Burma), is a female worker bee adapted for pollen gathering (Figure 15.15). Bee society already had a sophisticated structure.

## Later Stages in the Revolution

The changes to angiosperms, insects, and other animals did not all happen in the Cretaceous; the KTR began in the mid-Cretaceous but the full impacts of all the changes in terrestrial ecosystems were not felt until the Cenozoic. After the extinction of dinosaurs, angiosperms show several further hikes in importance. With dinosaurs gone, as we shall see in Chapter 17, mammals and birds diversified and many groups adopted new diets. Birds and small mammals, especially early primates and bats, all joined the seed- and fruit-eating guilds in Early Cenozoic times. Some tropical flowers today still rely for pollination on bats, small marsupials, or lemurs.

It's quite by accident that we happen to sense and appreciate the scents and colors of the flowers around us, because most of them were selected for the eyes and senses of insects. (We probably have color vision to help us choose between ripe and unripe fruit.) But we can gain a scientific as well as an aesthetic kick from looking at flowers if we admire their efficiency as well as their beauty.

Many predatory groups of animals benefited from the explosion in angiosperms, but especially from the explosion among insects. For example, spiders show a burst of evolution in the mid-Cretaceous, and their general purpose is not to terrify humans in their homes but to eat insects. Lizards and mammals arose in the Late Triassic (see Chapter 12) and both clades diversified substantially in the Cretaceous, as did birds (see Chapter 14). Many of these smaller animals were insect eaters, and there were diversifications both in the Cretaceous and at the beginning of the Cenozoic, as angiosperms and their coevolving insects showed further bursts of diversity. New mammal groups such as insectivores and bats became insect specialists around angiosperm forests (see Chapter 19).

## Angiosperm Chemistry

As we have seen, many angiosperms attract animals to themselves for pollination and seed dispersal. The plant usually pays a price in the production cost of substances such as nectar and in the cost of seeds eaten. Browsing animals and plant- and sap-eating insects often eat more plant material than they return in the form of services to the plant, and attracting such creatures results in a net loss of energy.

Angiosperms have therefore evolved an amazing variety of structures and chemicals that act to repel herbivores. These can be as simple and as effective as spines and stings, they can be contact irritants as in poison ivy and poison oak, or they can be severe or subtle internal poisons. Cyanide is produced by a grass on the African savanna when it is grazed too savagely. Many of our official and unofficial pharmacological agents were originally designed not for human therapy but as plant defenses. More than 2000 species of plants are insecticidal to one degree or another. Caffeine, strychnine, nicotine, cocaine, morphine, mescaline, atropine, quinine, ephedrine, digitalis, codeine, and curare are all powerful plant-derived chemicals, and it is not a coincidence that many of them are important insecticides or act strongly on the nervous, reproductive, or circulatory systems of mammals (some are even contraceptive and would act directly to decrease browsing pressure). Every day, 150 million pyrethrum flowers are harvested, to fill a demand for 25 000 tons of "natural" insecticide per year. A million tons of nicotine per year were once used for insect control, until it was found that the substance was extremely toxic to mammals (self-destructive humans still smoke it!). Other plant chemicals are powerful but can be used to flavor foods in low doses. All our kitchen flavorings and spices are in this category. Garlic keeps away insects as well as vampires and friends.

For paleobiologists, the problem of angiosperm chemistry is its failure to be preserved in the fossil record. Clearly, the increasing success of angiosperms in the Late Cretaceous and Early Cenozoic occurred in the face of intense herbivory by the radiating mammals and insects of that time. The chemical defenses of angiosperms probably evolved very early in their history.

## Reference

Vermeij, G.J. and Grosberg, R.K. (2010). The great divergence: when did diversity on land exceed that in the sea? *Integrative & Comparative Biology* 50: 675–682.

## Further Reading

Barba-Montoya, J., dos Reis, M., Schneider, H. et al. (2018). Constraining uncertainty in the timescale of angiosperm evolution and the veracity of a Cretaceous Terrestrial Revolution. *New Phytologist* 218: 819–834. [The definitive paper on phylogeny of angiosperms and other land plants].

Berendse, F. and Scheffer, M. (2009). The angiosperm radiation revisited, an ecological explanation for Darwin's 'abominable mystery'. *Ecology Letters* 12: 865–872. [Presents the evidence that angiosperms outgrow gymnosperms in most situations].

Condamine, F.L., Clapham, M.E., and Kergoat, G.J. (2016). Global patterns of insect diversification: towards a reconciliation of fossil and molecular evidence? *Scientific Reports* 6: 19208. [The fossil record of insects does not support the KTR].

Herendeen, P.S., Friis, E.M., Pedersen, K.R. et al. (2017). Palaeobotanical redux: revisiting the age of the angiosperms. *Nature Plants* 3: 17015. [The latest on the fossil evidence].

Lloyd, G.T., Davis, K.E., Pisani, D. et al. (2008). Dinosaurs and the Cretaceous Terrestrial Revolution. *Proceedings of the Royal Society B: Biological Sciences* 275: 2483–2490. [The KTR is named].

Peris, D., Pérez-de la Fuente, R., Peñalver, E. et al. (2017). False blister beetles and the expansion of gymnosperm-insect pollination modes before angiosperm dominance. *Current Biology* 27: 897–904. [Amazing beetle pollen preservation in amber].

Poinar, G.O. Jr. and Chambers, K.L. (2017). *Tropidogyne pentaptera*, sp. nov., a new mid-Cretaceous fossil angiosperm flower in Burmese amber. *Palaeodiversity* 10: 135–140. [Cretaceous flowers in amber].

Ren, D., Labandeira, C., Santiago-Blay, J.A. et al. (2009). A probable pollination mode before angiosperms: Eurasian, long-proboscid scorpionflies. *Science* 326: 840–847.

Sun, G., Ji, Q., Diulcher, D.L. et al. (2002). Archaefructaceae, a new basal angiosperm family. *Science* 296: 899–904. [Possibly the oldest decent fossil angiosperm].

Willis, K.J. and McElwain, J.C. (2014). *The Evolution of Plants*, 2e. Oxford: Oxford University Press. [A good overview of the entire history of plants].

## Questions for Thought, Study, and Discussion

1   The rise of the flowering plants coincided with a rise in many insect lineages. But plants and insects had both lived on Earth for many tens of millions of years. What was special about the parallel rise in flowering plants and insects?

2   Describe the science behind this limerick:
We're proud of humanity's powers,
But these potions and medicine of ours,
Coffee, garlic, and spices
Evolved as devices
So that insects would stop bugging flowers.

# 16

## The End of the Dinosaurs

| In This Chapter |
| --- |

The extinction of all dinosaurs (except birds), all plesiosaurs, all mosasaurs, and all pterosaurs at the end of the Cretaceous was a huge ecological disaster. (Many lineages survived but the tops of the food chain were destroyed.) In 1980, an astonishing paper suggested that the extinction was caused by the impact of a 6 km wide asteroid on to Earth. There was great resistance to the idea at first but evidence has built up now to the point where the impact and its size are not in doubt. The impact crater (in Mexico) has been found, along with dozens of sites where debris was scattered over the Earth. This chapter describes the effects of the impact and discusses a huge volcanic eruption that took place at the same time, this one centered on India. The paleontological evidence is mixed and sometimes contradictory. For example, while certain groups disappeared, others such as lizards, birds, and mammals survived. But new evidence shows that none of these survived comfortably – indeed, hundreds of species died out and some, such as birds, barely scraped through the crisis to become reestablished in the Paleogene.

## The Extinction at the End of the Cretaceous

Almost all the large vertebrates on Earth, on land, at sea, and in the air – all dinosaurs, plesiosaurs, mosasaurs, and pterosaurs – suddenly became extinct about 66 Ma, at the end of the Cretaceous Period. At the same time, most plankton and many tropical invertebrates, especially reef dwellers, became extinct and many land plants were severely affected. This extinction event was recognized in 1860 by John Phillips and he used it to mark a major boundary in Earth's history, the Cretaceous–Tertiary boundary, and the end of the Mesozoic Era. We now divide the Cenozoic into the Paleogene and Neogene periods, not the Tertiary and Quaternary, and so refer to this most famous of all catastrophes as the Cretaceous–Paleogene (or K–Pg) mass extinction.

The K–Pg extinction was worldwide, affecting all continents and oceans. There are still arguments about just how short the extinction event was. It was certainly sudden in geological terms and may have been catastrophic by anyone's standards. And it's not just the number of genera and families that became extinct. As we saw in Chapter 6, the K–Pg extinction was a huge ecological disaster, second only to the Permo-Triassic extinction.

Despite the scale of the extinctions, however, we must not be trapped into thinking that the K–Pg boundary marked a disaster for all living things. Most major groups of organisms survived. Insects, mammals, birds, and flowering plants on land and fishes, corals, and mollusks in the ocean went on to diversify tremendously during the Paleogene. The K–Pg casualties included most of the large creatures of the time but also some of the smallest, in particular the plankton that generate most of the primary production in the oceans.

There have been many inadequate theories to explain dinosaur extinctions. More bad science is described in this chapter than in all the rest of the book. For example, even in the 1980s, a new book on dinosaur extinctions suggested that they spent too much time in the sun, got cataracts, and because they could not see very well, fell over cliffs to their doom. In 2012, a team of reputable scientists proposed that dinosaurs farted themselves to death. The researchers noted that a cow produces 30–50 gal of methane gas every day from digestion of

*Cowen's History of Life*, Sixth Edition. Edited by Michael J. Benton.
© 2020 John Wiley & Sons Ltd. Published 2020 by John Wiley & Sons Ltd.

grass. With 1.5 billion cows on the planet, this amounts to a huge amount of gas, and methane is about 30 times more efficient as a greenhouse gas than carbon dioxide. The researchers estimated that all sauropod dinosaurs on Earth together, each 50 times the size of a cow, produced more than 500 million tons of methane per year, equivalent to the total modern methane production from cows and from human-caused pollution.

But no matter how convincing or how silly it is, any theory that tries to explain only the extinction of the dinosaurs ignores the fact that extinctions took place among land, sea, and aerial faunas and floras. In fact, it would probably be relatively easy to kill off the dinosaurs because they were so huge, but any killing model must also explain why plankton, ammonites, and many birds also disappeared. The K–Pg extinctions were a global event, so we look for globally effective agents to explain them: geographic change, oceanographic change, climatic change, or an extraterrestrial event (see Chapter 6). The most recent work on the K–Pg extinction has centered on two hypotheses that suggest a violent end to the Cretaceous: a large asteroid impact and a giant volcanic eruption. We have an enormous amount of data to assess these two hypotheses.

## An Asteroid Impact?

An asteroid hit Earth precisely at the end of the Cretaceous (Figure 16.1). The impact threw up a huge cloud of dust, which encircled the globe and blacked out the Sun. This cut out light and heat, and the surface of the Earth became freezing cold and, without sunlight, photosynthesis ceased. Since food chains on land and in the oceans are based on photosynthetic plants (diatoms among the plankton in the oceans), plant eaters died off, and then the predators. If the absence of food was not enough to kill the dinosaurs, plankton and ammonites, then the cold would have hit them all hard. What is the evidence for this model?

### Iridium and Glass Spherules

The suggestion of an impact was first made by Luis Alvarez and his colleagues in a remarkable paper in 1980. These researchers found that rocks laid down precisely at the K–Pg boundary contain extraordinary amounts of the metal **iridium**. It does not matter whether the boundary rocks were laid down on land or under the sea. In the Pacific Ocean and the Caribbean, the iridium-bearing clay forms a layer in ocean sediments (Figure 16.2); it is found in continental shelf deposits in Europe, and in much of North America it occurs in coal-bearing rocks laid down on floodplains and deltas. The dating is precise and the iridium layer has now been identified in several hundred places worldwide. This has allowed geologists to agree that the K–Pg boundary should be defined by the impact layer. Cretaceous rocks lie under it. The layer itself belongs to the Cenozoic or Paleogene (see Chapter 2, Figure 2.6). In marine sediments, the iridium occurs just above the last Cretaceous microfossils, and the sediments above it contain Paleocene microfossils from the earliest Cenozoic.

The iridium is present only in the boundary rocks and therefore was deposited in a single large **spike**: a very short event. Iridium occurs in normal sea-floor sediments in microscopic quantities, but the K–Pg iridium spike is very large. Iridium is rare on Earth. Chemical

Figure 16.1 Visualization of the end-Cretaceous asteroid just before it struck. *Source:* Still from a Discovery Channel simulation, available at: www.youtube.com/watch?v=bU1QPtOZQZU.

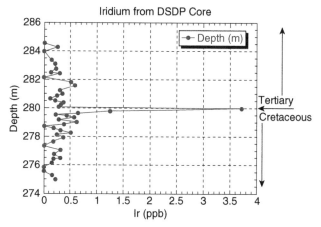

Figure 16.2 The iridium layer marking the Cretaceous-Tertiary boundary is often identified in drill holes in the ocean floor. This is a typical iridium anomaly or "spike" from a hole drilled during the Deep Sea Drilling Program. *Source:* Image from NASA.

processes in a sediment can concentrate iridium to some extent, but the K–Pg iridium spike is so large that it must have arisen in some unusual way. Iridium is much rarer than gold on Earth yet in the K–Pg boundary clay, iridium is usually twice as abundant as gold, sometimes more than that. The same high ratio is found in meteorites. The Alvarez group therefore suggested that iridium was scattered worldwide from a cloud of debris that formed as an **asteroid** struck somewhere on Earth.

An asteroid big enough to scatter the estimated amount of iridium in the worldwide spike at the K–Pg boundary may have been about 10 km (6 miles) across. Computer models suggest that if such an asteroid collided with Earth, it would pass through the atmosphere and ocean almost as if they were not there and blast a crater in the crust about 100 km across. The iridium and the smallest pieces of debris would be spread worldwide by the impact blast, as the asteroid and a massive amount of crust vaporized into a fireball.

If indeed the spike was formed by a large impact, what other evidence should we hope to find in the rock record? Well-known meteorite impact structures often have fragments of **shocked quartz** (Figure 16.3) and **spherules** (tiny glass spheres) (Figure 16.4) associated with them. Shocked quartz is formed when quartz crystals undergo a sudden pulse of great pressure, yet do not melt. The shock causes peculiar and unmistakable microstructures. The glass spherules are formed as the target rock is melted in the impact, blasted into the air as a spray of droplets, and almost immediately cooled and solidified. Over geological time, the glass spherules may decay to clay.

All over North America, the K–Pg boundary clay contains glass spherules, and just above the clay is a thin-

Figure 16.4 Tiny glass spherules picked out of the K–Pg boundary layer and glued on to a specimen card. Scale in millimeters. *Source:* Image from NASA.

ner layer that contains iridium along with fragments of shocked quartz. It is only a few millimeters thick but in total, there must be more than a cubic kilometer of shocked quartz in North America alone.

When the Alvarez team published their 1980 paper, no crater was known. The crater was not essential, as the iridium, shocked quartz, and glass spherules were strong enough evidence for the impact and the worldwide ash and the killing. Still, it was a bit frustrating for the proponents of the impact idea to have to say, "Well, the crater might have been under the ocean and lost to subduction." Critics demanded to see the smoking gun. Then the crater was found in 1991.

### Finding the Crater

Alan Hildebrand had been looking for the K–Pg impact crater as part of his doctoral studies. He located it by looking at borehole data from Pemex, the Mexican state petroleum company. In the 1960s, Pemex had made geophysical surveys and boreholes through an interesting circular structure near the village of Chicxulub on the Yucatán peninsula in south-eastern Mexico. At first, Pemex had thought this might be an oil trap but when they hit igneous rock in the middle of the structure, they abandoned their efforts. But Hildebrand recognized the igneous rock as evidence that rocks had been melted by high pressure and high temperature conditions caused by an impact. He had the crater – but it was entirely covered by younger sediments.

The **Chicxulub Crater** is a circular structure (Figure 16.5) about 180 km across, one of the largest impact structures so far identified on Earth. A borehole drilled into the

Figure 16.3 Shocked quartz: a crystal that has been caught up in a meteorite impact has characteristic shock marks in its crystal structure. *Source:* Image from NASA.

Figure 16.5 The location of the Chicxulub crater in the Yucatán peninsula of Mexico. It was discovered during drilling for oil and is completely covered by younger sediments, but its recognition as the K–Pg impact site was a stroke of insight. *Source:* Image from NASA.

Chicxulub structure hit 380 m (more than 1000 ft) of igneous rock with a strange chemistry that could have been generated by melting together a mixture of the sedimentary rocks in the region. The igneous rock contains high levels of iridium and its age is 66 Ma, exactly coinciding with the K–Pg boundary.

Geophysical survey of the Chicxulub site (Figure 16.6) confirms the circular shape of the crater and that there are multiple rings. This was expected. When a small asteroid hits the Earth, it forms a single-ring crater but a large asteroid, as here, hits and the Earth then rebounds, forming the circular outer crater ring. But in the center of a large crater, the floor of the crater shoots up in the air, like the rebound after a water droplet hits the surface of a sheet of water. Then the center of the crater falls back, leaving the inner ring.

On top of the igneous rock lies a mass of broken rock, probably the largest surviving debris particles that fell back on to the crater without melting, and on top of that are normal sediments that formed slowly to fill the crater in the shallow tropical seas that covered the impact area.

Well-known impact craters often have tektites associated with them as well as shocked quartz and tiny glass spherules. **Tektites** are larger glass beads with unusual shapes and surface textures. They are formed when rocks are instantaneously melted and splashed out of impact sites in the form of big gobbets of molten glass, then cooled while spinning through the air (Figure 16.7).

In Haiti, which was about 800 km from Chicxulub at the end of the Cretaceous, the K–Pg boundary is marked by a normal but thick (30 cm) clay boundary layer that consists mainly of glass spherules (Figure 16.4). The clay is overlain by a layer of **turbidite**, submarine landslide material that contains large

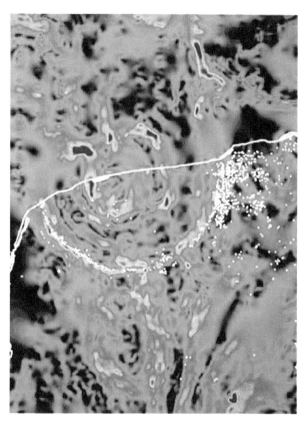

Figure 16.6 Gravity anomaly map of the Chicxulub crater, showing the inner and outer rings. The white line is the current coastline and white dots are sinkholes in the overlying limestone. *Source:* United States Geological Survey.

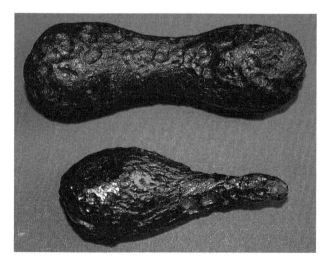

Figure 16.7 Two tektites, showing shapes molded while the tektites were still molten and spinning through the atmosphere. These are not from the K–Pg boundary. *Source:* Photograph by Brocken Inaglory (Wikimedia).

rock fragments. Some of the fragments look like shattered ocean crust, but there are also spherical pieces of yellow and black glass up to 8 mm across that are unmistakably tektites. The tektites were formed at about 1300 °C from two different kinds of rock, and they are dated precisely at 65 Ma. The black tektites formed from continental volcanic rocks and the yellow ones from evaporite sediments with a high content of sulfate and carbonate. The rocks around Chicxulub are formed dominantly of exactly this mixture of rocks, and the igneous rocks under Chicxulub have a chemistry of a once-molten mixture of the two. Above the turbidite comes a thin red clay layer only about 5–10 mm thick that contains iridium and shocked quartz.

## The Impact and Its Consequences

This evidence can be explained as follows: an asteroid struck at Chicxulub, hitting a pile of thick sediments in a shallow sea (Figure 16.1). The impact melted much of the local crust and blasted molten material outward from as deep as 14 km under the surface. Small spherules of molten glass were blasted into the air at a shallow angle and fell out over a giant area that extended north east as far as Haiti, several hundred kilometers away, and to the north west as far as Colorado. Next followed the finer material that had been blasted higher into the atmosphere or out into space and fell more slowly on top of the coarser fragments.

Other sites in the western Caribbean suggest that normally quiet, deep-water sediments were drastically disturbed right at the end of the Cretaceous, and the disturbed sediments have the iridium-bearing layer right on top of them. At many sites from northern Mexico and Texas, and at two sites on the floor of the Gulf of Mexico, there are signs of a great disturbance in the ocean at the K–Pg boundary. In some places, the disturbed sea-floor sediments contain fossils of fresh leaves and wood from land plants, along with tektites dated at 66 Ma (Figure 16.5). Around the Caribbean and at sites up the eastern Atlantic coast of the United States, existing Cretaceous sediments were torn up and settled out again in a messy pile that also contains glass spherules of different chemistries, shocked quartz fragments, and an iridium spike.

All this implies that a series of **tsunamis** or tidal waves affected the ocean margin of the time, washing fresh land plants well out to sea and tearing up sea-floor sediments that had lain undisturbed for millions of years. The tsunamis were generated by the shock of the impact, which has been estimated as the equivalent of a magnitude 11 earthquake, with 1000 times more energy than any "normal" earthquake recorded on Earth (about 100 million megatons, if you like measuring in terms of hydrogen bomb blasts!).

Geologists have now found hundreds of localities containing K–Pg boundary sediments, so we can put together the history of the impact, sometimes down to the minute. Larger fragments of solid rock and molten lava were blasted outward from the crater at lower angles, but not very far, and were deposited first and locally as mixed masses of rock fragments within 500 km of the crater: 100 m of rock rubble close to the crater rim, for example. Up to about 1000 km, round the Gulf of Mexico and in the Caribbean, we find meters of rock rubble with glass spherules and because these were coastal areas, they are mixed with tsunami debris. Up to 5000 km (3100 miles) away, we find the glass spherule layer that was blasted out at low angles (about 15 minutes travel time to Colorado, for example), and on top of them smaller fragments, including shocked quartz crystals and iridium, that had been lofted higher and fell more slowly (about 30 minutes to Colorado). Then, over the rest of the Earth we find the bulk of the mass in the fireball that had been vaporized to form molten debris high above the atmosphere. It was deposited last, slowly drifting downward as frozen droplets and dust particles to form a thin layer, now perhaps only 2 mm thick, that is usually made of clay rich in iridium.

Some scenarios have come to be extensively quoted. Thus, the dust that was blasted out into space and then fell back to Earth has been envisaged as forming millions of meteor-like trails in the atmosphere, which together would have heated the Earth's surfaces as if it was in a microwave oven. The heat was first envisaged as starting enormous forest fires worldwide, producing smoke, soot, and carbon dioxide in prodigious quantities. As the heat shock was absorbed into the ocean, it was thought that it would produce "hypercanes," gigantic hurricanes far larger than Earth's normal climate can generate. Sulfur-bearing aerosols were suggested as causing acid rain with the strength of battery acid. Overall, global ecological cycles would have been devastated, with primary production of plankton in the ocean and plants on land cutting off all food supplies to consumers on land and in the sea.

All these scenarios now look too extreme. There was no global forest fire, for example, and there is no evidence of hypercanes. The environmental insults that followed the impact were severe enough to cause global extinctions without us exaggerating their magnitude.

An immediate consequence of the impact was 100 000 years of global warming. Study of oxygen isotope records from fish debris at and immediately after the K–Pg event as recorded in Tunisia, North Africa, shows a temperature rise of 5 °C. The new result confirms other data from study of fossil leaves, paleosols and modeling but contradicts earlier suggestions that there was long-term cooling following the impact. Perhaps the cooling caused by the vast dust cloud lasted less than a year and after the dust cleared, the effects of global warming were more sustained. A temperature rise of 5 °C was much less than the 15–20 °C experienced after the Permian–Triassic mass extinction (see Chapter 6), but it would have still been enough to perturb paleogeographic distributions of plants and animals.

## A Giant Volcanic Eruption?

Exactly at the K–Pg boundary, a new volcanic plume (see Chapter 6) was burning its way through the crust close to the plate boundary between India and Africa. Enormous quantities of basalt flooded out over what is now the Deccan Plateau of western India to form huge lava beds called the **Deccan Traps**. Some of these flows are the longest and largest ever identified on Earth. A huge extension of that lava flow on the other side of the plate boundary now lies underwater in the Indian Ocean (Figure 16.8).

The Deccan Traps cover $500\,000\,km^2$ now (about $200\,000\,mile^2$), but they may have covered four times as much before erosion removed them from some areas. They have a surviving volume of 1 million $km^3$ ($240\,000\,mile^3$) and are over 2 km thick in places. The entire volcanic volume that erupted, including the underwater lavas, was much larger than this. A huge area of eastern India is dominated by these volcanic beds, which form a characteristic scenery (Figure 16.9), with deep valleys cutting through numerous layers. The separate basalt flows form steps on the hillsides, hence the name "Traps" (from Old German *treppe*, a step).

The date of the Deccan eruptions cannot be separated from the K–Pg boundary. The peak eruptions may have lasted only about 1 million years but that short time straddled the boundary. The rate of eruption was at least 30 times the rate of Hawaiian eruptions today, even assuming it was continuous over as much as a million years; if the eruption was shorter or spasmodic, eruption rates would have been much higher. The Deccan Traps probably erupted as lava flows and fountains like those of Kilauea, rather than in giant explosive eruptions like that of Krakatau. The Deccan plume is still active; its hot spot now lies under the volcanic island of Réunion in the Indian Ocean.

Thus the K–Pg boundary coincided with two very dramatic events. The Deccan Traps lie across the K–Pg boundary and were formed in what was obviously a major event in Earth history. The asteroid impact was exactly at the K–Pg boundary. The asteroid impact or the gigantic eruptions, or both, would have had major global effects on atmosphere and weather.

In considering the relative roles of the Chicxulub impact and the eruption of the Deccan Traps, it has commonly been argued that the impact was enough on its own to cause all the catastrophic extinctions. However, the Deccan Traps were comparable in size to the Siberian Traps, which triggered the even more catastrophic Permian–Triassic mass extinction (see Chapter 6), and so could have had a comparable effect, or at least should not simply be ignored. Indeed, on their own, the Deccan Traps might have been much less deadly than the Siberian Traps as they were erupted through more ancient crust, so there were not all the additional interactions between lava and sediments.

For a long time, geologists struggled to understand whether the eruptions on one side of the world and the impact on the other might in some way be linked. Paul Renne and colleagues presented evidence for a definite link. They noted that the Deccan Traps had been erupting in a modest way for $350\,000$ years before the impact, and then exactly at the time of the impact, the rate of eruption massively increased. They estimate that 70% of

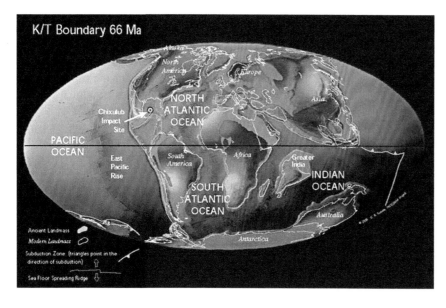

**Figure 16.8** World paleogeography at the K–Pg boundary. Gondwana and Laurasia are split into pieces, with Australia just leaving Antarctica. The Chicxulub impact is marked, and the great eruptions took place on and between India and the east coast of Africa. *Source:* Paleogeographic map by C.R. Scotese © 2012, PALEOMAP Project (www.scotese.com).

Figure 16.9 Satellite image of the Deccan Traps scenery showing multiple layers of basalt, Maharashtra, India. *Source:* Photography by Planet Labs, Inc. (Wikimedia).

all the Deccan lavas were erupted in the 50 000 years immediately after the impact. The eruptions eventually ended 500 000 years after the time of the Chicxulub impact.

There is evidence from various locations that some global warming may have begun in the last half a million years of the Cretaceous, triggered by the Deccan Traps eruptions, and there is evidence for early extinctions at this time. Then it seems the asteroid struck, causing the bulk of the extinctions and kicking off an intense and short spell of enhanced eruption. The Deccan Traps then carried on erupting for a further half a million years before ceasing.

## Paleobiological Evidence from the K–Pg Boundary

How do we assess the catastrophic scenarios that have been suggested for the K–Pg boundary? The paleontological evidence from the K–Pg boundary is ambiguous. Many phenomena are well explained by an impact or a volcanic hypothesis, but others are not. The fossils provide us with real evidence about the K–Pg extinction events and we can, for example, compare survivors and victims. On land, dinosaurs and pterosaurs died out, birds were quite hard hit and survived, whereas crocodiles, turtles, lizards, and mammals took a hit but survived quite successfully.

Regionally, there is little doubt that the North American continent would have been absolutely devastated by dust, hot toxic debris, and tsunamis. Globally, even a short-lived catastrophe among land plants and surface plankton at sea would drastically affect normal food chains. Pterosaurs, dinosaurs, and large marine reptiles would have been vulnerable to food shortage, and their extinction after a catastrophe seems plausible. Lizards and primitive mammals, which survived, are small and often burrow and hibernate; for a certain amount of time, they would have found plenty of nuts, seeds, insect larvae, and invertebrates buried or lying around in the dark.

In the oceans, invertebrates living in shallow water along the shore would have suffered greatly from cold or frost, or perhaps from $CO_2$-induced heating. But deeper water forms are insulated from heat or cold shock and have low metabolic rates; they therefore would be able to survive even months of starvation. High-latitude faunas in particular were already adapted to winter darkness, though perhaps not to extreme cold. Thus, tropical reef communities could have been decimated, but deep-water and high-latitude communities could have survived much better. All these patterns are either observed or under test.

The best-studied terrestrial sections across the K–Pg boundary are in North America. Immediately this is a problem, because we know that the effects of the asteroid impact were greater here than in most parts of the world. Perhaps this has given us a more catastrophic view of the boundary event that we would gather from, say, comparable careful research in New Zealand. Even so, it is obvious that life, even in North America, was not wiped out: many plants and animals survived the K–Pg event.

### Land Plants

North American land plants were devastated from Alberta to New Mexico at the K–Pg boundary. The sediments below the boundary are dominated by angiosperm

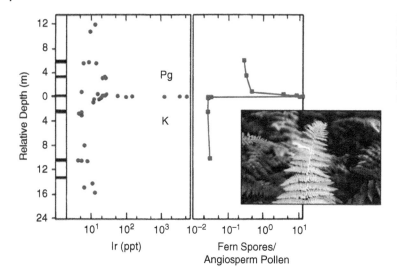

Figure 16.10  The iridium spike (*left*) and fern spike (*right*) from a borehole through the K-Pg boundary in the Raton Basin, New Mexico. The fern spike, showing the ratio of fern spores to angiosperm pollen, indicates a very short-lived but dramatic expansion of ferns immediately after the impact. *Source:* Image courtesy of David A. Kring.

pollen, but the boundary itself has little or no angiosperm pollen and instead is dominated by fern spores in a **fern spike** analogous to the iridium spike (Figure 16.10). Normal pollen counts occur immediately after the boundary layer. The fern spike therefore coincides precisely with the iridium spike in time and is equally intense and short-lived. A fern spore spike also occurs in New Zealand, suggesting that the crisis was widespread.

The fern spike could be explained by a short but severe crisis for land plants, generated by an impact or an eruption in which all adult leaves died off (some mix of lack of light, prolonged frost, or acid rain). Perhaps ferns were the first plants to recolonize the debris and higher plants returned later. This happened after the eruption of Krakatau in 1883, and of Mount St Helens volcano in 2008. Ferns quickly grew on the devastated ash-covered surfaces, presumably from wind-blown spores, but they in turn were replaced within a few decades by flowering plants as a full flora was reestablished.

Evidence from leaves confirms the data from spores and pollen. Land plants recovered from the crisis but many Late Cretaceous plant species were killed off.

Angiosperms were in the middle of a great expansion in the Late Cretaceous (see Chapter 15), and the expansion continued into the Paleocene and Eocene. Climate and plant diversity were fluctuating before the K–Pg boundary, but not to the extent that the plant crisis at the boundary can be blamed on climate. Yet there were important and abrupt changes in North American floras at the K–Pg boundary. In the Late Cretaceous, for example, an evergreen woodland grew from Montana to New Mexico in a seasonally dry, subtropical climate. At the boundary, the dominantly evergreen Late Cretaceous woodland changed to a largely deciduous Paleogene swamp woodland growing in a wetter climate. The fern spike marks a period of swampy mire at the boundary

itself. Deciduous trees survived the K–Pg boundary events much better than evergreens did; in particular, species that had been more northerly spread southward. More significantly, there were very few changes in the high northern (polar) floras of Alaska and Siberia.

These changes are probably best explained by a catastrophe that wiped out most vegetation locally, with recolonization from survivors that remained safe during the crisis as seeds and spores in the soil or even as roots and rhizomes. Other survivors came from larger refuges such as the high Arctic.

### Dinosaurs

There used to be a debate about whether dinosaurs died out at the end of the Cretaceous either with a whimper or a bang (Figure 16.11). In other words, were they already

Figure 16.11  Dinosaurs doomed by the Deccan Traps eruptions. *Source:* Image by Zina Deretzky for the National Science Foundation (in the public domain).

seriously in decline and the impact just finished them off, or were they at the peak of their success and then disappeared overnight? The discussion has reopened and two lines of evidence suggest that dinosaurs were in decline well before the end of the Cretaceous.

One piece of evidence comes from studies of evolutionary rates. Manabu Sakamoto and colleagues showed in 2016 that dinosaurs were evolving fast from the Late Triassic to Early Cretaceous, but that there was a turndown through most of the Late Cretaceous. Two new groups, the hadrosaurs and ceratopsians, were expanding throughout the Late Cretaceous but all other groups of dinosaurs were in slow decline. If the Chicxulub impact had not happened, Sakamoto and colleagues suggested that dinosaurs might very well have dwindled to extinction anyway, perhaps driven by slow climate change, especially cooling temperatures and break-up of the continents.

Another piece of evidence comes from a 2012 study by Jonathan Mitchell and colleagues, who looked at the community structure of dinosaur ecosystems through the last part of the Cretaceous. They found that food webs were initially quite robust to perturbation but then in the last 6 m.y. of the Cretaceous they became more unstable. Their models suggested that environmental changes would have greater destabilizing effects on these dinosaur communities.

It's easy to look at the very last dinosaurs on Earth, animals like *Tyrannosaurus rex*, *Triceratops*, and *Ankylosaurus* from the Hell Creek Formation of Montana (Figure 16.12), and assume that all was well. Here are a bunch of the most famous dinosaurs of all time, and they are diverse and successful. But cooling climates of the latest Cretaceous (see Chapter 17) might already have been making their lives more difficult, and speciation rates had fallen. Scientists used to speculate about what a modern dinosaur would look like if the Chicxulub impact had never happened; it's unlikely dinosaurs would have lasted more than another 10–20 million years into the Paleogene.

## Lizards, Birds, and Mammals

It is often said simply that lizards, birds, and mammals survived the K–Pg mass extinction, but this is entirely untrue. In fact, many species of lizards, birds, and mammals died out, and birds might have gone under completely. However, the crisis provided a remarkable opportunity for all three groups, and they exploded in diversity through the Cenozoic (see Chapter 19).

It was difficult to say what happened to lizards, other than that there were lizards throughout the Mesozoic, and they diversified in the Cretaceous, and there are lots of lizards today, so evidently they did just fine. However, close study of North American faunas by Nick Longrich and colleagues showed that 83% of species of lizards and snakes died out at the K–Pg boundary, and it took 10 m.y. of the Paleogene before they recovered their preextinction diversity levels. Larger lizards and snakes were selectively driven to extinction, and the survivors were all small.

Among birds, four key groups that had diversified in the Cretaceous (see Chapter 15) disappeared: the enantiornithines, comprising 80 species in the Cretaceous, the quite rare palintropiforms, and the marine hesperornithiforms and ichthyornithiforms. Only a few species of land-living ancestors of ducks and chickens survived through the K–Pg boundary. Dan Field and colleagues showed that tree-dwelling bird species were very hard hit by the impact, whereas ground dwellers survived better. Even though many modern birds of course live in trees, their ancestors at the end of the Cretaceous ran about on the ground. The suggestion was that forests were devastated by the impact and did not recover for

**Figure 16.12** Dinosaurs and pterosaurs from the Hell Creek Formation. From back to front: *Ankylosaurus*, *Tyrannosaurus*, *Quetzalcoatlus*, *Triceratops*, and *Struthiomimus* in the middle, with *Archeraptor*, *Pachycephalosaurus* and *Anzu*, from left to right in the front row. *Source:* Artwork by Durbed (Wikimedia).

1000 years, so the only birds that could survive were those that ran about on the ground feeding on seeds or burrowing bugs.

Mammals were affected too, with losses among all three main Late Cretaceous groups: the multituberculates, marsupials, and placentals (see Chapter 18). Indeed, there is evidence from the Hell Creek Formation that mammal faunas were already ecologically unstable 500 000 years before the K–Pg boundary, and then some 75% of species died out during the mass extinction. Maybe, as with the dinosaurs, climatic perturbation, perhaps triggered by the onset of the Deccan Traps, or perhaps not, was affecting terrestrial communities.

### Freshwater Communities

Some ecological anomalies at the K–Pg boundary are not easily explained by a catastrophic scenario. Freshwater communities were less affected than terrestrial ones. For example, turtles and other aquatic reptiles survived in North Dakota while dinosaurs were totally wiped out. Freshwater communities are fueled largely by stream detritus, which includes the nutrients running off from land vegetation. It has been suggested that animals in food chains that begin with detritus rather than with primary productivity would survive a catastrophe better than others.

### Marine Life

The K–Pg event was just as catastrophic for life in the sea. Famously, ammonites, belemnites, and rudist bivalves all died out. The ammonites and belemnites had been major swimmers in Mesozoic oceans (see Chapter 11) and their disappearance must have changed food chains massively. Rudist bivalves were unusual mollusks, often large and with massive, heavy shells (Figure 16.13). Millions of individuals clustered together to build great reefs in hot, shallow seas around the world. Rudists disappeared entirely but other reef builders, including corals, also suffered, with losses of 60% of genera. Perhaps all these marine invertebrates, reliant on skeletons composed of calcium carbonate, suffered from acidification of the oceans or loss of oxygen following global warming, even if just for a few years.

At the base of the food chain, it might have been expected that key components of the plankton would have shown severe extinction. However, there does not seem to have been a productivity collapse during the crisis. Extinctions of many plankton species have been noted but overall the foraminifera, which feed on photosynthesizing plant-like forms, showed little effect of the crisis. The microfossils show there was a rapid burst of ocean acidification, which caused stress for populations of species with calcareous shells, but this passed by

**Figure 16.13** Rudist bivalve from the Late Cretaceous of southern France. *Source:* Photograph by Wilson44691 (Wikimedia).

quickly and species numbers and global abundance recovered fast.

Fishes show losses also, with the extinction of seven families, and 84% of species of sharks and rays. Hardest hit were shallow-water forms, and their extinctions might have been caused by the loss of prey species as well as the sudden climatic shocks associated with the Deccan Traps eruptions and the Chicxulub impact. Among teleosts, the dominant group of bony fishes, large forms were particularly vulnerable to extinction, as were those which had special adaptations for fast, rather than powerful, jaw closing. However, small snappy fish and tooth grinders of all sizes survived.

At the top of the food chain, the marine reptiles (see Chapter 11) finally disappeared. Ichthyosaurs had died out 30 m.y. earlier, but long-necked plesiosaurs and mosasaurs were important marine predators right to the end of the Cretaceous, and they then disappeared.

### Where Are We?

The K–Pg impact was sudden and coincided precisely with the asteroid impact. The Deccan Traps eruptions were massive and lasted perhaps a million years, beginning before and ending after the impact. The clear implication is that the asteroid impact triggered the mass extinction yet the eruptions were not environmentally benign. The unusual ecological severity of the K–Pg extinction, and its global scope, may have happened because an asteroid impact and a gigantic eruption occurred when global ecosystems were particularly vulnerable to a disturbance of oceanic stability. However, that is a difficult argument to test, so most scientists naturally focus on the Chicxulub impact as the major cause of the K–Pg extinction.

There are interesting patterns among the survivors. Hardly any major groups of organisms became entirely extinct. Even the dinosaurs survived in one sense (as birds). Planktonic diatoms survived well, possibly because they have resting stages as part of their life cycle. They recovered as quickly as the land plants emerged from spores, seeds, roots, and rhizomes. The sudden interruption of the food chains on land and in the sea may well have been quite short, even if full recovery of the climate and full marine ecosystems took much longer. On one hand, climate modelers and paleobotanists have concluded that land plants recovered to full production in perhaps 10 years yet normal productivity took a few thousand years to reestablish in the oceans. However, it took about 3 million years for the full marine ecosystem to recover, probably because so many marine predators (crustaceans, mollusks, fishes, and marine reptiles) disappeared, and had to be replaced by evolution among surviving relatives.

We still do not have a full explanation for the demise of the victims of the K–Pg extinction, while so many other groups survived. Improvements in rock dating are helping geologists discriminate between truly fast events (years to centuries), and geologically fast events (less than 1 million years). Also, it is important to consider the global stage. Intensive study of case reports from North America provides good evidence of those regional effects, but impacts were less on other continents, and comparative studies of different groups of plants, animals, and microorganisms are needed from all parts of the world through the K–Pg crisis time.

## Further Reading

Alegret, L., Thomas, E., and Lohmann, K.C. (2012). End-Cretaceous marine mass extinction not caused by productivity collapse. *Proceedings of the National Academy of Sciences of the United States of America* 109: 728–732.

Alvarez, W. (2015). *T. rex and the Crater of Doom*, 2e. Princeton: Princeton University Press. [The best of many books on the extinction].

Alvarez, L.W., Alvarez, W., Asaro, F. et al. (1980). Extraterrestrial cause for the Cretaceous-Tertiary extinction. *Science* 208: 1095–1108. [The paper that started it all].

Field, D.J., Bercovici, A., Berv, J.S. et al. (2018). Early evolution of modern birds structured by global forest collapse at the end-Cretaceous mass extinction. *Current Biology* 28: 1825–1831.

Hildebrand, A.R., Penfield, G.T., Kring, D.A. et al. (1991). Chicxulub Crater: a possible Cretaceous/Tertiary boundary impact crater on the Yucatán Peninsula, Mexico. *Geology* 19: 867–871. [Recognition of Chicxulub as the K-T asteroid crater].

Johnson, K.R. and Ellis, B. (2002). A tropical rainforest in Colorado 1.4 million years after the Cretaceous-Tertiary boundary. *Science* 296: 2379–2383. [Forest changes following the impact].

Kring, D.A. (2007). The Chicxulub impact event and its environmental consequences at the Cretaceous–Tertiary boundary. *Palaeogeography, Palaeoclimatology, Palaeoecology* 255: 4–21.

Longrich, N.R., Bhullar, B.-A.S., and Gauthier, J.A. (2012). Mass extinction of lizards and snakes at the Cretaceous–Paleogene boundary. *Proceedings of the National Academy of Sciences of the United States of America* 109: 21396–21401.

MacLeod, K.G., Quinton, P.C., Sepúlveda, J. et al. (2018). Postimpact earliest Paleogene warming shown by fish debris oxygen isotopes (El Kef, Tunisia). *Science* 360: 1467–1469. [Evidence for 100 000 years of global warming following the impact].

Mitchell, J.S., Roopnarine, P.D., and Angielczyk, K.D. (2012). Late Cretaceous restructuring of terrestrial communities facilitated the end-Cretaceous mass extinction in North America. *Proceedings of the National Academy of Sciences of the United States of America* 109: 18857–18861.

Morgan, J.V., Gulick, S.P.S., Bralower, T. et al. (2016). The formation of peak rings in large impact craters. *Science* 354: 878–882. [Reports the latest geophysical surveys of Chicxulub].

Renne, P.R., Sprain, C.J., Richards, M.A. et al. (2015). State shift in Deccan volcanism at the Cretaceous–Paleogene boundary, possibly induced by impact. *Science* 350: 76–78. [Dating of the Deccan Traps, and possible link to the Chicxulub impact].

Sakamoto, M., Benton, M.J., and Venditti, C. (2016). Dinosaurs in decline tens of millions of years before their final extinction. *Proceedings of the National Academy of Sciences of the United States of America* 113: 5036–5040. [Evidence for long-term decline in dinosaurs before the impact].

Schulte, P., Alegret, L., Arenillas, I. et al. (2010). The Chicxulub asteroid impact and mass extinction at the Cretaceous-Paleogene boundary. *Science* 327: 1214–1217. [The case for impact as main driver of the mass extinction].

## Question for Thought, Study, and Discussion

One disputed calculation suggests that the asteroid that impacted the Earth at 66 Ma was formed in a major collision in the "asteroid belt" some time in the Jurassic. Tens of millions of years later, it arrived in Earth's orbit and collided with it. This is about as random as geological processes get. So the question is, what difference would it have made in the great sweep of life on Earth if it had arrived, say, 50 million years before or 50 million years after it did? This is a question asking for thoughtful speculation.

17

Changing Oceans and Climates

**In This Chapter**

In this chapter, we explore the link between climate and life in the ocean over the last 100 million years. We learn about how we drill into the ocean floor to obtain archives of its history and how we can reconstruct climate change using proxies. The chapter covers the most important climate changes over the last 100 million years, starting with the Cretaceous hothouse. We show how the origin of marine plankton in the Mesozoic affected the chemistry of the ocean. At the end of the Paleocene, an extraordinary warm climate changed life both on land and in the ocean for a short time. From the late Eocene, Earth started to cool, descending into an Icehouse. With geography changing all the time, South America drifted close enough to North America to limit water flowing between the Atlantic and Pacific Oceans and allowing animals to exchange. Our current climate is different from the rest of the last 100 million years. We have large temperature differences between ice ages and the intermittent warmer intervals and anthropogenic $CO_2$ emissions change the climate from the natural baseline. Understanding the deep-time history of climate change helps us understand the short-term changes that everyone, including even some politicians, worry about.

## Climate

Climate is one of the most important environmental factors for all organisms. Climate changes have taken place all through the history of the Earth, and they can directly affect the evolution of life. Plate tectonic movements can change oceanic and continental geography, and those geographic changes can modify climatic patterns and affect the ecology and evolution of organisms in major global events (see Chapter 6).

We are living through an ice age now, geologically speaking, and have been for the past 2.5 million years or so. Within this overall cooler time in Earth history, we happen to live during a warm stage. Great ice sheets expanded and covered much of the northern continents, then retreated again. Ice ages are not common events in Earth's history. There was a widespread ice age toward the end of the Precambrian around 600 Ma: Snowball or Slushball Earth (see Chapter 4). In Late Ordovician times, when northern Gondwana was over the South Pole, a great ice sheet spread over most of North Africa and probably further. Gondwana drifted across the South Pole during the rest of the Paleozoic, with a particularly important glacial period in South America at the end of the Devonian. Large-scale glaciation once again spread over most of Gondwana in the Late Carboniferous and Early Permian. Traces of this event, in the form of scratched rock surfaces and piles of glacial rock debris, are widespread around the world (see Chapter 10).

But afterward there was no major ice age for 250 m.y., until the present one began. The Cretaceous was particularly hot. Paleoclimatic evidence suggests that the Earth's surface cooled over the past 50 m.y., until finally the planet dropped into the present ice age (see Chapter 23).

Reconstructing past changes in climate such as these provides us with a challenge. We cannot simply stick a thermometer into the ocean to tell what the temperature was millions of years ago. We must use something that changes with temperature (or $CO_2$ concentrations or the strength of ocean currents) to understand the

*Cowen's History of Life*, Sixth Edition. Edited by Michael J. Benton.
© 2020 John Wiley & Sons Ltd. Published 2020 by John Wiley & Sons Ltd.

ocean of the past. We call these things a **proxy**, meaning a method of calculating one property by measuring another. If you have looked outside to see if people are wearing rainjackets or carrying umbrellas, then you have done this. You are not getting wet, but you still figure out it is raining.

Luckily for Earth scientists, there are many proxy systems to understand ancient climates: trees grow bands thicker or thinner depending how warm/cold and wet/dry a season was, glacial ice cores have bubbles of past atmosphere trapped in the ice, pollen in lakes can tell us what kind of plants were growing around the lake, finding charcoal can tell us how frequent forest fires were in prehistory, and so on. One of the most frequently used proxies for ancient ocean conditions is in the shells of foraminifers (Figure 17.1). These single-celled protists make their calcium carbonate ($CaCO_3$) shells (micropaleontologists refer to them as "**tests**") in seawater.

## Scientific Ocean Drilling and Proxies

The *R/V JOIDES Resolution* is possibly the best tool to aid in this endeavor today; this is a nearly 165 yards long, 200 ft tall ship that acts as a floating mix of hotel, science lab, and industrial complex (Figure 17.2). The *R/V JOIDES Resolution* is run by Texas A&M University and the International Ocean Discovery Program (IODP). The ship houses 125 people, roughly 60 of them scientists and technical staff, for two-month expeditions where the scientific participants work 12-hour shifts, seven days a week. The most important part of the ship is the long coring tool (like a big drill with a hole in the middle) which extends from roughly the middle of the ship to the bottom of the ocean. That coring tool has sometimes reached almost 5 km below the sea floor. It brings up tubes of sediments called cores that are often full of tiny fossils.

Scientific ocean drilling has a more than 70-year history and has produced a large amount of knowledge. Currently, there are 23 countries participating in the IODP, with hundreds of scientists involved around the globe. The goals of the IODP have expanded to understand the biosphere below the ocean floor, ocean hazards, plate tectonics, the history of the oceans, and past changes in climate.

Figure 17.1 These tiny fossils record the environments of past oceans in their form, the chemistry of their shell, and their abundance. *Source:* Photo by Daniela Schmidt.

Figure 17.2 The *R/V JOIDES Resolution* is one of the main ships operated by the International Ocean Discovery Program (IODP). The IODP is the Earth sciences' counterpart to the NASA Apollo program, bringing samples to science inaccessible before the program's inception. *Source:* Photo by Arito Sakaguchi and IODP/TAMU (Wikimedia).

We measure the isotopes and trace elements of carbonate in foraminifers to reconstruct past climates quantitatively. An **isotope** is an atom with a certain number of neutrons. For example, $O^{16}$ has eight neutrons (and eight protons: $8 + 8 = 16$) while $O^{18}$ has 10 neutrons. That difference in the number of neutrons means that $O^{16}$ and $O^{18}$ have slightly different properties.

Paleontologists working for oil companies have been studying foraminifera for decades, not for their paleoclimatic importance but because they evolve rapidly and can be exquisitely accurate indicators of the age of the sediment they are preserved in. In drilling for oil, paleontologists can sample rock fragments broken off by the drill bit and brought up the drill pipe, to allow them to know when they hit the target layer that (they hope) will yield oil or gas. But rocks on land hardly ever record continuous layers of sediment like those being deposited on the ocean floor. It was not until the late 1960s that we had the equipment and skills to take long samples drilled from the ocean floor and to interpret the foram tests as temperature recorders.

The $O^{16}$ isotope is lighter, so moves very slightly faster than the $O^{18}$ isotope. In warmer water, $O^{16}$ is taken up slightly faster than $O^{18}$ into the carbonate of the foram. So, if we can measure the ratio of the two isotopes, we have a thermometer for the seawater at the moment the test was crystallized. But if there are large ice caps, the ice traps slightly more $O^{16}$ as it freezes, so the ocean water has slightly more $O^{18}$ during a cold period. This alters the isotopes in forams that form at cold times (Figure 17.3).

The realization that oxygen isotopes can record ancient ocean temperatures, and their link to ice ages, has a long history. During the 1800s, astronomers were making links between climate cycles and the nature of the Earth's orbit, and these ideas were crystallized by the brilliant Serb mathematician Milutin Milankovitch (1879–1958) when he was a prisoner of war in Austria during the First World War (see later in the chapter). Then, Harold Urey and colleagues at the University of Chicago measured oxygen isotopes in carbonate shells of Cretaceous cephalopods and noted in 1951 that the proportion of $O^{16}$ to $O^{18}$ in carbonate records indicated the ocean temperature when the animal was growing (the shells recorded seasonal variation over several years, too).

In 1957, Cesare Emiliani of the University of Miami showed how to reconstruct ocean temperatures through long spans of geological time by using foraminifera from cores in ocean floor ooze. The stage was now set for building a series of drilling ships specifically designed for collecting long drill cores from the ocean floor. The first of these ships, the *R/V Glomar Challenger*, began drilling the ocean floor in 1968.

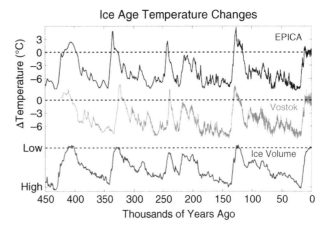

Figure 17.3 Analyzing the isotope composition, here of bubbles in ice cores, tells us about how regularly climate is changing on Earth and allows us to better understand the mechanisms. *Source:* Image by EPICA Community Members (Wikimedia).

## Changes in Marine Life over Time

As we learned in Chapter 15, life on land changed dramatically during the Cretaceous and similar revolutionary changes also happened in the ocean. The composition of marine plankton species living at the surface of the ocean also changed fundamentally (Figure 17.4). New groups of organisms suddenly appeared. Some used the light of the sun for energy (photosynthesis), such as coccolithophores which built our beautiful chalk cliffs (Figure 17.5). Some could use energy from the sun but also prey on other species, such as dinoflagellates. Lastly, some only lived by eating others, such as zooplankton with shells, the foraminifers, and radiolarians. The last addition to the modern plankton were diatoms, today the most abundant and diverse phytoplankton in the oceans. Having larger prey to eat affected the marine food chain and allowed larger predators to thrive.

The largest change to the chemistry of the ocean was that some of these groups make their shells out of carbonate, like mussels and corals. The main carbonate microfossils are the coccolithophores and foraminifers. Before that, carbonates were only grown by species living in shallow water shelves near the continents. Making carbonate in the open ocean and sinking these particles into the deep sea made the ocean a much better climate regulator. Carbonates are made of calcium, magnesium, and carbonate ions. Magnesium and calcium come from weathered rocks on land and are transported by rivers into the ocean, whereas carbonate ions come from atmospheric carbon dioxide in exchange with the ocean. Growing shells in the

Figure 17.4 Shells of marine plankton, including diatoms, radiolarians, foraminifers, ostracods, coccolithophores, and sponge spicules. *Source:* Photos by Hannes Grobe/AWI (Wikimedia).

Figure 17.5 The White Cliffs of Dover were built by billions and billions of microscopic carbonate plates of coccolithophores on the Cretaceous sea floor. *Source:* Photo by Immanuel Giel (Wikimedia).

surface ocean, away from the shelves, allowed the products of land-based rock weathering to be locked away in the deep sea, and once in place, the deep ocean carbonate storage acts as a great regulator for the ocean's chemistry. Some marine plankton, such as coccolithophores, are present in such huge abundance that their ocean surface "blooms" can be seen from space (Figure 17.6).

These changes in the "carbonate factory" caused changes in the Earth system. Before this regulator was in place, high $CO_2$ emissions, mainly from volcanoes such as the Siberian Traps at the end of the Permian and the Central Atlantic Magmatic Province at the end of the Triassic, were strongly associated with extinction (see Chapter 6). In contrast, the Deccan Traps eruption at the end of the Cretaceous had little impact on life because the deep-sea carbonates neutralized the $CO_2$ over thousands of years (see Chapter 16).

## The Cretaceous Hothouse

The hottest time in the last 100 million years was the middle Cretaceous. Vast amounts of carbon dioxide were produced by large undersea volcanic areas such as the Ontong Java Plateau and the Kerguelen Plateau. The carbon dioxide led to warmer air and oceans, resulting in temperatures much above our current ones. Deep ocean temperatures were around 15°C, compared to 2°C today, with temperatures in high latitudes like the southern tip of South America near 30°C. Such high polar temperatures are supported by findings of fossils, like vertebrates, which need warm temperatures to live.

The hot temperatures prevented the growth of ice, so sea level was very high during the Cretaceous, and large seaways flooded across the middle of Africa and the middle of North America, separating the East Coast from the West coast (Figure 17.7). At its largest, the Western Interior Seaway stretched from the Rockies east to the Appalachians, and even though it was not very deep, it still hindered the exchange of land animals across the continent. One prominent example of rocks deposited at this time is Monument Rocks in Kansas.

One problem with warmer water is that it cannot store as much gas, like oxygen, as cold water. Therefore, during the warm Cretaceous, less oxygen was available for life in the ocean. Warm climates may also have increased the weathering of rocks on land, so high levels of nutrients were transported by rivers to the ocean. High nutrient levels were great for plankton in the surface ocean and therefore they flourished, producing large amounts of organic matter and exporting it to the deep sea. This increase in organic matter within the surface ocean made more food available for other organisms, both those that prey on the plankton and the bacteria that decompose organic matter.

Consuming organic matter uses oxygen, something which was already scarce. As less and less oxygen was available to break down the organic matter, the organics accumulated in the deep sea, resulting in organic-rich mud. As a result, the deep oceans, and sometimes even parts of the shallow ocean, regularly lost most of their oxygen, indicated by characteristic black mudstones

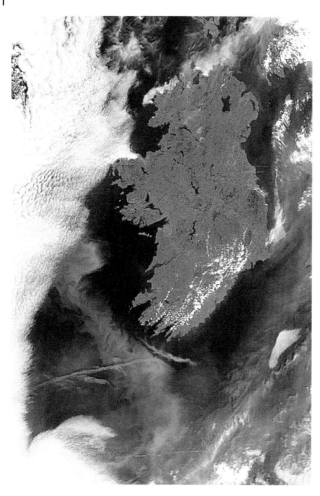

Figure 17.6 A coccolithophore bloom in green is seen to the south west of Ireland in this SeaWiFS satellite image. Coccolithophores can be seen with a powerful microscope (Figure 17.4) in rocks (Figure 17.5), and from space (this figure), but cannot be seen with the naked eye. *Source:* Image by the SeaWiFS Project, NASA/Goddard Space Flight Center (Wikimedia).

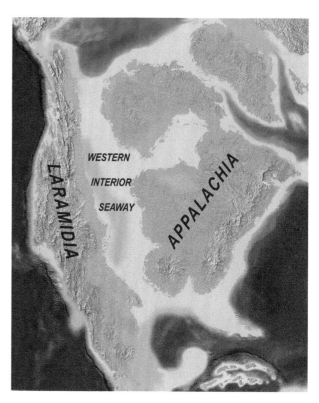

Figure 17.7 Map of North America with the Western Interior Seaway during the Campanian (Upper Cretaceous). *Source:* Image modified after Blakey (Wikimedia).

deposited around the world (Figure 17.8). We call those time intervals "ocean anoxic events," when low oxygen conditions affected species in the deep ocean, resulting in dramatic decreases in both diversity and abundance or sometimes even a total absence of deep-sea benthic foraminifers. But in the shallow waters of the ocean, exposed to sunlight, species flourished, diversified, and occupied a wide range of niches until the K–Pg mass extinction (see Chapter 16, Figure 16.9). The K–Pg event can be seen clearly in many ocean-floor boreholes, where a sudden, but short, change in sedimentation can be detected (Figure 17.9).

The effects of the K–Pg mass extinction on the marine plankton lasted for millions of years. Many of the surviving species lived in vast areas of the oceans, or near the coast. Most surviving species were tiny (Figure 17.10) and smaller organisms produce less carbonate in the surface ocean, so less carbonate reached the deep sea. There was a dramatic extinction among plankton with carbonate shells, such as foraminifers and coccolithophores, whereas species that could both photosynthesize and eat prey, like the dinoflagellates, and species with a skeleton made of opal, such as diatoms and radiolarians, were much less affected. Some evidence suggests that it took just a few hundred thousand years for enough food from the surface ocean to reach the deep ocean to support life there. It took millions of years, however, for carbonate to be produced in similar amounts as before the mass extinction. Carbonate production recovered roughly at the same time as micro- and nanofossil biodiversity recovered.

## Hyperthermals

Just when diversity started to reach values comparable to those before the K–Pg boundary, ocean ecosystems were exposed to a series of abrupt climate change events. The largest event occurred at the Paleocene–Eocene boundary and is called the Paleocene–Eocene Thermal Maximum, or PETM. The PETM was likely caused by a rapid emission of carbon but there is still debate about where exactly the carbon came from. Some scientists

342-U1407A-28H-1,
cm    0–53 cm

342-U1407B-24H-4,
cm    0–43 cm

342-U1407C-28H-1,
cm    72–93 cm

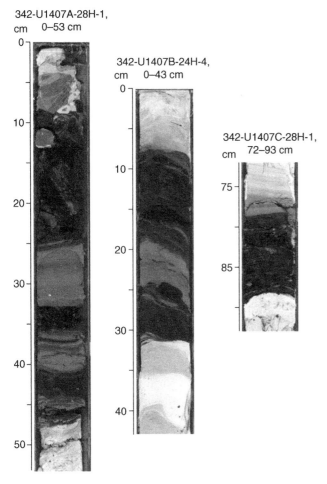

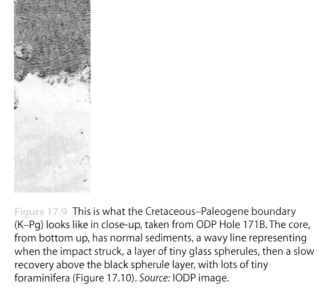

Figure 17.8 Boreholes from the ocean floor showing evidence for Ocean Anoxic Event 2, from off the coast of Newfoundland (Expedition 342). The lighter colored sediment is chalk, indicating normal sedimentation. The darker material in the middle is organic-rich sediment deposited during the Ocean Anoxic Event. *Source:* IODP image, from publications.iodp.org/proceedings/342/342bib.htm.

Figure 17.9 This is what the Cretaceous–Paleogene boundary (K–Pg) looks like in close-up, taken from ODP Hole 171B. The core, from bottom up, has normal sediments, a wavy line representing when the impact struck, a layer of tiny glass spherules, then a slow recovery above the black spherule layer, with lots of tiny foraminifera (Figure 17.10). *Source:* IODP image.

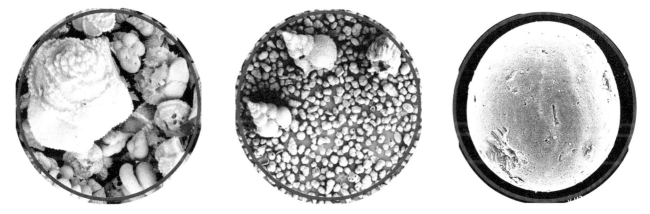

Figure 17.10 Collection of the tiny foraminifera from after the K–Pg extinction. *Source:* IODP image.

Figure 17.11 Core of deep-sea sediment during the PETM. The white is carbonate, while the red lacks carbonate. The sharp boundary at the base (top in this photo) of the red layer shows the change in the depth where carbonate was deposited. That sharp contact is the PETM, and the change in color is a record of ocean acidification in the deep sea. *Source:* Photo by Daniela Schmidt.

have suggested volcanoes, others deep-sea deposits of frozen methane or other sediments rich in organic matter heated by lava flows during the opening of the North Atlantic Ocean. The carbon moved into the atmosphere and ocean, possibly as $CO_2$ or maybe as methane rapidly converted to $CO_2$.

The input of carbon at the PETM may have occurred in pulses over a period of 10 000–40 000 years, with individual pulses occurring very rapidly geologically, in less than 1000 years. This input of carbon into our atmosphere led to an average global warming of 4–5 °C in the deep ocean and 5–8 °C in the surface ocean. Sediment cores suddenly turn from white to red (Figure 17.11). This change in color represents a change of the depth at which carbonate could be preserved. In some places, this depth rose by hundreds of meters, in others by nearly a mile, and this rise records the extent of ocean acidification. On land, the warmer climate also led to increased rain and erosion, which we see from changes in clay minerals and a larger amount of weathered materials in the ocean.

The amazing impacts of the climate change at the PETM can be found in both animals and plants on land and in the ocean. Deep-sea benthic foraminifera, which were not affected at all by the K–Pg mass extinction (see Chapter 16), lost between 30% and 50% of their species. The PETM changed the composition of plankton across the globe. Planktic foraminifers and coccolithophores in the surface of the ocean migrated to make use of the new warm areas of the ocean, while warm-water dinoflagellates were found suddenly at the North Pole. Terrestrial organisms migrated to high latitudes and changes found in fossil leaves indicate that the warming environment had an effect on plant life. Plant species were lost in some localities during the warm event and were replaced by new, often dry tropical species, but those local species returned when the temperatures dropped back down. Several new mammal lineages arose at this time, including the earliest horse (*Hyracotherium*) in North America (see Chapter 19).

The PETM was not unique, however (Figure 17.12). Layers in the deep ocean without carbonate, combined with proxy evidence of warming, have been identified in upper Paleocene and lower Eocene sediments. Scientists have not yet settled on a single explanation as to what was driving these events; some suggest that they happen at a specific orbital configuration, or that they are caused by greenhouse gas inputs, or changing ocean chemistry and circulation attributed to Milankovitch forcing (Box 17.1).

## Descent into the Icehouse

The mid-Cenozoic marked the beginning of a trend toward a cooler Earth, more like the one we know now. At the end of the Eocene (34 Ma), temperatures dropped and a large amount of ice started to develop at the South Pole.

There are two main ways in which climate can change dramatically on Earth: greenhouse gases such as $CO_2$ and methane and external factors that could generate major climate change are astronomical processes – changes in Earth's orbit or changes in solar radiation. Such changes occur but they are probably too small to generate major climate change by themselves. They cause fluctuations in climate, however (see Box 17.1).

It seems that we must look for mechanisms here on Earth for major climatic changes. The most important control on climate is the concentration of carbon dioxide ($CO_2$) and, to a lesser extent, methane ($CH_4$) in the atmosphere. Gases in the atmosphere, especially carbon dioxide and methane, are very effective in absorbing solar radiation (the greenhouse effect), and changes in the amounts of these gases could strengthen or weaken the effect of solar radiation or override it completely.

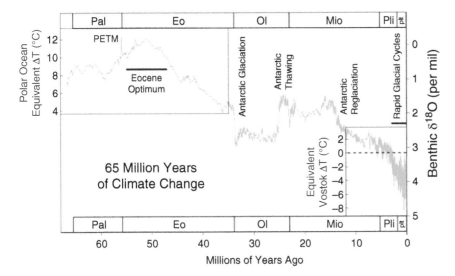

Figure 17.12 Climate change over the past 65 million years. For the PETM, note the rising temperatures during the Paleocene (Pal) that triggered the PETM at the beginning of the Eocene. An event as large, sudden, and short-lived as the PETM never happened again. *Source:* Diagram by Robert A. Rohde of the Global Warming Art Project (Wikimedia).

---

## Box 17.1  The Components of the Milankovitch Theory

**Tilt**. Increased or decreased tilting of the Earth's axis (Figure 17.13a), which varies between 22° and 24.5°, increases or decreases the effect of the seasons in a cycle of about 41 000 years (Figure 17.14).

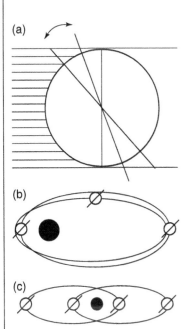

Figure 17.14 Earth's axis is tilted. As it orbits around the sun, solar radiation is concentrated on one hemisphere and then the other, giving Earth its seasons. As the seasons go by, we see the Sun gradually moving higher in the sky, then lower.

**Precession.** Earth's orbit is not a circle but an ellipse, with the sun at one focus (Figure 17.14). One pole is closer to the sun in its winter, while the other is closer in its summer. Thus, at any time, one pole has warm winters and cool summers, while the other pole has warm summers and cool winters. However, the slow rotation or precession of the Earth's orbit around the sun alternates the effect between the two poles in cycles of 19 000 or 23 000 years (Figure 17.13c).

**Eccentricity.** Earth's orbit varies so that it is more elliptical at some times than at others, strengthening or weakening the precession effect (Figure 17.13b). This variation in orbital eccentricity affects climate in cycles of about 100 000 years and about 400 000 years. Of course, when eccentricity is low (when the orbit is closer to being circular), the precession effect is much lessened.

These components add together to produce some of the climatic cycles that we see. Because of their effect on climate, and how climate controls the environments that produce rocks, we see cyclical rocks deposited in some areas (Figure 17.15). These changes happen on time scales much longer than modern climate change.

Figure 17.13 Some parameters of Earth's orbit can change over time, and as they do so they affect Earth's climate. (a) Greater or lesser tilting of Earth's axis causes stronger or weaker seasons. (b) The eccentricity of the Earth's elliptical orbit means that one pole almost always feels greater seasonal effects than the other; changes in eccentricity weaken or strengthen that effect. (c) The precession of the elliptical orbit alternates the eccentricity effect between the poles.

Figure 17.15 The influence of these cycles on climate results in the sometimes cyclical packages of rocks, here from Jurassic-aged Swiss rocks. Although the changes to our orbit are small, these alternating bands of lighter and darker sediment record the changing orbit of the Earth, so clearly they have a big effect! *Source:* Photo by Woudloper (Wikimedia).

One of the pieces of evidence we have for this is the change in carbon dioxide concentration during the glacial–interglacial cycles from the air bubbles trapped in the glacial ice. The amount of solar radiation that Earth retains can be altered by several processes. Some solar radiation is reflected back into space (the albedo effect), and a change in the amount of heat reflected would cool or warm the Earth. Albedo has to do with how much of the incoming light is absorbed by the surface. Think of bright snow reflecting light or dark soil absorbing the light.

The basic preconditions for climate change on Earth are simple. If there is a lot of snowfall in one area, it will build up rather than melt. Such a situation can occur if Earth's global geography is arranged in the right way. Changes in geography also act to vary the albedo of the Earth, the scale and activity of ocean currents, and the distribution of heat to different regions, all of which affect climate. In general, ice ages require large areas of land in high latitudes. The poles should be isolated from warm water. Finally, to lock the Earth into a long glacial period, there must be room for large continental ice sheets to spread out and provide high reflectivity over large regions.

Earth's major climate changes include distinct fluctuations. Dramatic advances and retreats of ice can occur even while Earth is overall in a cold phase. Huge areas of the northern continents are still covered by debris dropped by ice sheets during 40 or so glacial advances and retreats during the past 2 million years. This is corroborated by large temperature fluctuations which are recorded by microfossils in sea-floor sediments. World sea level has fluctuated up and down by more than 70 m (220 ft) as 5% of Earth's water has alternately been frozen into ice sheets and melted away. Such changes in sea level are recorded worldwide even in sediments far from the ice sheets. For example, islands and atolls in the Atlantic and Pacific Oceans have been repeatedly exposed and reflooded.

The astronomical theory of ice ages was proposed more than a century ago. It suggests that slight long-term variations in Earth's orbit around the sun and in the tilt of the Earth's axis make significant differences to climate (see Box 17.1). Milankovitch worked out the principles by hand during World War I, as we saw above. We now see evidence of the effects of "Milankovitch cycles" from Precambrian rocks and through the Pleistocene. In particular, Milankovitch cycles can trigger the advance and retreat of ice sheets, if conditions for an ice age are already present. Computer models of ice advances and retreats agree well with data from the geological record, especially the evidence gathered from foraminifera in deep-sea drill cores. The models and the data suggest that the present mild climate on Earth is very unusual for our geography. Interglacial periods with reduced northern ice sheets are very short in comparison with glacial periods with large ice sheets.

Although major ice growth began in the later part of the Eocene, ice grew to its largest extent in the Pliocene and Pleistocene.

## The Pliocene

During the Pliocene, the oceans changed to a configuration they never had before. Throughout most of Earth history, there were ocean gateways connecting the tropical oceans. The development of a land bridge in what is now Panama stopped tropical water exchanging between the Atlantic and Pacific for the first time in millions of years (Figure 17.16). Changing from an island chain, not dissimilar to the Antilles today, to a land bridge took millions of years, making it hard to determine the precise time when the water was completely cut off. At the same time, sea level dropped as ice started to grow in the northern hemisphere. The emergence of this land bridge allowed mammals, arthropods, reptiles, birds, and plants

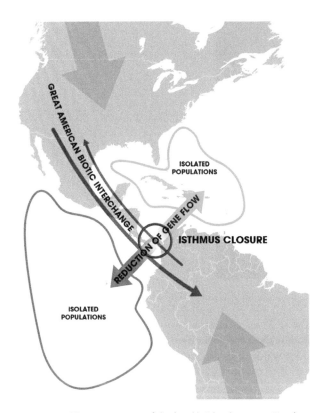

Figure 17.16 The emergence of the land bridge between North and South America stopped the exchange of marine species across the tropical Atlantic and Pacific and facilitated land species migrating between the continents. *Source:* Andrew Z. Colvin (Wikimedia).

to migrate more easily between North and South America (see Chapter 20).

The closure of the isthmus of Panama affected the oceans too, cutting the free movement of marine organisms between east and west. Species of fishes and mollusks in the Pacific and Caribbean have since diverged because they can no longer interbreed. Also, of course, the closure cut the great east-to-west equatorial ocean current that used to flow directly from the Atlantic to the Pacific. The currents flowing from west Africa now enter the Caribbean and are diverted north along the coasts of Mexico and the southern United States, and they then hug the east coast and shoot across the Atlantic as the warm Gulf Stream, leaving eastern Canada cold and warming western Europe.

Scientists have suggested that this change in ocean circulation led to the first significant ice build-up on the northern hemisphere (Figure 17.17), but there is a competing hypothesis. Change in $CO_2$ in the atmosphere has been suggested as an alternative, as atmospheric circulation changed due to the rise of the Rocky Mountains as well as tropical circulation similar to a permanent El Niño. How do we test these competing ideas? Combining both proxy data and climate models together shows that a change in El Niño has no impact on Greenland ice. The formation of the isthmus and the tectonic uplift of the Rocky Mountains only increase the ice a bit. To reach an amount of ice similar to what we have today requires that $CO_2$ must drop below modern levels. Proxy data from marine organisms support this. They show a drop in $CO_2$ at the same time as proxy evidence (e.g., drop stones from icebergs) that ice was increasing, thus leading the way to the large volumes of ice of today.

## The Ice Ages

The current climate we live in is exceptional for the last 100 Ma. The last 2.5 Ma have been dominated by glacial–interglacial cycles known popularly as the "ice ages." The interglacial we happen to be living in now is the warm stage of those cycles. Great ice sheets expanded and covered much of the northern continents, then retreated again. The effects of the ice ages have been most marked in the northern hemisphere, where huge new ice sheets grew, mostly around the north Atlantic Ocean (Figure 17.18).

Once ice sheets built up, they altered climatic patterns in the north Pacific and north Atlantic Oceans. Sea surfaces in the north Atlantic froze as far south as New York and Spain. Warm Gulf Stream waters were diverted eastward toward North Africa, instead of bringing a warm, moist climate to western Europe as far north as Scandinavia.

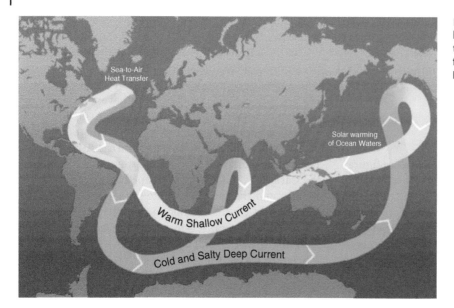

Figure 17.17 Simplified ocean circulation highlighting the impact of the closure of the circumtropical seaways. *Source:* Image from the US Global Change Research Program (Wikimedia).

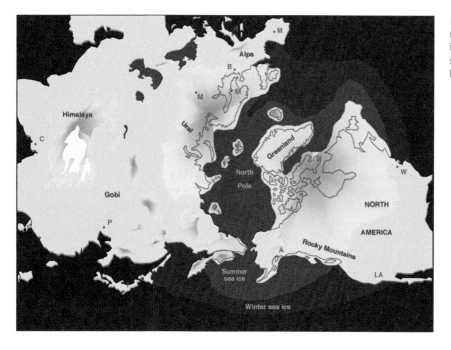

Figure 17.18 Great ice sheets around the northern hemisphere were a major influence on climates during the successive ice ages. *Source:* Map by Hannes Grobe (Wikimedia).

At its maximum, somewhere around 21 000 BP, Canadian ice advanced as far south as New York, St Louis, and Oregon (Figure 17.18). Ice scour removed great blocks of rock and transported them for hundreds of miles. The North American ice sheets diverted the jet stream and the main storm track southward. The western United States became much wetter than it is today, so that great freshwater lakes formed from increased rainfall and from meltwater along the front of the ice

sheet. River channels were blocked by ice to the north, and at the southern edges of the ice sheets, much of the melt water drained south to the Gulf of Mexico down a giant Mississippi River.

As the North American ice sheets began to melt and retreat, water draining from the melting ice sheet changed the seawater composition of the Gulf of Mexico as it poured southward down the Mississippi in enormous quantities, beginning about 14 000 BP, perhaps at

10 times its current flow. Finally, as the edge of the ice sheet retreated, the Great Lakes began to drain to the Atlantic instead, first down the Hudson River, then the St Lawrence, and finally north to the Mackenzie delta and the Arctic Ocean.

More subtle effects occurred in warmer latitudes. For example, increased rainfall in the Sahara during ice retreat formed great rivers flowing to the Nile from the central Sahara; they were inhabited by crocodiles and turtles, and rich savanna faunas lived along their banks.

## Climate Change

Climate has changed in the past, as we have seen, yet climate scientists are concerned about climate change happening today. Large changes in climate have been implicated in mass extinctions, from the glaciation at the end of the Ordovician (see Chapter 6) to the quick hot-to-nuclear winter transition at the demise of the dinosaurs (see Chapter 16).

We can look at these changes, and more recent ones, to understand links between climate change and how organisms on land and in the ocean respond. A key point is that global changes in climate have rarely, if ever, been as fast as they are right now. As we have seen, the PETM was the fastest carbon-driven climate change in the geological record, and yet it occurred over 10 times more slowly than the rate at which we are now putting carbon into the atmosphere by burning fossil fuels, making cement, and changing land use (Figure 17.19). The only time the Earth system was pushed faster was during the K–Pg mass extinction, which ended poorly for a large number of organisms.

Projecting into the future, climate change will lead to a warmer ocean. As we learned when we discussed the Cretaceous, a warmer ocean can hold less oxygen than a cool ocean, leading to the expansion of oxygen-poor areas in deep water (Figure 17.20). At the same time, the uptake of $CO_2$ by the ocean will lower its pH – in other words, make it more acid. These acidified waters will sink into the deep ocean and result in a shallowing of the depth at which $CaCO_3$ structures, such as shells, dissolve.

Climate change has become a much-debated topic, understandably interesting all citizens. If the scientific evidence is trustworthy, there will be rapid and profound changes in Earth's climate within tens and hundreds of years (Figure 17.21). On the other hand, if the scientific evidence is wrong, then perhaps, as some argue, nothing much will change. The public debate about climate change is a good thing – science should inform decision making by politicians – but some of the debate is poorly informed and provides a test of how well politicians and others are able to handle evidence, a theme sometimes called "data literacy" (Box 17.2).

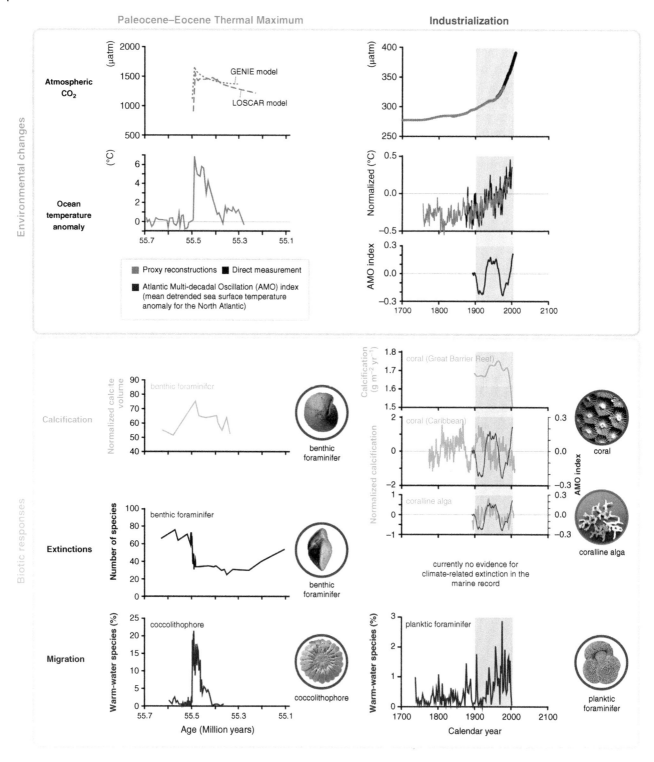

Figure 17.19 Environmental changes (*top*) and associated biological responses (*bottom*) for the PETM (*left*) and the industrial era (*right*). The PETM represents the best geological analog for the future ocean because of its rapid environmental change. Episodes of largest environmental change are indicated with darker bands. Note the different time scale between the two columns. Both time intervals are characterized by rapid warming both on land and in the ocean and increases in $CO_2$. Note the species-specific calcification responses to climate change with decreases, increases, and high variability. While there was extinction during the PETM, there is currently no evidence for climate-related extinction in the marine record. Warming led to migration of warm-water species into previous cold-water habitats. Pictures are examples of organisms highlighting the processes in each panel and are not to scale. Full reference for the sources in Pörtner et al. (2014). *Source:* IPCC Image.

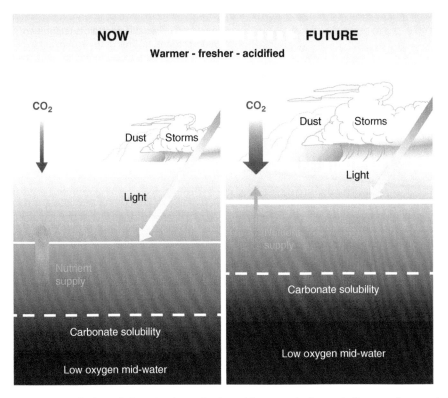

**NOW** **FUTURE**

Warmer - fresher - acidified

Figure 17.20 Projected alteration (magnitude and frequency) of oceanic fluxes and atmospheric events due to a changing climate in the coming decades. Ocean properties will be altered from the sunlit surface layer down to the mid-waters. In the surface ocean, the depth of the mixed layer (solid horizontal line) will be shallow, resulting in higher mean light levels. Increased density stratification (i.e., a strengthening seawater density gradient shown by the increasing thickness of the solid horizontal line) will reduce the vertical supply of nutrients for photosynthesizing organisms living in the mixed layer. Anthropogenic $CO_2$ will acidify, that is, lower the pH of, the surface ocean (note this happens in a pH range higher than 7 such that oceans will remain alkaline but less so due to acidification). The penetration of acidified waters to depth will result in a shallower depth (dashed horizontal line) at which $CaCO_3$ structures, such as shells, dissolve. At depth, the location of low-$O_2$ waters will progressively become shallower. In addition, changes in storm activity and dust deposition will influence ocean physics and chemistry, with consequent effects on ocean biota and hence ecosystems. Full reference for the sources in Pörtner et al. (2014). *Source:* IPCC image.

---

Box 17.2  Data Literacy

It is easy to find "information" online about any scientific issue, climate change included. It is not always easy, however, to tell whether the information is based on a careful and critical analysis of the data (the "scientific" approach) or whether it is presented in a selective manner to justify a particular case (the "nonscientific" approach). When evaluating evidence about climate change (or any topic), it is useful to keep in mind a few criteria of good data versus bad data.

**Bad Axes.** See Figure 17.21a. The axes do not go all the way to the present and they stop at 1997. We have records of temperature far longer than that. Context is important.

**Outliers.** Outliers need to be identified as such or put into a broader context. El Niño is an "event" that makes the global temperature much warmer. See how on Figure 17.21b the 1998 temperature is really high, even compared to everything around it for about a decade. Losing the context

of the 1970s, 1980s, and 1990s makes it look as if nothing is happening.

**Units.** Scientists are careful to note the units (feet, meters, degrees, etc.) on everything.

**Relevant credentials**. Look for whether the authors have relevant experience in the discussed field, best with a terminal degree (i.e. PhD).

**Reliable and Reproducible.** Did you find this on a reputable website? Is it from a trusted source? Even just having a copyright means that there was an extra amount of effort put into generating and protecting that piece of information, making it more trustworthy. Have you found similar data or claims somewhere else? Most importantly, scientists always cite their sources so check each piece of information back to the source, and the source should be a peer-reviewed journal or report, not another website written by a friend.

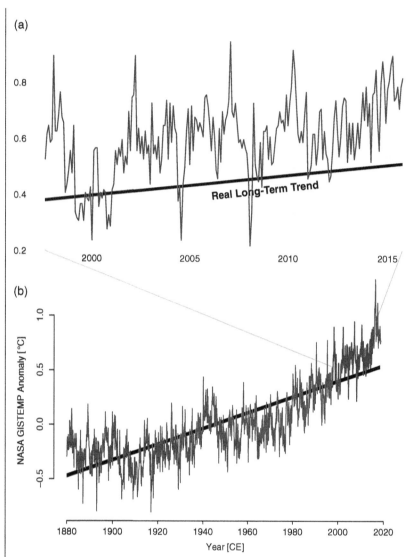

(a)

(b)

Figure 17.21 Two different plots telling two different stories. (a) This graph is similar to some that are shared around the internet. It is flawed in many ways and appears to be purposefully obscuring important aspects of the data. (b) This graph is from a scientific report. It has labeled axes, a larger context, and appears more professionally generated with attribution of the data.

**Motives.** Consider the motives of the person posting the information, as well as the source. What are they motivated by? Some highly professional websites on scientific topics such as evolution or climate change are funded by industries or religious groups that might benefit by persuading people about a particular point of view. At one time, for example, cigarette manufacturers funded publications and conferences that claimed cigarettes were not bad for your health, despite all the medical evidence. Humans tend to search for information that agrees with their previous ideas or strongly held personal beliefs, discounting information that does not conform. Read more in Oreskes and Conway (2010).

**Agreement.** Conflicts in science are interesting and fun to learn about, but scientists tend to agree on the important questions. For example, more than 97% of climate scientists agree that the world is warming and we are responsible. Climate science has a single mechanism and hypothesis to explain all our observations: the burning of fossil fuels adds $CO_2$ gas to our atmosphere, which has warmed the planet due to the greenhouse effect. The "skeptics" have many, many, *many* explanations. Their inability to agree on a mechanism is a strong argument for the conclusions of climate science.

## References

Oreskes, N. and Conway, E.M. (2010). *Merchants of Doubt: How a Handful of Scientists Obscured the Truth on Issues from Tobacco Smoke to Global Warming*. New York: Bloomsbury.

Pörtner, H.O., Karl, D.M., Boyd, P.W. et al. (2014). Ocean systems. In: *Climate Change 2014: Impacts, Adaptation, and Vulnerability. Part A: Global and Sectoral Aspects. Contribution of Working Group II to the Fifth Assessment Report of the Intergovernmental Panel on Climate Change* (ed. C.B. Field, V. Barros, D.J. Dokken, et al.), 411–484. Cambridge: Cambridge University Press.

## Further Reading

McInerney, F.A. and Wing, S.L. (2011). The Paleocene-Eocene Thermal Maximum: a perturbation of carbon cycle, climate and biosphere with implications for the future. *Annual Review of Earth and Planetary Sciences* 39: 489–516.

Ridgwell, A. (2005). A mid-Mesozoic revolution in the regulation of ocean chemistry. *Marine Geology* 217: 339–357.

Ruddiman, W.F. (2007). *Earth's Climate: Past and Future*, 2e. New York: W. H. Freeman.

Thomas, E., Brinkhuis, H., Huber, M. et al. (2006). An ocean view of the Early Cenozoic Greenhouse World. *Oceanography* 19 (4): 94–103.

Zachos, J., Pagani, M., Sloan, L. et al. (2001). Trends, rhythms, and aberrations in global climate 65 Ma to present. *Science* 292: 686–693.

## Questions for Thought, Study, and Discussion

1 The oxygen isotope curve (Figure 17.12) was constructed using data from multiple ocean drilling sites and is used to interpret Antarctic ice volume. Explain how many major steps in ice growth you can identify.

2 The PETM occurred 56 million years ago. How does this ancient example of global change compare with present conditions of increasing concentrations of atmospheric carbon dioxide and what can we learn from the geological record?

3 Sixty-five million years ago, a 10 km meteorite crashed into what is now Mexico's Yucatán Peninsula, creating a 177 km crater and causing mass extinctions across the globe. Using information from recent IODP drill expeditions, research why the location of the impact was important for the consequences.

4 Using information from the literature, consider the impacts that climate change in the next century may have on marine ecosystems. Link this to the environmental changes which are projected based on the current $CO_2$ emissions.

18

The Origin of Mammals

---

### In This Chapter

Mammals evolved from earlier synapsids at the end of the Triassic, and by the mid-Jurassic were the only survivors, already diversifying into several distinct lineages. By the Early Cretaceous, the modern lineages of mammals were established: monotremes, marsupials, and placentals. Much of mammal evolution is to do with the change to very small size in the transition from therapsid to mammal. We begin by summarizing the features of the earliest mammals, both completely new features and ones inherited from earlier synapsids, and then we consider how later mammals acquired their more derived features from this original mammalian condition. Mesozoic mammals were diverse in their ecologies and probably numerous as individuals, but remained small until after the extinction of the dinosaurs (the few "giants" among them were no bigger than a raccoon). Key questions are how these early mammals interacted with dinosaurs (if they did), and why they could not replace dinosaurs as key larger land animals in the Mesozoic.

---

## The Derived Features of Mammals

Living reptiles and birds (diapsids) and living mammals (synapsids) are very different, and these lineages diverged from each other some time in the mid-Carboniferous. But, as amniotes, the two lineages share a number of derived features absent from other vertebrates, in addition to the amniote egg described in Chapter 10: ventilation of the lungs via the ribs; an ankle joint with a hinge-like articulation; a keratinous (horny) outer layer of the skin that can form scales, hair, nails and horns; and more complex hearts and lungs than seen in fish and amphibians (although the complication of these organs may have evolved convergently in the two groups). There are many diapsids today, including turtles, lizards, snakes, crocodiles, and birds, and the earliest diapsids were probably somewhat like lizards in their anatomy and ecology. But only mammals survive from the synapsid lineage, which makes it challenging to try to understand the probable biology of the earlier forms (in the Permian and Triassic).

Although we usually consider mammals to be "more derived" than reptiles, in a few respects mammals are more like basal tetrapods: they have a glandular skin (lacking the hard type of keratin that is seen in reptile scales or bird feathers), and they excrete liquid urine that they can use for social functions such as scent marking (in contrast, most diapsids excrete a semi-solid paste of uric acid).

Mammals are endothermic (warm-blooded), although endothermy evolved convergently in birds (perhaps inherited from their dinosaurian ancestors) and they have hair, rather than scales or feathers. Mammals suckle their young, via milk-producing mammary glands in the females (a bonus of having a glandular skin). Mammals have muscles of facial expression, which include the muscles around the mouth that enable suckling.

The mammalian lower jaw consists of a single bone on either side, the **dentary**. The dentary is the tooth-bearing bone in all bony vertebrates, but in nonmammals there are several other bones behind the dentary, including the articular that forms the jaw joint with the quadrate in the skull. Mammals have a new jaw joint, between the dentary and a bone in the skull called the **squamosal** (Figure 18.1).

---

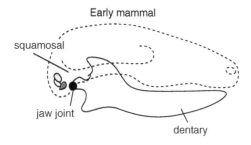

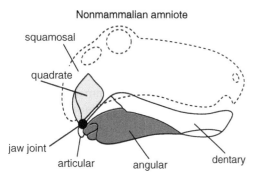

Figure 18.1 Jaw joints in (a) an early mammal and (b) an earlier synapsid (a pelycosaur). The ancestral condition in amniotes is to have several bones making up the lower jaw. The color coding shows the equivalent bones between the mammal and the pelycosaur. In mammals, the lower jaw is entirely formed by the dentary, and the quadrate (*blue*) and articular (*yellow*) are miniaturized into the incus and malleus respectively and included in the middle ear. The malleus in the mammalian ear has a small contribution from the angular bone (*pink*). The stapes is not shown. (See Figure 18.8 for the position of the three ear ossicles in a mammal.) *Source:* Artwork by Philcha (Wikimedia). (= old figure 15.1).

The bones that formed the original jaw joint are now contained within the middle ear cavity, joining the stapes (which forms the single middle ear bone seen in other tetrapods), and are used for hearing (see Figure 18.8). This "three-boned middle ear" (one on each side of the head, of course) is a distinctive and defining mammalian feature, giving mammals particularly acute hearing at high frequency so they can hear squeaks and insect buzzing.

While most vertebrates have simple teeth that all look the same, most mammals have a "differentiated" dentition, with different types of teeth (incisors, canines, premolars, and molars). In reptiles, teeth are replaced continuously during life, but in mammals, teeth are at best replaced only twice, as in ourselves, with our "milk" teeth and "permanent" teeth. Unlike most other vertebrates, mammals chew (masticate) their food and swallow a discrete chewed-up portion (or bolus). Muscular cheeks (part of the facial expression complex) help to contain the food during chewing.

The mammal brain is enlarged and specialized over the general amniote condition (birds have also enlarged their brain but in a different fashion). The forebrain contains a completely new structure, the **neocortex**, and there is improved sensitivity to hearing, smell, and touch.

It is impossible to imagine all these differences arising overnight; in fact, we see many of them evolving gradually within the therapsids that were the ancestors of mammals, as we will describe below (and as shown in Figure 18.2). The fossil record of the transition is richest in jaws and teeth which tell us not only about diet and food processing but also about the evolution of hearing.

## Cynodonts and Defining the First Mammals

At the beginning of the Triassic, the world was in recovery from the extinction at the end of the Permian (see Chapter 6), and throughout the Triassic the diapsids took over as the predominant large land vertebrates (see Chapter 12). The surviving therapsids were the dicynodonts (described in Chapter 10) and the cynodonts, relative newcomers who made their first appearance at the very end of the Permian. The dicynodonts were all herbivores and ranged up to the size of a cow. Cynodonts had a diversity of diets, some herbivorous, although most were carnivores of some sort and were generally smaller in size, with the largest being about the size of a medium-sized dog like a Labrador.

The Early Triassic radiation of cynodonts was of medium-sized generalist forms, such as *Thrinaxodon* (Figure 18.3). But by the Middle Triassic, cynodonts had split into two main lineages (Figure 18.2): the larger-sized Cynognathia (which included both predacious carnivores and herbivores, e.g., *Cynognathus*; Figure 18.4) and the smaller-sized Probainognathia (mostly generalized carnivores or insectivores, e.g., *Bonacynodon*; Figure 18.5). The probainognathians are known primarily from South America and they include animals very close to the ancestry of mammals. The most derived forms from the Late Triassic, such as *Brazilitherium*, were very similar to the earliest mammals and almost as small. The Cynognathia were extinct by the end of the Triassic, but a couple of lineages of probainognathians survived into the Jurassic along with the first mammals and may even have persisted until the Early Cretaceous.

There's a problem about defining the evolutionary transition from therapsid to mammal. Therapsids are a clade, cynodonts are a clade within therapsids, and mammals are a clade within cynodonts (Figure 18.2). But how does one define when cynodonts became mammals?

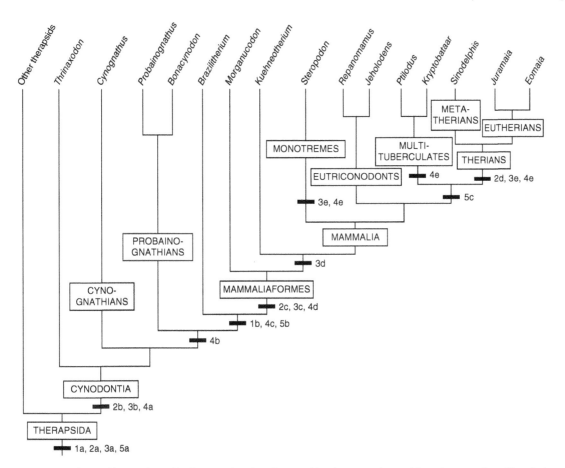

Figure 18.2  Synapsids mentioned in the text, showing the transition between therapsids and mammals and key features in the transition. (i) Hair: **1a** = at least some hair, especially whiskers; **1b** = probably fully furry. (ii) Posture and locomotion: **2a** = semi-erect posture; **2b** = loss of lumbar ribs (indicating presence of diaphragm); **2c** = more fully erect posture, backbone can flex up and down; **2d** = fully erect posture, mobile shoulder and bounding gait. (iii) Dentition: **3a** = teeth differentiated into incisors, canines, and postcanines; **3b** = postcanine teeth with three cusps; **3c** = postcanines divided into molars and premolars, maximum of two sets of teeth, precise occlusion; **3d** = triangular-shaped molars; **3e** = tribosphenic molars (not identical between therians and monotreme ancestors, lost in modern monotremes). (iv) Jaws and ears: **4a** = enlarged dentary, presence of masseter muscle; **4b** = dentary makes up most of lower jaw; **4c** = incipient mammalian (dentary/sqamosal) jaw joint alongside original jaw joint (quadrate + articular); **4d** = fully mammalian jaw joint (quadrate + articular) miniaturized and used for hearing, lower jaw can move sideways, chewing on one side of the mouth at a time; **4e** = quadrate + articular detached from jaw, incorporated into middle ear. (v) Reproduction: **5a** = at least some parental care (also evidenced in earlier synapsids); **5b** = evidence of lactation and suckling; **5c** = viviparity (may have evolved earlier, or later convergently between multiberculates and therians).

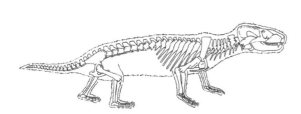

Figure 18.3  Body and skeletal outline of the early cynodont *Thrinaxodon* (about the size of a fox), known from the Early Triassic of South Africa and Antarctica. Note the expanded ribs, restricted to the thoracic region of the spine (so suggestive of the presence of a diaphragm). *Source:* Based on skeletal reconstruction by Farish Jenkins.

Figure 18.4  The carnivorous cynognathid cynodont *Cynognathus crateronotus* (about the size of a Labrador dog), known from the Middle Triassic of South America, Antarctica, and Argentina. (At this time, these three continents were connected together in the supercontinent Gondwana: Antarctica was ice-free and the environment was temperate.) *Source:* Artwork by Nobu Tamura (Wikimedia).

**Figure 18.5** The probainognathid cynodont *Bonacynodon shultzi* (about the size of an opossum) from the early Late Triassic of Brazil. *Bonacynodon* was a medium-sized probainognathid; others were as small as mice or as big as a corgi dog. *Source:* Artwork by Jorge Blanco in Martinelli et al. (2016) (Wikimedia).

Most paleontologists use a *crown-group* definition of Mammalia: the first member of Mammalia would be the latest common ancestor of all living mammals. This definition would exclude a lot of Triassic and Jurassic creatures that had many characters traditionally considered to be "mammalian," such as single lower jaw bone, jaw joint between dentary and squamosal, expanded brain, and so on. Nowadays, the term "Mammaliaformes" includes these critters plus the "crown-group" mammals (Figure 18.2), but we will use the informal term of "mammal" to refer to all of them.

## Teeth and Tooth Replacement

All therapsids differed from cynodonts in having some degree of tooth differentiation, possessing incisors (for food intake), canines (for food capture in carnivores), and postcanines (for food processing). In cynodonts, these postcanine teeth were more complex, sporting three cusps rather than a single one: this implies more complex food processing than in other therapsids. Reptiles replace their teeth often during life and although the process has some systematic pattern to it, any adult reptile has a mixture of larger, older teeth and smaller, newer teeth along its upper and lower jaws. This means that top and bottom teeth cannot be relied upon to meet precisely against one another, so that tooth functions are comparatively crude.

As we will discuss below, cynodonts had a more sophisticated set-up of jaw muscles than other therapsids, and probably better control of jaw movements with some degree of tooth occlusion (that is, teeth meeting each other, rather than simply chomping on food), and the rate of tooth replacement was reduced. However, it was not until the earliest mammals (or possibly their closest relatives among the cynodonts) that the postcanine teeth show what we would term "precise occlusion."

Mammals maintain this precise occlusion by having essentially a single set of teeth (at least as adults). Marsupials have only a single set of teeth, but in placentals, the young have an initial set of "milk teeth," consisting of incisors, canines, and premolars, later replaced by the adult dentition. There is only ever a single generation of molars, which fully erupt into the jaw somewhat later in life. (Our last molars are called our "wisdom teeth" because they do not erupt until our late teens or early twenties, by which time we are presumed to have acquired a degree of wisdom.)

Molars are usually complex in form and fit together in a precise interdigitating manner, allowing for efficient food processing. The molars of all mammals, including the earliest ones, are double-rooted (cynodont postcanines were single-rooted) and have a more durable ("prismatic") type of enamel than seen in cynodonts. These features correlate with the single generation of molars, that have to last a lifetime.

## Skulls and Jaws

The more derived therapsids, including cynodonts, developed a secondary palate, the bony division between the mouth and nasal passages that allows chewing and breathing at the same time.

Most therapsids retained the same set-up of jaw-closing muscles seen in other tetrapods today; the primary muscle was the adductor mandibularis, running from the top of the skull to the lower jaw. In cynodonts, this muscle became divided into two: the temporalis (in the position of the original muscle) and the **masseter**, which runs from the skull under the cheekbone to the outer side of the lower jaw (Figure 18.6). (Put your fingers on your cheek, clench your teeth and you will feel the masseter at work. Put your fingers just above and behind your eyes and you will feel the action of the temporalis.) The masseter probably originally served to hold the lower jaw in a cradle of muscles, and stresses acting on the jaw joint during chewing were much reduced. This was important, as the complex cheek teeth of cynodonts show they processed their food for longer, perhaps to fuel an increasingly higher metabolism. In addition, as discussed below, the bones forming the jaw joint were also involved in hearing. In mammals (but not in cynodonts, as we can

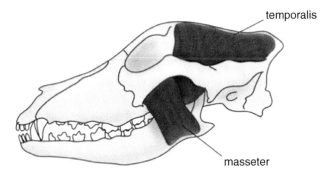

Figure 18.6 The masseter muscle runs from the cheek bone to the outside of the lower jaw, while the temporalis runs to the top of the lower jaw from the back of the skull in this dog.

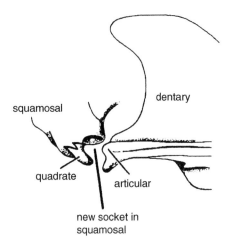

Figure 18.7 The structure of the back of the jaw in the derived cynodont *Probainognathus*. Only a small transition would be needed to change the hinge from the articular and quadrate, as normal in nonmammalian synapsids, to the dentary and squamosal, as in mammals. *Source:* After Romer with permission from the Museum of Comparative Zoology, Harvard University.

tell from wear patterns on the teeth), the masseter enabled the lower jaw to be moved sideways, and the earliest mammals chewed their food on one side of the jaw at a time, just as we do.

In the general tetrapod condition, the lower jaw is made of several bones; the most anterior of these bones, the dentary, is the one that bears the teeth (Figure 18.1). But in more derived therapsids, and especially in cynodonts, the dentary became larger and the other bones became smaller and were crowded back toward the jaw joint. Stresses on the jaw joint itself were reduced as the masseter evolved, and the bones forming the jaw joint on each side became specialized for transmitting vibrations to the stapes in the middle ear. Eventually, the dentary became the only bone in the lower jaw, a new jaw joint was formed between the dentary and the squamosal, and the old jaw joint bones (the articular from the lower jaw and the quadrate from the skull) joined the stapes in the middle ear.

It may seem incredible and unlikely that in evolution a jaw joint could change from one set of bones to another. But not only do we see this gradual change occurring in the fossil record, the same change happens in every mammal today during development. Mammals start off their embryonic life with developing upper and lower jaws that end in a quadrate and articular bone, respectively, and these bones later detach from the jaws and move into the developing middle ear. In the cynodont *Probainognathus* from the Middle Triassic of South America, the functional jaw joint is still between the quadrate and the articular but the dentary and squamosal are almost in contact (Figure 18.7). Still more derived cynodonts had a double jaw articulation. In the earliest mammals, the dentary–squamosal jaw joint became the primary articulation with a true ball-and-socket anatomy, although they still retained the quadrate–articular joint.

## Hearing

As we saw in the previous section, the evolution of the mammalian jaw is intimately associated with the evolution of the distinctive mammalian three-boned middle ear (Figure 18.8). Early synapsids had a rather large stapes and lacked the type of enclosed middle ear with an eardrum seen today in both mammals and diapsids (where it evolved convergently). They probably perceived only low-frequency vibrations. However, the quadrate and articular jaw joint bones were in contact with the stapes and were probably used for sensing vibrations in the earliest synapsids, perhaps with some type of eardrum positioned in the angular bone of the lower jaw. As the more derived therapsids began to use their jaws for more frequent food processing, a conflict arose between chewing and hearing, and reducing the size of these vibration-sensing bones would have made hearing more acute.

Clearly, air-borne sound became increasingly important to late cynodonts and early mammals. Perhaps they hunted insects at least partly by sound, or parental care became more important and they needed to hear the squeaks of their offspring. The middle ear bones were still linked to the jaw in earliest mammals, just as they are in mammalian embryos, although the bones were now very small and probably used primarily for hearing. But in later mammals (and independently in different lineages – see Figure 18.2), these bones became detached from the jaw and enclosed in a middle ear cavity, which probably resulted in improved hearing.

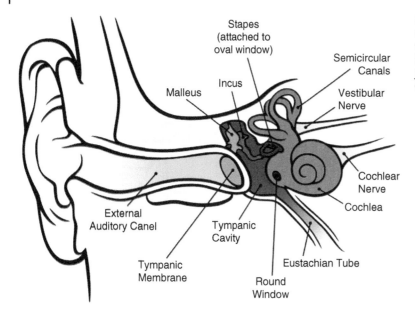

**Figure 18.8** Anatomy of the ear in a living mammal (human). Incoming sound is transmitted via the eardrum and then through three small bones (ossicles), two of which once formed the jaw joint in the ancestral synapsids. *Source:* Artwork by Chittka L, Brockmann (Wikimedia).

These middle ear changes were accompanied by changes in the inner ear that indicate greater hearing acuity, including lengthening of the cochlea that contains the cells that detect sound (Figure 18.8). However, only therian mammals (marsupials and placentals) have the highly coiled cochlea that we see in ourselves, and only therians have distinctive external ears. External ears are absent in monotremes and may have been absent in many or most mammals less derived than therians.

## Locomotion

All therapsids had a somewhat erect (i.e., less sprawling) posture, which would have helped them overcome "Carrier's constraint" and run and breathe at the same time (see Chapter 10). Cynodonts retained the general therapsid condition of a "semi-erect" posture with fore limbs that were somewhat more sprawling than the hind limbs, as did many of the early mammals, but cynodonts showed some more mammal-like characteristics, including a more sophisticated type of ankle joint. The backbone of earlier synapsids would have had some degree of lateral flexion, as seen in reptiles, although they were not as flexible as modern lizards. Cynodonts reduced or lost their posterior ribs, and their backbone was now divided into thoracic (rib-bearing) and lumbar (potentially more mobile) regions, as seen in mammals (Figure 18.3). However, the lumbar region of cynodonts was relatively stiff, and it was not until early mammals that up-and-down flexion of this part of the vertebral column became possible. The type of fully erect posture typical of many mammals today is probably a therian feature.

## Thermoregulation and Metabolic Level

Several lines of evidence suggest that therapsids, and cynodonts in particular, had at least some degree of endothermy by the Late Permian (see Chapter 10). The differentiated dentition and more sophisticated jaw muscles of cynodonts may reflect the need to process more food to fuel a higher metabolic rate. Mammals have a diaphragm as an important part of the breathing system. This sheet of muscle aids the ribs in expanding the chest cavity and filling the lungs with air. A diaphragm can work only when there are no ribs around the abdomen, and it may have evolved within early cynodonts, which lost their lumbar ribs (*Thrinaxodon*, Figure 18.3).

Whatever the metabolism of therapsids, they did not evolve the apparently athletic behavior of many dinosaurs but remained stocky with relatively short limbs. A few forms were perhaps arboreal but there were no fliers, specialized swimmers, or runners (either bipedal or quadrupedal). As therapsids evolved into mammals, they became smaller. A therapsid with endothermy would have found this difficult, because small bodies lose heat faster than large ones, even with hair/fur.

## Evolution of Hair

Hair rarely fossilizes directly, although some Mesozoic mammals have been preserved with impressions of a fur coat (see Figure 18.17). Some coprolites (fossilized feces) known from the Late Permian contain what appear to be

Figure 18.9 A CT scan of the skull of the early cynodont *Thrinaxodon*. The skull is about 7 cm long. The snout is pitted, which probably suggests it had whiskers (which are modified hair). *Source:* Courtesy of Professor Timothy Rowe and the Digimorph Project at the University of Texas at Austin.

hairs, which could only have come from the therapsids that were around at the time, so perhaps at least some hair was present in these beasts. *Thrinaxodon* had pits on the bones in its snout, which probably contained the roots of whiskers (Figure 18.9). That implies that it had hair, since whiskers are modified hairs.

Recent paleoneurological studies by Julien Benoit and colleagues, based on CT scans of skulls of derived probainognathian cynodonts, have thrown some interesting light on this issue. These scans showed the presence of a distinct canal for the nerve that supplies the snout, seen in mammals but not in earlier cynodonts, implying the presence of whiskers. Study of the brain showed certain derived features (e.g., an enlarged cerebellum, the motor area of the hindbrain) that in mammals today are under the control of the gene MS×2. This same gene also has a role in the development of both hair and mammary glands. Thus, the implication is that these neurological features are genetically and developmentally linked with the defining mammalian features of hair and mammary glands, providing evidence that these features of the soft anatomy evolved close to the origins of mammals.

## Mammalian Reproduction

A major difference between most modern mammals and most diapsids is that mammals are **viviparous** (i.e., give birth to live young), while diapsids usually lay eggs. Many species of lizards and snakes are viviparous but in these reptiles, the situation appears to be more fluid (among closely related species, one may be egg-laying and the other viviparous), whereas therian mammals are committed to viviparity. Of course, some mammals today lay eggs, the monotremes (echidnas and platypus), and this was likely the condition in most other synapsids, including the early mammals. But monotremes do not lay eggs in the fashion of most reptiles: the eggs are retained

within the mother for a long time; when laid, the eggs are small, thin-shelled, and contain little yolk; and the young are hatched at a very immature stage and are dependent on the mother's milk. In contrast, reptile hatchlings are usually able to fend for themselves right away, but of course, bird hatchlings depend on parental care (as, indeed, do crocodile hatchlings).

Why would mammals have evolved viviparity, and when did this trait appear in the mammalian lineage? Unfortunately, there is no simple answer to either of these questions. As discussed below, there is evidence that at least some multituberculates gave birth to live young, and viviparity may have arisen in the common ancestor of multituberculates and therians (as shown in Figure 18.2). Alternatively, viviparity might have evolved convergently in these two groups. Viviparity is advantageous for the young, as it affords them protection and insulation, but it places a lot of demands on the mother and requires modification to the immune system so that the developing young are not rejected as foreign bodies.

Parental care clearly evolved convergently in mammals and archosaurs (where it is also known in dinosaurs, as seen in Chapter 13). Colleen Farmer has speculated that the evolution of endothermy may be tied to the evolution of parental care: many of the hormones involved in thermoregulation are also involved with parental behavior, and endothermy allows the parent to control the incubation temperature of the young, thus promoting rapid growth.

When would parental care have commenced in the synapsid lineage? Surprisingly, there is some evidence that pelycosaurs had some parental care. Jennifer Botha-Brink and colleagues have described skeletons of a varanopid pelycosaur from the Middle Permian of South Africa that appear to represent an adult buried together with four juveniles. However, parental care does not mean that the adults provided food for the young as birds or mammals do; crocodiles care for their young but do not feed them.

## Lactation

Living monotremes may resemble reptiles in laying eggs, but they are like other living mammals in that they nourish their hatchlings by **milk** produced from mammary glands (**lactation**), rather than collecting food for them as birds do. This behavior has advantages: the parent does not have to leave the hatchling to search for suitable food for it, because any normal adult food can be converted into milk. The hatchling digests milk easily, and its parent is never far away, providing protection and warmth.

Charles Darwin suggested how suckling might have evolved in mammals, even before western science discovered monotremes, and his ideas have been extended and elaborated by present-day scientists such as Olav Oftedal. Let us assume that mammalian ancestors were already caring for eggs by incubating them. The synapsid glandular skin would have been capable of secreting fluids to keep the eggs moist and to protect them from bacteria (milk even today has antibacterial properties). Hatchlings that licked the mother's skin would benefit by gaining fluids in addition to the food brought back by the parents, and this would set the stage for selection of nutritive substances to be added to the fluid and the evolution of a source of nutrition. At this point, the mother's excursions for food could be reduced and the hatchlings benefited even more by her increased attendance. In living therian mammals, the mammary glands have nipples to which the young attach to suckle, but nipples are absent in monotremes. Does this mean that early mammals also lacked nipples? Possibly, but note that monotremes appear to have lost the ability to suckle (see discussion below), so nipples may have been present in the earliest mammals but lost in monotremes.

Could lactation of some sort have been a primitive feature for synapsids in general, in the way that parental care may have been? It seems unlikely that lactation predated the origin of mammals by any great extent. As previously mentioned, genetic evidence suggests that mammary glands only appeared in the most mammal-like of the cynodonts. Independent evidence for this time frame comes from the work of Fuzz Crompton and colleagues. Suckling milk requires a sophisticated swallowing system that involves pressing the back of the tongue against the secondary palate (i.e., the roof of the mouth); the mammalian anatomy that allows for this behavior was not present until highly derived probainognathids like *Brazilitherium*. Interestingly, modern monotremes appear to have lost this palatal anatomy, and the young lick the milk from the mother's fur. (Perhaps it's difficult to suckle anyway if your mouth forms a toothless beak, as it does in extant monotremes.)

## Early Mammaliaformes

The earliest mammals were tiny, about the size of a shrew, being around 10 cm long and weighing no more than 25 g, and their fossils are rare and difficult to collect except by washing and sieving enormous volumes of soft sediment. But after years of effort, we now have fragments (mostly teeth) from many localities in many continents, beginning close to the Triassic–Jurassic boundary. Until relatively recently, we knew little about these animals apart from their teeth, but fossils from China now provide more detail about their skeletons.

Most early mammals had small pointy teeth and were obviously little carnivores, probably eating insects, worms, and grubs. They had relatively longer snouts and much larger brains than cynodonts. The skeletons show that they were relatively agile, probably with a generalized "scampering" type of lifestyle. The molar teeth had three cusps in a line, so the name **triconodont** is used for this structure. Cynodonts also had postcanine teeth with three cusps but in mammaliaforms these teeth now precisely interdigitate (Figure 18.10): vertical faces shear up and down past one another, resulting in a zigzag cut like that of pinking shears in dressmaking.

The latest Triassic to early Jurassic mammaliaforms include the **morganucodonts** (Figure 18.11), named after *Morganucodon*, which were found almost worldwide. A contemporaneous form from Europe was *Kuehneotherium*, although this animal was more closely related to crown-group mammals, as revealed by its teeth. The molars of *Kuehneotherium* and later mammals were shaped like triangles, interlocking between upper and lower tooth rows and providing an extended

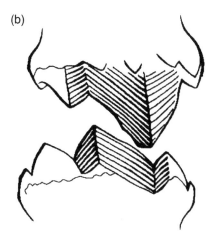

**Figure 18.10** (a) Triconodont teeth have three cusps in a row. *Source:* After Simpson. (b) They had an action rather like that of pinking shears; *Source:* After Jenkins.

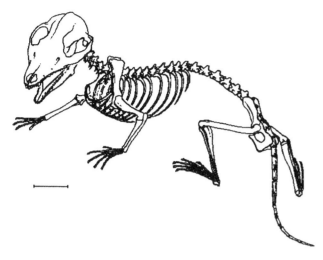

Figure 18.11 Skeleton of an early mammaliaform, the morganucodontid *Megazostrodon*, known from the Early Jurassic of South Africa. The scale bar is 2 cm. *Source:* After Jenkins and Parrington.

shearing surface, compared with the simple three cusps in a line of the earliest mammaliaforms.

Pam Gill and colleagues studied the biomechanics of the jaws of *Morganucodon* and *Kuehneotherium*, which were found in the same fossil deposits in South Wales. They showed that *Morganucodon* had the more powerful crushing bite and probably ate harder prey such as beetles, while *Kuehneotherium* preferred softer food such as caterpillars.

There were some other lineages of Mesozoic mammaliaforms, comprising slightly larger (mousetorat size) forms, that are also not included within the crown-group mammals. These were the docodonts, which included beaver-like swimmers, and the haramiyids, which included herbivorous gliders like present-day flying lemurs.

## Early Crown-Group Mammals

The three living clades of mammals are the **monotremes**, **marsupials**, and **placentals**. The easy distinction between them today is reproductive: monotremes lay eggs and suckle their young, while marsupials and placentals have live birth. Marsupials and placentals are classed together as therian mammals, the **Theria**, and they have many features in common besides their mode of reproduction.

Therians have **tribosphenic** molar teeth. These are more complex in shape than the triconodont molars and add a crushing function to the original shearing. Tribosphenic molars are particularly well suited for puncturing and shearing, and especially for grinding, superbly fitting mammals for a diet of insects and high-protein seeds and nuts. (Interestingly, similar, but not identical,

types of molars were seen in some Mesozoic mammals more closely related to monotremes.) Therians also have a shoulder girdle with a mobile scapula (shoulder blade). The classic bounding type locomotion common in mammals today, such as seen in squirrels, is probably a therian feature as the mobile scapula allows the fore limbs to reach forwards together during the bound.

The earliest monotreme is the diminutive *Steropodon* known only from teeth from the Early Cretaceous of Australia. Monotremes have only been found in Australasia (Australia plus New Guinea), except for one platypus-like form from the Paleocene of Patagonia. There was a moderate diversity of small monotremes in the Mesozoic of Australia, but only ancestors of the modern types (platypus and echidnas) are known from the Cenozoic. Monotremes were always a group with limited diversity, and they do not represent a relic of a more widely spread radiation (unlike, for example, the sphenodontid tuatara in New Zealand; see Chapter 12).

A number of other lineages of Mesozoic mammals are considered to be more closely related to therians than are monotremes (see Figure 18.2), although they did not possess these therian specialties. We will discuss only the two major ones.

**Eutriconodonts** were successful into Early Cretaceous times. Several well-preserved specimens come from the same remarkable rocks in China that also yielded feathered theropod dinosaurs (Chapter 13), many early birds (Chapter 14), the earliest therians, and the earliest flowering plants (Chapter 15). *Repenomamus giganticus* was the size of a raccoon, the largest Mesozoic mammal so far discovered. One specimen was found to have a baby dinosaur inside it, presumably its last meal (Figure 18.12). *Jeholodens* was a smaller eutriconodont with grasping hands and a prehensile tail, indicating that it was arboreal (Figure 18.13).

**Multituberculates** were successful in the Late Jurassic, Cretaceous, and early Cenozoic. They often make up

Figure 18.12 *Repenomamus*, a raccoon-sized eutriconodont from the Early Cretaceous of China, shown capturing a baby dinosaur. *Source:* Artwork by Nobu Tamura (Wikimedia).

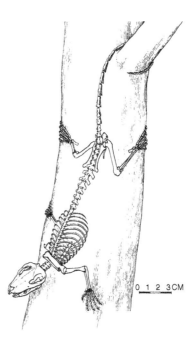

Figure 18.13 *Jeholodens*, a mouse-sized triconodont from the Early Cretaceous of China. *Source:* Artwork by Nobu Tamura (Wikimedia).

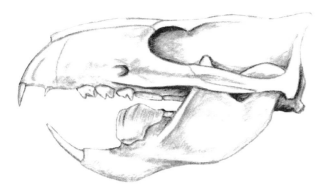

Figure 18.14 The skull of *Ptilodus*, a squirrel-sized multituberculate from the Paleocene of North America. *Source:* Artwork by Nobu Tamura (Wikimedia).

Figure 18.15 Reconstruction of *Ptilodus* as a tree-climbing multituberculate. *Source:* Courtesy of Professor David Krause of Stony Brook University.

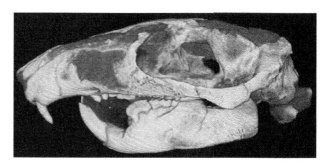

Figure 18.16 CT scan of the skull of the multituberculate mammal *Kryptobataar*, from the Cretaceous of Mongolia. Skull about 3 mm long. *Source:* Courtesy of Professor Timothy Rowe and the Digimorph Project at the University of Texas at Austin.

more than half the mammals in Late Cretaceous faunas. Multituberculates evolved superficially rodent-like teeth, so are sometimes called the "rodents of the Mesozoic." The range in tooth style and body size (mouse- to rabbit-sized) indicates a fairly wide ecological range. They survived the great extinction at the end of the Cretaceous but by the end of the Eocene, they had been replaced by more modern mammals, especially the true rodents.

The incisor teeth of multituberculates were usually specialized for grasping and puncturing, rather than gnawing as in rodents. The very large, sharp-edged premolars were designed for holding and cutting, while the molars were grinding teeth. The dentition appears well suited for cropping and chewing with a back-and-forward jaw action (Figure 18.14), although most were probably omnivorous rather than strictly herbivorous.

Some later forms from the early Cenozoic were clearly tree dwellers. *Ptilodus* had a prehensile tail and squirrel-like hind feet that could rotate backwards for climbing downward (Figure 18.15). *Kryptobataar*, from the Late Cretaceous of Mongolia (Figure 18.16), most likely had live birth. It had a narrow, rigid pelvis that was incapable of widening during birth to more than a few millimeters. Thus, the animal could not have laid any reasonable-sized egg, but it could have borne a very small fetus (newborn marsupials weigh about 1 g).

## Therian Mammals

Formally, therians include **metatherians** and **eutherians**. Metatherians consist of the common ancestor of living marsupials and all its descendants; eutherians consist of the common ancestor of living placentals and all its descendants.

Therians had diverged into separate marsupial and placental clades by the Early Cretaceous, but we do not know when their styles of reproduction diverged.

Marsupial and placental styles of reproduction are now quite distinct, but they probably both evolved from a state of delivering small, helpless young, in the manner of a modern marsupial. The egg yolk is no longer present in therians (although some of the genes responsible for its production are retained in nonfunctional forms, or pseudogenes). Both marsupials and placentals carry the young for a time in the uterus, but the special adaptation that placentals have evolved is the ability to combat the reaction of the mother's immune system and to retain the young for a longer period of time so that it can be born at a more mature stage.

Both marsupials and placentals care for their young for the same amount of time, but a good way of thinking about the difference between them is that marsupials have a short period of gestation and a long period of lactation, while placentals have a long period of gestation and a short period of lactation (at least relative to marsupials). Many, but not all, marsupials carry their young in a pouch, but in all marsupials the newborns clamp onto a nipple for both attachment to the mother and nutrition. The fetal-stage marsupials must climb unaided from the birth canal to the nipple: newborn marsupials have precociously developed fore limbs to accomplish this daunting task.

The term "placental mammal" is in fact a misnomer, as all therian mammals have a placenta of some sort, crafted from the general amniote extraembryonic membranes. In monotremes, as in other egg-laying amniotes, these membranes (which are enriched with blood vessels) help to contain the yolk and transfer nutrition to the embryo (later becoming a fetus), and to transfer oxygen from the outside to the developing young. In therians, the extraembryonic membranes now form the embryo's side of the placenta, while the other side is formed from the mother's uterine tissues.

The placenta is used to supply the developing young with nutrition, oxygen, and hormones, and to pass waste products from the fetus to the mother for disposal. In all therians, there is an initial placenta that is formed in part from the yolk sac membrane; this is a short-lived structure in all placentals, but is the only placenta seen in most marsupials. In placentals, the yolk sac placenta is replaced by one formed from the chorion and allantoic membranes (a handful of marsupials form a transitory **chorioallantoic placenta** at the end of their short gestation period).

None of this means that marsupials are inferior to placentals. A marsupial mother who experiences a natural crisis can easily abandon her young while they still essentially at the fetal stage, because she already carries them as an external litter. She may be ready to breed again quickly. A few placentals can absorb their fetuses but most placental mothers must carry their fetuses to term

for a comparatively long gestation period, even during a flood, drought, or harsh winter, often at risk to their own lives. Marsupial females can delay embryonic development after implantation, whereas placental females rarely can.

The marsupial reproductive system stresses flexibility in the face of an unpredictable environment, so it may sometimes be superior to the placental system. This may be an advantage in the arid and unpredictable environment of Australia. However, it must be remembered that Australia has only been arid for the past few million years, and that a great many marsupials live today in the equable tropical forests of South America and New Guinea.

Other major differences between marsupials and placentals today are in thermoregulation and metabolic rate. Size for size, placentals thermoregulate at slightly higher temperatures and have slightly higher metabolic rates. This need not affect reproduction because female marsupials increase their metabolic rate during pregnancy and lactation, up to placental levels. The brain grows faster in fetal placental mammals than in marsupials, probably because it is easier and more efficient to transfer nutrients across the placenta than via the milk. Not all placentals are large-brained but almost no marsupials are, and placentals have an additional brain difference in the specialized nerve tract, the corpus callosum, that connects the two hemispheres.

In spite of the metabolic differences, however, there is no systematic difference in at least one vitally important aspect: locomotion. Marsupials can run at about the same maximum speeds as equivalent placentals, and they have about the same stamina. However, marsupials may be limited in the locomotor modes they can develop; the need for the newborn to have fore limbs that can aid it to climb may limit how specialized the adult fore limb can become. There are no marsupials with fore limbs that have a greatly reduced number of digits, as seen in one-toed horses or two-toed antelope. Rather, the fast locomotors among marsupials are the kangaroos, which hop on specialized hind limbs but retain five-fingered fore limbs.

Also, no marsupials have made their fore limbs into flippers, as seen in whales. Of course, there are other problems with a marsupial evolving an aquatic lifestyle: the tiny helpless young could not be born underwater, and immature young in the pouch (or affixed to a nipple) would drown. Only one marsupial is even semi-aquatic, the yapok (or water opossum) of South America, which tightly seals its pouch during brief underwater forays.

The lineages leading to marsupials and placentals diverged by the Early Cretaceous. The first definitive eutherian is *Eomaia*, from the Early Cretaceous of China about 125 Ma. It was tiny, probably weighing less than 25 g, less than an ounce (Figure 18.17). An earlier potential eutherian is *Juramaia*, from the Jurassic of China at about 160 Ma, although both its age and its eutherian identity have been disputed. The Chinese formation that contains *Eomaia* has representatives of both basal placental and marsupial mammals. *Sinodelphys* is about the same size as *Eomaia*, but it has been identified as a metatherian (although its identity has recently been questioned, and it may in fact be a stem eutherian). Notice that if *Juramaia* is definitely on the placental line, there must be marsupial "ghost ancestors" for *Sinodelphys* stretching back at least 35 Ma. This underlines how rare early mammaliaforms are, compared with contemporary archosaurs (but, of course, their much smaller size means that they are less likely to preserve as fossils).

These three earliest therian mammals would not have been greatly different ecologically, and their skeletons provide evidence of a "scansorial" lifestyle, both climbing trees and spending time on the ground, much like a modern squirrel. Early placental mammals would still have had tiny, helpless young. The evolution of precocious young such as colts, calves, and fawns, which are large and can run soon after birth, had to wait until placental mammals reached large size. Placental mammals may well have little or no advantage over marsupials when both are small, but large precocious young are not an option for marsupials, while they are for placentals.

Figure 18.17 The little (mouse-sized) mammal *Eomaia* from the Early Cretaceous of China is an early fossil in the eutherian line: that is, the lineage leading to placental mammals. This specimen is preserved with its hair intact. *Source*: Photograph by Laikayu (Wikimedia).

## The Inferiority of Mammals

If cynodonts were moderately warm-blooded, why did they evolve to smaller size in both the latest probainognathans and the earliest mammals? Given our ideas about dinosaur biology and physiology (see Chapter 13), it was probably because of competition from diapsids. Ecologically squeezed between the first dinosaurs (fast-moving predators with sustained running) and the small, lizard-like reptiles of the Triassic (running on cheap solar energy with a low resting metabolic rate), the later cynodonts may have escaped extinction only by evolving into a habitat suitable for small, warm-blooded animals and no-one else: the night. In doing so, they underwent the radical changes in body structure, physiology, and reproduction that resulted in the evolution of mammals.

Some evidence for a shift to a nocturnal way of life comes from the visual abilities of modern mammals. Mammals have poor color vision; they appear to have lost the receptor cells in the eye (and the genes that code for them) that enable other vertebrates to see at least four different basic colors. Most mammals are dichromatic – they can only perceive two different types of colors (like a red-green color-blind person), although most humans and many other primates have duplicated an existing gene to give us trichromatic color vision (red, green, and blue). However, mammals are much better at seeing in the dark than most other vertebrates.

By the end of the Triassic, archosaurs had replaced and probably outcompeted the therapsids, driving them underground and/or into nocturnal habits all over the world. And as the few surviving therapsids (i.e., mammals) were confined to tiny body size, the dinosaurs evolved into one of the most spectacular vertebrate groups of all time.

Burrowing in the dark, the early mammals lived in a habitat that required much greater sensitivity to hearing, smell, and touch. This requirement may have selected for a relatively large, complex brain and sophisticated intelligence. So why did they not take over the Cretaceous world? It may have depended not only on competition with other vertebrates, including dinosaurs, but also on the nature of their vegetational habitat.

With the spread of flowering plants (angiosperms) in the Early Cretaceous (see Chapter 15), herbivorous dinosaurs, insects, and mammals all increased in diversity. The increase in food in the form of insects, seeds, nuts, and fruits provided a great ecological opportunity for small mammals. Mammals did increase in diversity through the Cretaceous but only at small body sizes, and probably in environments that do not yield many fossils, such as the forest canopy.

Although forest ecosystems had flourished since Carboniferous times (see Chapter 8), the types of forests inhabited by mammals today are dominated by angiosperms. Such habitats, whether tropical forests or temperate woodlands, provide a multi-storied canopy and lush vegetational undergrowth suitable for both arboreal and small terrestrial mammals. Contrast this with the forests today dominated by gymnosperms, such as pine trees, which are much sparser in mammal inhabitants. Mammal evolution in the Mesozoic may have been held back not so much by competition with dinosaurs as the lack of suitable habitats for their radiation.

Angiosperms did not come to dominate forest habitats until the latest part of the Cretaceous, and the diversity of both therians and multituberculates tracked the development of these habitats. Although no larger mammals appeared until the Cenozoic, it seems possible, even likely, that even in the absence of dinosaur extinction, these new habitats would have encouraged the evolution of a broader diversity and size range of mammals, perhaps one that would have come to challenge at least the smaller dinosaurs.

Of course, we cannot replay the tape of life and see what would have happened if the K–Pg mass extinction (discussed in Chapter 16) had never happened. However, immediately after the end of the Cretaceous and the disappearance of the dinosaurs, mammals began a tremendous radiation into a diversity of body sizes and ways of life. The inverse relationship between the success of Mesozoic archosaurs, especially dinosaurs, and Mesozoic therapsids and mammals is probably not a coincidence. It may reflect some real inability of the synapsid lineage to flourish in open terrestrial environments at the time, due to either competition or environmental factors. The extinction at the Cretaceous–Paleogene boundary that finally seems to have "released" the evolutionary potential of mammals is covered in the next few chapters.

## Reference

Martinelli, A.G., Bento Soares, M., and Schwanke, C. (2016). Two new cynodonts (Therapsida) from the Middle-Late Triassic of Brazil and comments on South American probainognathians. *PLoS One* 11 (10): e0162945.

## Further Reading

Bajdek, P., Qvarnström, M., Owocki, K. et al. (2017). Microbiota and food residues including possible evidence of pre-mammalian hair in Upper Permian coprolites from Russia. *Lethaia* 49: 455–477.

Brawand, D., Wahli, W., and Kaessman, H. (2008). Loss of egg yolk genes in mammals and the origin of lactation and placentation. *PLoS Biology* 6: e63.

Benoit, J., Manger, P.R., and Rubidge, B.S. (2016). Palaeoneurological clues to the evolution of defining mammalian soft tissue traits. *Scientific Reports* 6: 25604.

Bi, S., Zheng, X., Wang, X. et al. (2018). An Early Cretaceous eutherian and the placental-marsupial dichotomy. *Nature* 558: 390–395.

Botha-Brink, J. and Modesto, S.P. (2007). A mixed-age classed 'pelycosaur' aggregation from South Africa: earliest evidence of parental care in amniotes? *Proceedings of the Royal Society B: Biological Sciences* 274: 2829–2834.

Crompton, A.W., Owerkowicz, T., Bhullar, B.A.S. et al. (2017). Structure of the nasal region of non-mammalian cynodonts and mammaliaforms: speculations on the evolution of mammalian endothermy. *Journal of Vertebrate Paleontology* 37: e1269116.

Farmer, C. (2000). Parental care: the key to understanding endothermy and other convergent features in birds and mammals. *American Naturalist* 155: 326–334.

Gill, P.G., Purnell, M.A., Crumpton, N. et al. (2014). Dietary specializations and diversity in feeding ecology of the earliest stem mammals. *Nature* 512: 303–307.

Grossnickle, D.M. and Newham, E. (2016). Therian mammals experience an ecomorphological radiation during the Late Cretaceous and selective extinction at the K-Pg boundary. *Proceedings of the Royal Society B: Biological Sciences* 283: 20160256.

Hu, Y., Meng, J., Wang, Y. et al. (2005). Large Mesozoic mammals fed on young dinosaurs. *Nature* 433: 149–152. [*Repenomamus*].

Kim, J.-W., Yang, H.J., Oel, A.P. et al. (2016). Recruitment of rod photoreceptors from short-wavelength-sensitive cones during the evolution of nocturnal vision in mammals. *Developmental Cell* 37: 520–532.

Lee, M.S.Y. and Beck, R.M.D. (2015). Mammalian evolution: a Jurassic spark. *Current Biology* 25: R753–R773.

Luo, Z.-X. (2007). Transformation and diversification in early mammal evolution. *Nature* 450: 1011–1019.

Maier, W. and Ruf, I. (2016). Evolution of the mammalian middle ear: a historical review. *Journal of Anatomy* 228: 270–283.

Meng, J. (2014). Mesozoic mammals of China: implications for phylogeny and early evolution of mammals. *National Science Review* 1: 521–542.

Oftedal, O.T. (2011). The evolution of milk secretion and its ancient origins. *Animal* 6: 355–368.

Renfree, M.B. (2010). Review: marsupials: placentals with a difference. *Placenta* 31 (suppl A, vol 24): S21–S26.

## Questions for Thought, Study, and Discussion

1  As you know now, the first mammals were small (tiny) and not very fast. They must have evolved into a world that already had efficient small animals on land: lizards, for example, and other small reptiles that are now extinct. Discuss how mammals might have been able to survive and flourish in the face of this competition.

2  We humans have only two sets of teeth: "baby teeth" and "permanent teeth." Reptiles grow new teeth all their lives. Old mammals may starve to death as their teeth wear out, especially if they grind food in their molars. So why have mammals not evolved to replace their "permanent teeth" with another new set?

3  Reptiles usually have a homodont dentition (teeth look all the same), while most mammals usually have a heterodont (differentiated) dentition. But dolphins have lower jaws that are long and thin with lots of little pointy teeth and no differentiation into different types. If you found a dolphin jaw by the seaside, how would you know it belonged to a mammal and not to a marine reptile (perhaps a living ichthyosaur!)

4  Describe the science behind this limerick:
   Early mammals all suckled their brood
   They breathed in and out as they chewed
   Their molar tooth facets
   Were masticatory assets
   And processed varieties of food.

# 19

# Cenozoic Mammals

## In This Chapter

One of the immediate results of the K–Pg mass extinction was the radiation of the mammals, which filled at least some of the ecological niches vacated by the dinosaurs. Paleontologists study this radiation using both fossil and molecular data, and we explore how these studies provide evidence of how evolutionary radiations work in general. Mammals radiated on continents that had been separated by continental drift, and sometimes similar adaptations arose through convergent evolution in different lineages on different continents. We quickly review the history of the early Cenozoic. The Cenozoic is marked by climatic change, moving from an overall tropical world early on to one that became increasingly cooler and drier. As the climate changed, so did the vegetation and the types of animals present. One enlightening episode is the way that different mammals on different continents reacted to the appearance of large grasslands called savannas. Finally, as an example of evolutionary trends among mammals, we outline the history of two lineages that made major transitions in their modes of life: bats, evolving to be fliers, and whales, evolving from land dwellers to the giant blue whale that is the largest vertebrate of all time.

## Evolution Among Cenozoic Mammals

The end of the Cretaceous Period was marked by so many changes in life on the land, in the sea, and in the air that it also marks the end of the Mesozoic Era and the beginning of the Cenozoic Era (see Chapter 17). Survivors of the Cretaceous extinctions radiated into an impressive and varied set of organisms, beginning in the Paleocene epoch, the first 10 Ma of the Cenozoic. In the marine fossil record, the Cenozoic is dominated by mollusks, especially by bivalves and gastropods, the clams and snails of beach shell collections.

On land, the Cenozoic is marked by the dominance of flowering plants (angiosperms), insects, and birds, and in particular by the radiation of the mammals from insignificant little insectivores into dominant large animals in almost all terrestrial ecosystems. Cenozoic mammals have a very good fossil record, with thousands of well-preserved skeletons, and we understand their evolutionary history very well.

Evolution is the result of environmental factors acting on organisms through natural selection. But it is easier to understand evolutionary processes if we can isolate some of the different aspects involved. In this chapter and the next, we describe how successive groups of mammals evolved to replace dinosaurs and marine reptiles and discuss some of the major evolutionary events of the Cenozoic. In so doing, we consider not only how mammals evolved various adaptations for different modes of life, but we also think about the ecological context of these events, and how both geography and climate have influenced mammalian evolutionary history.

Much of the turnover in the fossil record consists of the ecological replacement of one group of animals by another. An older group may disappear, for various reasons, offering an ecological opportunity for a new set of species that evolves and replaces the older set. Sometimes a new group outcompetes an older group, driving it to extinction, but more usually, as we saw in

*Cowen's History of Life*, Sixth Edition. Edited by Michael J. Benton.
© 2020 John Wiley & Sons Ltd. Published 2020 by John Wiley & Sons Ltd.

the case of the replacement of dinosaurs by mammals (see Chapter 18), the first group is killed off by some catastrophe and the replacing group only then has its chance. The replacement group often evolves much the same adaptations as its predecessor, providing wonderful examples of **convergent evolution**; certain body patterns are apparently well suited for a particular way of life, so they evolve again and again in different continents at different times.

Then we look at a smaller-scale phenomenon, **evolution by adaptive change**. We often see changing morphology through time within a lineage. These evolutionary changes can often be interpreted as a series of increasingly good adaptations to the characteristic way of life, or as a set of alternative adaptations within the general way of life. However, such changes may reflect changing morphologies in response to changing environmental conditions, including climatic change, rather than simple "progression." In a successful, long-lived lineage, one can trace the various adaptations that eventually led to the derived characters of the survivors.

Obviously, one must first have a good idea of the evolutionary relationships within the group (a reliable phylogeny, in other words). In almost all cases, the evolutionary and adaptive pattern of a group is not a straight line but a winding pathway through time. But the attempt to trace a lineage through the complexity of evolution can be instructive, showing how the adaptations correspond to environmental opportunities.

## Morphological Evidence

The surviving major groups of terrestrial creatures after the Cretaceous extinction were mammals and birds (which are highly derived dinosaurs, of course). Crocodilians were amphibious rather than terrestrial. Most Mesozoic mammals had been small insectivores or omnivores, probably nocturnal, many of them tree dwellers or burrowers, and with limbs adapted for agile scurrying rather than fast running. Flying birds must be small but there is not the same constraint on terrestrial birds. There was probably intense competition between ground-dwelling birds and mammals in a kind of ecological race for large-bodied ways of life during the Paleocene, with crocodilians playing an important secondary role in some areas.

The fossil record suggests that there was an "explosive radiation" among mammals in the early Cenozoic. Dinosaurs had been dominant in terrestrial ecosystems, worldwide, for over 100 million years, and may have suppressed any ecological radiation of Mesozoic mammals

larger than the size of a raccoon. With the disappearance of the (nonavian) dinosaurs, new ecological roles suddenly became available for mammals (and birds), and dramatic adaptive radiation was a predictable response to that ecological opportunity.

Disappointingly, it has been hard to understand the pattern of origin of the major groups of placental mammals. This might seem surprising when we consider the numbers of biologists who have worked on modern mammals for 200 years, and the fact that fossils of mammals are increasingly commonly found in the Late Cretaceous, and especially in the Paleocene. Repeated study of the anatomy of fossil and modern mammals showed some elements of the pattern of relationships among the various orders, such as a natural grouping of the South American sloths, anteaters, and armadillos (xenarthrans, based on their shared unusual vertebrae), and a grouping of elephants, sea cows, and hyraxes (paenungulates, based on shared features of the skull, teeth, and wrist bones), but little else was clear. This is frustrating as we would like to understand the true cladogram of relationships of modern mammals, not least where our own mammal group, the primates, sits.

There are two main ways of determining the relationships of organisms: through studying either their morphology (i.e., anatomy) or their molecular characters. If we cannot determine the nearest relatives among mammals of the carnivores, primates, or perissodactyls, perhaps a better approach is to consider **molecular evidence**.

## Molecular Evidence

Understanding of placental mammalian phylogeny was revolutionized in 1997 when evidence for the clade Afrotheria was published. Molecular analyses showed strong evidence for the pairing of elephants, and other paenungulates, with an odd selection of insect-eating mammals, including the tiny golden moles and African shrews, as well as the strange aardvarks. This exciting proposal seemed at first crazy to the anatomists – how on Earth could we say elephants and African moles were more closely related to each other than, say, elephants to hippos or African moles to European moles? However, the result stood firm after repeated analyses using different molecular data and by different laboratories.

The molecular methods date back to 1953, when the structure of DNA (deoxyribonucleic acid) was established, and 1963, when the idea of the molecular clock was proposed (see Chapter 4). Since 1963, many molecular studies of the phylogeny of mammals have been published. Early results were often quite different from the paleontological evidence, and there were many arguments, but now the paleontological and molecular data are usually read together – after all, you need well-dated fossils to date the molecular trees – and the molecular results have become much more compatible with the fossil record. We shall see a particularly good example of that in the primate fossil record (see Chapter 21).

So let us return to the question of the timing of the mammalian radiation. The initial Paleocene radiation of mammals apparently occurred very rapidly, obscuring the branching events that resulted in the great morphological diversity of the major groups of mammals alive today. This is why anatomists could not identify deep relationships of modern mammalian orders. The combined evidence of fossils, anatomy, and molecules shows that the main divisions of mammals (monotremes, marsupials, and placentals) split in the Jurassic and Cretaceous, and the modern orders, such as primates, bats, and whales, split in the latest Cretaceous, or

immediately after the K–Pg mass extinction, in the early Paleocene. Figure 19.1 shows the current consensus of how the lineages of living mammals are related to each other.

We saw in Chapter 18 that there was a southern (Gondwana) origin of monotremes and a northern (Laurasian) origin of therians (marsupials and placentals). But molecular methods have established another set of landmarks in mammal history. By the early Cenozoic, marsupials and placentals had arrived in Gondwana as it was breaking up and founded lineages in the southern hemisphere that evolved in those regions separately from mammalian evolution elsewhere. These include two lineages of placentals, the afrotheres in Africa and the xenarthrans in South America, and essentially all of the marsupials, which radiated in South America and Australia. We will consider marsupials in more detail in Chapter 20.

The **Afrotheria** forms an African clade, including elephants, sirenians (manatees and dugongs, or sea cows),

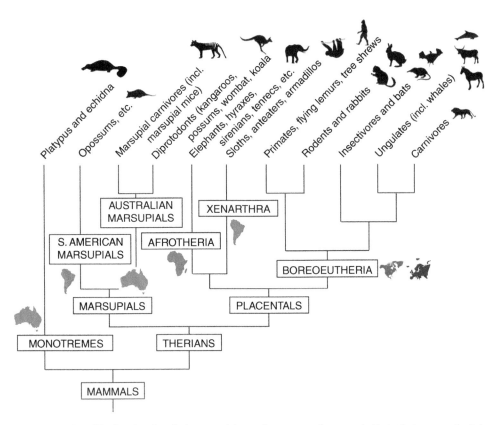

Figure 19.1 Simplified molecular phylogeny of the major groups of mammals. Note that some scientists would put either Afrotheria or Xenarthra as more closely related to Boreoeutheria, or would place Chiroptera (bats) inside the grouping of ungulates and carnivores. Continental outlines (Australia, South America, Africa, North America-Europe-Asia) indicate the major biogeographic splits early in the history of the modern mammals.

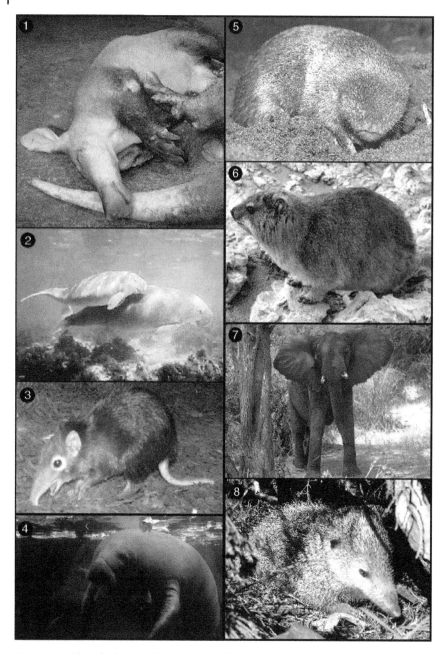

Figure 19.2 The Afrotheria, a clade of mammals that are grouped together with certainty by the DNA that they share. The skeletal and ecological differences between them are great, and only a few morphological features can be found to unite them. (1) Aardvark. (2) Manatee. (3) Rufous elephant shrew. (4) Dugong. (5) Golden mole. (6) Rock hyrax. (7) African elephant. (8) Tenrec. *Source:* Collage of Wikimedia images assembled by Esculapio.

hyraxes, aardvarks, elephant shrews, tenrecs (found today in Madagascar rather than on the African mainland), African shrews, and golden moles: a tremendous array of mammals with different body plans, sizes, and ecologies (Figure 19.2). The clade Afrotheria indicates that Africa became isolated in the Late Cretaceous or early Cenozoic, carrying a cargo of early placentals that evolved to fill all these ecological roles, separated from

evolution on other continents, in an astounding case of an adaptive radiation.

The **Xenarthra** is a South American clade of placental mammals that has long been recognized based on morphology and genomic data. Xenarthrans (sometimes known as edentates, because of their tendency to reduce or lose their teeth) include living sloths, anteaters, and armadillos, as well as many

**Figure 19.3** The Xenarthra, a South American clade of mammals, again showing a wide variety of ecologies after a long time of radiation on the continent. Clockwise from the top left: *Megatherium*, a Pleistocene giant ground sloth; giant anteater; two-toed tree sloth; armadillo. *Source:* Collage of Wikimedia images assembled by Xvazquez.

extinct mammals such as the giant ground sloths and tank-like glyptodonts (Figure 19.3).

The rest of the placentals comprise a northern clade, the **Boreoeutheria** (meaning "northern eutherians). Molecular evidence splits this lineage into two groups: one the ancestors of ungulates, including whales, carnivores, insectivores, and bats, the other the ancestors of primates and related forms plus rabbits and rodents (see Figure 19.1).

When did these branching events take place? Some molecular results indicate that the Afrotheria became separate at perhaps 105 Ma and the Xenarthra perhaps 95 Ma. These estimates coincide roughly with the major break-up of Gondwana to form the southern continents. There were apparently major differences between molecular estimates for the origins of the major mammal groups within these clades and the first fossils, but recent work by Mario dos Reis and others has shown good agreement between molecular and paleontological dates – the modern mammal orders all arose, or at least diversified, shortly before the K–Pg mass extinction or immediately after it.

So, the debate between morphological and molecular approaches to reconstructing phylogeny did not end in a stalemate, or with one side winning and the other side losing. In fact, paleontologists are lucky that they have access to two sorts of data they can use to try to understand the life of the past. The fossils are great when they are well preserved and complete but in some cases,

such as in the case of the mainly tiny Mesozoic mammals, there are not enough specimens to be sure about large-scale patterns. Molecules can plug the gaps in knowledge. Equally, fossil data are needed to understand the meaning of the molecular results. We end up with both sides collaborating, and a much richer field of study.

## The Paleocene

It is important to remember, when trying to understand mammal evolution and diversification, that in the early Cenozoic (the Paleocene and Eocene), the world was a much more equable place than today. There was no ice on the poles and the vegetation in the higher latitudes was like that of tropical forests and woodlands today, extending even to within the confines of the Arctic circle. Thus, the morphological diversity of mammals in the early Cenozoic reflected these types of habitats. Large grazers and fast runners did not evolve until later, when the world became cooler and drier, and habitats like grasslands were prevalent.

By Paleocene times, mammals included recognizable ancestors of a great many living groups, including opossum-like marsupials, carnivores, elephants, rodents, and forms related to modern primates. The ancestors of the peculiar South American ungulates and marsupial carnivores were already isolated on that continent. However, there were also many lineages that were termed "archaic

Figure 19.4 *Phenacodus,* a Paleocene mammal that looked a little like a sheep (and was about the same size) but had a more omnivorous diet. *Source:* Artwork by Heinrich Harder, in the public domain.

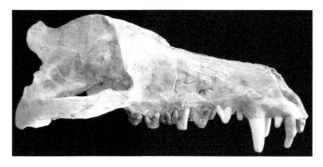

Figure 19.5 *Andrewsarchus,* the largest terrestrial mammalian scavenger/carnivore that has ever lived, from the Paleocene of Mongolia. The skull is about 1 m long. *Source:* Photograph by Ryan Somma (Wikimedia).

mammals" that did not survive past the mid-Cenozoic, and without molecular data it is difficult to determine how they were related to the mammals alive today.

Among all this diversity, the dominant group of Paleocene mammals was a set of generalized, rapidly evolving early "ungulates" (hoofed mammals), most of them generalized herbivores of various sizes, such as *Phenacodus* (Figure 19.4). But arctocyonids had low, long skulls with enlarged canines and molars indicative of an omnivorous diet and were probably rather like raccoons in their ecology. *Chriacus* had much the same size and body plan as the tree-climbing coati, but *Arctocyon* itself was the size of a bear and probably had much the same omnivorous ecology. Mesonychids were probably carnivores or scavengers, but some of them were running predators on land. For those interested in the largest of anything, the *Andrewsarchus* from the Eocene of Mongolia (a relative of the mesonychids) was the largest terrestrial carnivore/scavenger among mammals, with a skull nearly 1 m (3 ft) long (Figure 19.5).

Paleocene mammals are generally primitive in their structure, but after a drastic turnover at the end of the epoch, many new groups appeared in the Eocene that survive to the present.

## The Eocene

The turnover at the end of the Paleocene is partly related to a chance event: climatic change briefly allowed free migration of mammals across the northern continents of Eurasia and North America (more detail in Chapter 20). Roughly the same fossil faunas are found across the northern hemisphere in North America and Eurasia. In contrast, South America, Africa and Arabia, India, and Australasia were island continents to the south of this great northern land area (Figure 19.6).

Many modern groups of mammals appeared early in the Eocene in the northern hemisphere, including true primates, bats, and modern artiodactyls (even-toed ungulates, including antelope, deer, giraffe, pigs, and hippos) and perissodactyls (odd-toed ungulates, including horses, rhinos, and tapirs). Small at first, both groups of ungulates later evolved members with long, slim, stiff legs and other adaptations for fast running.

The Eocene faunas record the initial diversification of many groups of mammals as specialist herbivores of all sizes. Many of these early herbivores were small or medium-sized, including the earliest known horses which were dog-sized at this time, but soon there were large-bodied herbivores that ranged up to 5 tons (the size of a living elephant). In North America, the large herbivores were uintatheres, followed by brontotheres, which also radiated in Asia; in South America they were astrapotheres (Figure 19.7; see also Figures 19.12 and 19.13), and in Africa, they were arsinoitheres and early proboscideans (elephants and relatives).

By the end of the early Eocene, digging, running, climbing, leaping, swimming, and flying mammals were well established at a variety of body sizes. Whales were evolving from land mammals, along the southern coasts of Eurasia. Proboscideans and sea cows (afrotheres) evolved along the African shores of the tropical ocean that spread east–west between Africa and Eurasia (Figure 19.6). Many other herbivores evolved in isolation in South America. Some lineages of these endemic South American ungulates survived into the Pleistocene, and genomic analysis shows they are related to perissodactyls. However, it is not known if the other lineages (including the astrapotheres) were part of the same clade or an unrelated separate invasion of the South American continent.

Some early carnivorous mammals, the mesonychids, arctocyonids, and creodonts, were probably the equivalents in size and ecology of hyenas, coyotes, and wolves.

Figure 19.6 Paleogeography of the Eocene. The arrangement of the continents was an important controlling influence on the distribution of mammals (see text). *Source:* Paleogeographic map by C.R. Scotese © 2012, PALEOMAP Project (www.scotese.com)

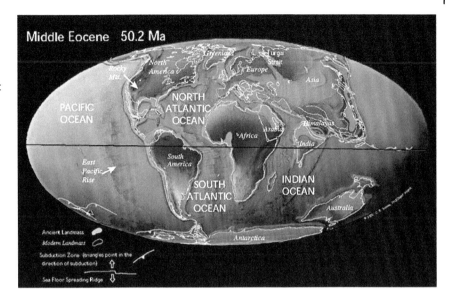

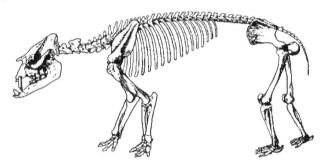

Figure 19.7 Various megaherbivore (>1 ton) mammals evolved in the early Eocene on different continents. This is *Astrapotherium*, which lived in South America. *Source:* After Riggs.

Ancestors of modern carnivores were present, but at this time they were small, playing the ecological roles of weasels and civets. They did not evolve into larger-sized predators until the late Eocene, by which time the earlier radiation of carnivorous mammals was largely extinct.

In South America, the predominant carnivorous mammals were marsupials, from an extinct lineage known as sparassodonts (also sometimes called borhyaenids). However, they were outclassed in body size by large, flightless birds with massive heads and impressive beaks, the phoruschacids (see Chapter 14) that were likely the dominant predators in this ecosystem. Similar large terrestrial birds were also found across the northern hemisphere, the diatrymas; these were originally thought to be carnivorous but recent research has shown that they were in fact herbivorous, with their large beaks used for crushing plant material rather than shearing flesh (see Chapter 14). At the same time, some crocodiles became important predators on land; for example, the pristichampsid crocodiles of Europe and North

America evolved the high skulls, serrated teeth, and rounded tails of terrestrial carnivorous reptiles. One Eocene crocodile evolved hooves!

## The End of the Eocene

As Antarctica became isolated and began to refrigerate, the Earth's climate began to cool on a global scale (see Chapter 17). It seems that the cooling took place in sharp steps, occasionally reversing for a while, so that there may have been a series of climatic events, each of which set up stresses on the ecosystems of the various continents. These climate changes resulted in the cooling and drying of the higher latitudes and the once-tropical forests gave way to seasonal woodlands able to resist winter frosts. The end-Eocene extinction was the largest one of the Cenozoic, but it was nowhere as great in magnitude than the K–Pg extinction, and it was gradual rather than catastrophic.

Toward the end of the Eocene, many families of "archaic mammals" became extinct and were replaced by those of more modern aspect, that were better adapted the new environmental conditions. Among the more modern mammals, those adapted to more tropical habitats, such as primates, either became locally extinct or, if migration was possible (it was not for the primates in the then-isolated continent of North America), they moved toward the Equator, following the retreating forests.

In North America, the waves of extinction commenced in the later Eocene. In Europe, the extinctions were concentrated at the end of the Eocene, an event that has been called La Grande Coupure – "the great cut-off." In Europe, the global climatic events were made worse by the drying up of the sea (the Turgai Straits) that had previously separated Europe from Asia, and the European mammals were inundated by an invasion of Asian ones, which replaced many of the Europeans.

## The Later Cenozoic

In the Miocene, the refrigeration of the Antarctic deepened and its ice cap grew to a huge size, affecting the climate of the world. Vegetation patterns changed, creating more open environments (i.e., with few or no trees), and a major innovation in plant evolution produced many species of grasses that colonized the open plains. The mammals in turn responded and a grassland ecosystem evolved on many continents, continuing with changes to the present. The "Savanna Story" receives separate treatment later in this chapter.

Climatic and geographical changes allowed exchanges of mammals between continents, often in pulses as opportunities occurred. A favorite example is *Hipparion*, a horse that migrated out of North America at around 11 Ma, where horses had originally evolved and spent most of their evolutionary history. Hipparions radiated in the Old World but were replaced in the Pleistocene by a second immigration from North America of the modern horse genus *Equus*.

By the end of the Miocene, the mammalian fauna of the world was essentially modern. Two further events demand special attention: the great series of ice ages that have affected the Earth over the last few million years (see Chapter 23) and the rise to dominance of animals that greatly changed the faunas and floras of Earth – humans (see Chapters 22 and 23).

## Ecological Replacement: The Guild Concept

Ancient mammal communities may have included some strange-looking animals but nevertheless, certain ways of life are usually present in different types of ecosystems. For example, in a tropical ecosystem, which would have been the case for much of the world in the early Cenozoic, plant life is abundant and varied, and provides food for herbivores, usually medium to large in size. Small animals feed on high-calorie fruits, seeds, and nuts. Pollen and nectar feeding is more likely to support really tiny animals. Carnivores range from very small consumers of insects and other invertebrates to medium-sized predators on herbivorous mammals; scavengers can be any size up to medium. There may be a few rather more specialized creatures, such as anteaters, arboreal or flying fruit eaters, or swimming mammals.

Figure 19.8 A Nubian woodpecker from Kenya. *Source:* Photograph by Brad Schram (Wikimedia).

Easily categorized ways of life that have evolved again and again among different groups of organisms are called **guilds**, occupied by organisms of similar **ecomorphology**, and their recognition helps to make sense of some of the complexity of evolution on several continents over more than 60 Ma.

For example, the **woodpecker guild** includes many creatures that eat insects living under tree bark. Woodpeckers do this on most continents. They have specially adapted heads and beaks for drilling holes through bark (Figure 19.8), and very long tongues for probing after insects. But there are no woodpeckers on Madagascar, where the little lemur *Daubentonia*, the aye-aye, occupies the same guild. It has ever-growing chisel-like incisors, and instead of using beak and tongue like a woodpecker, it gnaws with its teeth and probes for insects with an extremely long middle finger (Figure 19.9).

In the tropical forests of New Guinea (the large island just north of Australia, see Figure 19.6) and the northwestern tip of Australia, where there are no primates and no woodpeckers, the marsupial *Dactylopsila*, the striped possum, has also evolved chisel-like incisors and a very long finger (but the fourth finger) for the same reasons (Figure 19.10). The adaptations to the "woodpecker guild" seen in living mammals enable us to interpret the morphology of extinct species, whose behavior we cannot directly observe. In the Paleocene and Eocene of the northern hemisphere, there was a group of small insectivore-like mammals called apatemyids. They had similar teeth to the aye-aye and the striped possum and, as seen in one individual where the skeleton was preserved (Figure 19.11), the second and third fingers on each hand

Figure 19.9 The lemur *Daubentonia*, the aye-aye of Madagascar. Its long fingers and chisel-like teeth make it a member of the woodpecker guild there. *Source:* Photograph of a mounted specimen by Matthias Kabel (Wikimedia).

Figure 19.10 The marsupial *Dactylopsila*, the striped possum of New Guinea and northeastern Australia. Its long fingers and chisel-like teeth make it a member of the woodpecker guild there. *Source:* Painting by Joseph Wolf in 1858, in the public domain.

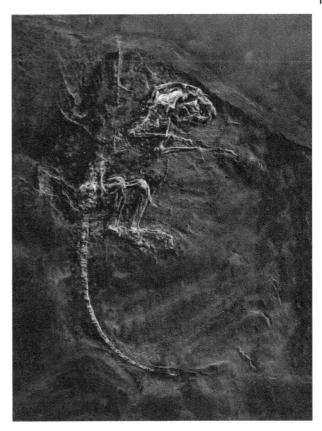

Figure 19.11 *Heterohyus*, from the Eocene of Germany, had similar teeth and fingers to both the aye-aye and the striped possum, allowing it to be identified as a member of the woodpecker guild. *Source:* Photo by Ghedoghedo, of a fossil at the Museum of Natural History, Karlsruhe (Wikimedia).

include rock hyraxes in Africa, pikas (small relatives of rabbits) in Asia and North America, rock cavies (chinchilla-like rodents) in South America, and even rock wallabies in Australia.

## Cenozoic Mammals in Dinosaur Guilds

All Mesozoic mammals were small, mostly no bigger than a rat, the largest the size of a raccoon. Mammals with small bodies can play only a limited number of ecological roles, mainly insectivores and omnivores. But when dinosaurs disappeared at the end of the Cretaceous, some of the Paleocene mammals quickly evolved to take over many of their ecological roles, particularly as large carnivores and herbivores. Others continued to occupy the same small-bodied guilds that Mesozoic mammals had occupied for 100 million years. Even today, 90% of all mammal species weigh less than 5 kg (11 pounds).

Dinosaurs dominated many guilds in the Cretaceous, including that of large herbivores. Most of them, such as the ceratopsians, hadrosaurs, and iguanodonts, weighed

are very long. This allowed paleontologists to identify them as earlier members of the woodpecker guild. Interestingly, apatamyids became extinct about the time that true woodpeckers evolved!

Some mammalian guilds are less exotic. For example, there is a recognizable guild of small herbivorous mammals that live among rocks, found on several different continents, that look alike and behave alike. These

around 2–7 tons as adults. The K–Pg extinction wiped out all these creatures, and it was not until the late Paleocene that the guild was occupied again, by large mammals.

Although some large flightless birds are herbivores (e.g., ostriches, rheas, and cassowaries), mammals are the dominant large herbivores today. Even at the very beginning of the Paleocene, mammal faunas were dominated by the largely herbivorous early ungulates. Herbivorous mammals make their originally tribosphenic teeth more complex by adding an extra cusp, and the teeth are now flattened for grinding and shredding food, rather than for cutting and piercing it. Herbivorous mammals also move their jaw sideways to a larger extent when chewing than other mammals in order to thoroughly grind up tough vegetation, and this is reflected in their larger masseter muscles (see Chapter 18, Figure 18.6). A great advantage of herbivory is that mammals can now bulk process vegetation and grow to huge sizes.

There seems to be something special about the 2–7-ton range for large land herbivores, possibly related to metabolic efficiency with a diet of low-calorie forage. This size range applied to all dinosaurs except for sauropods, and it has apparently applied to almost all mammals since, including living elephants and rhinos. The 2–7-ton size was approached by different mammalian groups in the different continents of the Paleocene and Eocene (Figure 19.5). The best record is in North America, where uintatheres (Figure 19.12) and later the brontotheres (also known as titanotheres) (Figure 19.13) were the first large mammalian herbivores. The larger members in both lineages evolved bony horns on their heads. (In contrast, the horns of rhinos are not made of bone but of keratin.) Brontotheres became extinct at the end of the Eocene and it was not until the early Miocene that this guild was filled by modern types of rhinos and proboscideans.

**Figure 19.13** Brontotheres evolved to very large body size between the early Eocene and the end of the epoch. (i) An early Eocene small brontothere, the pig-sized *Eotitanops*, and (ii) a gigantic late Eocene one, *Megacerops* (formerly known as *Brontotherium*) in the act of displaying. The huge double horns may have been used for wrestling or ramming other members of the species. *Source:* From Osborn.

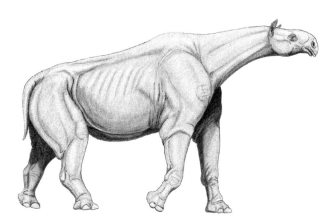

**Figure 19.14** The gigantic Oligocene rhinoceros *Paraceratherium* (also known as *Indricotherium* and *Baluchitherium*). It weighed up to 15 tons, far beyond the "usual" size range for large herbivores. *Source:* Artwork by Dmitry Bogdanov (Wikimedia).

However, the largest herbivorous land mammal that ever lived is known from the Oligocene of Asia; this was *Paraceratherium* (Figure 19.14), a hornless rhinoceros (but belonging to a different family to the modern diversity of rhinos), which weighed as much as 15 tons. *Paraceratherium* had relatively long legs and a long neck, which is why it is sometimes called the "giraffe rhino."

## The Savanna Story

An image of the diversity of large mammals today would almost inevitably be one of the Serengeti, the savanna habitat of East Africa, teeming with zebra, antelope, giraffe, rhinos, elephants, and, of course, carnivores like lions and cheetahs. Savanna is the term given to habitats, mostly tropical today, that combine a basic covering of

**Figure 19.12** The rhino-sized *Uintatherium*, which gives its name to the uintatheres, a clade of large late Paleocene and Eocene North American herbivores. They had ridged molar teeth for shredding vegetation, but also sported large canine teeth modified into sabers which were probably for fighting between adults. *Source:* Reconstruction by Bob Guiliani. © Dover Publications Inc., New York. Used by permission.

grass with a scattering of trees and bushes. This contrasts with the largely treeless grasslands seen in the higher latitudes, variously named prairie, steppe, or pampas.

As might be expected from the diversity of vegetation, savannas represent a more productive ecosystem than treeless grasslands, and support a greater diversity of large mammals, including both grazers (eating grass, e.g., zebras) and browsers (eating the leaves of trees and bushes, e.g., giraffe). Grazers tend to live in herds, while most browsers are solitary. Most savanna herbivores are actually mixed feeders, including both grass and browse in their diet.

Savanna ecosystems produce a great deal of edible vegetation, even though grasses have high fiber and low protein. Grasses are adapted to withstand severe grazing; they recover quickly after being cropped because they grow from the base instead of mainly at the growing tip: the more grass is cut off from the top, the more it tends to grow from the base, as anybody responsible for mowing the family lawn knows only too well.

Furthermore, with a seasonal and local variation in food supply, it is easy to envisage the evolution of a guild of grazing species, each specializing in a different part of the available food. In the Serengeti, for example, three different grazers eat grass and herbs. Zebras eat the upper parts of the blades of grass and the herbs, wildebeest follow up and eat the middle parts, and the Thomson's gazelle eats the lower portions. The teeth and digestive systems of each animal are specialized for its particular diet. Thus, a succession of animals grazes the plain at different times, each species modifying the plants in a way that (by chance) permits its successor to graze more efficiently. A great diversity of grazers is encouraged – today there are 10 separate tribes of bovid antelopes on the savannas of Africa.

Similar principles probably apply to browsers too. Obviously, the rules will be rather different, because the defense of many plants against browsing is to grow tall quickly.

And finally, herbivores, whether they are grazers or browsers, are a food resource for predators and scavengers. The animals of the African savannas are in a delicate and interwoven ecological network.

Although today savannas are tropical in their distribution (and only the African savannas contain this type of large mammal diversity), savannas first appeared in higher latitudes in the early Miocene, with the spread of grasslands. The African savannas were not established until the Plio-Pleistocene, some 5 Ma, by which time the cooling and drying of the higher latitudes had turned their savannas into prairies.

A major climatic change in the Miocene was apparently triggered by the refrigeration of the Antarctic and the growth of its huge ice cap. The cooler climate encouraged the spread of grasslands in subtropical and temperate latitudes, at the expense of thicker forests and woods. The new feature of Miocene open country was the spread of grasses, with their high productivity. Savanna ecosystems became prominent in North America, Eurasia, and Argentina, each filled with their own unique diversity of mammals that evolved from the preexisting fauna.

Although mammalian faunas evolved to fill the new savanna ecosystems, this response was not immediate. Work by Caroline Strömberg and colleagues has shown that, in North America, the grasslands spread at the start of the Miocene, around 23 Ma, but it was not until 5 Ma later that the ecosystem "took off" in terms of a diversity of mammals adapted to this habitat, not only grazers but also browsers and mixed feeders. A particular feature of open habitats is the evolution of high-crowned, or **hypsodont**, cheek teeth, as seen today in horses, antelope, camels, and elephants. The larger savanna animals also showed changes in size and locomotion consistent with life in open country. They became taller and longer-legged, well adapted for both running fast and traveling economically around large home ranges.

Hypsodont cheek teeth were long thought to be a specific adaptation for grazing, as grass blades contain tiny silica fragments, or **phytoliths**, that make them tough to chew. But further study of both living and fossil mammals has revealed that the underlying reason for hypsodonty is simply to resist abrasion, whether from silica in grasses or other sources, such as dust and grit on the leaves of both grass and browse. A larger source of abrasion than silica for grazing mammals appears to be the soil that they inevitably ingest along with the plants. Hypsodont teeth last a lifetime of wear: the tall crowns mean that most of the tooth is actually retained within the jaw itself, and as the top portion wears down, the reserve crown becomes exposed in turn, much like the way that a long piece of lead renews the working tip in a mechanical pencil.

Figure 19.15 gives a glimpse of the North American Miocene savanna ecosystem, showing animals that look like those in the Serengeti but which are not closely related. This early Miocene scene would in reality have been more wooded than shown here; none of these animals has adaptations for grazing and the horses have not yet become hypsodont. The radiation of hypsodont horses into a diversity of grazers and mixed feeders as great as that of the antelope in Africa today happened a few million years later, and they were accompanied by a radiation of large browsing horses, camelids, ruminant artiodactyls (pronghorns and deer-related forms), rhinos, and proboscideans.

The North American Miocene savanna reached its peak in the mid-Miocene, from around 15 to 12 Ma. But following this, the climate became progressively cooler

Figure 19.15 A reconstruction of the savanna ecosystem of the early Miocene of North Miocene savanna, around 18 Ma. The big (buffalo-sized) pig-like animal in the middle is the carnivorous/omnivorous *Daeodon*, a member of the extinct artiodactyl family Entelodontidae. In the lower left-hand corner is a herd of the pony-sized horse *Parahippus*. Both horses and camels had their original evolutionary radiation in North America, and the camels shown here are the gazelle-like *Stenomylus* (*bottom right*) and the lama-like *Oxydactylus* (*middle, top*). The horned animal below *Daeodon* is *Syndyoceras*, distantly related to camels. True rhinos made their first appearance in North America at this time, and the tapir-sized *Diceratherium* can be seen on the middle right and middle left. *Source:* From a painting by Jay Matternes (Wikimedia).

and dryer, and as the grasslands turned from savannas to prairie, the diversity of mammals decreased and the specialized browsers became extinct. Of the original diversity of North American herbivorous ungulates, only one species remains on that continent today, the pronghorn "antelope." The remaining large native ungulates (horses, camels, mastodons) all went extinct at the end of the Pleistocene (see Chapter 23). The large ungulates on the prairies and woodlands of North America today, buffalo, sheep, and deer, represent Plio-Pleistocene immigrants from Eurasia.

## Evolutionary Change over Time

The fossil record of mammals is so good that we can trace related groups of mammals through long time periods, and often across large areas and across geographic and climatic barriers. In many cases, we can see considerable evolutionary change in the groups, and because we understand the biology of living mammals rather well, and we also know of accompanying climatic and environmental changes, we can interpret the morphological changes with confidence. Often the

changes can be linked with specific biological functions and can be seen as allowing the animals to perform those functions in a more effective way: that is, becoming "better adapted."

The notion of improvement over time, or "evolutionary progress," is a popular one. But while animals can, in retrospect, be seen to be "improving" in one way or another, this must still be viewed in the context of changing situations. For example, the evolution of horses is often portrayed as a progressive trend toward better adaptations for feeding (hypsodont cheek teeth) and for rapid locomotion (longer legs with only a single toe). These morphological features clearly make horses well adapted for their current habitat. But we must remember that ancestral horses, such as the early Eocene *Hyracotherium*, lived in tropical forest-like habitats, where such morphology would either be of little value (hypsodont teeth) or a liability (stiff, inflexible limbs). In the evolution of horses from *Hyracotherium* to *Equus*, it was not that later horses sought out new environments; rather, they stayed where they had always lived but the environments changed around them, the tropical forests of the Eocene giving way to the savannas of the Miocene and then to the prairies of the Plio-Pleistocene.

Two particular examples of progressive adaptations to changing environments can be seen in mammals that rapidly altered their choice of environment to inhabit, rather (unlike the horses) than having environmental change imposed on them by changing climatic conditions. These include the bats, which took to the air, and the whales, which returned to the sea.

## Bats

Bats (order Chiroptera) were the last of the vertebrates to evolve flapping flight, following pterosaurs in the Triassic and birds in the Jurassic. In all bats, the wing is stretched between arm, body, and leg, with the fingers of the hand splayed out in a fan toward the wingtip (Figure 19.16). The wing membrane has little strength of its own but it is elastic, and tension has to be maintained in it by muscles and tendons. The hind leg is used as an anchor for the trailing edge of the membrane, which means that the limb is not free for effective walking and running. Bats therefore are forced into unusual habits, which include roosting in inaccessible places where they hang upside down.

Bats have evolved special adaptations to maintain flight during pregnancy and nursing and, unusually for small mammals, often only have one young at a time. Bats also live a long time in comparison with other small mammals; this may be related to their small litter size – if they had short lives they might not produce enough offspring to ensure the survival of the species.

Another unique feature of bats is their ability to use sonar to echolocate; they produce extremely high-pitched sounds, and the echo received from surrounding objects allows them to navigate in the dark. Bat sonar presumably evolved from the acute hearing of little, nocturnal, insect-hunting mammals in the forest canopy of the Late Cretaceous. (Some ability to echolocate is probably a general mammal feature; humans can do this to a limited extent, and blind people rely on it.)

Echolocation is lacking in the fruit bats. Fruit bats (also termed mega-bats) are a separate family from the other bats (micro-bats), known from the Old World tropics (the micro-bats are known worldwide). As the terms "mega" and "micro" suggest, the fruit bats are generally larger than the other bats, up to the size of a rabbit (i.e., around 5 pounds). The evidence now suggests that fruit bats had an echolocating ancestry but came to rely more on vision and smell to locate the fruit on which they feed. The large size of fruit bats and their fox-like faces (they lack the gargoyle-like faces of micro-bats, which are associated with their production of ultrasound) have led them to be erroneously cast as vampire bats in movies. In the house of Count Dracula, these bats could only suck oranges, not blood!

The earliest bats, *Onychonycteris* and *Icaronycteris*, are known from a few extremely well-preserved fossils from Early Eocene lake beds in Wyoming (Figure 19.17).

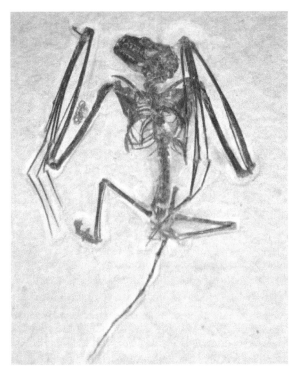

Figure 19.17 *Onychonycteris*, the earliest known bat, from the Eocene of Wyoming. *Source:* Photograph of a replica of the only specimen, taken by Arvid Aase for the US National Park Service.

Figure 19.16 The bat *Corynorhinus townsendii*, showing the arm and hand bones supporting the wings. Note also the tail membrane or uropatagium, extending between the hind limbs and supported by a long tail. By coordinating movements of the hind limbs and tail, the membrane can be contracted or expanded, and raised or lowered. In other words, it is an active component of the flight system. *Source:* Image from the Bureau of Land Management; US government image, in the public domain.

*Onychonycteris* is the most primitive bat yet known: it retained claws on all its (relatively short) fingers, had relatively long hind legs, and evidence from the preserved ear region indicates that it did not have the powers of echolocation (although there remains dispute about this issue). In contrast, *Icaronycteris* has the wing proportions and the reduced number of claws (two) seen in modern bats, and appears to have been capable of echolocation.

Obviously, bats must already have had an eventful evolutionary history before the Eocene, but we still have to find the fossils that will tell us precisely who their ancestors were, and how they evolved flight. It has often been assumed that bats evolved flying from a gliding ancestry; many mammals glide today, including two different types of gliding squirrels, several different types of gliding possums, and flying lemurs (which are not true lemurs but a group in their own order, Dermoptera). But recent work highlights the importance of the tail in bat flight, with the implication that gliding was not an intermediate form of locomotion for these animals.

A number of small bats today begin flight by flapping the tail membrane as well as the wings as they take off. The coordination of the two systems is exact, and at very slow speed the tail is a vital component of bat flight, very unlike the action of the tail in a gliding mammal. Rather, bats used the legs to jump into the air, gradually evolving longer and faster flights under better and better control from the actions of the skin membranes of the wings and tail membrane (uropatagium). It may be important that the earliest bat, *Onychonycteris*, also had a long bony tail that could have supported and operated a relatively large uropatagium.

## Whales

Whales, dolphins, and porpoises (order Cetacea) are well-known marine mammals today. Modern whales consist of two distinct lineages: **odontocetes** (toothed whales, including dolphins and porpoises) and **mysticetes** (the baleen whales, including the 120-ton blue whale). Odontocetes are the more numerous lineage in terms of species, and are also usually smaller in size than the mysticetes (the sperm whale is an exception). Odontocetes hunt large prey, and they perfected the echolocation that is such an important factor in their ability to locate prey in dark or turbid water. Some of them (e.g., the sperm whale) have physiological adaptations for deep diving. Mysticetes are filter feeders and strain small planktonic organisms (such as the tiny crustaceans called krill) through plates of keratinous baleen that hang from their upper jaw. Mysticetes lack echolocation, but these are the whales that sing. Both types of whales have moved their nostrils to the top of their heads to act as blow-holes. The form of the blow-hole is slightly different in the two lineages, suggesting a degree of parallel evolution; however, the beginning of the retraction of the nostrils can be seen in more primitive fossil whales.

Whales are beautifully adapted to carnivorous ways of life in the sea, with streamlined bodies, fore limbs turned into flippers, hind limbs lost, and tail flukes that act for propulsion. Whales give birth underwater; the tail of the infant is delivered first and the head last. In this way, the whale infant can still get some oxygenated blood through the umbilical cord and does not drown as it is pushed out into the water. Once born, its mother helps it to swim rapidly to the surface to take its first breath of air.

The relationship of whales to other placental mammals remained a mystery for many years, but genomic evidence now shows that their ancestors were artiodactyls. Specifically, among the artiodactyls, whales appear to be most closely related to hippos. The fact that hippos are semi-aquatic today is probably a coincidence; the common ancestor of whales and hippos, some time in the early Eocene, was likely a terrestrial animal. Paleontologists were originally skeptical about whales being descended from an artiodactyl ancestry. However, when a greater diversity of early whale fossils came to light, including some with functional hind legs, it could be seen that whales had the same peculiarity of the ankle joint as seen in artiodactyls (specifically, a "double-pulley astragalus"), and so the morphology could be seen to back up the molecular data.

In the Eocene, there was a group of extinct whales that was more primitive than the modern groups, the archaeocetes (the name means "ancient whale"). Some of these spent much of their lives on land, and others were more fully adapted to life in the water. All archaeocetes retained hind legs, although these were tiny in the most derived ones, the basilosaurids. *Basilosaurus*, from the late Eocene (Figure 19.18), was recognized as a huge fossil whale over 150 years ago; it was termed "saurus" (meaning "lizard") because its elongated body meant it was originally mistaken for a sea serpent. Below are descriptions of some of the archaeocetes, roughly in morphological and temporal order, since they show transitions between land- and water-based ecologies.

*Indohyus* (Figure 19.19), from the early and middle Eocene of India, is not classified as a whale, but is the artiodactyl that is most closely related to whales. It shares with whales a characteristic feature of a thickening of the portion of the skull housing the inner ear bones. It was of a similar size, and probably similar in its ecology, to the water chevrotain, a browsing deer-like artiodactyl about

Figure 19.18 *Basilosaurus* was recognized as a huge fossil whale as early as the 1840s. Known from late Eocene rocks, it ranged up to 18 m (60 ft) in length. *Source:* Reconstruction by Pavel Riha (Wikimedia).

Figure 19.19 Reconstruction of *Indohyus*, from the Eocene of India. *Source:* Artwork by Nobu Tamura (Wikimedia).

Figure 19.20 Reconstruction of *Pakicetus*, from the Eocene of Pakistan. *Source:* Illustration by Carl Buell. Used by permission.

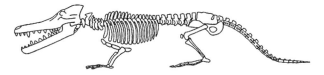

Figure 19.21 Reconstruction of *Ambulocetus* from the Eocene of Pakistan. The body was alligator-sized, about 12 ft long. *Source:* Artwork courtesy of Hans Thewissen, and used by permission.

Figure 19.22 Reconstruction of *Kutchicetus*, from the Eocene of India, a whale that looked like a giant otter. *Source:* Illustration by Carl Buell, used by permission.

the size of a medium-sized dog, which lives today in central Africa and escapes predators by diving into nearby water. We can tell that *Indohyus* spent significant amounts of time in the water as its bones are dense, as they are today in hippos, which make walking in shallow water more stable.

*Pakicetus* (Figure 19.20), from the middle Eocene of Pakistan, looked superficially like a wolf and was of a similar size, but its skeleton showed that it was an excellent swimmer and it probably spent much of its time in the water. Its predatory types of teeth are more like those of a fish eater than like those of a meat eater, and it lived in fresh waters or near-shore environments.

*Ambulocetus* (Figure 19.21), the "walking whale," is also from the middle Eocene of Pakistan. It was much larger than *Pakicetus*, up to 12 ft long, and looked superficially

like an alligator, with short powerful limbs and a long flat head with fish-eating teeth. It clearly could swim powerfully, though slowly, probably with an otter-like vertical bending of the spine. But it could also walk perfectly well on land, and studies of the chemical composition of its bones show that, like *Pakicetus*, it drank fresh water.

*Kutchicetus* (Figure 19.22) was a whale from the middle Eocene of India that probably behaved like a giant otter, even better adapted to feeding and swimming in water than previous whales. It had a long beak-like set of jaws, with the skull mounted on a relatively stiff neck. The tail was relatively long and powerful and may have been the major power producer in swimming, and its inner ear region showed that it could hear underwater, like all later whales.

The next whales we see in the fossil record are distributed in a great tropical belt from southern Asia westward to North America, and that means that whales had become long-distance ocean-going swimmers. The protocetids include more than a dozen genera. They are varied in size and represent a radiation of early whales away from their origins in the Indian subcontinent. There are two particularly interesting protocetids, partly because they are very well preserved and partly because of the persuasive reconstructions by the artist John Klausmeyer. *Rodhocetus* (Figure 19.23), from the middle Eocene of Pakistan, still has strong limbs, although its skull is very whale-like, with the nostrils starting their retraction up the snout (in modern whales the nostrils migrate to the top of the head in development, forming the blow-hole). If one had to search for a living comparison in ecological terms, one might choose sea lions, which spend most of their time at sea but still go onto land for mating and giving birth.

Figure 19.23 The Eocene protocetid whale *Rodhocetus*. *Source:* Reconstruction by John Klausmeyer, used courtesy of the University of Michigan Museum of Natural History.

*Maiacetus*, from the middle Eocene of Pakistan, is known from an almost complete skeleton, allowing a reconstruction in swimming position (Figure 19.24). The sea lion analogy seems reasonable for this animal, too. But the most compelling aspect of this specimen is that it was a female with a late-stage fetus inside the body cavity (Figure 19.25). It is clear that the head of the fetus faces backward, as it does in large mammals that give birth on land but unlike the condition in modern whales. The evidence from this single specimen of *Maiacetus* tells us that these whales gave birth on land (Figure 19.26), as we could deduce indirectly from the fact that they had strong limbs.

But *Basilosaurus*, known from the late Eocene, was different. As we have seen, it was a huge whale and it had lost any functional hind limbs, although it retained tiny, but complete, ones that may have been used in mating. Its fore limbs were now completely flipper-like, with no bend at the elbow, and all the propulsion came from the tail (the tail vertebrae allow us to infer the presence of a tail fluke) (Figure 19.18). There is no way that it could have emerged on to land – it spent its entire life at sea. *Basilosaurus* and a related smaller genus, *Dorudon*, had fearsome fish-eating teeth set in a long jaw (Figure 19.27).

The archaeocetes went extinct at the end of the Eocene, and both lineages of modern whales appeared at the start of the Oligocene. Erich Fitzgerald has shown that the first whales that are recognizably mysticetes (from aspects of their skull anatomy) still retained teeth, and they were dolphin-sized – they did not become filter-feeding giants until the mid-Miocene. It seems that early mysticetes used a widely opening mouth to suck in dozens of fish at a time, and it could be that the filter feeding of later baleen whales evolved from that sort of prey capture. Overall, the evolution of whales from small herbivores on land to carnivorous predators at sea is the most spectacular, and now one of the best understood, transitions in the mammalian fossil record.

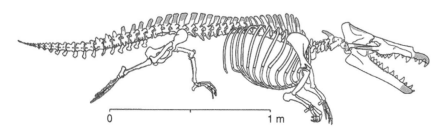

0                    1 m

Figure 19.24 The swimming position of the protocetid *Maiacetus*. *Source:* Part of figure 1 in Gingerich et al. (2009): drawn by Bonnie Miljour (Wikimedia).

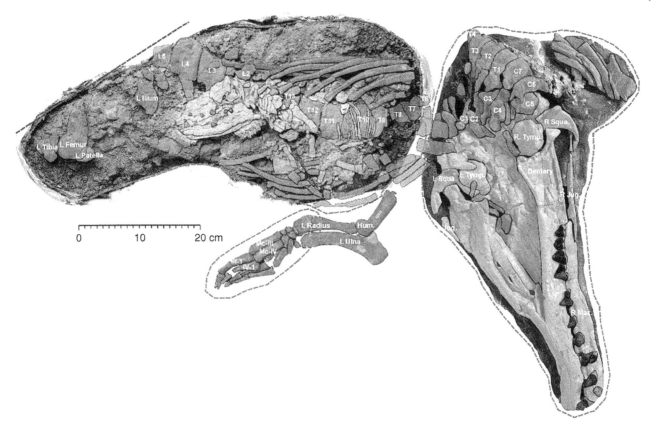

Figure 19.25  The single fossil of *Maiacetus*, showing the fetus (in blue) inside the body cavity. *Source:* Figure 2 in Gingerich et al. (2009) (Wikimedia).

Figure 19.26  A female *Maiacetus* and her newborn pup. *Source:* Artwork by John Klausmeyer, used courtesy of the University of Michigan Museum of Natural History.

Figure 19.27 *Basilosaurus* and a smaller relative *Dorudon*. *Source:* Photograph by Philip Gingerich, used courtesy of the University of Michigan Museum of Natural History.

## Reference

Gingerich, P.D., von Koenigswald, W., Sanders, W.J. et al. (2009). New protocetid whale from the middle Eocene of Pakistan: birth on land, precocial development, and sexual dimorphism. *PLoS One* 4: e4366. [*Maiacetus*].

## Further Reading

dos Reis, M., Inoue, J., Hasegawa, M. et al. (2012). Phylogenomic datasets provide both precision and accuracy in estimating the timescale of placental mammal phylogeny. *Philosophical Transactions of the Royal Society of London. Series B, Biological Sciences* 279: 3491–3500.

Fitzgerald, E.M.G. (2012). Archaeocete-like jaws in a baleen whale. *Biology Letters* 8: 94–96.

Foley, N.M., Springer, M.S., and Teeling, E.C. (2016). Mammal madness: is the mammal tree of life still not resolved? *Philosophical Transactions of the Royal Society of London. Series B, Biological Sciences* 371: 20150140.

Goswami, A. (2012). A dating success story: genomes and fossils converge on placental mammal origins. *EvoDevo* 3: 18. [A readable short review of the often dense literature on molecules versus morphology in mammal evolution].

McGowen, M.R., Gatesy, J., and Wildman, D.E. (2014). Molecular evolution tracks macroevolutionary transitions in Cetacea. *Trends in Ecology and Evolution* 29: 336–345.

Phillips, M.J. and Fruciano, C. (2018). The soft explosive model of placental mammal evolution. *BMC Evolutionary Biology* 18: 104.

Pyenson, N.D. (2017). The ecological rise of whales chronicled by the fossil record. *Current Biology* 27: R558–R564.

Simmons, N.B., Seymour, K.L., Habersetzer, J. et al. (2008). Primitive Early Eocene bat from Wyoming and the evolution of flight and echolocation. *Nature* 451: 818–821.

Strömberg, C.A.E. (2011). Evolution of grasses and grassland ecosystems. *Annual Review of Earth and Planetary Science* 39: 517–544.

Thewissen, J.G.M., Cooper, L.N., George, J.C. et al. (2009). From land to water: the origin of whales, dolphins, and porpoises. *Evolution: Education and Outreach* 2: 272–288.

Uhen, M.D. (2010). The origin(s) of whales. *Annual Review of Earth and Planetary Science* 38: 189–219.

## Questions for Thought, Study, and Discussion

1 What are the strengths and weaknesses of reconstructing the evolution of mammals based only on molecular and genetic data, compared with the strengths and weaknesses of reconstructing the evolution of mammals based only on the skeletons of living and fossil mammals? What happens when the results give different answers?

2 The concept of a guild is rather simple in principle. And it's an easy question to give some examples of guilds alive on Earth today. But it's more difficult (so try it) to pick out guilds that once lived on Earth but no longer do. And it's even more difficult to ask why those guilds disappeared (so answer it for the examples you chose).

3  Whales are not the only mammal group that lives successfully on the sea. For example, seals, sea lions, sea otters, manatees, and even an extinct group of sloths all evolved to take on a successful life in the sea. Find out more about these other animal groups and then point out the features that make whales different, perhaps better adapted to marine life. Do you think that whales are different because they "got there first"?

On the heels of the dread K–Pg
Mammals flourished immediately
Through climatic changes
They shifted their ranges
Some even returned to the sea

20

## Geography and Evolution

**In This Chapter**

Here we look at ways that changing geography influenced evolution during the Cenozoic. Australia was an isolated continent with a limited original fauna. The marsupials that had reached Australia radiated into an amazing array of creatures that occupied many of the ecological niches in Australia that placental mammals occupied elsewhere, and Australia's reptiles are equally impressive. New Zealand became isolated and while bats flew there, birds became the dominant land animals, including the recently extinct flightless giant moas. South America had a radiation of its own placental mammals, the Xenarthra and many lineages of now-extinct ungulates, but the Cenozoic ecosystem had marsupials and flightless birds as the top predators. In the Pliocene, South America drifted close enough to North America that a land bridge formed and animals were exchanged, and many of the endemic South American animals became extinct. Africa had suffered a similar fate in the early Miocene, when it joined with Eurasia. Madagascar, a large island technically part of Africa, had its own independent evolution, with a unique endemic fauna and flora today. Smaller islands provide interesting microcosms of isolated evolution, where small mammals tend to become larger and larger ones become smaller: the Miocene island of Gargano is a particularly interesting case.

## Mammals and Geography

Huge changes took place during the Cenozoic, with the final steps in the break-up of the Pangea supercontinent (see Chapter 17), as the Atlantic Ocean continued to open, driving the Americas and the Old World further apart. The southern continents moved to their present locations, Antarctica moving south, Australia east, and India northeast to continue driving into southern Asia. At the same time, world temperatures were broadly cooling down continuously, with occasional warming episodes (see Chapter 17). These changes in the layout of continents also changed ocean circulation patterns and the changing temperatures led to major changes in vegetation, with the spread of grasslands in the Oligocene and Miocene (see Chapter 19). All groups of plants and animals were affected but we focus here on mammals, and a little on the other tetrapods like amphibians, reptiles, and birds, and look at how the break-up of continents has led to increasing endemism (see Chapter 6).

## Australia

Australia is linked in people's minds with exotic creatures such as kangaroos, jillaroos and koalas, but they are only a part of the story of evolution on this isolated continent. Australian plants, insects, amphibians, reptiles, birds, and mammals are all unusual. Australia and New Zealand were part of Gondwana in Cretaceous times, joined to Antarctica in high latitudes (see Chapter 16, Figure 16.8). The climate was mild, however, and pterosaurs, dinosaurs, and marine reptiles have been found there. In the early Cenozoic, the two land masses broke away from Antarctica and began to drift northward and diverge. Both Australia and New Zealand became isolated geographically and ecologically from other land masses, and evolution among their faunas and floras led to interesting parallels with other continents.

Among amphibians, Australia has (or had) at least two species of frogs that brood young in their stomachs. Instead of the colubrid snakes and vipers that are

*Cowen's History of Life*, Sixth Edition. Edited by Michael J. Benton.
© 2020 John Wiley & Sons Ltd. Published 2020 by John Wiley & Sons Ltd.

abundant elsewhere, Australia had a radiation of elapid snakes (cobras and their relatives) into 75 species, all of them virulently poisonous. The largest Australian predators are the salt-water crocodiles (the world's largest surviving reptiles), which lurk along northern rivers and shorelines, and giant monitor lizards the size of crocodiles, related to the Komodo dragon of Indonesia. Monitors are ambush predators, the largest living Australian monitor being 2 m (over 6 ft) long. Smaller monitors dig for prey like the badgers of larger continents. In contrast, most large Australian mammals are herbivores.

Extinct Australian reptiles included giant horned tortoises that weighed up to 200 kg (450 pounds) and a monitor 7 m (23 ft) long that weighed perhaps a ton, and competed with large terrestrial crocodiles of about the same size and weight. The giant snake *Wonambi* was 6 m (19 ft) long and weighed around 100 kg (220 pounds). Extinct Australian birds included *Dromornis*, the heaviest bird that has ever evolved.

Australia is the only continent with living **monotremes**. They have been in Australia since the Early Cretaceous (see Chapter 18), but teeth from an early Cenozoic monotreme from Argentina shows that they once ranged more widely over Gondwana. The surviving monotremes are egg-laying mammals, including the platypus of Australia and the echidnas (or spiny anteaters) of Australia and New Guinea. Many aspects of monotreme biology are bizarre: not only are they more primitive in certain respects, such as laying eggs and

lacking external ears (see Chapter 18), but they have their own unique specializations, such as a beak that can sense electric currents produced by other animals. The semi-aquatic platypus uses its rather duck-like beak to hunt for invertebrates underwater, while the echidnas use their long thin beaks to detect ants and termites in the leaf litter.

Marsupials had originally evolved in Asia, and there was a modest radiation in the Cretaceous of Asia of rather opossum-like animals that were not closely related to the living marsupials. Figure 20.1 shows a phylogeny of living marsupials. Marsupials reached both North and South America and radiated there in the Cenozoic (the South American marsupials are considered later in this chapter). In the Eocene, small opossum-like marsupials (Figure 20.2) migrated to the Old World from North America via Greenland (see Chapter 19), and although they were always rare, they were found in fossil deposits in Europe, Africa, and Asia. However, all of these northern hemisphere marsupials were extinct by the mid-Cenozoic, probably due to climatic changes. The Virginia opossum (Figure 20.3) found in North America today is not a relic of the North American radiation but a relatively recent immigrant (early Pleistocene) from South America. Marsupial fossils have now been discovered in Eocene rocks in Antarctica and Australia, so it is likely they reached Australia from South America across Antarctica when the region was much warmer than it is now, well before the refrigeration of Antarctica in Oligocene times (see Chapters 17 and 23).

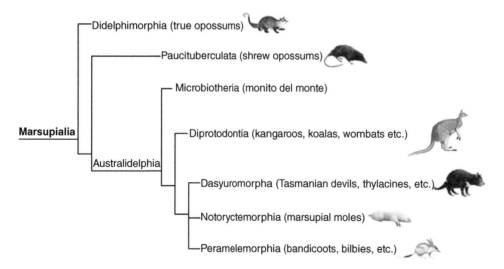

Figure 20.1 A phylogeny of extant marsupials. The opossums, shrew opossums, and microbiotheres are all known from South America. However, the microbiotheres (only one species alive today) are more closely related to the Australian marsupials than the other South American lineage and are included with them in the Australidelphia.

Figure 20.2 *Mimoperadectes*, a little (mouse-sized) opossum-related marsupial from the Eocene of North America. *Source:* By Jorge González (Wikimedia).

| | |
|---|---|
| kangaroos | antelope |
| wallabies | rabbits |
| wombats | marmots |
| phalangers | squirrels |
| koala | sloths |
| "mice" | cats and weasels |
| "moles" | moles |
| numbats | anteaters |
| Tasmanian devil | wolverine |
| †diprotodonts | rhinos, tapirs |
| †marsupial "lion" | large cats |
| †Tasmanian "wolf" | dogs |

Figure 20.4 A gallery of Australian marsupials, each of which has a placental ecological counterpart on other continents. Extinct forms marked with a dagger (†).

Figure 20.3 The Virginia opossum (*Didelphis virginiana*), a relatively recent immigrant to North America from South America, now found as far north as southern Canada. Opossums (not to be confused with the Australian herbivorous possums) are omnivorous and are semi-arboreal (the Virginia opossum has a prehensile tail, not clearly seen in this photo). *Source:* By Drcyrus (Wikimedia).

By the late Cenozoic, marsupials had evolved to fill most of the ecological roles in Australia that are performed by placental mammals on other continents (Figure 20.4). Wallabies and kangaroos are terrestrial long-legged herbivores comparable with antelope and deer, wombats are large burrowing "rodents" rather like marmots, and koalas are slow-moving browsers like sloths. The cuscus and other Australian possums are rather like lemurs, and the numbat is a marsupial anteater, resembling small placental anteaters like the African bat-eared fox rather than the big South American giant anteater. There are marsupial "cats" – quolls, actually more generalized carnivores like civets; marsupial "moles," which are more like the sand-swimming African golden moles than the European soil-shoveling moles; marsupial "mice" – dasyurids, actually more like shrews as they are mainly carnivorous rather than omnivorous; and half a dozen types of gliding possums that can be compared with flying squirrels. The honey possum *Tarsipes* is the only nonflying mammal that lives entirely on nectar and pollen, which it gathers with a furry tongue (Figure 20.5). The possum *Dactylopsila* of New Guinea has evolved specialized teeth and a very long finger to become a marsupial "woodpecker" (see Figure 19.10). The thylacine (Tasmanian wolf or Tasmanian tiger: Figure 20.6) and the Tasmanian devil are marsupial carnivores comparable in size and ecology to coyote and wolverine, respectively. They both once ranged over the main continent of Australia. The Tasmanian devil is now confined to Tasmania and the thylacine, only known from Tasmania in historical times, is probably extinct. The last captive thylacine, called Benjamin, died in the Beaumaris Zoo in Hobart in 1936.

Figure 20.5 *Tarsipes*, the honey possum, which feeds on pollen and nectar with a long hairy tongue. *Source:* Painting by John Gould (1863), in the public domain.

Figure 20.6 The thylacine (*Thylacinus cynocephalus*), portrayed in a painting from 1850 by Joseph Wolf (Wikimedia).

The fossil record of extinct Australian marsupials is even more impressive. Entire families of marsupials are now extinct. Many were very large, including giant kangaroos (see Chapter 23, Figure 23.16) and a giant wombat that each weighed 200 kg or more (450 pounds). *Thylacoleo* was a Pleistocene carnivore whose name means the marsupial lion. It was the size of a leopard and had efficient stabbing and cutting teeth. Steve Wroe has shown that it was better adapted for processing chunks of flesh than any living carnivore, and its bite was one of

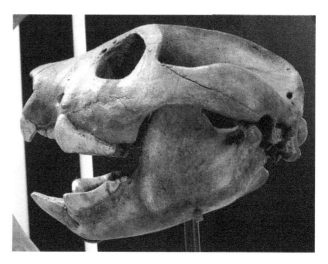

Figure 20.7 *Thylacoleo*, an extinct Australian marsupial carnivore the size of a leopard. *Source:* By Ghedoghedo (Wikimedia).

the most powerful for its size ever evolved by a mammal (Figure 20.7) (see also Chapter 23, Figure 23.15). It also had the unusual feature, not seen in any placental carnivore, of a huge claw on a rotatable thumb, which it may have employed to dispatch its prey.

Diprotodonts were quadrupedal herbivorous marsupials. Miocene forms were sheep-sized and one species foraged in the trees rather than on the ground (Figure 20.8). By the Pleistocene, diprotodonts became very large, and were the largest marsupials ever. *Diprotodon optatum* (see Chapter 23, Figure 23.15) was the size of a rhino and would have weighed between 2 and 3 tons. Discoveries by Mike Archer and colleagues of enormous numbers of Miocene bats and marsupials at Riversleigh, in Queensland, now provide an excellent description of the earlier radiation of these Australian mammals.

People often talk of marsupials as primitive and inferior to placentals, and it's true that today they are outclassed in diversity and range by placentals. But marsupials are in no way inferior (see Chapter 18). For example, a kangaroo may appear rather clumsy as it locomotes slowly around on the ground, using its tail as an extra limb in what is really a five-footed movement. It does use more energy than a placental at this speed, but at high speed a kangaroo is not only very fast (up to 60 kph, or 40 mph), but its incredibly long leaps are much more efficient than the full stride of a four-footed runner of the same weight.

Dromornithids (mihirung in aboriginal legend) were giant extinct Australian birds that evolved large body size and flightlessness (Figure 20.9). They are distantly related to ducks and geese; at one time they were thought to be carnivorous, and Mike Archer called them "the Demon Ducks of Doom." However, subsequent research has shown that they were actually herbivorous, like their living relatives. They are the largest known birds; standing up to 10 ft tall and weighing up to half a ton, they were twice as tall as an emu and more than 10 times the weight. The living Australasian emu and cassowary are large ground-running ratites, related to the ratites on other remnants of Gondwana (see Chapter 14).

The isolated position of Australia has meant that only a few placental mammals reached there after initial Eocene invasion by marsupials. Bats arrived in the Eocene, rodents in the Pliocene (before the mice and rats brought later by humans), and humans in the late Pleistocene. All of these invasions appear to have been from the Asian continent. Interestingly, one type of placental (a small ungulate) appears to have migrated to Australia with the marsupials in the Eocene, from South America via Antarctica, but the lineage soon became extinct.

The original human migrants to Australia, the aboriginals, arrived there around 50 000 years ago (see Chapter 23). It is often thought that they brought the dingo (a kind of dog) with them, but dingoes did not reach Australia until a few thousand years ago, and they may have made their own way there from Asia. Europeans traveling to Australia over the past few hundred years have brought with them a host of invaders, such as rats, cats, dogs, sheep, cattle, rabbits, horses, camels, cactus, fish, and cane toads, with serious results for the Australian ecosystem.

## New Zealand

New Zealand was part of Gondwana until the Cretaceous, and it had a fauna like the rest of the world at that time. But after New Zealand became separated, it underwent a turbulent geological history, being almost completely submerged at times, and much of its current fauna and flora represents descendants of the original inhabitants. Animal old-timers of New Zealand include a diversity of insects, including huge flightless insects – enormous weevils and wetas (giant grasshoppers), four species of frogs (primitive forms that lack a tadpole stage), and the tuatara, the only surviving member of the ancient sphenodontid reptiles (see Chapter 12).

There are only a few native lizard species, which appear to be Cenozoic immigrants: 11 geckos that all have live birth and 18 skinks, of which 17 have live birth. New Zealand had a bigger diversity of reptiles in the past than today, including dinosaurs in the Mesozoic and crocodiles and turtles in the Miocene.

Today, many mammals flourish on New Zealand but they are almost all immigrants brought by humans (Europeans) in the past few hundred years. Mammals may have been part of the original fauna – a Miocene fossil may represent a Mesozoic survivor. But the only truly native mammals are a couple of species of bats, which are known since the Oligocene and must have arrived by flying from Australia. Interestingly, these bats are now quite terrestrial in their habits, unlike bats elsewhere in the world.

Figure 20.8  The late Miocene diprotodont *Nimbadon lavarackorum*, from the Riversleigh National Heritage site in Queensland, shown foraging in the trees accompanied by a juvenile. *Source:* By Peter Schouten (Wikimedia).

Figure 20.9  *Dromornis*, the heaviest bird ever known, from the Miocene of Australia. *Source:* By McBlackneck (Wikimedia).

The dominant fauna of New Zealand today, and for much of the Cenozoic, consists of birds. There is a large diversity of birds there today, including many different types of parrots, and the emblematic animal of the country, the kiwi. Kiwis are the smallest of the ratites (about the size of a chicken) and are unusual among birds in being nocturnal in their habits, foraging for insects with their long beak. Until only a few hundred years ago, very large ratites, the moas, were also present. Like other large ratites, moas were herbivorous. The largest moa (females were much larger than males) was 3.5 m (11 ft) in height (see Figures 20.10 and 23.18). Both the moas and their likely predator, the enormous Haast's eagle (Figure 20.10), twice the size of a modern eagle, went extinct shortly after the arrival of the Maoris (the indigenous native people of New Zealand), around 1300 CE.

Moas coevolved with New Zealand plants so that 10% of the native woody plants have a peculiar branching pattern called *divarication* – they branch at a high angle to form a densely growing plant with interlaced branches that are difficult to pull out or break. There are few leaves on the outside and the largest, most succulent leaves are on the inside. But nine species of divaricating plants that grow more than 3 m (10 ft) tall look more like normal trees once they reach that height, and other divaricating species grow more normally on small offshore islands. The only reasonable explanation of divarication is that it evolved as a defense against browsing moas.

## South America

South America split away from Africa in the Late Cretaceous (around 80 Ma) to become an island continent (Figure 20.11). We know more about mammalian evolution on that continent than on Australia because there is a rich fossil record from many places, for much of the Cenozoic.

In Cretaceous times, the South American mammals and dinosaurs included unique forms belonging to basal Jurassic groups. Examples include the giant dinosaur *Megaraptor*, a large sphenodont, and early mammals such as eutriconodonts and multituberculates (see Chapter 18). However, therian mammals (marsupials and placentals) were not present. Therians first appear in the South American fossil record in the early Cenozoic, presumably immigrants from North America. There was apparently an opportunity for migration between North and South America in the early Cenozoic but after this, faunal interchange was much more limited, until the late Miocene. The early Cenozoic arrivals included opossum-related marsupials, xenarthrans, and various primitive kinds of ungulates. These ungulates originally included some primitive northern forms, but these gave rise to a new, uniquely South American clade, the Meridiungulata, which diversified into five different and distinct lineages.

The climatic changes in the late Eocene coincided with the arrival of new immigrants into South America:

Figure 20.10 A Haast's eagle (*Harpagornis*), related to living eagles but about twice the size, attacking a couple of moas (*Dinornis*). *Source:* By John Megahan (Wikimedia).

Figure 20.11 South America drifted away from Africa, first west and then west-north-west during the Cenozoic, and for most of that time it was an island continent accessible only to lucky or mobile immigrants until it joined up to North America via the isthmus of Panama around 3 million years ago.

Figure 20.12 The capybara is the largest rodent today, weighing around 50 kg. Its scientific name, *Hydrochoerus*, means "water pig": these animals are about the size of a big pig, and they spend much of the time in the water. But unlike pigs, they are grazing herbivores rather than omnivores. *Source:* Photograph taken at Bristol Zoo by Adrian Pingstone (Wikimedia).

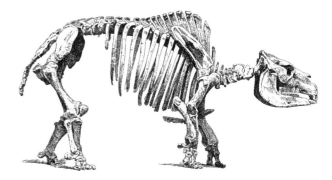

Figure 20.13 *Toxodon*, a rhino-sized notoungulate from the Pleistocene of South America. *Source:* Image in the public domain.

rodents and monkeys, tortoises, and colubrid snakes. The mammals, at least, appear to have reached South America from Africa, as that is where their closest relatives are found. Rodents made the journey a little earlier than primates: the first South American rodent is known from 41 Ma and the first primate from 36 Ma. Both groups radiated widely and remain diverse and successful today. The primates radiated into the distinctive New World monkeys (including marmosets), evolving habits and characters convergently with gibbons and Old World monkeys (see Chapter 21). The rodents were the caviomorphs, a distinctive clade rather different from rodents elsewhere in the world. Commonly known caviomorphs are guinea pigs and chinchillas, that are kept as pets, but there are a number of larger ones, ranging from the size of a cat to a large dog, that are specialized herbivores, almost like mini ungulates. The largest of these is the capybara (Figure 20.12) but capybaras would be dwarfed by some Miocene giants: *Josephoartigasia* from the Pliocene of Uruguay weighed 1 ton (about 2200 pounds)!

After these early Cenozoic migrations, evolution in South America took place more or less in isolation for around 30 m.yr. (although there is growing evidence for some limited exchange in the early Miocene). The strange South American mammals in particular are well known, and they divided up available ecological roles as did mammals elsewhere. Charles Darwin noticed peculiar fossil mammals in Argentina during his voyage on the *Beagle*, and later expeditions to Argentina have found hundreds of beautifully preserved Cenozoic fossils.

From early Cenozoic times, the South American marsupials took on the roles of small omnivores and insectivores

(and still do). There is a living semi-aquatic marsupial, the water opossum or yapok, with webbed feet and a watertight pouch. *Argyrolagus* was a rabbit-sized marsupial from the Pliocene of Patagonia that looked like a giant kangaroo rat. It hopped and had ever-growing molars for grazing coarse vegetation. The arrival of the placental rodents did not affect these small marsupials; indeed, as noted, many of the rodents diversified into roles taken by ungulates elsewhere.

There were some large, heavy-bodied South American ungulates in the early Cenozoic but by the middle Miocene, only two of the original five lineages remained. The notoungulates were more chunky in appearance, ranging from the size of a rabbit to the size of a rhino (Figure 20.13). Many notoungulates had hypsodont or even ever-growing molars, indicating an abrasive diet. The litopterns were more slender, with longer legs and longer faces. Some of them looked rather like horses, and

**Figure 20.14** A scene from the middle Miocene of Patagonia. A pony-sized *Theosodon*, a litoptern, defends her young from the Labrador dog-sized marsupial carnivore, *Borhyaena*. *Source:* From a painting by Charles R. Knight (Wikimedia).

some rather like camels (Figure 20.14); one horse-like litoptern even paralleled true horses in reducing the number of toes to a single digit. Litopterns retained low-crowned molars and presumably had a less abrasive, browsing type of diet.

The South American ungulates radiated on the savannas that dominated the landscape of Patagonia (southern South America) in the Miocene. However, in the cooling climate of the Pliocene, these savannas turned into less productive pampas grassland, and their numbers and diversity declined. Nevertheless, one litoptern (*Macrauchenia*) and one notoungulate (*Toxodon*) survived until around 10 000 years ago, and their remains were well enough preserved that scientists were able to extract the protein collagen from their bones and incorporate the results into a phylogenetic analysis with extant mammals. These South American endemics were shown to be related to perissodactyls.

Xenarthrans, today comprising armadillos, sloths, and anteaters, are also uniquely South American mammals (although today they can also be found in Central and North America). Armadillos and their relatives evolved heavy body armor for protection and became highly successful opportunistic omnivores. The Pleistocene armadillo relative *Glyptodon* was a very large herbivore 1.5 m (5 ft) long. It had a thick armored skullcap as well as body armor, and some glyptodont species had a spiked knob at the end of the tail (see Chapter 23, Figure 23.11). Glyptodonts were certainly too big to burrow like the smaller armadillos.

Sloths now live only in trees, eating leaves and moving with painful slowness, but the usual type of sloth for most

of the Cenozoic were ground dwelling, ranging from the size of a pony (around 250 kg, or 550 lb) to huge giants like *Megatherium*, which was the size of an elephant (around 4 tons) (see Chapter 19, Figure 19.3). One type of ground sloth, *Thalassocnus*, was even semi-aquatic.

Anteaters are related to sloths but are now specialized to an amazing extent for eating ants and termites. The giant anteater tears apart termite nests with tremendously powerful clawed forearms, while smaller anteaters, like the tamandua, live in trees and eat the ants that live there.

The most impressive South American creatures were the larger carnivores, which were all marsupials known as sparassodontids. Sparassodonts are technically metatherians, rather than marsupials, as they fall outside the crown group of living marsupials. The best known of the group are the bear- or wolf-like borhyaenids (Figure 20.14), mainly known from the Miocene. Other sparassodonts, many of them earlier forms, were smaller and more like ferrets or civets, possibly more omnivorous than the borhyaenids. Thylacosmilids were saber-toothed forms, paralleling the saber-toothed cats of the northern hemisphere. *Thylacosmilus* (Figure 20.15) looks like a fearsome predator but it had a number of strange features suggesting that it did not have quite the same ecology as the placental saber-tooths; for example, it lacked incisors and its bite was less powerful. As with the native ungulates, the diversity of the carnivorous marsupials declined in the Pliocene.

These amazing marsupials had unusual competitors for mastery of the carnivorous guild, the phorusrhacids: flightless, ostrich-sized birds equipped with very

Figure 20.15 The South American sparassodont marsupial saber-tooth *Thylacosmilus*, from the Pliocene of Argentina. *Source:* By Dmitry Bogdanov (Wikimedia).

Figure 20.16 The phorusrachid *Llallawavis*, an ostrich-sized flightless predatory bird from the Pliocene of Argentina. *Source:* By Rextron (Wikimedia).

powerful tearing beaks as well as foot talons (Figure 20.16). Many large flightless birds of the Cenozoic that were once thought to have been carnivorous (e.g., diatryma-forms, dromornithids) have since been shown to be herbivorous, but phorusrachids were definitely carnivorous. Other South American birds included the teratorns,

related to the New World vultures, including some of the largest flying birds known. Teratorns are also known from the Pleistocene of North America (see Chapter 23, Figure 23.10).

South America had its own group of crocodiles, the sebecids. They apparently evolved in Gondwana in the Cretaceous, survived the K–Pg extinction and radiated in the early Cenozoic in South America to become powerful terrestrial predators. Unlike aquatic crocodiles, they had high, deep skulls and snouts. Other crocodilians in South America also evolved into unusual morphologies; for example, a duckbilled caiman is known from the Miocene of Colombia. The largest turtle of all time, *Stupendemys*, lived along the north coast around 5–6 Ma.

This unique ecosystem suffered four tremendous shocks in 10 million years and has almost completely disappeared. First, Antarctica froze up, with the result that the Humboldt Current, flowing most of the way up the west coast of South America, became much colder and stronger. Second, tectonic activity along the Pacific coast raised the Andes as a major mountain chain. Together, these two events drastically lowered rainfall over most of the continent, and much of the area turned from forest and well-watered plain to dry steppe. This led, in the later Miocene, to the extinction of many animals, including the terrestrial crocodiles and especially the large-bodied savanna herbivores and the marsupial carnivores that fed on them.

Third, South America drifted northward toward Central and North America (Figure 20.11). By about 6 Ma, the gap was small enough to allow a few animals to cross it, more or less by accident. North American raccoons crossed to the south, while two kinds of ground sloths crossed to the north. Finally, at about 3 Ma, the last important sea barrier was bridged and animals could walk from one continent to the other.

The "Great American Biotic Interchange" happened after 3 Ma (Figure 20.17), with the formation of the isthmus of Panama linking the two continental land masses. The South American immigrants to North America flourished there for a while, but many met their demise in the Pleistocene extinctions (see Chapter 23) in both North and South America (e.g., ground sloths, glyptodonts, giant birds). Overall, the North American species that moved south were more successful, but even some of them suffered in the Pleistocene extinctions (e.g., horses and gomphothere proboscideans). However, today half of the mammalian genera in South America are of North American origin. While some of the original fauna remains successful and diverse (smaller xenarthrans, monkeys, caviomorph rodents, and opossums), many of the larger forms are extinct (sparassodont carnivorous marsupials, native ungulates, phorusrachid birds).

Figure 20.17 Some of the animals involved in the Great American Biotic Interchange. South American creatures that went north, olive green; North American creatures that went south, blue. The South American migrants to North America included ground sloths, anteaters (limited to Central America), armadillos, glyptodonts, notoungulates (*Toxodon*), opossums, capybaras (now extinct in the US), monkeys (limited to Central America), and teratorn and phorusrachid birds. The North American migrants to South America included horses (now extinct there), gomphothere proboscideans (now extinct everywhere), tapirs, peccaries, llamas, deer, bears, raccoons, skunks, weasels, dogs, cats, peccaries, shrews, rabbits, and murid rodents (rats and mice).
*Source:* Diagram by Woudloper (Wikimedia).

It's difficult to know how much of this replacement of the South American fauna was due to competition from "superior northerners," as is often supposed. As previously noted, the environmental changes in cooling climates of the late Miocene and Pliocene resulted in the loss of most of the ungulates and large carnivores before the interchange took place. Many of the groups that are highly successful and diverse in South America today, including the dogs and cats, are tropical species and we have little knowledge of what lived in the earlier tropics, as such habitats are rarely preserved in the fossil record. And we must not forget that many of the original South American animals are doing well today in the tropics of Central America, which is of course part of the North American continent.

The geographical changes that had allowed the interchange also altered the currents in the Atlantic Ocean, and this in turn caused drastic changes in the land ecology of North and South America as the northern ice ages began in earnest around 2.5 Ma. Both North and South American faunas suffered a catastrophic extinction in the late Pleistocene, and we shall examine this in Chapter 23.

## Africa

Africa (plus Arabia, = Africarabia) was part of Gondwana until the Cretaceous, when it broke away from South America on the west and Antarctica and India on the east (see Chapter 16, Figure 16.8). From Late Cretaceous times onward, Africa and South America, and the animals and plants living there, had increasingly different histories. Africa lay south of its present position, bounded on the north by the tropical Tethys Ocean, not yet joined to Eurasia (see Figure 20.20).

Africa had dinosaurs much like those of the rest of the world during the Late Cretaceous, but there is no record of any Cretaceous mammal on the African mainland. There are a few fossils of Mesozoic mammals on the island of Madagascar, off the southeast coast of Africa (see Figure 20.20), but these are all basal forms, unrelated to the therians that later radiated on the African mainland.

Our first good look at the fossil Cenozoic life of Africa comes from the Eocene rocks of Egypt, laid down on the northern edge of the continent. Shallow warm seas teemed with microorganisms whose shells formed the limestones from which the Pyramids and the Sphinx were carved.

Here we find early whales and sea cows, which probably evolved adaptations for marine life in swamps and deltas along the shores of Tethys. Moeritheres were amphibious basal proboscideans (Figure 20.18). Other Eocene fossils from North Africa include some early primitive carnivores, the creodonts (also known from other continents), and a handful of early primates.

By Oligocene times, Egypt was the site of lush deltas where luxuriant forest growth housed rodents, primates, and bats, all recent Eurasian immigrants. Pig-like anthracotheres had crossed from Eurasia but there were African groups too, including a diversity of early proboscideans. Hyraxes are rabbit-sized herbivores that look much like rodents today (see Chapter 19, Figure 19.2), but in the Oligocene hyraxes were much more diverse and many were much larger, the size of pigs. *Arsinoitherium* was a large-bodied browser related to proboscideans (Figure 20.19).

Eocene and Oligocene African mammals are a mixture of native African groups and a few successful immigrants

Figure 20.18 *Moeritherium*, a hippo-like (but the size of a tapir or a pygmy hippo) early proboscidean browser from the Eocene of Egypt. *Source:* Artwork by Heinrich Harder, about 1920, now in the public domain.

Figure 20.19 *Arsiniotherium*, a large-bodied browser from the Oligocene of northern Africa. *Source:* Image by Trent Schindler for the National Science Foundation, in the public domain.

Figure 20.20 Africarabia drifted slowly northeast during the Cenozoic. Finally it collided with western Asia in the Early Miocene along a line that is now the Zagros Mountains. The African continent then rotated slightly clockwise, splitting away from Arabia. At the end of the Miocene the northwest corner of Africa collided with western Europe to close off the Mediterranean Sea as a vast lake that quickly dried up. Meanwhile, the Red Sea opened up as Africa split away from Arabia, and the great African Rift Valley was formed. Madagascar is the island off the southeast coast.

from Eurasia. Even in the late Oligocene, the large African mammals were still arsinoitheres and a diverse set of proboscideans and hyraxes. But shortly before the start of the Miocene, Africarabia drifted far enough north to bring it close to Eurasia, and finally the two continental edges collided around 24 Ma (Figure 20.20). There were important times of uninterrupted migration between the two land masses. The interchanges affected animal life throughout the Old World, almost on the same scale as the Great American Interchange.

Twelve families of small mammals appeared in Africa in the early Miocene, mostly insectivores and rodents from Eurasia. Antelope and pigs largely replaced the hyraxes at medium sizes, and rhinos and the first giraffes were large-sized invaders. Cats arrived and began to replace the older creodonts. Going the other way, proboscideans walked out of Africa into Eurasia, in at least two major lineages: mastodons and elephants. Hyraxes also migrated to Asia and can be found today in the Middle East, as the "conies" of the Bible. Some large creodonts even reinvaded Eurasia from Africa. In a second exchange around 15 Ma, a new set of African animals, including apes, quickly spread over the woodlands and forests of Eurasia. Hyenas and shrews migrated into Africa.

By the end of the Miocene, more immigrants had appeared in Africa: small animals, including many bats, the three-toed horse *Hipparion*, and dogs that replaced many of the hyenas. Meanwhile, hippos evolved in Africa, and the antelope and cattle that had arrived earlier evolved into something close to the incredible diversity we see today in the last few game reserves.

Africa and Eurasia have been connected by land since the Miocene but this does not automatically imply free exchange of animals. For example, the Mediterranean Sea dried into a huge salty desert like a giant version of Death Valley, around 6 Ma. Only a few animals could have crossed this barrier. Later, the development of desert conditions in the Sahara formed another fearsome barrier to animal migration for most of the past few million years. Today, North African animals are more like those of Eurasia than those of sub-Saharan Africa.

The Early Pleistocene saw a large extinction in Africa but the extinct species were replaced by newly evolving species, so total diversity remained high. Africa apparently did not feel the effects of the ice ages too drastically, and there were few extinctions at the end of the Pleistocene, in contrast to the rest of the world (see Chapter 23). Today Africa, along with southern Asia to a lesser extent, remains as the place where the former worldwide diversity of large mammals still persists.

## Madagascar

Madagascar is the fourth largest island in the world today, and it has sometimes been termed the "eighth continent." Its habitats range from tropical forest to desert. Although Madagascar is considered to be part of Africa, it had an independent geological and biogeographical history. Madagascar was part of Gondwana in the Mesozoic and as that supercontinent broke apart, it became isolated in the Indian Ocean around 88 Ma. Its present-day fauna and flora are highly endemic, with 90% of the species unique to the island. Some of these species are descended from lineages from Gondwanan times, but many of them (including all of the mammals) evolved from immigrants that must have arrived there via rafting from Africa.

There is a good fossil record from the Cretaceous of Madagascar, including a giant frog, several dinosaurs, a terrestrial herbivorous crocodile, as well as other reptiles, birds, and a few mammals. Strangely, these animals were not similar to contemporaneous African ones, but rather to animals from South America and India, revealing complex biogeographic interconnections within the Gondwanan continent. Unfortunately, the Madagascan Cenozoic fossil record is fragmentary: there is a good selection of Holocene subfossil animals but little record of their Cenozoic history.

Although all of the fauna and flora of Madagascar are unique and interesting (for example, it is home to two-thirds of the world's species of chameleons), we will focus on the mammal fauna. Lemurs are primates unique to Madagascar, and perhaps its emblematic mammals. We will consider them further in Chapter 21 and discuss here the mammalian groups that are less well known.

Three nonprimate lineages of terrestrial mammals are tenrecs (afrotheres), rodents, and carnivorans. There is also a diversity of bats (46 species, including fruit bats) and marine mammals around the coasts (including a dugong, a seal, and around 36 species of whales). A few thousand years ago, there were also several kinds of pygmy hippos and a weird aardvark-like mammal probably distantly related to tenrecs.

The tenrecs, related to the African otter shrews, arrived on Madagascar between 42 and 25 Ma. The 30 or so species of tenrecs take the role of small omnivores and insectivores (see Chapter 19, Figure 19.2) and include species that resemble hedgehogs and small otters.

The rodents belong to the family Nesomyidae, which is also known from Africa, and they arrived on Madagascar 24–20 Ma. The 39 endemic species are mostly small forms resembling mice and rats, but there is a large (rabbit-sized) form – the Malagasy giant jumping rat (Figure 20.21).

The carnivores are the most recent immigrants, arriving in Madagascar 26–19 Ma. Many of them look like mongooses but they are not in the same family; rather, the 10 species of Malagasy carnivores are in their own family (albeit mongoose related), the Eupleridae. The most spectacular of these carnivores is the fossa (Figure 20.22), a fierce lemur-eating predator that has converged on the cat-like ecological niche.

Figure 20.21 The Malagasy giant jumping rat, *Hypogeomys*. *Source:* Photograph taken at Bristol Zoo by Adrian Pingstone (Wikimedia).

Figure 20.22 The fossa (*Cryptoprocta*), a carnivore unique to Madagascar. This animal is very cat-like in appearance and is a little bigger than a large house cat in size. It evolved into a cat-like ecological niche on an island where no cats were present. However, it is more arboreal in its habits than any type of cat. *Source:* Photograph taken at Frankfurt Zoo by MM (Wikimedia).

The unique ecosystem of Madagascar is one of the most threatened in the world today, and many of the mammal species can only be found in protected areas and national parks. Clearance of forest by humans leaves barren landscapes subject to soil erosion, with devastating effects on the native flora and fauna.

## Islands and Biogeography

Strict geographic barriers prevent land plants and animals living on islands from moving easily to other land areas, and potential invaders also must cross barriers. This means that island faunas and floras tend to evolve in greater isolation than those with wider and more variable habitats. Of course, this is true at any scale, whether we look at small islands or continent-sized ones. Islands past and present can teach us a great deal about evolution. It is no accident that Darwin was particularly enlightened by his visit to islands like the Galápagos, and Wallace by his years in Indonesia.

We have seen some of the vagaries of continental faunas over a time scale of tens of millions of years, but it is worth looking at cases where smaller-scale events on smaller islands over smaller lengths of time show the rapidity and power of natural selection in isolated populations.

There is a general "island rule," whereby large mammals on islands tend to get smaller, while smaller ones get larger. Larger mammals probably become smaller because food resources on islands are limited. Smaller mammals may become larger if the island is free of predators and they no longer need to be small enough to hide in holes. Classic examples of larger mammals becoming smaller on islands are the dwarfed elephants and hippos in the Pleistocene fossil record of the Mediterranean islands. The late Miocene record of Gargano, in Italy, provides an interesting case of island evolution, including numerous examples of smaller mammals becoming larger.

In 1969, three Dutch geologists were exploring the Mesozoic limestones of the Gargano Peninsula in southern Italy (Figure 20.23). Some time in the early Cenozoic, this block of land was raised above sea level and caves and fissures formed in the limestone. In early Miocene times, the Gargano area was cut off from the mainland by a rise in sea level to form an island in the Mediterranean Sea, isolating the animals living there. Over a few million years, animals occasionally fell into fissures in the limestone, where they were covered by thin layers of soil and preserved as fossils. Today, the limestone is quarried for marble and the fossil bones can be found in the pockets of ancient soils exposed in the quarry walls.

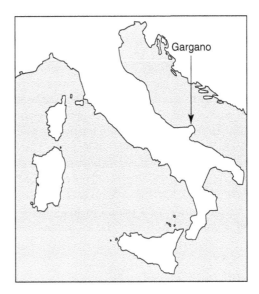

Figure 20.23 The Gargano Peninsula in southern Italy. It was a small rocky offshore island in Miocene times.

(a)

(b)

Figure 20.24 Some of the endemic mammals from Gargano. (a) The five-horned artiodactyl *Hoplitomeryx*. *Hoplitomeryx* was related to deer but was not a true deer. (b) The giant hedgehog *Deinogalerix*, shown besides a regular European hedgehog for comparison. *Source:* artwork by Mauricio Antón, by permission.

No large animals were isolated on Gargano as it was cut off. The only large reptiles were swimmers (turtles and crocodiles) and the only mammalian carnivore was also a swimmer, a large otter with rather blunt teeth that probably ate shellfish most of the time and would not have hunted on land.

Small mammals evolved quickly into larger forms, including pikas (nonhopping relatives of rabbits), dormice, true mice, and hamsters. One particularly well-known "giant" form was the hedgehog *Deinogalerix* (Figure 20.24b). Some of the Gargano mice grew to giant size, with skulls 10 cm (4 in) long, and many evolved fast-growing teeth as complex as those of beavers. They probably chewed very tough material. *Hoplitomeryx* was a small deer-like artiodactyl which evolved horns instead of antlers (Figure 20.24a). If there were no cats, dogs, or other terrestrial carnivores, how were the small mammals kept under control? And why did *Hoplitomeryx*

evolve spectacular horns, if there were no carnivores to fight off? The horns were too lethal to have been used for fighting between individuals of the species.

The answer to these questions seems to have been raptors – birds of prey. A giant buzzard, *Garganoaetus*, was as large or larger than a golden eagle. Presumably it hunted by day. It would have been perfectly capable of taking a small or young *Hoplitomeryx*, and the horns may have evolved to protect the back of the neck against raptors (horns exist year-round while antlers are shed seasonally). Normally, small deer hide in vegetation but Gargano was a bare, limestone island, with no cover by day. At night, the owls took over; the largest barn owl of all time evolved on Gargano.

## Further Reading

Bacon, C.D., Molnar, P., Antonelli, A. et al. (2016). Quaternary glaciation and the Great American Biotic Interchange. *Geology* 44: 375–378.

Black, K.H., Archer, M., Hand, S.J. et al. (2012). The rise of Australian marsupials: a synopsis of biostratigraphic, phylogenetic, palaeoecologic, and palaeobiogeographic understanding. In: *Earth and Life* (ed. J.A. Talent). New York: Springer.

Bond, M., Tejedor, M.F., Campbell, K.E. Jr. et al. (2015). Eocene primates of South America and the African origin of New World monkeys. *Nature* 520: 538–541.

Buckley, M. (2015). Ancient collagen reveals evolutionary history of the South American ungulates. *Proceedings of the Royal Society B: Biological Sciences* 282: 20142651.

Carrillo, J.D., Forasiepi, A., Jaramillo, C. et al. (2015). Neotropical mammal diversity and the Great American Biotic Interchange. *Frontiers in Genetics* 5: 451.

Croft, D.A. and Simeonovski, V. (2016). *Horned Armadillos and Rafting Monkeys: The Fascinating Fossil Mammals of South America*. Bloomington: Indiana University Press. [A popular account of the extinct South American mammals, with wonderful illustrations].

Faurby, S. and Svenning, J.-C. (2016). Resurrection of the Island Rule: human-driven extinctions have obscured a basic evolutionary pattern. *American Naturalist* 187: 812–820.

Flannery, T.F. (1995). *The Future Eaters: An Ecological History of the Australasian Lands and People*. New York: George Braziller. [Part I is the story of Australasian life before the arrival of humans].

Kappelman, J., Rasmussen, D.T., Sanders, W.J. et al. (2003). Oligocene mammals from Ethiopia and faunal exchange between Afro-Arabia and Eurasia. *Nature* 426: 549–552.

Sánchez-Villagra, M.R. (2013). Why are there fewer marsupials than placentals? On the relevance of geography and physiology to evolutionary patterns of mammalian diversity and disparity. *Journal of Mammalian Evolution* 20: 279–290.

Samonds, K.E., Godfrey, L.R., Ali, J.R. et al. (2012). Spatial and temporal patterns of Madagascar's vertebrate fauna explained by distance, ocean currents, and ancestor type. *Proceedings of the National Academy of Sciences of the United States of America* 109: 5352–5357.

Sen, S. (2013). Dispersal of African mammals in Eurasia during the Cenozoic: ways and whys. *Geobios* 46: 159–192.

Strömberg, C.A., Dunn, R.E., Madden, R.H. et al. (2013). Decoupling the spread of grasslands from the evolution of grazer-type herbivores in South America. *Nature Communications* 4: 1478.

Worthy, T.H., Tennyson, A.J.D., Archer, M. et al. (2006). Miocene mammal reveals a Mesozoic ghost lineage on insular New Zealand, southwest Pacific. *Proceedings of the National Academy of Sciences of the United States of America* 103: 19419–19423.

Wroe, S., McHenry, C., and Thomason, J. (2005). Bite club: comparative bite force in big biting mammals and the prediction of predatory behaviour in fossil taxa. *Proceedings of the Royal Society B: Biological Sciences* 272: 619–625. [*Thylacoleo*].

## Questions for Thought, Study, and Discussion

1 Find three examples where mammals did manage to cross ocean barriers and colonize another continents. Think of ways in which ocean crossing could have happened, and try to think of characters that such pioneering animals had in common.

2 Choose a good example of convergent evolution in two mammals or mammal groups in different continents. Describe the similarities, try to identify differences, and explain how the similarities are related to their way of life.

Were marsupials inferior critters?
Well, no, in some ways they were fitter
In a tough situation
Unconfined by gestation
They could simply get rid of their litter

21

Primates

---

**In This Chapter**

We are particularly interested in our own ancestry. It has always seemed reasonable that the earliest primates would have been small tree dwellers, and that has now been confirmed by multiple fossil discoveries. Larger primates with a greater diversity of lifestyles started to radiate around 25 million years ago. The diversity of living primates has a long and complex history. An early diverging branch comprised lemurs, bushbabies, and lorises, found in Africa and Asia. The Asian tarsier is rather like the African bushbabies but is actually more closely related to apes and monkeys (anthropoids). The earliest anthropoids were Asian but the major radiation was African (living Asian anthropoids are descended from the African ones). One branch of anthropoids managed to cross the Atlantic to South America to become the South American monkeys. The branch remaining in Africa split into monkeys and apes. There was a large radiation of generalized, medium-sized apes, which diversified into Europe and Asia in the Miocene warm period. The apes alive today, including ourselves, are the specialized (and mostly large) remnants of a once more generalized and diverse radiation.

## Primate Characters

Most living primates are small to medium-sized (rat-size to dog-size), tropical, tree-dwelling animals; they are mostly herbivorous, eating fruit and soft leaves, although the smallest ones (mainly among the more basal prosimians) also eat insects. Primate ancestors likely searched for insects, fruit, seeds, or nectar on small branches, high in trees and in smaller bushes, as is true of their closest living relatives, the tree shrews and the dermopterans (also known as colugos or "flying lemurs") (Figures 21.1 and 21.2). The group containing all these mammalian orders is the Euarchonta, a division of Boreoeutheria (see Chapter 19, Figure 19.1).

Living primates have large eyes, turned forward to give excellent stereoscopic vision. This anatomical adaptation may have evolved in primates – as also in owls and fruit bats – to help them search for food by sight rather than smell, because it allows the animal to judge the distance of a food item without moving its head. Primates also have a reduced sense of smell in comparison with most other mammals. Stereoscopic vision promotes agility and coordination, especially when an animal has hands and feet adapted for grasping and fine manipulation, with nails rather than claws, and sensitive finger pads, as seen in living primates. Grasping feet and hands allow primates to forage along narrow branches, and live prey or other food can be reached or seized by a hand rather than by a lunge with the whole body and head.

Primate fetuses show rapid growth of the brain relative to the body, so they are born with relatively larger brains than other mammals. Primate gestation time is long for their body size and primates have small litters of young that develop slowly and live a long time. Primates evolved high learning capacity, complex social interactions, and unusual curiosity. The evolution of curiosity is useful in searching for food, and high learning capacity, memory, and intelligence help individuals to make correct responses in a complex, ever-changing environment.

Living primates are often divided into two groups: smaller-brained and usually small-bodied animals called

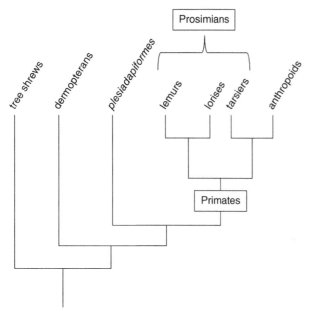

Figure 21.1 Cladogram of the Euarchonta, including the clades of mammals close to primates, and also showing the major groups of living primates. The term "lorises" also includes pottos and bushbabies (galagos). Extinct groups are in italics.

many early primates that expand our picture of very early primate evolution.

## The Earliest Primates

The Plesiadapiformes, an important group of animals found mainly in the Paleocene of North America and Europe, are the extinct sister group of primates (Figure 21.1). They were not a distinct clade but rather a paraphyletic grouping of stem primates and included forms that ate insects, fruit, gum, and soft leaves. Plesiadapiforms in general looked and probably lived like large rodents (Figure 21.2), they may have competed to some extent with Paleocene multituberculates (see Chapter 18).

Larger (cat-sized) plesiadapiforms were rather heavy in build, ecologically like squirrels or marmots, small brained and rather small-eyed. The mouse-sized *Carpolestes* (Figure 21.3) had grasping hands and feet and the big toe had a nail (like a primate) while the other toes had claws (like other plesiadapiformes). But *Carpolestes* did not have stereoscopic vision and apparently did not leap, as do similar small primates today (e.g., bushbabies and small lemurs).

The first true primates are known from the earliest Eocene of North America and Eurasia and were of a similar size to small monkeys today. Adapids looked rather like small lemurs and probably moved in the same way. They were mostly fruit and leaf eaters. *Darwinius*

prosimians, and the relatively large-brained anthropoids (monkeys and apes). However, prosimians contain two clades, each with a long evolutionary history. The lorises and lemurs form one clade (Strepsirrhini or "wet noses,"), while tarsiers form another, linked with anthropoids in the Haplorhini ("dry noses") (Figure 21.1). But there are

Figure 21.2 The groups of the Euarchonta. Clockwise from the upper left: the plesiadapiform *Plesiadapis*; the tree shrew (order Scandentia) *Tupaia*; the baboon (order Primates) *Papio*; and the colougo (order Dermoptera) *Cynocephalus*. *Source:* By Esculapio (Wikimedia).

Figure 21.3 The derived plesiadapidiform *Carpolestes* from the Eocene of North America. After Bloch and Boyer (2002). *Source:* Image by Sisyphos23, based on a drawing by Mateus Zica (Wikimedia).

Figure 21.5 The beautifully preserved skeleton of *Darwinius* from the Eocene of Germany. *Source:* Photograph from Franzen et al. (2009), published in PLoS One (Wikimedia).

Figure 21.4 Life reconstruction of the adapid *Darwinius* by Nobu Tamura (Wikimedia).

Figure 21.6 Life restoration of *Tetonius*, a small Eocene omomyid from North America, as an alert tarsier-like animal. By L. Kibiuk under the supervision of K. D. Rose. *Source:* Courtesy of Kenneth D. Rose of Johns Hopkins University.

(Figures 21.4 and 21.5) is the best-preserved adapid, from the Eocene of Germany. Omomyids were small, alert, active nocturnal insect eaters in the forest (Figure 21.6). Ecologically, they were probably like tarsiers, and in terms of phylogenetic relationships, they were probably near the base of the anthropoid/tarsier clade, or the sister group of tarsiers (Figure 21.1).

The northern continents slowly cooled as they drifted northward during the Eocene, and ice began to form over Antarctica (see Chapter 23). Finally, by the end of the Eocene, with a few stragglers persisting a little later, primates disappeared from northern latitudes and did not return until the world warmed again in the early Miocene (see Chapter 20).

## The Living Prosimians

Living lemurs are confined to Madagascar and genomic data show that they must have reached that island from Africa by the mid-Eocene. There is little fossil record for the Cenozoic and fossils of lemurs are not found until the Pleistocene. However, lemurs evidently flourished in their island refuge; today there are five families of lemurs, containing nearly 100 different species.

Most living lemurs are medium-sized (weighing a few pounds) and are mainly herbivorous, eating fruits and leaves, although some smaller ones are insectivorous, as is the cat-sized peculiarly specialized aye-aye (see Chapter 19, Figure 19.11). Some lemurs are specialists at vertical clinging and leaping, in which the front limbs are used for manipulating, grasping, and swinging, while the hind limbs are powerful for pushing off. A few lemurs are small: *Microlemur* the mouse lemur (Figure 21.7) weighs only 50 g or so (about 2 oz). Most prosimians, lemurs included, are nocturnal (active at night) but some of the larger lemurs are diurnal (active during the day).

Until recently (a couple of thousand years ago or less), there was a diversity of relatively giant lemurs that paralleled the diversity of apes in Africa and Asia. The largest lemur, *Archaeoindris*, reached around 200 kg, the size of a gorilla (Figure 21.8) and became extinct only recently; as an adult it must have been a ground dweller. *Palaeopropithecus* was adapted for moving slowly in the forest canopy in much the same way as the South American sloth, while *Megaladapis* was probably rather like the Australian koala in its ecology.

Figure 21.8 The largest lemur, *Archaeoindris*, from the Pleistocene of Madagascar, now extinct. *Source:* Artwork by Smokeybjb, based on a reconstruction by Stephen Nash (Wikimedia).

Figure 21.9 Tarsier from the Philippines. *Source:* Photograph by Roberto Verzo and used under Creative Commons license.

Lorises and bushbabies are small, nocturnal hunters of insects. Slow-moving lorises live in African and Southeast Asian tropical forests, while agile, leaping bushbabies (or galagos) are exclusively African. Tarsiers are also nocturnal small, agile insectivores (Figure 21.9), found today only in Southeast Asia. Tarsiers have huge eyes, each weighing as much as the brain. Tarsier fossils are known since the Eocene, and tarsiers once had a broader distribution throughout southern Asia and possibly also Africa. Both anatomical (e.g., the dry nose) and molecular data show that they are more closely related to anthropoids than are lemurs or lorises. Tarsiers diverged from anthropoids so long ago that the two groups share little similarity today.

Figure 21.7 The gray mouse lemur, *Microlemur*, from Madagascar. *Source:* Photograph by Gabriella Skollar, edited by Rebecca Lewis (Wikimedia).

## The Origin of Anthropoids

The living higher primates, or anthropoids (monkeys and apes), have evolved into a variety of life styles and habitats that extends from the huge herbivorous gorillas to the tiny gum-chewing marmosets of South America. Some of the Eocene anthropoids were even smaller, and although anthropoids are today mostly larger than prosimians, they came from a stock of very small primates. Anthropoids differ from prosimians in a variety of ways: they are generally larger-brained, with a shorter snout, and even more forward-directed eyes with the back of the eye socket walled in by bone. Extant anthropoids are almost all diurnal, the only exception being the night monkey (*Aotus*, also known as the owl monkey) of South America.

The shift from prosimian to anthropoid may have originally matched a shift from a nocturnal habit to a diurnal one, and living anthropoids have better color vision than prosimians. All of the Old World anthropoids have trichromatic vision (i.e., they can see three basic colors, while most mammals can only see two), while New World monkeys are somewhat variable in this ability. Only some of the New World monkeys have trichromatic vision (and it seems to be more prevalent in females than

in males) but they have retained a better sense of smell than the other living anthropoids. Color vision in mammals in general is discussed in Chapter 18.

While most living prosimians are arboreal (tree-living; a few lemurs are more terrestrial), anthropoids exhibit a diversity of locomotor types. Many Old World monkeys are terrestrial quadrupeds (e.g., macaques and baboons). Among the arboreal forms, smaller anthropoids may leap from branch to branch with their bodies held vertical, and larger anthropoids may practice suspensory locomotion (i.e., move by hanging below the branch). A highly specialized form of locomotion, brachiation (or swinging from branch to branch), has evolved convergently in two groups of anthropoids: in the spider monkeys of South America (which also employ a prehensile tail in this effort) and the gibbons and siamangs of Asia.

Living anthropoids are divided into two main evolutionary groups: platyrrhines (= flat-nosed), the New World radiation in South America, and catarrhines (= narrow nosed), the Old World radiation in Africa and Asia. Catarrhines can be further divided into the cercopithecoids (Old World monkeys) and the hominoids, which include gibbons, great apes, and humans (Figure 21.10).

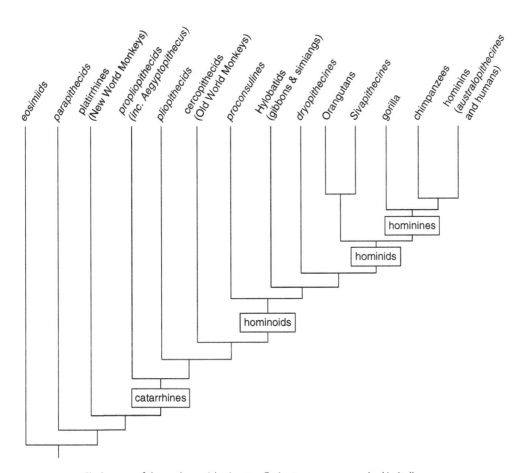

Figure 21.10 Cladogram of the anthropoid primates. Extinct groups are marked in italics.

## The Eocene and Oligocene Radiation of Anthropoids

The earliest known and most primitive of the anthropoids were the eosimiids (= dawn monkeys) from the late Eocene and early Oligocene of India, Pakistan, and China (Figure 21.11). However, the main radiation of early anthropoids is known, from the same time, from the Fayum district of Egypt, not far from Cairo. Thousands of fossilized tree trunks, some of them more than 30 m (100 ft) long, show that tropical forests of mangroves, palms, and lianas grew along the levees of a lush, swampy delta. Water birds such as storks, cormorants, ospreys, and herons were abundant, as they are today around the big lakes of central Africa. Fishes, turtles, sea snakes, and crocodiles lived in or around the water, and early relatives of elephants and hyraxes foraged among the rich vegetation. The primates presumably ate fruit in the trees. The same fauna has been discovered as far south as Angola, so the Fayum animals were widespread around the coasts of Africa in Eocene and Oligocene times.

More than 2000 specimens of at least 19 species of fossil primates have now been collected from the Fayum deposits, including prosimians as well as anthropoids, all of which look like tree-climbing fruit and insect eaters. Many of the Fayum primates had some advanced characters but were more generalized than living monkeys or apes. They are placed into a basal stem anthropoid group called parapithecids (Figure 21.10). Parapithecids were small, weighing only up to 3 kg (7 pounds). This miscellaneous group probably had a basic style of primate ecology, eating fruit in the trees.

Those Fayum anthropoids that are well enough known to compare individual sizes are sexually dimorphic. Males are larger and had much larger canine teeth than females, implying that males displayed or fought for rank, and that the animals had a complex social life that included groups of females dominated by a single male. Among living primates, it is generally the larger-bodied species that have these characters, especially in the Old World. However, the Fayum anthropoids show that size is not important in evolving these sex-linked characters, and they also suggest that these traits may well be basic to anthropoids. Elwyn Simons and his colleagues, who collected many of these fossils, suggest that these canines arose when anthropoids became active in daylight; group defense may be linked with the social structure.

*Aegyptopithecus* (Figure 21.12) is the best known of the Fayum anthropoids, belonging to a basal group of catarrhines called the propliopithecids (Figure 21.10). It was a larger, powerfully jawed, monkey-sized primate with an adult weight of 3–6 kg (7–14 pounds). Its heavy limb bones suggest that it was a powerfully muscled, slow-moving tree climber, ecologically like the living howler monkey of South America. *Aegyptopithecus* is often incorrectly termed an "early ape," which underscores the fact that many of these early anthropoids looked more like small apes than monkeys. Because we are apes, we tend to think of apes as being the derived form and monkeys being the more primitive ones. But, in fact, monkeys are highly specialized in their own right, and while the living apes are also specialized in their own

Figure 21.11 *Eosimias*, from the late Eocene of China, was one of the earliest anthropoids. It was tiny – at only 12 g it was much smaller than a mouse. *Source:* Drawing by DiBdg (Wikimedia).

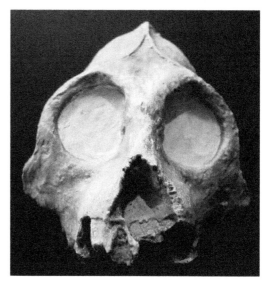

Figure 21.12 One of the fine casts on display at the Museum of Anthropology at the University of Zürich. This is *Aegyptopithecus* from the Fayum deposits of Egypt. *Source:* Photograph by Nicholas Guérin, and used under the Creative Commons Attribution-Share Alike 3.0 Unported license.

ways, a general ape-like body plan may be the basic one for the catarrhine anthropoids.

Another group of catarrhines, more closely related to the living radiation than *Aegyptopithecus*, was the plio-pithecids. This was the first group to leave Africa, being known from the early to late Miocene of Europe and Asia. They were medium-sized primates, about the size of small dog, and are interesting because at least some of them evolved limbs rather like those of gibbons, with the implication that they may have been brachiators.

## The New World Monkeys (Platyrrhines)

Primates reached South America by the late Eocene and evolved there in isolation, probably from African immigrants that crossed the widening Atlantic. No prosimian or ape-like primate has ever been found in South America. Instead, the New World primates evolved to fill the ecological niches that monkeys and gibbons occupy in Old World forests. They evolved a number of typically monkey-like features in parallel with the Old World primates, such as a larger brain.

There are three main lineages of living South American primates: the pitheciids (sakis and titi monkeys), cebids (squirrel monkeys and marmosets), and atelids (spider monkeys [Figure 21.13], howler monkeys, and wooly monkeys). The atelids are the largest of these primates (large howler monkeys are similar in size to a medium-sized dog), and they are also the only primates that have prehensile tails. While the lemurs evolved terrestrial forms, now all extinct, similar to the African baboons and great apes, none of the South American monkeys ever abandoned the trees. Perhaps other types of South American mammals, such as the small ground sloths, took the roles that terrestrial primates have elsewhere in the world.

## The Old World Monkeys

The Old World monkeys are the cercopithecids and the split between them and hominoids (including apes and humans) has been dated by molecular studies to around 28 Ma, which broadly coincides with the first fossil record appearance of both groups. Cercopithecids were originally an African radiation; the small fossil *Victoriapithecus* from the Miocene of East Africa is an early member of this group and predates the radiation of the living forms. Interestingly, it is a rather small monkey, with an estimated weight of around 7 kg (15 pounds), and seems to have had a semi-terrestrial ecology rather than the tree-dwelling habit that one might have expected.

Monkeys were rather rare in the early and middle Miocene of Africa but by the late Miocene, they were more abundant and had diverged into the two living groups: the colobines and the cercopithecines. By the end of the Miocene, both lineages had members that left Africa and radiated into Europe and Asia. Colobines (e.g., langurs and colobus monkeys) are the more arboreal and more specialized leaf eaters, while cercopithecines (e.g., macaques, baboons, and the patas monkey [Figure 21.14]) are more terrestrial and more omnivorous. Both lineages of Old World monkeys are known from both Africa and Asia but the colobines are more numerous in Asia, and the cercopithecines are more numerous in Africa.

Figure 21.13 Brown-headed spider monkey (*Ateles fusciceps*), a New World monkey, showing the very long arms (used for brachiating) and the long prehensile tail used as a "fifth limb." *Source:* Photograph taken at the Cleveland Metropark Zoo by Lea Maimone (Wikipedia).

Figure 21.14 The patas monkey (*Erythrocebus patas*), an Old World monkey, is the most terrestrial of the cercopithecines and the only monkey that gallops. *Source:* Photograph taken in Senegal by Charles R. Sharp (Wikimedia).

## Emergence of the Hominoids

Living hominoids include hylobatids (gibbons); pongids or Asian apes (only orangutans survive); the African apes (chimps and gorillas); and hominids (only humans survive). African apes are sometimes grouped together as the panids but this is not an evolutionary grouping; molecular data show that chimpanzees are more closely related to humans than are gorillas (Figure 21.10). The physical, molecular, and genetic structure of living hominoids has been studied closely. Humans, gorillas, and chimps are very similar in genetic make-up and protein chemistry, much closer than they are in body structure, but the orangutan differs significantly and gibbons even more.

Traditionally, humans were classified in their own family (Hominidae), separately from any of the apes, but the genetic similarities between us and the other great apes has led to the extension of that family to include the orangutans and African apes. Humans and their immediate fossil relatives have now been "demoted" to the level of the tribe Hominini (i.e., hominins). Chimps, gorillas, and humans are grouped together as hominines (Figure 21.15). Molecular data show that the split between gibbons and hominids took place around 17 Ma, orangutans branched off from the hominines around 13 Ma, gorillas diverged around 9 Ma, and chimpanzees and humans diverged around 6.6 Ma. These dates fit well with the earliest emergence of hominins in the fossil record (see Chapter 22).

## Early and Middle Miocene Hominoids

About 20 Ma, Africarabia formed a single land mass that lay south of Eurasia and was separated from it by the last remnant of the Tethys Ocean. African animals were evolving largely in isolation from the rest of the world and some groups, including the hominoids, were confined to Africarabia at this time, although they were widespread across it.

Early Miocene faunas of Africa were dominated by elephants and rhinos at large body size, antelope and giraffids at medium size, and insectivores were common at small sizes. The environment was forest, broken by open grassland and woodland. Primates of all kinds flourished, although it is difficult to describe their ecology and habits because body skeletons are not as well known as skulls. But prosimians and monkeys were rare, while hominoids were diverse and abundant. We have over 1000 hominoid fossils from the Early Miocene of Africa, most dating from 19 to 17 Ma and most from East Africa.

The dominant hominoids were the ape-like proconsulines, the most basal of the hominoid lineages (Figure 21.10). Like living African apes, they had relatively small cheek teeth with thin enamel, implying a soft diet of fruits and leaves, and a way of life foraging and browsing in trees like most living monkeys, and their brain size was similar. (True monkeys were scarce at this time, remember.) Proconsulines varied in weight from a large species of *Proconsul* in which males weighed up to 50 kg (110 pounds, the size of a large dog) down to *Micropithecus* at about 4 kg (9 pounds, the size of a cat). Proconsulines flourished during the early and middle Miocene but went extinct in the late Miocene when the true monkeys radiated.

There were several species of *Proconsul* known by 18 Ma. *Proconsul* had advanced hominoid characters of the head and jaws but the skeleton showed a mixture of generalized characters indicating a rather basic quadrupedal, tree-climbing, fruit-eating primate (Figure 21.16) that could have been ancestral to all later hominoids. A large *Proconsul* may have spent a lot of time in deliberate climbing or on the ground, like a living chimp. Like them, it was probably versatile in its movements and capable of occasional upright behavior.

*Morotopithecus* is a large proconsuline from Uganda, probably as old as 20 Ma. Although we do not have a skeleton as complete as *Proconsul*, *Morotopithecus* is clearly large (40–50 kg, or 100 pounds), and its skeleton shows evidence of suspensory climbing, like an orangutan. It was probably a rather heavy slow climber, hanging in trees and eating fruit.

Another lineage of hominoids was the dryopithecines, which were more derived than the proconsulines and more closely related to the living great apes than are gibbons (Figure 21.10). They ranged in size from a large monkey to a small chimpanzee, and probably represent

Figure 21.15 Pictoral cladogram of the Homininae.
*Source:* By Merrilydancingape (Wikimedia).

**The
Great
Apes**

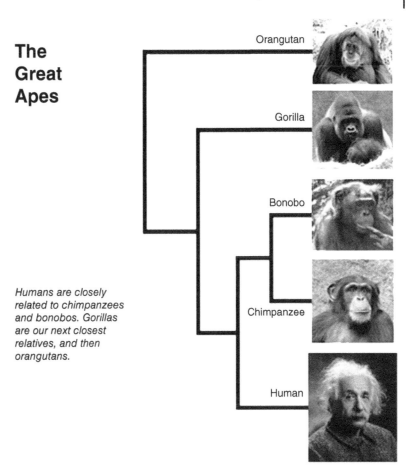

Orangutan

Gorilla

Bonobo

*Humans are closely
related to chimpanzees
and bonobos. Gorillas
are our next closest
relatives, and then
orangutans.*

Chimpanzee

Human

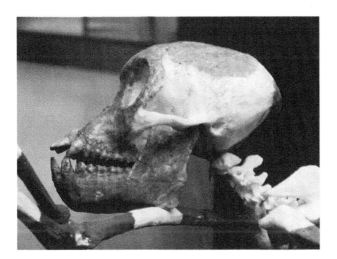

Figure 21.16 Another of the fine casts on display at the Museum
of Anthropology at the University of Zürich. This is a reconstructed
skeleton of *Proconsul* from the Miocene of East Africa.
*Source:* Photograph by Nicholas Guérin, and used under the
Creative Commons Attribution-Share Alike 3.0 Unported license.

a hominine stem group rather than a distinct clade
(evolutionary lineage).

Dryopithecines left Africa and radiated in Eurasia in
the middle to early late Miocene (around 13–8 Ma).
They were especially numerous in southern Europe
(Spain, Italy, Greece, Hungary) when the mid-Miocene
higher global temperatures (see Chapter 20) made that
area a more habitable place for primates than it was
previously or is today. Dryopithecines were leaf eaters
with thin-enameled molars; most had suspensory
types of arboreal locomotion (Figure 21.17) but one
species, *Oreopithecus*, may have evolved a type of
bipedal locomotion convergent with that of hominins
(although it was rather short-legged with a very diverging big toe, and probably stood to forage rather than
for locomotion).

Living alongside the dryopithecines in the Miocene
was another lineage of ape, the sivapithecines, known
only from Asia. Sivapithecines, especially the genus
*Sivapithecus* (formerly known as *Ramapithecus*), were
once thought to be human ancestors but they are now

Figure 21.17 *Dryopithecus*, from the late Miocene of Europe, was one of the more northern apes that radiated in the mid-Miocene warm period. *Source:* Drawing by DiBdg (Wikimedia).

## Late Miocene to Pleistocene Hominoids

As the world cooled in the late Miocene (see Chapter 20), the northern latitudes returned to a more temperate climate. The woodlands that had been the habitats of the more northern apes turned to bushland and grassland, and both dryopithecines and sivapithecines dwindled and disappeared. Meanwhile, in Africa, the cercopithecid monkeys embarked on their successful radiation and the numbers of African ape-like forms (like the proconsulines) largely went extinct. The surviving African apes, the ancestors of chimpanzees and gorillas, probably evolved at this time into the specialized tropical forest forms that they are today.

However, one lineage of sivapithecines persisted. *Gigantopithecus* lived in southern and eastern Asia from about 7 Ma well into the Pleistocene. It had huge grinding teeth and weighed several hundred pounds. It probably lived on very coarse vegetation, as an ecological equivalent of the giant ground sloth of the American Pleistocene, the Asian giant panda, or the African mountain gorilla. *Gigantopithecus* survived in Asia as recently as 300 000 or 250 000 years ago. It was certainly contemporaneous with *Homo* in eastern Asia and its bones, teeth, and jaws may be responsible for Himalayan folklore about the abominable snowman or yeti (Figure 21.18).

known to be stem group orangutans. Sivapithecines appeared to be more terrestrial than the dryopithecines and were a little larger in size (comparable to living orangutans), with a more omnivorous type of diet.

Sivapithecines had thick tooth enamel and powerful jaws, suggesting that their diet required prolonged chewing and great compressive forces on the teeth. In living primates with thick enamel, such as orangutans or mangabeys, teeth and jaws like these are correlated with a diet of nuts or fruits with hard rinds. One can hear an orangutan cracking nuts 100 m away! Perhaps sivapithecids diverged from the earlier hominoid diet of soft leaves and fruits to exploit a food source that had so far been available only to pigs, rodents, and bears. Either the different diet or the area of Eurasia inhabited may explain why sivapithecines persisted for longer than the dryopithecines, in the face of late Cenozoic cooling (see Chapter 20), the main radiation lasting until the end of the Miocene and one lineage (*Gigantopithecus*) surviving until almost the present day.

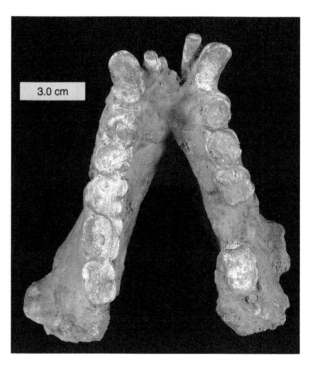

Figure 21.18 The massive lower jaw and teeth of the huge sivapithecine *Gigantopithecus*. *Source:* Photograph by Dr Mark Wilson of the College of Wooster (Wikimedia).

After about 11 Ma, migration between Africa and Eurasia was essentially cut off. The hominoid groups evolved independently in Eurasia and Africa, eventually leaving the sivapithecines, gibbons, and orangutans in Asia and the hominines in Africa. The late Miocene African fossil record of hominines is horribly incomplete, probably because the ancestors of gorillas and chimpanzees were living in tropical rainforests, where fossilization is virtually unknown. A few fossils, mostly isolated teeth, have been found. *Chororapithecus* and *Nakalipithecus* are possible basal gorillas from ~8 to 10 Ma in Kenya, but no evidence of chimpanzees is known until the Pleistocene. However, starting around 4.5 Ma, we begin to pick up evidence of the radiation of the hominins, humans and their fossil relatives. This radiation will be covered in the next chapter.

## References

Bloch, J.I. and Boyer, D.M. (2002). Grasping primate origins. *Science* 298: 1606–1610, and comment, pp. 1564–1565; arguments, Science 300: 741.

Franzen, J.L., Gingerich, P., Habersetzer, J. et al. (2009). Complete primate skeleton from the middle Eocene of Messel in Germany: morphology and paleobiology. *PLoS One* 4: e5723.

## Further Reading

Begun, D. (2010). Miocene hominids and the origins of the African apes and humans. *Annual Review of Anthropology* 39: 67–84.

Bloch, J.I., Silcox, M.T., Boyer, D.M. et al. (2007). New Paleocene skeletons and the relationship of plesiadapiforms to crown-clade primates. *Proceedings of the National Academy of Sciences of the United States of America* 104: 1159–1164.

Ciochon, R., Long, V.T., Larick, R. et al. (1996). Dated co-occurrence of *Homo erectus* and *Gigantopithecus* from Tham Khuyen cave, Vietnam. *Proceedings of the National Academy of Sciences of the United States of America* 93: 3016–3020.

Fleagle, J.G. (2013). *Primate Adaptation and Evolution*, 3e. Philadelphia: Elsevier.

Kay, R. (2015). New World monkey origins. *Science* 347: 1068–1069.

Silcox, M.T. and López-Torres, S. (2017). Major questions in the study of primate origins. *Annual Review of Earth and Planetary Sciences* 45: 113–137.

Simons, E.L., Plavcan, J.M., and Fleagle, J.S. (1999). Canine sexual dimorphism in Egyptian Eocene anthropoid primates: *Catopithecus* and *Proteopithecus*. *Proceedings of the National Academy of Sciences of the United States of America* 96: 2559–2562.

Williams, B.A., Kay, R.F., and Kirk, E.C. (2010). New perspectives on anthropoid origins. *Proceedings of the National Academy of Sciences of the United States of America* 107: 4797–4804.

## Questions for Thought, Study, and Discussion

1   If humans are so wonderful, how is it that dozens of other primate species also flourish all across the warm latitudes of the world? Choose a few varied primate species and find out how they make a living.

2   In the mid-Miocene warm period, primates (both monkeys and apes) migrated from Africa into Eurasia, and they persist there today in the more tropical areas of southern Asia. But primates do not appear in the American subtropics, like Panama and Costa Rica, until around 2 million years ago. Why is this the case (you might need some help from Chapter 20 to figure this one out!)

   We think that as apes, we are derived
   But this viewpoint is rather contrived
   Monkeys' specializations
   Caused us complications
   We are lucky to have even survived

22

# Evolving Toward Humans

### In This Chapter

The African clade containing humans (*Homo*) and fossil relatives, the hominins, has a patchy fossil record. The australopithecines are a group of maybe a dozen species, all with large brains (compared to chimpanzees) and the ability to walk upright. Some species were clearly omnivorous but others had huge teeth and jaws and probably ate tough vegetation for most of their diet. Beginning about 2.5 Ma, we begin to find primitive stone tools associated with australopithecines, and there is little doubt that they made and used such tools for butchering game animals. There were numerous species in Africa, many of which have been claimed to be ancestral to the earliest *Homo* species. After 2 Ma, we see the first definitive evidence of *Homo*, with *Homo erectus* displaying a further increase in brain size, a more upright human-like stature, larger body size, and associated with more sophisticated tools that could well have been used for hunting. *H. erectus* and its close relatives spread all over south Asia into east Asia and further north than any previous hominin. A branch of hominins later migrated into Europe and evolved there within the last 1 million years, eventually giving rise to the Neanderthals, *Homo neanderthalensis*. They were formidable Ice Age hunters. *Homo sapiens* evolved in Africa over 300 000 years ago and within the last 100 000 years, a population of *sapiens* migrated out of Africa to take over most of the Old World. Older species became extinct during this time, although there is clear evidence of a small amount of interbreeding with Neanderthals. The study of ancient DNA is rapidly changing our understanding of the later stages of human evolution – the "Denisovans" are known almost exclusively from their DNA. Natural selection continues to produce evolutionary change in *H. sapiens*.

## The Earliest Hominins

We know practically nothing of the evolution of the hominoid lineages that led to gorillas and chimpanzees. Molecular and genetic evidence suggests that our closest living relatives are chimpanzees, with gorillas a little further away. Our own lineage, the hominins, began separating from that of chimpanzees as early as 12 Ma (Scally and Durbin 2012). Even so, our DNA is more than 95% identical to that of chimpanzees. Obviously, the 5% that is different reflects very important evolutionary changes in our bodies, brains, and behavior.

Over time, there have been at least a dozen species of hominins but we, as *Homo sapiens*, are the only surviving one. As many as six earlier species of *Homo*, ranging back to about 2 Ma, have become extinct and at least as many again are placed in a group called **australopithecines**, which ranges back to about 4 Ma and contains the ancestor of *Homo*.

This simple picture is changing rapidly as more fossils are found and new techniques to study them become available. In many areas, paleoanthropologists continue to debate different hypotheses. For identifying the earliest hominins, the evidence we have is very scarce.

One candidate is *Sahelanthropus*, dated close to the base of the hominins, at 6 Ma or 7 Ma (see Chapter 21, Figure 21.15). It is from Chad, far to the west of the "classical" East African sites. It is known

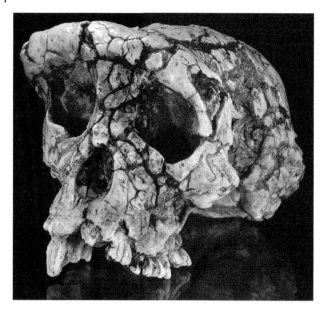

Figure 22.1 A cast of *Sahelanthropus*, from the Miocene of Chad. *Source:* Photograph by Didier Descouens (Wikimedia).

from skull pieces and shows a puzzling combination of very "primitive" characters (small brain, for example) with "advanced" characters such as eyebrow ridges. However, the single skull was badly crushed (Figure 22.1) and a different reconstruction of it might allow a different interpretation.

*Orrorin* dates from perhaps 6 Ma in East Africa. It is known mostly from a few pieces of limb bones so is difficult to place. Rivals of the discovery team suggest that it may in fact be an ancestral gorilla or chimpanzee rather than a hominin. It is difficult to resolve with only fragmentary remains and no pieces of the skull.

Appearing later are two other candidate species of early hominin: *Ardipithecus kadabba* in Ethiopia dated at 5.5 Ma and *Ardipithecus ramidus*, also from Ethiopia, and dated at 4.3–4.4 Ma. *Ardipithecus* retains primitive (that is, ape-like) traits and was probably a forest dweller. As with other early hominins, the position of *Ardipithecus* is strongly disputed but it looks as if it was a careful climber in trees, and perhaps it could move bipedally on the ground.

With these earliest "hominin" fossils, it is very hard to know whether they are genuinely hominins – ancestors on the branch toward *Homo* – or if they fit elsewhere, perhaps as ancestors on the lines to chimpanzees or gorillas, or as other lineages with a mixture of traits that ultimately became dead ends (Wood and Harrison 2011). Since we cannot yet give a reasonable story of very early hominin evolution, we

move to the australopithecines, the hominin species that predate *Homo*. They lived in Africa, south of the Sahara Desert, from perhaps as early as 4.2 Ma to about 1.4 Ma or a little later. Overall, we are reasonably sure of the position of australopithecines in hominin evolution and have enough evidence to reconstruct a vivid picture of australopithecine life.

### Early Australopithecines

The earliest species of australopithecine is *Australopithecus anamensis*, known from rocks in Kenya dated at 4.1–4.2 Ma. This species has an ape-like jaw but its arm and leg bones suggest some level of upright (bipedal) posture and locomotion. Bipedalism seems to be the key trait that separates these early hominins from more ape-like relatives. It would make a good ancestor for later *Australopithecus. A. anamensis* is rather large, perhaps 50 kg or 110 pounds.

The East African Rift splits East Africa from Ethiopia to Zambia and Malawi. Among its unusual geological features are volcanoes that sometimes erupt carbonatite ash, which is composed largely of a bizarre mixture of calcium carbonate and sodium carbonate. One of these volcanoes, Sadiman, stood near the Serengeti Plain in northern Tanzania. After carbonatite ash is erupted, the sodium carbonate in it dissolves in the next rain and as it dries out, the ash sets as a natural cement. Any animals moving over the damp surface in the critical few hours while it is drying will leave footprints that can be preserved very well. As long as the footprints are covered up quickly (for example, by another ash fall), rainwater percolating through the ash will react with the carbonate to make a permanent record.

Sadiman erupted one day about 3.6 Ma, toward the end of the dry season. Ash fell on the plains near Laetoli, 35 km (20 miles) away, and hominins walked across it, leaving their footprints along with those of other creatures. The vital point about the tracks is that the hominins were walking upright (Figure 22.2), long before the hominin jaws, teeth, skull, and brain reached human proportions, shape, or function.

Why would a hominin become bipedal? Most suggestions are related to carrying things with the hands and arms (infants, weapons, tools, food), food gathering (seeing longer distances, foraging over greater ranges, climbing vertically, reaching high without climbing at all), defense (seeing longer distances, throwing stones, carrying weapons), better resistance to heat stress (less sweat loss and better cooling), or staying within reach of rich food resources by migrating with the great plains

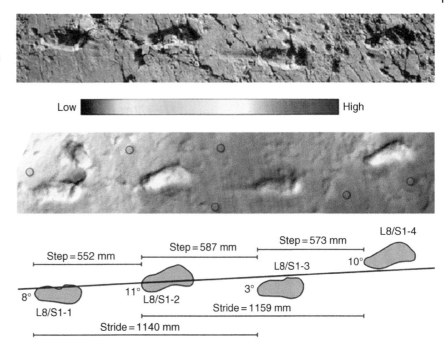

Figure 22.2 One of the trackways of australopithecine footprints from Laetoli, laser scanned. You can see four prints, each showing the heel and toe prints, with a separated big toe and a distinct arch to the foot. All these are features of an efficient, erect walking action.
*Source:* Original figure by Fidelis T. Masao and colleagues (Wikimedia).

animals (carrying helpless young over long distances). These are all reasonable suggestions but all are difficult to test.

The footprints at Laetoli were made by australopithecines that were walking upright. All australopithecines are similar below the neck, apart from size differences, so they all probably moved in much the same way. Their movements were probably not exactly like ours and their leg and hip bones retain primitive features that indicate that they walked with something of a "bent-hip, bent-knee" posture. At the same time, the upper limb joints and toes suggest that they spent a lot of time climbing in trees as well as walking upright on the ground.

Probably the trend toward the use of the fore limbs for gathering food and the hind limbs for locomotion began among tree-dwelling primates long before *Australopithecus*. Chimpanzees themselves, who primarily move by knuckle-walking, can move bipedally over short distances. Australopithecines were probably practicing an extension on these types of behaviors – moving bipedally quite often but only as part of wide range of locomotor behaviors. The final achievement of erect and fully committed bipedality on the ground was probably an extension of previous locomotion and behavior, rather than something completely different, and is not seen until *Homo erectus*.

Most environmental reconstructions of the habitats in which the australopithecines are found suggest quite open, mixed environments but where trees were still heavily present, allowing the australopithecines to exploit a range of types of locomotion. Only with *H. erectus* do hominins start using fully modern bipedalism to exploit much more open savanna environments.

Australopithecines were smaller than most modern people. They varied around 40 kg (90 pounds) as adults, but their bones were strongly built for their size. Their skulls were also very robust and quite different from ours. The relative brain size was less than half of ours, even allowing for the smaller body size of *Australopithecus*, but the jaw was heavy and the teeth, especially the cheek teeth, were enormous for the body size. The canine teeth were large and projecting. The whole structure of the jaws and teeth suggests strength.

The australopithecines had a small brain size relative to us, but it represented a small increase relative to more ancestral species (or in comparison to chimpanzees). This may have affected the birth canal, and the pelvis of australopithecines is wide from side to side but narrow from front to back. In *Homo* (from *H. erectus*), to accommodate the passage of a baby with a much larger head (and brain), the birth canal is rounder in shape. The evolution of a larger brain is closely connected to, and constrained by, these features of skeletal pelvic anatomy (Figure 22.3).

### Australopithecus afarensis

The best-known collections of early australopithecines are from Laetoli and from Hadar in Ethiopia. Each

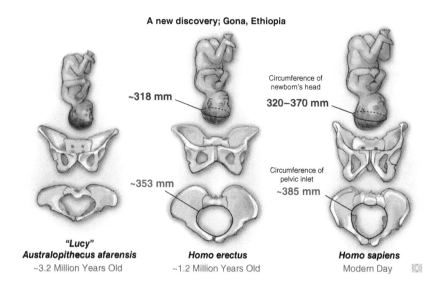

**A new discovery; Gona, Ethiopia**

~318 mm

Circumference of newborn's head

320–370 mm

~353 mm

Circumference of pelvic inlet

~385 mm

**"Lucy"**
***Australopithecus afarensis***
~3.2 Million Years Old

***Homo erectus***
~1.2 Million Years Old

***Homo sapiens***
Modern Day

**Figure 22.3** Australopithecines have a pelvis that is relatively narrow from front to back. Modern humans, and *Homo erectus* from more than a million years ago, have a rounded pelvis, probably related to giving birth to a large-brained baby. *Source:* Artwork by Zina Daretsky of the National Science Foundation, in the public domain.

district has produced spectacular finds. At Laetoli, there are the footprints, plus remains of at least 22 individuals; at Hadar, bone fragments from at least 35 individuals are preserved. All the specimens belong to one species, *Australopithecus afarensis*, which was very closely related to *A. anamensis* and was probably ancestral to all the later species of *Australopithecus* and to *Homo* as well (Figure 22.4).

Hadar lies in the Afar depression, a vast arid wilderness in northeast Ethiopia. At 3–4 Ma, it was the site of a lake fed by rivers tumbling out of winter snowfields high on the plateau of Ethiopia. The australopithecines lived and are fossilized along the lake edges. Delicately preserved fossils such as crab claws and crocodile and turtle eggs suggest that the australopithecines had rich protein foods available to them, and skeletons of hippos and elephants suggest that there was rich vegetation in and around the lake edges. All the Hadar specimens are dated at about 3.2 Ma, so they are considerably later than the Laetoli australopithecines.

The best-preserved Hadar skeleton is the famous **Lucy**. Lucy was small by our standards, a little over 1 m (42 in.) in height. She was full-grown, old enough to have had arthritis. Her brain was small at about 385 cc, compared with 1300 cc for an average human. Her large molar teeth suggest that *A. afarensis* was a forager and collector eating tough fibrous material.

*Australopithecus afarensis* was no more dimorphic than modern humans: males were bigger than females but not to the extent seen in baboons, for example. Across primates, extreme male size is correlated with intense physical competition between males for females, whereas monogamy is associated with very

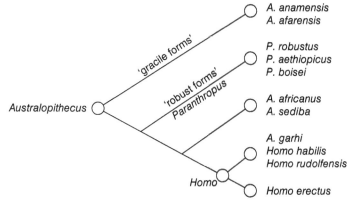

*Australopithecus*

'gracile forms'

'robust forms'
*Paranthropus*

*Homo*

A. anamensis
A. afarensis

P. robustus
P. aethiopicus
P. boisei

A. africanus
A. sediba

A. garhi
Homo habilis
Homo rudolfensis

Homo erectus

**Figure 22.4** Simplified cladogram of australopithecines. Similar species are grouped together: the early primitive species *afarensis* and *anamensis*, for example; the "robust" australopithecines; and the "gracile" South Africans species *africanus* and *sediba*. Note that if this cladogram is correct, *Australopithecus* is not a clade unless you include *Homo* in it. This "problem" is not really a problem: the aim of cladograms is to portray evolution; the naming schemes are simply convenient. The close relationships within the groups are also unclear – for example, the "robust" species may be closely related to a single ancestor or may appear similar because of convergent evolution on a similar morphology driven by selection for larger teeth and jaw musculature.

low levels of dimorphism. The reduction in body size dimorphism would suggest differences in the social structures of these hominins – they were likely living in small groups with close kin relationships and cooperative behaviors.

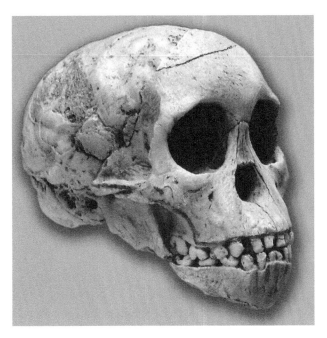

Figure 22.5 The first *Australopithecus* ever found was the fossil of a child, *Australopithecus africanus*. *Source:* Photograph by José-Manuel Benito Álvarez (Wikimedia).

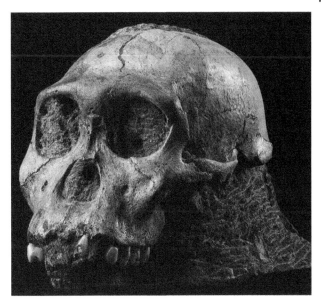

Figure 22.6 *Australopithecus sediba* from South Africa. *Source:* Photograph by Brett Eloff, courtesy of Professor Lee Berger and the University of the Witwatersrand (Wikimedia).

## *Australopithecus* in South Africa

Isolated caves scattered over the high plains of South Africa are mined for limestone, and hominin fossils have been found encased in the limestone. But cave deposits are difficult to interpret and date accurately. Roof falls and mineralization by percolating water have disturbed the original sediments, and few of the radioactive minerals in cave deposits allow absolute dating. Thus, there have been problems in relating South African hominin fossils to their well-dated East African counterparts.

One of the best-known early australopithecines from South Africa is *Australopithecus africanus* (Figure 22.5). Although it was about the same body weight as *A. afarensis*, *A. africanus* was taller but more lightly built and had a slightly larger brain, perhaps 450 cc. The teeth and jaws were large and strong, suggesting that the diet remained mainly vegetarian, and tough, fibrous or hard foods dominated. However, evidence from isotopes in the teeth suggest that *A. africanus* ate animals as well, possibly catching small animals or scavenging meat from carcasses. The arms were relatively long compared with *A. afarensis*, suggesting that *A. africanus*, although perfectly able to walk upright on the ground, spent a good deal of time in trees.

A newer species from South Africa, *Australopithecus sediba* (Figure 22.6), was discovered in 2008 and is dated to 2 Ma, so was contemporary with the earliest *Homo*. The species is known from several individuals found at Malapa Cave and shows a strange mixture of characters, some primitive and others more derived (closer to *Homo*). It was first argued to be a transitional species from *A. africanus* to later *Homo* (*erectus*), raising the idea that the origins of *Homo* might have been in South Africa and not, as often assumed, East Africa. *A. sediba* retains a small brain size and long arms and appears to be most like its putative ancestor *A. africanus*, but it also shows signs of "*Homo*-like" pelvis, ankle joint and quite modern hand morphology capable of a precision grip. The mixture of characters and relatively late date (2 Ma) suggests *sediba* may be an example of a late-surviving *Australopithecus* population, but not directly ancestral to later *Homo*. There are plenty of other candidates to consider for this transition, as we shall see below.

## Robust Australopithecines

All australopithecines had heavily built skeletons relative to modern humans but the australopithecines can be divided into two groups: the **robust** australopithecines had extremely robust skulls and extremely large molar teeth, which distinguish them from those with lightly built skulls (such as *A. afarensis*) which are called **gracile**. The best example of a robust skull is the oldest one, the so-called Black Skull (Figure 22.7) from the Turkana Basin of northern Kenya dating from about 2.5 Ma. It is much heavier and stronger than *A. afarensis*, although the brain was no larger and the body was not very different. The jaw extended further forward, the face was broad and dish-shaped, and there was a large crest on the top of the skull for attaching very strong jaw muscles

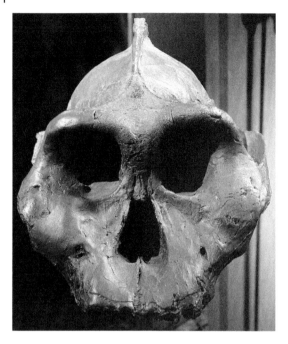

Figure 22.7 Skull of *Australopithecus aethiopicus*, from East Africa, the "Black Skull." One of the casts on display at the Museum of Anthropology at the University of Zürich. *Source:* Photograph by Nicholas Guérin, in the public domain.

Figure 22.8 Skull of *Australopithecus robustus* from South Africa. One of the casts on display at the Museum of Anthropology at the University of Zürich. *Source:* Photograph by Nicholas Guérin, in the public domain.

(Figure 22.7). The molar teeth of the Black Skull are as large as any hominin teeth known, about four or five times the size of ours. Yet the front teeth of robust australopithecines are small. This morphology of the skull suggests robust australopithecines were specialized on a particular diet, exploiting either hard or very tough foods (perhaps underground tubers) that required a strong bite or lots of chewing. This specialization when compared to the more generalized *Australopithecus* species has led most researchers to now class the robust australopithecines in a different genus: *Paranthropus.* The Black Skull is attributed to *Paranthropus aethiopicus.*

Later robust forms have been found all over East and South Africa between 2.5 Ma and 1.4 Ma. In South Africa, they are called *Paranthropus robustus* (Figure 22.8), and in East Africa they are usually called *Paranthropus boisei* (the famous Zinj of Louis Leakey) (Figure 22.9). There are enough fossils to suggest that robust australopithecines changed over the million years of their history, evolving a larger brain (perhaps 500 cc rather than 400 cc) and a flatter face.

The robust australopithecines are certainly similar ecologically. The large jaw and huge molars, with their very thick tooth enamel, were adaptations that indicate great chewing power and a diet of coarse fiber. However, almost all the characters that are used to define robust australopithecines are connected with the huge teeth, and the modifications of the jaws and the face during

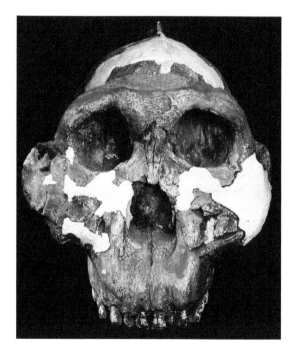

Figure 22.9 Skull of *Australopithecus boisei* from East Africa. One of the casts on display at the Museum of Anthropology at the University of Zürich. *Source:* Photograph by Nicholas Guérin, in the public domain.

growth that are required to accommodate the teeth. So, any australopithecine population that evolved huge teeth would have come to look "robust." Therefore, the robust australopithecines may not be an evolutionary clade.

They could be three separately evolved species, or three related species, or they could be variants of the same species (which would have to be called *robustus*).

## Transitional Species and Early Tool Use

In the period from 2.5 to 1.5 Ma, various hominin remains show transitional features from the gracile australopithecine traits to the more derived and undisputed *Homo* characteristics (larger brains and body size, gracile skulls and smaller dentition, taller stature, modern hip morphology, and fully modern upright bipedalism). Many of these species have been argued to be the direct ancestors of *Homo* or the first of the *Homo* species, but increasingly the picture is of a melting pot of different species – perhaps different lineages of hominins finding different ways to adapt to the changing landscapes. Knowing which (if any) of these was a direct ancestor to later *Homo* is a difficult task!

We have already seen the mosaic traits shown by *A. sediba*, and another (even more recently discovered) South African species with a similar story is *Homo naledi*. *H. naledi* has some more heavily derived characters (particularly in the hands and feet), placing it in the genus *Homo*, but also with primitive features such as a very small (for *Homo*) brain size (450–550 cc). Before directly dating the remains, *naledi* was also thought to date to this critical transitional time-period around 2 Ma but now we know the fossils actually date to 250 000 ya – remarkably late! So *naledi* can be ruled out as an ancestor to later *Homo* but it adds to the diversity of species we know and shows that relict populations persisted much later than thought alongside more derived later *Homo* (Spoor 2011).

The most familiar transitional species is *Homo habilis* or "handy man" (Figure 22.10). Originally *habilis* was thought to belong to the genus *Homo* because of its association with stone tools. Early *Homo* was small by modern standards, perhaps just over a meter (4 ft) tall, but was at least as heavy as contemporary robust australopithecines at about 30–50 kg (65–110 pounds). The brain size was about 650 cc, larger than the brain of an australopithecine. Perhaps, then, early *Homo* is marked by a new level of brain organization. *Homo rudolfensis,* also known from East Africa around 2 Ma, has been given a separate name because its brain size was even larger – above the upper range for *habilis* (at about 800 cc).

We have a good record of the tools that were used by *H. habilis.* They are called Oldowan tools because they were first identified by the Leakeys in Olduvai Gorge. They are often large and clumsy-looking objects with simple shapes, and not all of them were useful tools in themselves (Figure 22.11). Instead, many objects may be

Figure 22.10 Replica of the skull of *Homo habilis* from East Africa. *Source:* Photograph by Daderot (Wikimedia).

Figure 22.11 Oldowan chopper tools from Ethiopia, about 1.7 Ma. *Source:* Photograph by Didier Descouens (Wikimedia).

the discarded centers (cores) of larger stones from which useful scraping and cutting flakes had been removed by hammering with other stones.

Oldowan tools demonstrate the use of stone in a deliberate, intelligent way, and the flakes were probably made and used for cutting up food items. Experimental Oldowan-style artifacts have been reproduced from East African rock types and these show the toolmaker was sophisticated in selecting appropriate rocks and making the most of them. The experiments also show that Oldowan axes, flakes, and cores are excellent tools for slitting hides, butchering carcasses, and breaking bones for marrow.

*Homo habilis* was not the only tool user in Africa at this time. In Ethiopia, at 2.5 Ma we have evidence from two or three skeletons to allow the description of another species, *Australopithecus garhi*. *A. garhi* is a normal gracile australopithecine, except that it has very large teeth for the size of its jaw and skull. The skull is far too primitive for it to belong to *Homo*, and its brain size is only about 450 cc. But there is also evidence of the use of stone tools for butchering meat and smashing bones. Some of the animal bones in the same rock bed had been sliced and hammered in ways that betray intelligent butchering. Most likely, the butchers used their tools carefully, because there were no suitable rocks nearby, and all tools had to be carried in (and carried out for further use). Were early *Homo* hunting or scavenging carcasses from other predators? Evidence from Turkana and Olduvai suggests that early *Homo* was mainly scavenging on large carcasses but was perhaps hunting small- and medium-sized prey.

Before 1999, it was generally thought that the defining characters of *Homo* versus *Australopithecus* included a larger brain and the use of tools. The evidence now suggests that *A. garhi* was making, carrying, and using tools effectively. So, the tool use predates the large increases in brain size, which is not so surprising after all because chimpanzees and other primates also use tools in certain situations. Perhaps the great ecological advantages gained by the invention of butchering tools encouraged exactly those changes in the *A. garhi* lineage that ultimately led to increased brain size, reduced tooth size, and the status of first *Homo*.

Once again, apparently major transitions disappear as we collect more fossils; we have seen this for the transition between birds and dinosaurs, between cynodonts and mammals, and now between australopithecines and *Homo*. As we see it now, perhaps as early as 2.4 Ma, hominins with increased brain size and reduced teeth and jaws appeared in Africa. They are sufficiently like ourselves in jaws, teeth, skull, and brain size to be classed as *Homo*. But because one genus always evolves from another, there is always room to argue just where to draw the line, and this is happening as we try to decide which species actually was the first *Homo*. Increasingly, we realize that there is a great difference between early, transitional forms, and later species that everyone agrees belong to *Homo*.

## *Homo erectus* and the Modern Body Plan

After 2 Ma, some extraordinary changes took place in the African plains ecosystem. We are also blessed with the appearance of another new species of human, *H. erectus*, exploiting these environments. ("*Homo ergaster*" is a name also used for the early African fossils that may well be the same species as the later fossils from Eurasia that are called *H. erectus*.)

An excellent specimen of *H. erectus* was discovered in 1984 west of Lake Turkana in Kenya, in sediment dated about 1.6 Ma. The body had been trampled by animals, so the bones were broken and spread over 6 or 7 m, but careful collecting recovered an almost complete skeleton. The skeleton came from a boy (the "Turkana Boy") who was 8 years old and stood 1.6 m (65 in.) high (his stature is equivalent more to an 11- or 12-year-old modern human, as early *Homo* likely had a shorter period of growth and development relative to modern humans). Adult males were probably close to 1.8 m (6 ft). *H. erectus* was strongly built and was a specialized walker and runner with large hip and back joints capable of taking the stresses of a full running stride. There is less evidence of tree-climbing ability than there is in early *Homo*, and the overall body plan of *H. erectus* is much more like ours. *H. erectus* is also advanced in skull characters (Figure 22.12). The skull is thick and heavy by our standards, but brain size had increased to around 900 cc.

Quite suddenly, at about 1.4–1.5 Ma, all over East Africa, *H. erectus* is found associated with a completely new set of stone tools. The **Acheulean** tool kit is much more effective than the older Oldowan, but experiments by Nicholas Toth show that Acheulean tools required much greater strength and precision to make and use than Oldowan tools. Acheulean craftsmen shaped their stone cores into heavy axes (Figure 22.13) and cleavers at the same time as they flaked off smaller cutting and scraping tools. Most Acheulean tools are well explained as heavy-duty butchering tools. It is tempting to correlate all these events with the achievement of some dramatically new level of intellectual, physical, and technical ability in *H. erectus*. *H. erectus* was much bigger than any preceding human. Most paleontologists believe that the evidence from anatomy, from tools, and from animal remains found with *H. erectus* suggests that this was the first effective human hunter of large animals. Alan Walker suggested that the entire ecosystem of the African savanna was reorganized as *H. erectus* came to be a dominant predator instead of a forager, scavenger, and small-scale hunter. African kill sites with butchered animals suggest a sophisticated level of achievement.

The physical stature, large jump in brain size, and ecological impact of this new species of *Homo* are reasons why some experts suggest we should redefine the origin of *Homo* to the appearance of *H. erectus*, perhaps placing *H. habilis* and the other transitional forms back into *Australopithecus*.

We know from the shape of the pelvis that *H. erectus* babies were born as helpless as modern human babies are (Figure 22.3), and it is clear that the brain grew a lot after birth, as our brains do. This implies a long period of

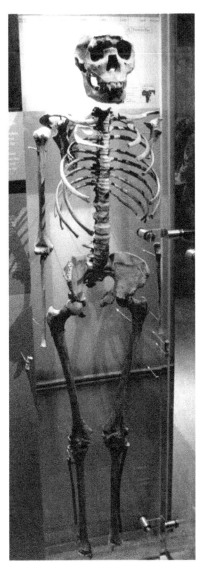

Figure 22.12 The "Turkana boy," an early example of *Homo erectus* ("*Homo ergaster*") from Kenya. *Source:* Photograph by Claire Houck (Wikimedia).

Figure 22.13 An Acheulean biface (hand ax) made of flint. This is actually from the locality of St Acheul, in France. *Source:* Photograph by Didier Descouens of specimen from the Museum of Toulouse, as part of Projet Phoenix (Wikimedia).

care for a baby that probably could not walk for several months. That is an enormous price to pay for a larger brain and would only have been evolutionarily worthwhile (selected for) if there was a large pay-off in learning and intelligence. The longer periods of growth and development also suggest a shift to greater care-giving from other members of the community in bringing up offspring. Such "grandmothering" behaviors, and exploiting rich plant resources such as tubers, may have had an even greater role in the evolutionary success of *H. erectus* than any hunting abilities (Hawkes and Coxworth 2013).

All these lines of evidence imply a complex and stable social structure for *H. erectus*, although details are certainly not available. It is difficult to pin down where the origins of language fit into the evolutionary picture but it is around this time, with greater social complexity and interaction among individuals, that language may have had its origin.

## Into Asia

The first fossils to be named *H. erectus* were in fact collected 100 years ago on the island of Java, in Indonesia. The earliest of the Java specimens of *H. erectus* may be as old as 1.8 Ma, although this date is contested. Specimens of *H. erectus* have been discovered in the southern Caucasus, in Georgia, and they date to around 1.7– 1.8 Ma. It seems likely that the expansion of *H. erectus* from Africa to Asia occurred soon after the species evolved. It was rapid, it extended across the warm regions of southern Asia from the Middle East to Indonesia, and it happened before the Acheulean tool kit was invented in Africa.

The migration of *H. erectus* left a corridor of hominin activity that stretched from South Africa to eastern Asia. All these populations evolved larger body size and more advanced skull characters, and all made new tool kits. Given the mobility of humans, there was no necessary dramatic or long-lived separation between these pantropical populations. There was one founding, and dominant, early human species, *H. erectus*, with locally variable anatomy and culture, just like *H. sapiens* today.

Other specimens of *H. erectus* from China are compatible with this story. The Chinese specimens have an age around 1.0 Ma and younger. *H. erectus* may have reached as far east as the island of Flores, in Indonesia, before 750 000 BP (years before present), a feat which involved two sea crossings, of 15 and 12 miles. There are no fossils, only a few tools, but the story fits with the fact that three major animals became extinct quite suddenly on Flores around 900 000 BP: a pygmy elephant, a giant tortoise, and a giant lizard related to the Komodo dragon.

*Homo erectus* from Java had a brain size just under 1000 cc, but brain size had reached 1100 cc by the time of "Peking Man," who occupied caves outside Beijing between 500 000 and 300 000 BP. The successful long-term occupation of north China by these people indicates that they had solved the problems of surviving a challenging northern winter. The Asian populations of *H. erectus* had their own versions of stone tool-making styles. Some Asian *erectus* made tools out of rhino teeth, since they were living in an area without good tool-making stone.

It looks as if *H. erectus* was the first species to control fire. There is very good evidence for fires in a South African cave dating from at least 1 Ma.

## After *Homo erectus*

After *H. erectus*, the story of human evolution starts becoming very messy again as anthropologists argue about the origin of modern humans. As fossil and molecular evidence has accumulated, anthropologists have accepted finer subdivisions for species and the majority opinion is that several species of *Homo* have evolved during the last million years, with all but one becoming extinct.

It appears that in different regions, local populations of *H. erectus* evolved separately and significantly. The name *Homo antecessor* was proposed for a group of well-preserved specimens found in Spain, dated between 1.2 Ma and 800 000 BP. Obviously, *H. antecessor* must have had ancestors somewhere in Africa (*H. erectus*?). *H. antecessor* left evidence for cannibalism: some *antecessor* bodies were processed – by human tools – to provide convenient chunks of meat, leaving characteristic cut marks on the skeletons (Bermúdez de Castro et al. 2017).

Perhaps dating as far back as 500 000 BP, a population, perhaps descended from *H. antecessor* or *H. erectus* and known from central and western Europe, is called *Homo heidelbergensis*. Around 400 000 BP, a number of *H. heidelbergensis* skeletons were deposited in the same caves in Spain associated with *H. antecessor* 400 000 years

before. In an act of complex behavior, a *heidelbergensis* placed a beautiful pink quartzite hand ax on a pile of skeletons in the cave. At about the same time, *H. heidelbergensis* in Germany were making beautifully crafted wooden hunting spears from spruce and pine. These are throwing spears, up to 2.5 m long (8 ft), carved to angle through the air like modern javelins, and the spears are associated with butchered horses and other bones from elephant, rhino, deer, and bear. Modern athletes have thrown replicas of these spears 70 m (over 200 yards).

*Homo heidelbergensis* in turn seems to have evolved into the Neanderthals, *Homo neanderthalensis*. They were strongly adapted for life in cold climates along the fringes of the Ice Age tundra from Spain to Central Asia.

We have only patchy evidence from Africa during this time but we assume that *H. erectus* continued to live across that continent. Meanwhile, *H. erectus* thrived in eastern Asia and we have populations from Indonesia ("Java Man") and China ("Peking Man") in the age range 1 Ma to 500 000 BP.

## The Origin of *Homo sapiens*

For some years now, the majority story has been that around 200 000 BP, an African *Homo* species (*heidelbergensis*/archaic *sapiens*) gradually developed a distinctive new type of stone tool technology which we call Middle Stone Age/Middle Paleolithic technology – MSA for short. And gradually, one of these African populations evolved into fully modern *H. sapiens*.

This story was strengthened when a new set of fossils was described from Ethiopia in 2003. Three skulls, dating to about 160 000 BP, were identified as *H. sapiens*. They still have some "archaic" features such as robust skulls and big eyebrows that would allow us to recognize them as "different" from modern humans, so they are called *H. sapiens idaltu*, a separate subspecies. Their tools were on the boundary between Acheulean and MSA.

Recent finds and dating at the site of Jebel Irhoud in Morocco have now established that MSA tools associated with remains of *H. sapiens* were already present there at 300 ka – the earliest known evidence for *H. sapiens*. These individuals also show some archaic features (Figure 22.14) and are robust compared to modern humans, but they suggest an earlier origin, and wider geographical range, for the early populations of *H. sapiens* than had previously been thought (Stringer and Galway-Witham 2017).

Where exactly in Africa we trace back the origins of our species is now hotly contested, but it is becoming clear that our species did not evolve only in a single

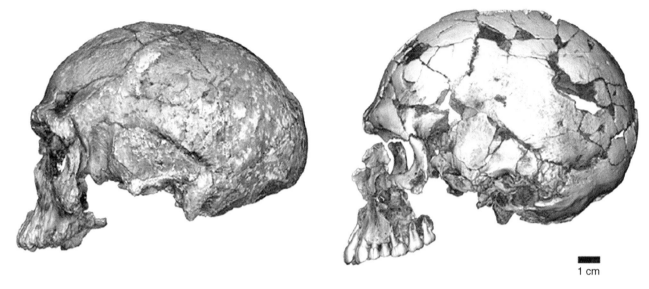

1 cm

Figure 22.14 Digital reconstructions of the early archaic *Homo sapiens* skull found at Jebel Irhoud (300 ka), with a later, more modern *H. sapiens*, Qafzeh, from 95 000 BP. *Source:* Image from Scerri et al. (2018), Creative Commons license.

region within Africa. Rather, the evidence points to complex, subdivided populations of (archaic) *H. sapiens* spreading and persisting across the continent, distinct from one another but likely with some gene flow among them (Scerri et al. 2018). It is from within this complex history, over 200 000 years, that we eventually see the emergence of the first fully modern humans: *H. sapiens sapiens*. However, we know about these first *H. sapiens sapiens* more from their tools and their traces in our genes than from clear fossil evidence.

Modern humans are known from evidence in South Africa by the time of the last interglacial (around 120 000 BP). There were initial early forays of populations pushing beyond the boundaries of Africa (into the Middle East), but later thereafter a small population of pioneering or refugee *H. sapiens* must have left Africa and successfully expanded into and over the Old World (Bae et al. 2017). The expansion and dispersal of populations were rapid and modern humans may have traversed around the shores of the Indian Ocean to reach Southeast Asia and Australia by 60 000 BP (Figure 22.15). There were also other species coexisting across the Old World at the same time. *H. neanderthalensis* was firmly entrenched in the colder areas of Europe and western Asia, with late *H. erectus* persisting in eastern Asia. In the Middle East, there is an alternating pattern of occupation, with Neanderthals reoccupying areas as the climate cooled again around 70 000 BP.

From 60 000 BP, modern humans spread rapidly across from the Old World, and other hominin species were soon to be extinct. Eventually, *H. sapiens* was to spread into parts of the world previously not occupied by any hominin, including the Americas and Polynesia. All non-African living humans are thus descended from what must have been a comparatively small original population of *H. sapiens* that emigrated from Africa (Figure 22.15).

This scenario was first favored in the late 1980s, when it was called the **Out-of-Africa hypothesis**. For a long time, debates persisted between proponents of this Out-of-Africa model and the alternative "multiregional" model of human evolution. The multiregional model made claims that fossil evidence showed transitional stages between *erectus* and *sapiens* in different regions – Indonesia, China, Africa, and the Middle East – and favored gradual evolution of *sapiens* simultaneously across the Old World.

These arguments have now been settled once and for all by masses of new data that have become available over the last 15 years, all strongly supporting the Out-of-Africa hypothesis. This is evidence not only from fossils but also from the genetics of modern humans. The DNA of living humans shows much greater genetic variation among African populations than there is among all non-Africans. But in general, the variation in modern human DNA is actually very restricted. Populations of chimpanzees in the Taï forest in Africa have more variability than does the entire human species today. Taken together, this suggests very strongly that all modern humans are descended from an ancestral population that was not only small – say 10 000 or so – but was both small for a long time and quite recently (within the last 200 ka). The only possible conclusion is that all living humans are descended from a small ancestral *sapiens* population

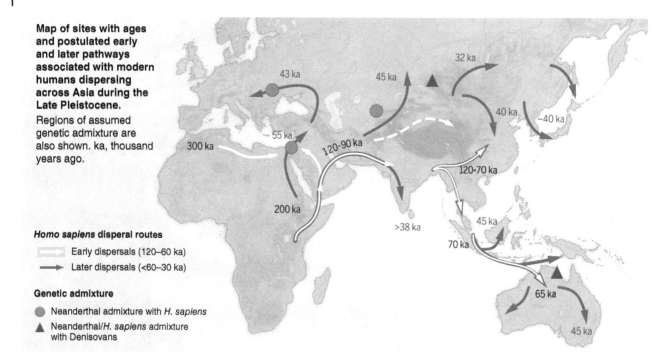

**Map of sites with ages and postulated early and later pathways associated with modern humans dispersing across Asia during the Late Pleistocene.**

Regions of assumed genetic admixture are also shown. ka, thousand years ago.

*Homo sapiens* disperal routes

Early dispersals (120–60 ka)

Later dispersals (<60–30 ka)

**Genetic admixture**

● Neanderthal admixture with *H. sapiens*

▲ Neanderthal/*H. sapiens* admixture with Denisovans

Figure 22.15 The dispersals of *Homo sapiens* out of Africa, with possible locations of admixture with archaic species. *Source:* Bae et al. (2017).

who evolved in a restricted region and spread from there throughout the Old World, replacing the earlier hominins in each region (*erectus* in Asia, the Neanderthals in Europe).

The difference that remains today between "African" and "non-African" DNA is explained if a small **founder population** left Africa, carrying with them only a small sample of the genetic variation that had by then evolved across Africa. These founder populations expanded as they occupied Eurasia, growing into a large population but with relatively low variation in their DNA structure. The pattern repeats again later, when a small subset of East Asian humans crossed the Bering Strait and populated the Americas with people who had even less genetic variation.

## The Neanderthals

The people we call Neanderthals (*H. neanderthalensis*) diverged from the modern human lineage over 400 000 BP and were a species living in Europe and the Middle East (but also as far east as the Altai Mountains) and surviving until as recently as 30 000 BP in some regions of southern Europe. They are named after a site in the Neander Valley in Germany, where some of the first Neanderthal fossils were discovered. Most Neanderthal fossils are found in deposits laid down in the harsh climates of the last two glacial periods.

Neanderthals had pushed into more northerly latitudes than any previous hominin, and over time they

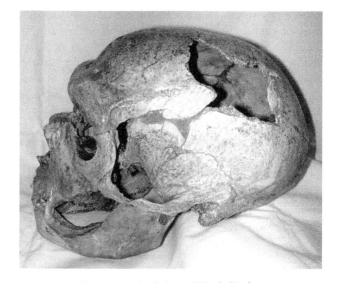

Figure 22.16 *Homo neanderthalensis*. This skull is from La Chapelle aux Saintes in France. *Source:* Image published in *PLoS Biology*, in the public domain.

became strongly adapted to life in the cold climates of western Europe and central Asia. The geographical isolation allowed unique Neanderthal characters to evolve, becoming visibly different from both *heidelbergensis* (earlier in southern Europe) and from the *sapiens* populations that were evolving in Africa.

Neanderthals differ from living humans in having robust skulls, big faces with large brow ridges, large front teeth, and little or no chin (Figure 22.16). Their skeletons show their body type was quite short but heavily built and stocky relative to modern humans, who are more gracile in comparison. There is variation through time, however, and the classic Neanderthal body type is observed most prominently in later populations. Their short, stocky stature shows they were well adapted for cold climates: a large body mass and shorter distal limbs act to minimize the surface area to body mass ratio and thus reduce heat loss to the surrounding environment. Their strong, stocky bodies with very robust bones may also reflect a lifestyle that required great physical strength. Studies of their arm bones have shown strong asymmetry in some individuals – as much as 50% stronger on the right-hand side. Researchers have hypothesized that this may relate to behaviors that load heavily onto one arm, for example spear throwing or directly thrusting spears into prey, or (perhaps less dramatically) to daily chores such as continual scraping of animal hides to make clothing (Shaw et al. 2012).

Neanderthal brain size, at 1450–1500 cc, was at least equal to that of living humans and sometimes greater, and Neanderthals had a way of life that was distinctly sophisticated in living sites, tools, and behavior.

Most Neanderthal tools are made in a style called **Mousterian**. They include scrapers, spear points, and cutting and boring tools (Figure 22.17) made from flakes carefully chipped off a stone core. Marks on Neanderthal teeth suggest that they stripped animal sinews to make useful fibers by passing them through clenched teeth. But perhaps the most enlightening Neanderthal finds are their ceremonial burials. Bodies were carefully buried, with grave offerings of tools and food. Enormous quantities of pollen were found with the body of Shanidar IV, a Neanderthal man buried in Iraq. The pollen came from

seven plant species, all of which have brightly colored flowers, all bloom together in the area in late April, and all have powerful medicinal properties. It is difficult to avoid the conclusion that Shanidar IV was carefully buried with garlands of healing herbs chosen from early summer flowers.

In the Middle East, between approximately 100 000 and 50 000 BP, Neanderthals seem to have alternated with *H. sapiens*, with a fluctuating border between them. Both peoples in the region were making the same Mousterian-style tools, which have been identified as far south as the Sudan. The population ranges reflect the climatic trends, with Neanderthals living in the Middle East in cooler, wetter times, while *H. sapiens* lived there in hotter, drier times. Each was fitted to a particular climatic zone in which the other could not compete, and their ranges expanded or contracted following the climatic oscillations.

European Neanderthal sites typically contain less standardized tools, made only from local stone and flint, but the last Neanderthals in western Europe showed a distinctly more advanced culture, with some similarities to the Aurignacian tools that the newly arrived, fully modern humans were using at about the same time in western Europe. The last Neanderthal sites in France also contain simple ornaments, and it is tempting to suggest that Neanderthals may have copied some of the Cro-Magnon technology and art. The earliest modern humans in Europe are known as the Cro-Magnon people (Figure 22.18).

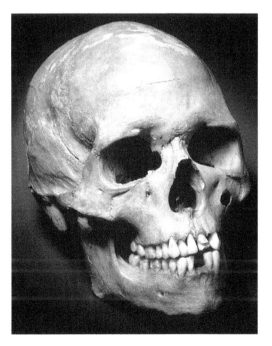

Figure 22.18 *Homo sapiens* (Cro-Magnon skull from France). *Source:* Photograph by Laténium (Wikimedia).

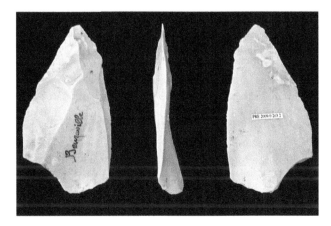

Figure 22.17 Mousterian tools made from flint, from Beuzeville, France. *Source:* Photograph by Didier Descouens of specimens from the Museum of Toulouse, as part of Projet Phoenix (Wikimedia).

Cro-Magnon sites yield richer and more sophisticated art and sets of tools, and more complex structures and burials than those typical of Neanderthals. In particular, Cro-Magnons made stone tips for projectiles. Neanderthals may have had arrows and spears but if they did, they were not stone tipped. Mousterian tools are more frequently wood-working tools than those of Cro-Magnons, who worked more with bone, antler, and stone.

Neanderthals disappeared from the Middle East about 45 000 BP, then from eastern and central Europe, and finally from northwest Europe (France and northern Spain). The last dated surviving Neanderthals existed in upland France until about 38 000 BP and in southern Spain and Portugal until about 30 000 BP.

Did modern humans drive Neanderthals extinct? Neanderthals do seem to decline at the same time as Cro-Magnon sites were spreading across Europe. However, the picture is not so simple, and it is also a time of increasingly cold climate, approaching the last glacial maximum at 18 000 BP. Yes, Neanderthals were a "cold-adapted" species but the last evidence for Neanderthals is in "glacial refugia" in southern Europe, which suggests Neanderthals may have been pushed to their limit in terms of the conditions in which they could survive, whereas the early modern humans perhaps had a greater cultural toolkit to withstand these same challenges.

Cro-Magnons were also responsible for the magnificent cave paintings (before 30 000 BP) of extinct Ice Age animals (Figure 22.19), drawn by people who saw them alive, and they made (and presumably played) bone flutes. The Gravettians, populations of modern humans living in Europe during an increasingly cold time between 22 000 and 33 000 BP, were firing small terracotta figurines in kilns.

Figure 22.19 A bison, painted on the wall of the cave at Altamira, Spain, well over 10 000 years ago. *Source:* Photograph by Ramessos (Wikimedia).

As Cro-Magnon people took over the cold and forbidding European peninsula, they were only a local population on the northwest fringe of the human species, but they are important because they have yielded us the best-studied set of fossils, tools, and works of art from the depths of the last Ice Age (Figure 22.19).

Modern Europeans today may share some ancestry with these first human occupants of Europe, but over the subsequent 20 000 years Europe was to see numerous further population movements and replacement. Most notable of these was the development and impact of agriculture – which took hold in the Middle East first, leading eventually to an influx of Neolithic people (and new ways of life) into central Europe, and from there further west. These dispersals were not always population replacement, but at times either dispersal just of cultural ideas or integration of knowledge (and people) into existing communities.

## Ancient DNA

Ten years ago, there was agreement about a model for the evolution of *H. sapiens*: modern humans evolved in Africa, a small population then migrated out of Africa (around 60 000 BP), spreading across the Old World and replacing any and all archaic species – either directly in competition or perhaps because they were better able to survive harsh climatic conditions as the last glacial maximum approached.

The advent of techniques to extract and sequence ancient DNA (aDNA) has dramatically altered what we know of the history of these last few hundred thousand years (Slatkin and Racimo 2016). The first full draft of the Neanderthal genome was published in 2010 by a group led by the geneticist Svante Pääbo (Green et al. 2010). Contrary to some earlier genetic evidence, this showed that there may have been interbreeding between Neanderthals and modern humans, because Neanderthals share slightly more genetic variants with non-African modern humans than with Africans. Eurasian *H. sapiens* that mated with Neanderthals and produced viable offspring thereby introduced Neanderthal genes into the *sapiens* population. It is suggested that the contribution to modern human (non-African) DNA after dispersals out of Africa is about 1–4%.

The Neanderthal DNA (recovered from the original fossil from the Neander Valley) is distinctly different from that of any living human population, and different from samples of early modern human aDNA. So, the evidence shows Neanderthals were evolving separately from other groups, but at times there was some limited interbreeding between *H. sapiens* and Neanderthals.

Figure 22.20 Denisova cave. The site in the Altai mountains is of great importance to understanding modern human evolution, and aDNA from both Denisovans and Neanderthals has been recovered there, showing it as a location where different hominin species interacted. *Source:* Photograph by Demin Alesksey Barnaul (Wikimedia).

Surprisingly, the Neanderthal gene sequences are found (at low frequency) in both European and Asian genomes (but not in African genomes) and so the likely location for interbreeding is perhaps the Levant or Middle East, but not from the vicinity of later sites in Europe where the archaeological evidence had previously hinted at close interaction.

Even more sensationally, aDNA has identified a whole new group of hominins of which nothing was known before and who we know almost exclusively from their genes. In 2010, the tip of a little finger bone from Denisova cave (Figure 22.20) in the Altai Mountains was sequenced, and the results were quite different from modern humans or Neanderthals – this bone must belong to another species of hominin, now known as the "Denisovans." They are more distantly related to modern humans than Neanderthals and so likely reflect a group that had reached the Altai from an earlier migration.

Denisova cave is a unique site because at varying times it has been occupied by both Neanderthals and Denisovans. From sequencing more material discovered at Denisova cave, researchers have now, most unexpectedly, come across the genome of an individual that is the child of a Neanderthal mother and Denisovan father (Slon et al. 2018).

It is clear these archaic species could interbreed with one another, and the consequences of these encounters are preserved in the genomes of all non-African modern humans (Figure 22.21). Overall, however, the levels of interbreeding required to explain the data are quite low, and in most cases Neanderthals and modern humans probably remained isolated from one another. But at key sites, particularly on the fringes of the species' ranges (such as at Denisova), interaction and interbreeding were perhaps more common than we thought. Ancient DNA has changed considerably our understanding of interactions between hominin groups and the evolution of our species.

## Evolution among Humans Today

Humans today have expanded to inhabit almost the full range of environments available on planet Earth (and beyond!). We are thus an extreme example of a generalist species, not specialized to a single unique environmental niche but rather highly adaptable to our environment.

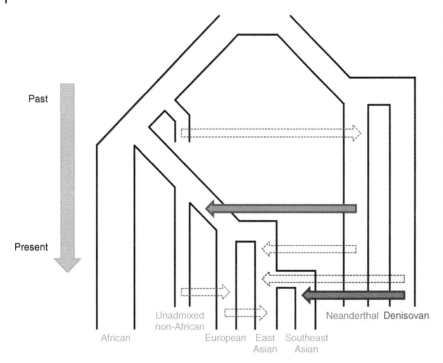

Past

Present

Unadmixed
non-African

African          European  East      Southeast        Neanderthal  Denisovan
                          Asian     Asian

Figure 22.21 A simplified model of DNA admixture from archaic species (Neanderthals and Denisovans) into the *Homo sapiens* lineage. There is a strong case for two definite gene flow events (*solid arrows*), and other cases of admixture, either among the archaic species or more recently among modern human populations, are also likely (Wolf and Akey 2018).

Evolution by natural selection depends on there being competition and environmental pressures giving rise to differences in reproductive success. A classic example from recent human evolution is the development of lactose tolerance in certain human populations following the origins of pastoral forms of agriculture. The ability to consume milk from domesticated animals into adulthood would have offered a significant survival advantage to individuals with specific mutations in the genes that control lactase production – and we can retrace how strong this selective force was from signals preserved in the genes of modern humans. Similarly, in tropical regions where malaria poses a risk to health, there continues to be selection for the sickle cell allele, which causes resistance to malaria in its heterozygous state but leads to the genetic disorder sickle cell anemia in the homozygous condition (Elguero et al. 2015).

Humans have unique abilities to buffer against environmental pressures very effectively. Adapting through cultural traits, changing behavior, or physiological responses (plasticity) are all routes to buffer external stresses before they can exert pressure for genetic adaptation (via natural selection). This is not uniquely human – many animals also buffer external pressures through some or all of these routes, but humans arguably have the most sophisticated and greatest range of responses available. Humans can modify their behaviors and develop their cultural toolkit in difficult environments based on a wealth of knowledge and experience built upon over generations – cumulative cultural change.

These various routes to adaptation can be seen in the success of populations in high-altitude environments, where reduced partial pressure of oxygen can lead to hypoxia (reduced oxygen supply to the body's tissues). Physiological responses to the reduced oxygen pressure take place (e.g., increased hemoglobin production) while culturally, clothing reduces exposure to cold, and behaviorally we can adjust (e.g., by reducing activity or taking time to acclimatize gradually). But despite all these options, sometimes only incomplete adaptation is possible and in these cases, natural selection may still act upon us. Tibetan populations, with a long history of inhabiting high-altitude environments in the Himalayas, possess unique genetic mutations that have been selected for because they provide an advantage over the typical physiological response individuals from low altitude experience when exposed to hypoxic environments.

Our understanding of humans today is closely entwined with our knowledge of the trajectory and history of human evolution. Understanding the interactions of early hominins with their environments, how natural selection favored increasing brain size, or the ways differing hominin species met and interacted allows us to discover and understand more about our own unique species today.

## References

Bae, C.J., Douka, K., and Petraglia, M.D. (2017). On the origin of modern humans: Asian perspectives. *Science* 358: eaai9067.

Bermúdez de Castro, J.M., Martinón-Torres, M., Arsuaga, J.L. et al. (2017). Twentieth anniversary of *Homo antecessor* (1997–2017): a review. *Evolutionary Anthropology* 26: 157–171.

Elguero, E., Delicat-Loembet, L., Rougeron, V. et al. (2015). Malaria continues to select for sickle cell trait in Central Africa. *Proceedings of the National Academy of Sciences of the United States of America* 112: 7051–7054.

Green, R.E., Krause, J., Briggs, A.W. et al. (2010). A draft sequence of the Neandertal genome. *Science* 328: 710–722.

Hawkes, K. and Coxworth, J.E. (2013). Grandmothers and the evolution of human longevity: a review of findings and future directions. *Evolutionary Anthropology* 22: 294–302.

Scally, A. and Durbin, R. (2012). Revising the human mutation rate: implications for understanding human evolution. *Nature Reviews in Genetics* 13: 745–753.

Scerri, E.M.L., Thomas, M.G., Manica, A. et al. (2018). Did our species evolve in subdivided populations across Africa, and why does it matter? *Trends in Ecology and Evolution* 33: 582–594.

Shaw, C.N., Hofmann, C.L., Petraglia, M.D. et al. (2012). Neandertal humeri may reflect adaptation to scraping tasks, but not spear thrusting. *PLoS One* 7: e40349.

Slatkin, M. and Racimo, F. (2016). Ancient DNA and human history. *Proceedings of the National Academy of Sciences of the United States of America* 113: 6380–6387.

Slon, V., Mafessoni, F., Vernot, B. et al. (2018). The genome of the offspring of a Neanderthal mother and a Denisovan father. *Nature* 561: 113–116.

Spoor, F. (2011). Malapa and the genus Homo. *Nature* 478: 44–45.

Stringer, C. and Galway-Witham, J. (2017). On the origin of our species. *Nature* 546: 212–214.

Wood, B. and Harrison, T. (2011). The evolutionary context of the first hominins. *Nature* 470: 347–352.

## Further Reading

Anton, S.C., Potts, R., and Aiello, L.C. (2014). Evolution of early *Homo*: an integrated biological perspective. *Science* 345: 1236828.

Asfaw, B., White, T., Lovejoy, O. et al. (1999). *Australopithecus garhi*: a new species of early hominin from Ethiopia. *Science* 284: 629–635.

Berger, L.R., de Ruiter, D.J., Churchill, S.E. et al. (2010). *Australopithecus sediba*: a new species of Homo-like australopith from South Africa. *Science* 328: 19–204.

Klein, R.G. (2009). *The Human Career: Human Biological and Cultural Origins*, 3e. Chicago: University of Chicago Press.

Lordkipanidze, D., de León, M.S.P., Margvelashvili, A. et al. (2013). A complete skull from Dmanisi, Georgia, and the evolutionary biology of early *Homo*. *Science* 342: 326–331.

Marzke, M.W. (2013). Tool making, hand morphology and fossil hominins. *Philosophical Transactions of the Royal Society of London. Series B, Biological Sciences* 368: 20120414.

Raichlen, D.A., Gordon, A.D., Harcourt-Smith, W.E.H. et al. (2010). Laetoli footprints preserve earliest direct evidence of human-like bipedal biomechanics. *PLoS One* 5 (3): e9769.

Stewart, J.R. and Stringer, C.B. (2012). Human evolution out of Africa: the role of refugia and climate change. *Science* 335: 1317–1321.

Wolf, A.B. and Akey, J.M. (2018). Outstanding questions in the study of archaic hominin admixture. *PLoS Genetics* 14 (5): e1007349.

Wood, B. (2010). Reconstructing human evolution: achievements, challenges, and opportunities. *Proceedings of the National Academy of Sciences of the United States of America* 107 (Supplement 2): 8902–8909.

## Questions for Thought, Study, and Discussion

1  The more hominin fossils we find, the more we find intermediate specimens and species along the various branches of hominin evolution. Explain how this makes it easier for a paleontology instructor to describe human evolution, and how at the same time it makes it harder to teach. Summarize your arguments in a few short questions that need to be answered. (They would all start with "Why?")

2 Traditionally it was thought the origins of *Homo* were closely associated with the first use of stone tools for butchering and an increase in hunting or scavenging carcasses. Is this still thought to be the case? What other hypotheses might explain the morphological changes we see in *H. erectus* compared to the earlier australopithecines?

3 Summarize the evidence that modern humans evolved in Africa and spread across the world from there.

4 There is a growing consensus that there was some interbreeding between *H. sapiens* and *H. neanderthalensis*. Do you agree that ancient DNA studies have revolutionized our understanding of later human evolution? Describe how our models of human evolution have changed.

## 23

## Life in the Ice Age

**In This Chapter**

The last 2.5 million years of Earth's history took place against a background of climate change known as the Ice Age, although the term is often used to refer to individual glacial episodes within this timeframe. Climate change caused plants and animals to migrate to remain in their preferred habitats. Toward the end of the last glacial episode, there was a significant extinction of large mammals, with the disappearance of such iconic animals as saber-tooth cats, mammoths, and mastodons, giant ground sloths, and many more. In many cases, these extinctions coincided with the arrival of *Homo sapiens* in their ecosystem but climatic and environmental changes likely also played an important role. Understanding these extinctions has relevance today, as human populations increase, climate change warms the planet, and ecosystems are increasingly devastated.

## Ice Ages and Climatic Change

Climate is one of the most important environmental factors for all organisms, and climatic changes have been major factors affecting the evolution of life. Polar ice is not a usual feature of Earth's climate. We now live in a time where polar ice is prominent but until relatively recently, the Earth was ice free since the Paleozoic. (Chapter 17 discusses the phenomenon of past "ice ages" in more detail.) The Antarctic ice cap has been present (with some fluctuations) since the late Eocene but the Arctic ice cap is more recent, only becoming established at around 5 Ma, at the start of the Pliocene.

We usually think of an "ice age" as a time when ice covered a greater area of the Earth than it does today but these are really episodes of glaciation, periods when the ice sheet is advancing. We are living through an ice age now and have been for the past 2.5 million years or so, marking the inception of the Pleistocene epoch. We happen to live during a warm stage (i.e., an interglacial period) but there is no sign that the overall ice age is over. In fact, this interglacial period might be cooler than some earlier ones: in the last interglacial, hippos lived as far north as England.

Great ice sheets expanded and covered much of the northern continents, then retreated again to approximately their present position, at least 17 times in the past 2.5 million years. The end of the last glaciation, around 12 000 years ago, marks the end of the Pleistocene and the beginning of the Holocene epoch, which we still inhabit. (Together, the Pleistocene and Holocene epochs comprise the Quaternary Period.)

As we saw in Chapter 17, it was not until the later Cenozoic that climatic cooling, related to continental movements, led to the build-up of ice at the poles. Cooling alone is not sufficient for ice build-up; continental movements must also result in large areas of land at high latitudes (the current position of Antarctica over the South Pole being a case in point). To lock the Earth into a long glacial period, there must be sufficient high-latitude land area for large continental ice sheets to spread out and provide high reflectivity over large regions.

The advance and retreat of the ice sheets are triggered by Milankovitch cycles, changes in the Earth's axis and orbit around the Sun that were discussed in Chapter 17. Huge areas of the northern continents are still covered by debris deposited by ice sheets during numerous glacial advances and retreats during the past 2.5 million years.

*Cowen's History of Life*, Sixth Edition. Edited by Michael J. Benton.
© 2020 John Wiley & Sons Ltd. Published 2020 by John Wiley & Sons Ltd.

## The Present Ice Age

The effects of the current ice age have been most marked in the northern hemisphere, because that is where the current high-latitude land mass is located. The southern continents also cooled but they were far enough away from Antarctica that ice sheets could not spread to cover them: only the very tip of South America experienced any glaciation. At its maximum, Arctic ice advanced as far south as New York, St Louis, and Oregon (Figure 23.1), and sea surfaces in the North Atlantic froze as far south as New York and Spain. The western United States became much wetter than it is today, so that great freshwater lakes formed from increased rainfall and from meltwater along the front of the ice sheet. More subtle effects occurred in warmer latitudes. For example, increased rainfall in the Sahara formed great rivers flowing to the Nile from the central Sahara; they were inhabited by crocodiles and turtles, and rich savanna faunas lived along their banks.

## Life and Climate in the Ice Ages

The severe fluctuations of climate greatly affected ice age plants and animals. Major shifts of range, extinctions, and evolutionary adaptations happened throughout the Pleistocene, although the basic types of animals present remained relatively constant. A burst of extinction, however, was concentrated at the end of the last period of glaciation (i.e., the end of the Pleistocene), and we will discuss later how much these were due to climatic change (versus the effects of humans). Glacial advances and retreats were rapid on a geological time scale but were slow enough to allow animals to migrate north and south with the ice sheets and the climatic zones and weather patterns affected by them. However, these migrations were individualistic in nature, rather than all of the animals in an ecological community moving together, so there was considerable episodic ecological disruption. Tropical rainforests were very much reduced but the habitat did not disappear, and their fauna and flora survived well.

Further interesting effects were controlled by changes of sea level that occurred with every glacial advance and retreat. As water was removed from the oceans into ice sheets, each major glaciation dropped world sea level by 120 m or so (about 400 ft), exposing much more land area and joining land masses together. Each new melting episode reflooded lowlands to recreate islands.

Most continents carry examples of organisms whose ranges were shifted by the climatic changes. For example, today the Sahara Desert is a hot and dry inhospitable habitat but ancient rock paintings of giraffes and antelope provide evidence of a moist climate there 2000 years ago, and there are a few surviving cypress trees perhaps 2000 years old. Some populations were literally stranded by climate warming after the last glaciation: certain plant species, formerly widespread across

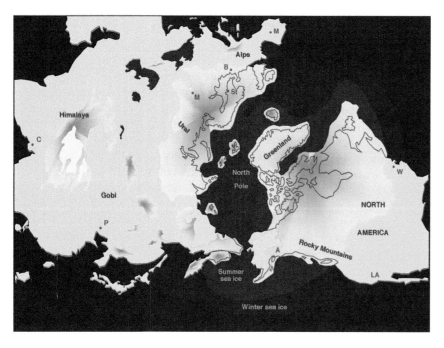

Figure 23.1 Great ice sheets (shown in light blue) around the northern hemisphere were a major influence on climates during the successive ice ages. *Source:* Map by Hannes Grobe (Wikimedia).

Europe, now live as isolated populations at high altitude in the Alps.

A few creatures were trapped in geographical cul-de-sacs and wiped out. Advancing ice sheets, not St Patrick, wiped out snakes from Ireland, and snakes did not recolonize before the postglacial rise of the Irish Sea cut off the island. More commonly, species' ranges contracted due to climate change: trees require adequate warmth to grow and the forests of western Europe were replaced by largely open habitats, except for Mediterranean refugia and a few stragglers that hung on near the coast of Norway. After the ice sheets melted, western Europe was eventually recolonized by deciduous hardwoods although elsewhere, in North America, Scandinavia, and Siberia, the great boreal (northern) forests are dominated by conifers.

We have good evidence of the plant and animal life of the Pleistocene. Enormous bone deposits in Alaska and Siberia and fossils found in river deposits, caves, sinkholes, and tar seeps have provided excellent evidence of rich and well-adapted ecosystems on all continents.

As the result of sea-level change, land areas sometimes showed dramatic changes. For example, at the peak of the last glaciation (21 000 years ago), Alaska and Siberia were joined across what is now the Bering Strait, and Greenland was joined to North America, to form one giant northern continent (Figure 23.1). Australia was joined to New Guinea, and Indonesian seas were drained to form a great peninsula jutting from Asia, incorporating islands such as Borneo and Sumatra in a land mass with the mainland. Madagascar and New Zealand, however, remained as separate islands.

## Faunal Changes

On major continents, the larger birds and mammal faunas of the Pleistocene were much more diverse than those of the present day. North America had mastodons (browsing proboscideans), mammoths (grazing proboscideans), giant bison, ground sloths, saber-tooth cats, dire wolves, moose, reindeer (caribou), horses, camels, and dozens of other large mammals. In some cases, the species were much larger than their present-day relatives (e.g., giant ground sloths in the Americas, giant marsupials in Australia). Some of these mammal groups had been there for much of the Cenozoic (horses, camels, mastodons, and dogs ancestral to wolves), while others (including the remainder of those mentioned above) first appeared in the Plio-Pleistocene (i.e., within the last 5 million years). Eurasia had many of these same mammals (not sloths, despite their presence there in the movie "Ice Age"!), plus giant deer and wooly rhinos (Figure 23.2). The moas of New Zealand and the elephant birds of Madagascar are well known (see Chapter 20), but Australia had giant ground birds as large as these and a dozen giant marsupials. All these creatures are now extinct.

The catastrophic extinctions occurred at different times on different continents. For example, in North America, perhaps 20 genera of varying sizes became extinct in the 2 million years or so before the start of the last glaciation, around 100 000 years ago, but these animals were usually replaced by the evolution of other, related, mammals (e.g., one kind of saber-tooth cat would be replaced by another). But thereafter, around 35 genera of large mammals were lost within a much shorter

Figure 23.2 Wooly mammoths foraging in a European Ice Age winter, among horses, lions, reindeer, and a wooly rhino. Mammoths were grazing proboscideans, closely related to modern elephants. Mastodons were another type of proboscidean but with teeth showing that they were browsers, and despite their superficial similarity to mammoths they were much more distantly related to modern elephants. *Source:* Painting by Mauricio Antón (Wikimedia).

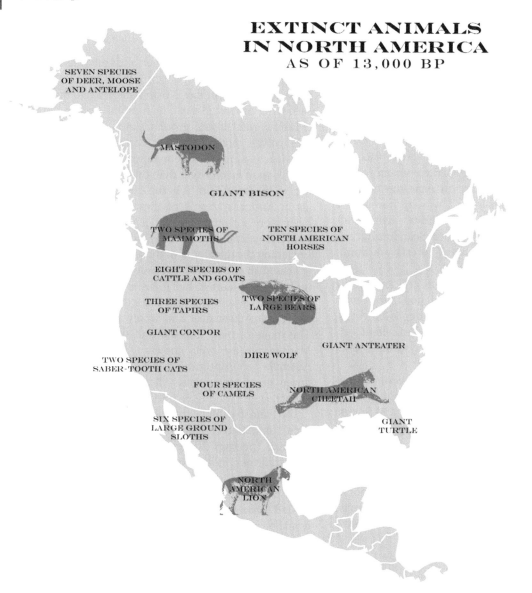

## EXTINCT ANIMALS IN NORTH AMERICA
### AS OF 13,000 BP

SEVEN SPECIES
OF DEER, MOOSE
AND ANTELOPE

MASTODON

GIANT BISON

TWO SPECIES OF
MAMMOTHS

TEN SPECIES OF
NORTH AMERICAN
HORSES

EIGHT SPECIES OF
CATTLE AND GOATS

THREE SPECIES
OF TAPIRS

TWO SPECIES OF
LARGE BEARS

GIANT CONDOR

GIANT ANTEATER

TWO SPECIES OF
SABER-TOOTH CATS

DIRE WOLF

FOUR SPECIES
OF CAMELS

NORTH AMERICAN
CHEETAH

SIX SPECIES OF
LARGE GROUND
SLOTHS

GIANT
TURTLE

NORTH
AMERICAN
LION

Figure 23.3 The astounding number of large North American mammals that became extinct about 12 000 years ago. *Source:* Artwork © 2012 Mike Hansen, and used by permission.

time period, most of them within a few thousand years toward the end of the glacial period 12 000 years ago (Figure 23.3). Note that, in talking about Pleistocene "megafaunal" extinctions, a "large" mammal is not necessarily an elephant or a rhino: it is defined as anything over 44 kg (~100 lb), so bigger than a large dog.

Some high-latitude ice-age animals, such as the wooly mammoth and wooly rhinoceros, were much hairier than their living relatives (as shown in cave paintings; Figure 23.4), and were specifically adapted to life in cold climates. Wooly mammoths sometimes died by falling into ice crevasses. Their bodies have been found still frozen in permafrost in Siberia, preserved well enough to tell us quite a lot about their way of life (Figure 23.5).

Frozen stomach contents show that wooly mammoths ate mainly sedges and grasses and browsed dwarf tundra trees such as alder, birch, and willow. Mammoths shifted their range in response to ice advances and retreats through the Pleistocene, and their molar teeth became progressively better adapted for chewing coarse forage (Figure 23.6).

Large Pleistocene mammals had certain advantages that allowed them to survive episodes of severe climate. Their large body sizes enabled them to bulk-process low-quality food, and to store sufficient fat reserves in the summer to last through winter food shortages. As adults, they were largely free from the danger of predation by carnivores. Yet many large mammals and birds

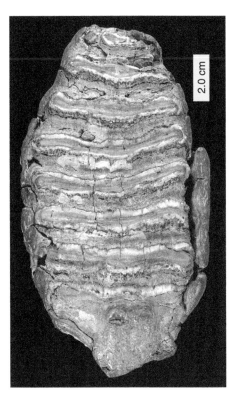

Figure 23.4 We do not have to rely on the inferences of paleontologists to describe the megaherbivores in Pleistocene ecosystems. The wooly mammoth and the wooly rhinoceros were observed and illustrated by competent European ecologists of the time.

Figure 23.6 A mammoth's molar, showing multiple ridges for shearing vegetation. *Source:* Photograph by Dr Mark A. Wilson of the College of Wooster (Wikimedia).

## Causes of the End-Pleistocene Extinction: Human Hunting

Figure 23.5 A baby wooly mammoth (called "Dima"), as it was found frozen into the permafrost in Siberia. It is now preserved and on display in St Petersburg. *Source:* Photograph by NOAA, in the public domain.

became extinct at the end of the Pleistocene, while smaller species did not suffer as much. The plants the large mammals ate are still living, and so are the small birds, mammals, and insects that lived with them. In the oceans, nothing happened to large marine mammals.

The problem of Pleistocene extinctions has been debated ever since ice-age animals were discovered, and there is still continuing major disagreement. In North and South America, many of the extinctions took place in a short time around the end of the last glaciation. There is no question that climatic change around 12 000 years ago was rapid, but the very same species of animals had already survived a dozen or more similar events. The previous ice retreat, about 125 000 years ago, resulted in relatively few extinctions. Was there something unique about climatic change at the end of the last glaciation or were there other factors involved? We know that the end-Pleistocene was a time of both rapid climatic change and human arrival in the Americas; as noted by David Meltzer, the problem is to distinguish a signal of human effects that is separate from the noise of the climatic and ecological changes.

Humans were long thought to have had at least some role in the extinctions, and several decades ago Paul Martin, of the University of Arizona, put forward the idea of the **overkill hypothesis**, the notion that human hunting was responsible for the extinction of the large mammals and birds. "Overkill" could take place over thousands of years by gradual attrition, or in less than a thousand years. This latter, more sudden, extinction was termed the Blitzkrieg hypothesis (with the overtones of human destruction by warfare), and Martin proposed this for the apparently rapid extinctions of large mammals in the Americas.

Martin noted that the extinctions were much less profound in areas of the world where humans had been present throughout the Pleistocene: that is, in Africa and to a lesser extent Eurasia. In other parts of the world, such as North and South America, and Australia, and particularly on islands such as Madagascar and New Zealand, the extinctions were much more profound and appeared to coincide (or to occur shortly after) humans arrived into these regions. Martin contended that human hunting did not have such a great effect in Africa and Eurasia because the animals there had coevolved with humans and learned to become wary of them, while elsewhere in the world they were naive to human behavior and by the time they learned to fear and avoid them, it was all too late.

While some scientists enthusiastically embraced this new hypothesis for the end-Pleistocene demise of large mammals, others were more skeptical. Could humans with primitive technologies and likely small population sizes really have exterminated so many species over the entire continent? And if human hunting caused such demise of so many large mammals, then why were there not more extinctions later on in the Holocene? Bison survived the end-Pleistocene extinctions and continued to survive intensive hunting by native North Americans, and then by commercial hunters with rifles in the nineteenth century (although this latter large-scale slaughter did bring the bison close to extinction). Reindeer and horses also survived extensive hunting in northern Eurasia.

It has also been noted that archeological evidence for human hunting in North America is not extensive and there is even less for South America. Out of 126 archaeological sites interpreted as "kill sites" (and that in itself is often contentious), only 15 of them involve extinct taxa, these consisting of only five of the genera that became extinct. However, Martin argued that these events were so rapid that we would not expect the evidence to be preserved. The extinction of carnivores, not hunted by humans, would follow from the loss of their prey or in some cases perhaps from competition for space (caves especially) by humans.

Although more recent supporters of "overkill" do not generally accept the sudden "Blitzkrieg" hypothesis, there have been a number of recent attempts to demonstrate Martin's overkill model through modeling of human hunting on megafauna. Early analyses suggested that, under reasonable assumptions of human population growth and hunting frequency, humans entering North America could have caused the extinction of a number of megafaunal species within 2000 years. However, the modeling did not take account of the rapid learning and behavioral avoidance by mammalian prey that is evident from present-day experience. Also, humans, like other predators, would have switched to other prey when one species became rare, thus were unlikely to drive any one species all the way to extinction.

More recently, models have attempted to test the impact of human hunting by examining the timing of extinction versus the time of arrival of humans in different areas. However, such studies have been criticized because the extinction date of the majority of species is still very poorly known, especially for those species that may have gone extinct before the limit of accurate radiocarbon dating (about 40 000 years ago). Such earlier extinctions would predate the arrival of humans in most areas.

## Causes of the End-Pleistocene Extinction: Climatic and Environmental Change

Many scientists see a much larger role for climatic change, or at least some combination of climatic change and human action, with habitat loss being at least as important as people killing the animals by hunting them. In a combined viewpoint, human arrival might have been the "last straw" for a fauna already stressed by environmental events. At the end of the Pleistocene, there were violent oscillations of global climate, starting around 15 000 years ago with a sudden warming event as the glaciers retreated, then pronounced cooling (an episode known as the Younger Dryas), and finally a warming event in the early Holocene. With each change, global temperatures shifted between 5° and 10°, sometimes over a period of only a few decades.

These sudden climatic changes might well have had a debilitating effect on the fauna, in many cases due to habitat loss rather than the effects of temperature per se. There were major environmental changes, including the replacement of the mammoth steppe type of vegetation (see description below), that covered much of the high latitudes of the northern hemisphere during the Pleistocene, by tundra or forest. John Stewart and colleagues note that both the fossil record and molecular data show that Late Pleistocene and Early Holocene environmental changes did not simply cause shifting of entire animal communities to more hospitable climes; rather, different species had individual responses, resulting in the break-up of previously established communities. Whether such events followed other end-glacial episodes is not clear, as we have much less detailed data.

Considerations about the role of climatic change in the extinction of Pleistocene mammals may conjure up images of mammoths trapped in ice floes. But the effects of environmental change may have been more subtle, shifting species' distributions over thousands of years. Adrian Lister and Tony Stuart have made extensive studies of the Eurasian Pleistocene megafaunal mammals over the period of time from the last glacial maximum to the present interglacial (from around 25 000 to 10 000 years ago), and present examples of species extinctions that appear to have been largely climate mediated.

Lister and Stuart showed that those species that eventually became extinct, such as mammoths and giant deer, shifted their ranges in response to vegetational change. As their ranges became smaller, the remaining populations became concentrated in refugia, and ancient DNA studies reveal a reduction in genetic variation. These species continued to survive for thousands of years in these refugia but eventually went extinct. In some cases, mammals limited to refugia were able to later radiate and recolonize larger areas, but many were not.

While it's possible that humans may have been the final cause of the demise of Eurasian mammoths and giant deer, environmental changes alone might have been sufficient. Large mammals are likely to be more sensitive to these types of environmental effects than smaller ones, because of their smaller population sizes and slower rates of reproductive turnover.

Ancient DNA studies also show that many late Pleistocene mammals were losing genetic diversity. In North America, this began long before the appearance of the first humans; the invading humans may have merely been finishing off the surviving members of genera that were already in severe decline.

Most scientists would agree that extinctions on islands were the result of human invasion, resulting either from human hunting, human environmental disturbance, or a mixture of the two. These island extinctions took place during the Holocene, many within the past few thousand years. However, debate continues about the role of humans in extinctions on continental land masses, which either took place during the Late Pleistocene (e.g., Australia) or at the transition between Pleistocene and Holocene around 12 000 years ago (e.g., the Americas). Part of the problem lies in obtaining accurate dates to establish the correlation of human arrival and local extinctions.

We will look at these patterns of extinction on the three major continents that were colonized suddenly by humans, North and South America, and Australia, as well as extinctions in northern Eurasia and on islands.

## North America: Human Arrival and Human Hunting

Humans crossed into North America from Siberia at a time when the Bering Strait region was a dry land area, Beringia. During the last glacial period, Beringia (including northeast Siberia, Alaska, and the Yukon), and wider areas of northern Eurasia and northern North America, supported a productive habitat called the "mammoth steppe." This high-latitude, largely treeless environment was nonetheless rich in grasses and other herbaceous plants, supporting a fauna of large ice-age mammals, resembling that of the present-day African savannas, including wooly mammoths, wooly rhinos (in Eurasia only), horses, bison, reindeer, moose, musk oxen, wolves, and cave lions. Further south, in what is now the continental US, the megafauna also included large mammals such as xenarthrans (ground sloths and glyptodonts, relatively recent immigrants from South America, see Chapter 20), plus camels, pronghorns, mastodons, and saber-tooth cats.

Did humans reach the Americas only as the last glacial period ended or did they arrive long before? There is now compelling evidence from a number of sites in North and South America that humans had arrived by 15 000 years ago. Reports of even earlier arrival remain contentious. Most likely, these people moved south of the Bering region by boat along the western American coast. When Europeans arrived on the Alaska coast in the eighteenth century, they found very proficient fishers and hunters there, using a variety of boats for hunting, transportation, and trade along the coast, possibly reflecting the lifestyle of the original inhabitants, as documented by scattered evidence from sites in British Columbia, southern California, and Peru. These people seem to have eaten shellfish, seabirds, and fish. As far as we can tell, they had very little effect on American continental ecosystems, and may not even have ventured inland across the mountain barriers that lie behind the entire west coast. American continental ecosystems did not receive full human impact until the arrival of big game hunters in the interior around 13 000 years ago.

For decades, we have been gathering evidence of a short-lived, distinctive tool and weapon culture, the **Clovis culture**, which was widespread across North America from Washington to Mexico. Dates for the earliest Clovis sites cluster around 13 000 years ago, approximately coinciding with a number of dated major

**Figure 23.7** A collection of Clovis spear points from a site in Iowa. The skillful preparation of the edges is clear. Once the point was made, two careful blows chipped off a central smooth area on each side of the point so that it could be hafted on to the shaft of the spear. The scale is in centimeters. *Source:* Photograph by Billwhittaker@en.wikipedia.

mammalian extinctions and supporting Martin's Blitzkrieg notions. However, as mentioned above, many of these extinctions are poorly dated and various of these now extinct mammals may have disappeared pre- or even post-Clovis (the culture lasted only a few hundred years).

The trademark of the Clovis culture is a large lethal spear point made of obsidian or chert. These are weapons made to kill large mammals (Figure 23.7). There is a strong implication that the precursors of the Clovis people were already skillful hunters of large mammals across the far northern plains of Asia and Beringia, before they reached the open plains of North America. A mastodon skeleton was found in Washington State with the tip of a long large spear point made of mastodon bone or tusk embedded in a rib (it must have been driven a foot [30 cm] deep into the body to reach this far). The mastodon kill is dated nearly a thousand years before Clovis. In addition, evidence of mammoth hunting around the Great Lakes at around the same time is likely to have been the work of pre-Clovis hunters who used bone to arm their spears, rather than stone.

However, well-dated extinctions have not been demonstrated before the arrival of the Clovis people. Note that several species of birds and reptiles also went extinct around this time, and that there is evidence that the Clovis people hunted not only proboscideans but also many of the species that survived, such as bison, musk oxen, moose, and caribou (reindeer).

We know that the Clovis people hunted mammoths and mastodons. There are cut marks on mastodon bones found around the edge of the ice sheet near the Great Lakes, and it seems that humans butchered the carcasses into large chunks and cached them for the winter in the frigid waters under shallow, ice-covered lakes, just as Inuit do today in similar environments. We can tell that the favorite hunting season for mastodons was late summer and fall, whereas natural deaths occurred mainly at the end of winter when the animals were in poor condition. A mammoth skeleton from Naco, Arizona, had eight Clovis points associated with it. Two juvenile mammoths

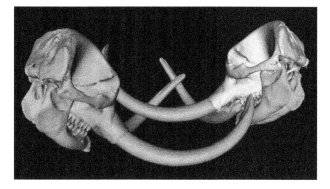

**Figure 23.8** Reconstruction of fighting between two male American mastodons. The reconstruction is based on massive damage to some fossil male skulls, which indicates lethal or sublethal strikes on a target area in which the tusk would smash under the cheekbone toward the braincase. *Source:* Courtesy of Daniel Fisher of the University of Michigan. © Daniel C. Fisher.

and seven adults were found with Clovis tools near Colby, Wyoming, and the way the bones are piled suggests meat caching there too. However, whether the animals were hunted or scavenged as already-dead carcasses can be difficult to establish in such cases.

It cannot have been easy to kill these proboscideans in any direct attack; they were lethally effective in using their tusks against one another (Figure 23.8), and that skill would easily have carried over into effective defense against their natural predators. But Clovis people were another story: they were armed with formidable weapons, traps, poisons, and intelligent group hunting tactics.

However, one of the problems with assessing the validity of Martin's Blitzkrieg hypothesis in relation to the Clovis people is as follows. Despite the fact that many Late Pleistocene large mammals were extinct by 10 000 years ago, we do not know how many of these extinctions were actually simultaneous and occurred shortly after the arrival of the Clovis people. Only about half of these species are known to have been still present during the time that the Clovis people were in North America. For the others, they may have been present or they may have gone extinct before the Clovis people arrived; the current data are insufficient to determine that. We certainly know that the Pleistocene extinctions in Eurasia were staggered over tens of thousands of years, and the extinctions in the Americas might not have been as rapid as assumed from current fossil record evidence. We need to know if, how, and when these megafaunal species began to decline in numbers, as well as the date of their final extinction. These data are becoming available as more radiocarbon dates are collected, and more studies of the DNA of these animals are accomplished.

## North America: Patterns of Megafaunal Extinction

By around 10 000 years ago, 37 species of North American large mammals known from the Late Pleistocene (representing 69% of the total megafauna) were extinct. Only nine megafaunal mammal genera survived, the largest being the American bison. Most of those extinctions that have been accurately dated appeared to have happened in the time window of 13 500–12 500 years ago. However, a number of megafaunal mammals made their latest recorded appearance in the fossil record several thousand years before the arrival of the Clovis people, including several ground sloths, the glyptodont, the American cheetah, and a saber-tooth cat (*Homotherium*).

Predator species such as the saber-tooth cats (Figures 23.9 and 23.10) and the North American lion

(a)

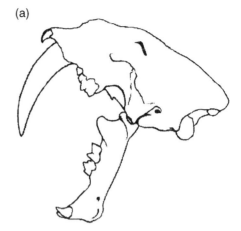

(b)

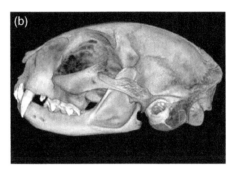

Figure 23.9 (a) The North American saber-tooth cat *Smilodon*. (b) A CAT scan of a cat (sorry, but it's true!). The skull structure is much the same, and the two cats differ mainly in the extravagant size of the upper canine in *Smilodon*. *Source:* Courtesy of Tim Rowe and the Digimorph Project at the University of Texas: www.digimorph.org.

Figure 23.10 A reconstruction of a scene from the La Brea tar pits, from the Late Pleistocene of California. The saber-tooth cat *Smilodon* menaces some giant ground sloths (*Paramylodon*), while the giant condors (*Teratornis*) look on. In the background are some Columbian mammoths (*Mammuthus*). *Source:* By Charles R. Knight (Wikimedia).

could have been reduced to dangerously low levels by the removal of their prey; there is no need in many cases to think in terms of the direct, systematic overkill of predator species that modern humans often carry out, although this has been suggested for cave bears and hyenas in Europe, as they competed with humans for precious living space (caves). In turn, scavengers may also depend on populations of large mammals to provide the carcasses they feed on. For example, the giant teratorn bird known from the La Brea tar pits (Figure 23.10) is extinct, and the so-called "California" condor once nested from the Pacific coast to Florida. Pleistocene caves high on vertical cliffs in the Grand Canyon of Arizona contain bones, feathers, and eggshells of this condor, along with the bones of horses, camels, mammoths, and an extinct mountain goat. The condor vanished from this area at the same time as the large mammals did, presumably because its food supply largely disappeared.

What about the surviving large mammals in North America? It turns out that many of them were originally Eurasian and crossed into the Americas late in the Pleistocene. Thus, the living species of bear, moose, musk oxen, and reindeer (caribou) had been previously exposed to humans in Eurasia, so may have had behaviors that reduced their vulnerability to human hunting. There were no North American extinctions after around 8000 years ago. There were separate regional cultures in the Americas by this time, but there were no new significant extinctions even though tools and weapons had improved.

Bison were a special case. They arrived in North America in the Early Pleistocene, earlier than these mammals that might have been exposed to human hunting in Eurasia. Although an immense long-horned species of bison became extinct, the smaller "American bison" (the extant *Bison bison*) survived in great numbers. Perhaps the removal of larger competitors encouraged its success. Moreover, bison survived in the face of intense and wasteful hunting by Native Americans and European colonists later in the Holocene and into recent times. Dale Guthrie, a supporter of climatic and vegetational effects, argued that the "mosaic" vegetation of the Pleistocene supported a wide array of herbivorous mammals, while the monotonous, widespread zones of vegetation in the Holocene support a few specialists (e.g., musk ox and reindeer in the tundra, bison in the grasslands).

A contributory factor to the extinctions of the smaller megafaunal mammals might have been the removal of **megaherbivores** (very large herbivorous mammals more than 1000 kg [1 ton] in weight) from the ecosystem. On the African savannas today, the largest animals, elephant and rhino, can have drastic effects on vegetation. Elephants destroy trees and open the woodland, opening up clearings in which smaller browsing animals multiply. They then migrate to another woodland habitat, leaving the trees to recover in a long-term ecological cycle that can take decades to complete. White rhinos graze high grass so effectively that they open up large areas of short grassland for smaller grazing animals.

Thus, in the long run, megaherbivores keep open habitats in which smaller plains animals can maintain large populations. Where elephants have been extinct for decades, the growth of dense forest is closing off grazing areas, and smaller animals are also becoming locally extinct. Many of the problems in African national parks today occur because they are not large enough to allow these cycles of destruction and migration to take place naturally.

But what would happen if megaherbivores were completely removed from an ecosystem – by hunting, for example? Norman Owen-Smith proposed that the disappearance of Pleistocene megaherbivores soon led to the overgrowth of many habitats, reducing their populations of smaller animals too. According to this theory, even if early hunters hunted or drove out only a few species of megaherbivores, they could have forced ecosystems so far out of balance that extinctions would then have occurred among medium-sized herbivores too, especially if hunters were forced to turn to the latter as prey when the megaherbivores had gone.

There may be more subtle effects of removing large herbivores. Plants sometimes coevolve with herbivores that disperse their seeds. Very large herbivores are likely to encourage the evolution of large, thick-skinned fruits, and a sudden extinction would leave the fruits without dispersers. Even today, guanacaste trees of Central America produce huge crops of large fruits, most of which lie and rot. Daniel Janzen suggested that these fruits coevolved with gomphotheres, proboscideans that lived in the Americas before the arrival of mammoths, which became extinct with the other large American mammals.

## South America

The South American megafauna consisted both of endemic species and relatively recent immigrants from the Great American Biotic Interchange (see Chapter 20). Unlike North America, South America suffered from very little glaciation, mainly limited to mountain glaciers in the Andes. The endemic species consisted of a diversity of ground sloths (see Chapter 19, Figure 19.3), glyptodonts (Figure 23.11) and related forms, as well as native ungulates such as *Toxodon* (see Chapter 20, Figure 20.13). The immigrants that did not survive the Pleistocene

Figure 23.11 Two Native Americans in North America size up their chances with a glyptodon (*Glyptodon*) (probably not very good!). Glyptodon was a South American species that migrated up to North America in the Pleistocene, along with its xenarthran relatives, the ground sloths. Glyptodons were extinct in South America before humans arrived there. *Source:* By Heinrich Harder, in the public domain.

included horses, gomphothere proboscideans, and the large saber-toothed cat *Smilodon* (Figures 23.9 and 23.10). Around 52 genera went extinct, 80% of the original megafauna, more than on any other continent.

There are as yet too few radiocarbon dates from South America to establish a reliable chronology for megafaunal extinctions or to correlate them with human arrival, but as far as our existing knowledge shows, here the extinctions were spread out over a longer period of time. The largest group, including gomphotheres (proboscideans), horses, and several genera of ground sloths, have latest known dates between 18 and 11 ka and their demise might be connected with the arrival of humans, first recorded around 14 500 years ago. Several genera (such as the giant armadillo relative *Glyptodon*) were apparently extinct earlier, before 18 ka. A third category continued longer, well into the Holocene (including *Smilodon* and the giant ground sloth *Megatherium*). Although there is little archeological evidence of humans killing the megafaunal mammals, and humans did not coexist with the megafaunal mammals throughout South America, scientists propose that the South American extinctions were due to some combination of human influence and severe climatic events at the Pleistocene/Holocene transition.

## Old World Continents

Africa and southern Asia are the two regions of the world where megaherbivores still survive. Most of the megafaunal extinctions in the Old World were in northern Eurasia, and this region has been subject to the most detailed study of any continent. Environmental change has been well documented here too, especially the disappearance of the so-called mammoth steppe habitat at the end of the Pleistocene. Modern humans reached the area around 45 ka, replacing the previous hominin inhabitants, the Neanderthals.

Around 18 out of 49 megafaunal mammals (including at least one human species) went extinct, representing 37% of the total. But most of the mammals over 500 kg disappeared, including three species of proboscideans and four of rhinos. The evidence clearly demonstrates that extinctions were spread out over a significant period of time. While several extinctions in northern Eurasia (such as wooly rhinoceros and "cave" lion) took place at the end of the Pleistocene (roughly in the interval 15–12 ka), other species disappeared between 40 and 25 ka (e.g., cave bear and the huge one-horned rhinoceros *Elasmotherium*) and others still as early as 100–50 ka (including two species of interglacial rhinoceros). Finally, some megafaunal mammals survived well into the Holocene in restricted areas, in particular wooly mammoths and giant deer.

Utilization of megafaunal resources by people in the last glaciation of northern Eurasia is well documented. For example, the Gravettian people (see Chapter 22) used mammoth bones as resources. At Předmostí in the Czech Republic, a site dating from just before the coldest period of the last glaciation (28000–22 000 BP) contains the bones of at least 1000 wooly mammoths. These people buried their dead with mammoth shoulder bones for tombstones. However, this intensive use of mammoth resources predates regional extinction of the species by several thousand years.

The giant deer *Megaloceros* (Figure 23.12) is sometimes called the Irish elk, partly because it is best known from Ireland – it is even depicted on the coat of arms for Northern Ireland (Figure 23.13). It was not an elk but a deer the size of a moose, with the largest antlers ever

Figure 23.12 *Megaloceros*, reconstruction by Pavel Riha (Wikimedia). This giant deer has several species across Pleistocene Eurasia but the largest and last surviving was the "Irish elk."

Figure 23.13 The coat of arms of Northern Ireland. Like all such devices, it is brimming in symbolism. The lion and the crown represent Great Britain, and the Red Hand is an ancient Celtic symbol adopted by the O'Neill kings of Ulster. The Irish elk, along with the harp, are perhaps the only pan-Irish symbols in the entire device.

evolved, more than 3 m (10 ft) in span. It was adapted for long-range migration and open-country running, and its diet was the high-protein willow vegetation on the edges of the northern tundra. It once ranged from Ireland to southern Siberia and there were related species in the Far East, but it did not reach North America.

The giant deer flourished in northwest Europe in a warm period until about 12 000 years ago, but it then died out in a brief cold period, probably because of a sharp reduction in vegetational productivity. However, a population survived until around 7700 years ago in central Russia, where it may have been wiped out by Neolithic farmers.

Wooly mammoths contracted their range progressively over a period of 20 000 years or more. They were gone from Europe by 14 ka, coincident with the spread of forest over former grassland habitats. Some wooly mammoths survived in the permafrost areas of northern Siberia until around 10 000 years ago, in an area where humans arrived late. Even then, there were still some remaining mammoth refuges, including Wrangel Island, a small low-lying island off the north coast of Siberia. Forage was poor and the last mammoths were small, perhaps 2 tons instead of the 6 tons of their ancestors. The last wooly mammoths died out on Wrangel Island only 3000–4000 years ago, at a time when there were large cities in the ancient civilizations of Eurasia and some of the Egyptian pyramids had already been built. We are not sure what killed off these last survivors of the great mammoths, but humans reached Wrangel Island about that time.

The situation was somewhat different in southern Eurasia (i.e., the Indian subcontinent and southeast Asia), where many fewer megafaunal animals became extinct. Although the dating of the fossils is still very poor for this region, Late Pleistocene extinctions included a couple of species of proboscideans and large species of hyena, tapir, and panda. The giant ape *Gigantopithecus* disappeared in the Middle Pleistocene. Many megafaunal mammals remain today, including elephants, rhinos, large apes such as the orangutan, and large carnivores such as lions and tigers. However, most of these species are today endangered.

In sub-Saharan Africa, most of the Late Pleistocene megafauna survived, although several species of large antelope went extinct, as did a large species of zebra.

## Australia

At the time of the lowered sea levels of the Late Pleistocene, Australia, New Guinea, and Tasmania formed a single land mass. This land mass had very limited glaciation but was subjected to fluctuating wet and dry periods corresponding to interglacial and glacial episodes elsewhere. Australia suffered more severe extinctions, in terms of percentage of the previous megafauna, than any other continental-sized land mass. Around 40 species became extinct, representing ~90% of the megafauna. The only animals larger than 44 kg in Australia today are a couple

Figure 23.14 The skull of *Megalania*, the largest terrestrial lizard of all time, from the Pleistocene of Australia. This skull is 30 in. long, and the largest individuals were over 20 ft long, the top terrestrial carnivores in the ecosystem. *Source:* Photograph by Stephen G. Johnson (Wikimedia).

of species of birds (emu and cassowary) and kangaroos (red and gray kangaroos; but many smaller kangaroo species survived), and the saltwater crocodile.

The first humans are estimated to have arrived, via Southeast Asia, some time between 62 to 43 ka, while all, or nearly all, of the extinct Australasian megafauna had gone by about 46 ka. However, the precise timing poses a problem for accurate dating of events as this is just outside the zone where radiocarbon dating can be employed, and there is intense debate on whether humans or natural climate change were responsible for the extinctions. A few species of megafaunal mammals are thought to have survived alongside humans at the site of Cuddie Springs, dated at around 30 000 years ago, and this has been argued to contradict the idea of "overkill." As with South America, there is no direct archeological evidence that humans hunted the megafaunal animals, although the eggs of the giant bird *Genyornis* show some evidence of being charred from campfires.

Australia lost every terrestrial vertebrate larger than a human. It lost a giant horned turtle as big as a car and its giant birds, the dromornithids (see Chapter 20, Figure 20.9) (see Chapter 14). It lost its top predators, including *Megalania*, the largest terrestrial lizard that ever evolved, 7 m (24 ft) long. *Megalania* was closely related to the living Komodo dragon but weighed more than eight times as much (Figure 23.14). Other predators were the marsupial "lion" (Figure 23.15; see also Chapter 20, Figure 20.7), a huge terrestrial crocodile and a 5 m (16 ft) python. Australia lost about 20 species of large marsupials, including all the diprotodonts, huge wombat-related herbivores (Figure 23.15; see also Chapter 20, Figure 20.8), the Pleistocene ones as large as a small rhino, and an entire family of giant kangaroos, the sthenurines. Sthenurines were short-faced browsers (Figure 23.16), ranging from the size of a large modern kangaroo to *Procoptodon*, which stood over 2 m high. Only a few large animals survived in Australia but small animals were less affected.

Figure 23.15 A marsupial lion (*Thylacoleo*) attacking a *Diprotodon. Source:* From Rom-diz, Russian Wikipedia, in the public domain.

Figure 23.16 *Sthenurus stirlingi*, a giant extinct kangaroo, around 6 ft tall. Some sthenurines could have been too big to hop and may have walked bipedally. *Source:* By Brian Regal (Wikimedia).

Figure 23.17 *Genyornis*, a giant ground-running bird from the Pleistocene of Australia, over 6 ft tall. *Source:* Artwork by Nobu Tamura (Wikimedia).

The slow-running giant dromornithid bird *Genyornis* (Figure 23.17) disappeared from habitats where fast-running emus survived, but there seems to be a memory of it in aboriginal legend as the mihirung. Some of the oldest Aboriginal rock art seems to show *Genyornis*, perhaps the last ones!

The extinctions coincide roughly with a change in vegetation associated with increased burning, as shown by the presence of charcoal in the fossil record, possibly generated by the early Australians. They were migrating into a dry country ecosystem that was unfamiliar to them because they came from the moister tropical ecosystems of southeast Asia. They had to learn slowly how to adapt to drier Australian conditions, just as European colonists had to do tens of thousands of years later.

One of the easiest ways of clearing Australian vegetation is to burn it; burning makes game easier to see and hunt, and Australian aborigines today have complex timetables for extensive seasonal brush burning that has dramatic effects on regional ecology. The Australian extinctions may have been the direct result of human invasion, through the introduction of large-scale burning as well as hunting. Others propose that natural cycles of dry climate during glacial periods expanded the arid core of the continent so that the lusher coastal vegetation, where many megafauna lived, became greatly reduced in area, leading to extinctions.

## Island Extinctions

New Zealand serves as a case study for island extinctions. A thousand years ago, it was an isolated set of islands without land mammals (except for two species of

Figure 23.18 One of the moas of New Zealand. *Source:* After George Edward Lodge.

bats). Birds were the dominant vertebrates and the largest were the moas, huge flightless browsing birds the size of ostriches (see Chapter 14) (Figure 23.18). The moas and other native creatures survived as glacial periods came and went, yet they became extinct within a few hundred years of the arrival of the Polynesian Maori people after 1000 CE.

There seem to have been two main reasons for the extinctions, both connected with human arrival. First, evidence of hunting is clear. Midden sites that extend for acres are piled with moa bones, with abundant evidence of butchering. The bones are so concentrated in some places that they were later mined to be ground up for fertilizer. Second, the arrival of people was accompanied by rats, which ate insects directly, killed off reptiles by eating their young, and exterminated birds by robbing their nests. The tuataras (sphenodontid reptiles; see Chapter 12), the giant flightless wetas (insects that had been the small-bodied vegetarians of New Zealand; see Chapter 20), and many flightless birds including the only flightless parrot, the kakapo, were practically wiped out by rats. There were many other more subtle ecological changes. A giant eagle that may have preyed on moas died out with them, for example.

Half of the original number of bird species in New Zealand was extinct before Europeans arrived in the late eighteenth century, and the new settlers only acted to increase the changes in New Zealand's landscape and biology. Forests were cleared even faster and new mammals were introduced. European rats were the worst offenders against the native birds, but cats, dogs, and pigs were also destructive, rabbits destroyed much of their habitat, and deer competed with browsing birds. The tuatara now lives only on a few small, rat-free islands and the kakapo survives precariously in remote areas where it is threatened by feral cats. Bird populations are still dropping in spite of efforts to save them.

As a microcosm of the problem, consider the Stephen Island wren, the only flightless songbird that has ever evolved. This species had already been exterminated from the main islands of New Zealand by the Polynesian rat before European arrival. The entire remaining population of this species, which was by then confined to one small island, was caught and killed by Tibbles, a cat brought to the island in 1894 by the keeper of a new lighthouse.

On Madagascar, large lemurs (some as large as gorillas; see Chapter 21, Figure 21.8), giant land tortoises, two species of pygmy hippos, and the huge flightless elephant birds (see Chapter 14) disappeared after the arrival of humans somewhere between 0 and 500 CE. Here too, large forest areas were cut back and burned off to become grassland or eroded, barren wasteland. No native terrestrial vertebrate heavier than 12 kg (25 pounds) survived after 1000 CE. Humans took a long time to penetrate the forest of this large island, and the extinction may have taken 1000 years instead of being sudden. It's clear that human arrival was part of a "recipe for disaster" (as David Burney has called it). The erosion and poverty of much of the countryside of Madagascar today underline the fact that humans are still involved in deforestation, in spite of the evidence all around them of its destructive after-effects.

Europeans killed off the dodo on Mauritius (Figure 23.19) and several species of giant tortoises there and in the Galápagos; deliberate burning, and the goats, pigs, and rats they brought, completed a great deal of destruction of native plants, birds, and animals. They killed off the great auk of the North Atlantic, the huia and other small birds of New Zealand, and an unknown number of species of birds of paradise in New Guinea, in many cases to satisfy the demands of egg and feather collectors. They drove many fur-bearing mammals close to extinction worldwide.

New discoveries of extinct flightless birds in Hawaii suggest that devastating extinctions followed the arrival of Polynesians. The Hawaiian Islands are famous for honeycreepers, which evolved there into many species like Darwin's finches on the Galápagos Islands. But there were 15 more species of honeycreepers before humans arrived. Two-thirds of the land birds on Maui (the second largest of the Hawaiian islands) were wiped out, probably by a combination of hunting, burning, and the arrival of rats. As in New Zealand, the extinctions that followed the European arrival were severe but not as drastic, probably because the bird fauna was already so depleted.

The same process is recorded on almost all the Pacific islands in Melanesia, Polynesia, and Micronesia. Almost all

Figure 23.19 The dodo, playing a cameo role in *Alice in Wonderland*, by Lewis Carroll. This famous image by Sir John Tenniel from 1869 shows an outmoded reconstruction of the dodo, perhaps based on overfed captive specimens.

of them, apparently, had species of flightless birds that were killed off by the arriving humans and/or their accompanying rats, dogs, pigs, and fires. As many as 2000 bird species may have been killed off as human migration spread across the ocean before the arrival of Europeans. It may not be an accident that Darwin was inspired by the diversity of the Galápagos Islands; these were never occupied before European discovery in 1535 and human impact was relatively slight until whalers arrived in strength around 1800.

More mammals went extinct in the Caribbean over the last 20 000 years than anywhere else on Earth. Extinction has dramatically transformed these islands with an entire lost world of monkeys, ground sloths, large and small rodents, and insectivores. Today, out of more than 130 original Caribbean mammal species, only 13 terrestrial mammals and 60 bats remain. An initial pulse of extinction, mainly focused on several genera of ground sloths, followed the first arrival of humans around 5000 years ago. Many monkeys, rodents, and insectivores were able to coexist alongside multiple human cultures over thousands of years, succumbing only in the last 500 years or so to agricultural clearance and the spread of domestic animals.

Several islands in the Mediterranean Sea (Cyprus and Crete are good examples) held fascinating evolutionary experiments during the ice ages. There were pigmy elephants and pigmy hippos, giant rodents, and dwarf deer. These mammals had evolved on these isolated islands in much the same way as did the fauna of Gargano during Miocene times (see Chapter 20). Some of the island animals disappeared as Neolithic peoples discovered how to cross wide stretches of sea and colonized the islands several thousand years ago. The timing and causes of extinction of other species are less clear.

## The World Today

As shown by the above detailed account, the causes of Late Quaternary extinctions are still debated for many parts of the world. It seems likely, however, that it was a combination of pressures that led to the extinction of so many large vertebrates. Severe environmental changes reduced and shifted the habitats to which animal species had become adapted. Some may have succumbed from this cause alone. For others, the added impact of human hunting tipped the balance.

In 1876 Alfred Russel Wallace wrote with respect to the Pleistocene megafaunal extinctions: "We live in a zoologically impoverished world from which all the hugest and fiercest and strangest forms have recently disappeared." But how much more depressing is it to consider that the end-Pleistocene extinctions may merely represent the "first pulse" of the Late Quaternary extinctions, and that the worst is yet to come?

The extinctions of the Late Quaternary, being restricted mainly to large terrestrial mammals, do not in themselves rank alongside the five great mass extinctions of the Paleozoic and Mesozoic (see Chapters 6 and 16), when numerous groups of vertebrates, invertebrates, and plants were decimated across multiple habitats. However, the events of the Late Quaternary can be seen as just the beginning of a true mass extinction – the so-called "Sixth Extinction" – that is in progress at the present day and shows every indication of accelerating into the future. Scientists have proposed a new name for the time period we currently live in, that is following the Industrial Revolution in the eighteenth century, as the Anthropocene, which emphasizes the changes in the world wrought by humans.

In today's world, human actions are affecting the natural world in multiple ways that act synergistically to multiply the threat.

- Global warming, driven by the burning of fossil fuels, is turning formerly fertile habitats into deserts (for example, in Australia), and melting Arctic and Antarctic ice sheets that immediately affect polar animals and pose further threats as low-lying lands become flooded.
- More than half of the Earth's ice-free land has been taken over by humans for their own use. The resulting destruction of massive areas of natural habitats through the spread of urbanization and agriculture has severely affected many animal populations. We know that the tropical regions of the world are a treasure house of species, many of them valuable to us and many of them undescribed. The stripping of tropical forest from hillsides not only removes the plants and animals that are best adapted to life there but results in erosion that removes the few nutrients left in the soil.
- Many animal species, such as tigers, gorillas, orangutans, chimps, elephants, rhinos, and whales, are still threatened by direct hunting, whether for food or for products such as ivory and horn.
- Tens of thousands of species have been transported all over the world, so that almost every ecosystem has been contaminated by invasive species. The introduction of plants and animals to new continents and islands, without proper ecological analysis of their possible impact, often has unintended consequences for the native species.
- Pollution, especially of the oceans, continues to pose an existential threat to many ecosystems.

As a result, species and populations of animals have been going extinct at elevated rates over the past few centuries. The Living Planet Report of 2016 (http://awsassets.panda.org/downloads/lpr_living_planet_report_2016_summary.pdf) revealed that, overall, global populations of fish, birds, mammals, amphibians, and reptiles declined by 58% between 1970 and 2012. Since 1900, around 500 species are known to have become extinct, and this is likely to greatly underestimate the true figure, as it is largely focused on vertebrates alone.

One can argue that humans at 13 000 BP, perhaps even at 500 BP, did not know enough ecology, did not have enough recorded history, certainly did not know enough archeology or paleontology, and did not have enough of a global perspective to realize the consequences of their impact on an ecosystem. But that is not true today. We have the theory and the data to know exactly what we are doing. The question is, do we have the will and foresight to halt and reverse the destruction of the biosphere in which our species originated and which it needs to survive?

## Further Reading

Barnosky, A.D. and Lindsey, E.L. (2010). Timing of Quaternary megafaunal extinction in South America in relation to human arrival and climate change. *Quaternary International* 217: 10–29.

Burney, D.A. and Flannery, T.F. (2005). Fifty millennia of catastrophic extinctions after human contact. *Trends in Ecology & Evolution* 20: 395–401.

Cafaro, P. (2015). Three ways to think about the sixth mass extinction. *Biological Conservation* 192: 387–393.

Cooper, A., Turney, C., Hughen, K.A. et al. (2015). Abrupt warming events drove Late Pleistocene Holarctic megafaunal turnover. *Science* 349: 602–606.

Diamond, J. (2011). *Collapse: How Societies Choose to Fail or Succeed*. London: Penguin Books. [Jared Diamond has written extensively about the interaction between human history, nature, and geography].

Flannery, T.F. (1995). *The Future Eaters*. New York: George Braziller. [The history of Meganesia (Australasia and associated islands). This book has much deeper significance than simply a regional history].

Guthrie, R.D. (1990). *Frozen Fauna of the Mammoth Steppe: The Story of Blue Babe*. Chicago: University of Chicago Press.

Johnson, C.N., Alroy, J., Beeton, N.J. et al. (2016). What caused extinction of the Pleistocene megafauna of Sahul. *Proceedings of the Royal Society B: Biological Sciences* 283: 20152399. [Sahul is the name given to the combined landmasses of Australia, New Guinea and Tasmania].

Lister, A.M. and Stuart, A.J. (2008). The impact of climate change on large mammal distribution and extinction: evidence from the last glacial/interglacial transition. *Comptes Rendus Geoscience* 340: 615–620.

Martin, P.S. (2005). *Twilight of the Mammoths*. Berkeley: University of California Press.

Meltzer, D.J. (2015). Pleistocene over-kill and North American mammalian extinctions. *Annual Review of Anthropology* 44: 33–53.

Nagaoka, N., Torben, R., and Wolverton, S. (2018). The overkill model and its impact on environmental research. *Ecology and Evolution* 2018: 1–14.

Owen-Smith, N. (1987). Pleistocene extinctions: the pivotal role of megaherbivores. *Paleobiology* 13: 351–362.

Price, G.J., Louys, J., Faith, J.T. et al. (2018). Big data little help in megafauna mysteries. *Nature* 558: 23–25.

Stewart, J.R., Lister, A.M., Barnes, I. et al. (2010). Refugia revisited: individualistic responses of species in space and time. *Proceedings of the Royal Society of London B: Biological Sciences* 277: 661–671.

Stuart, A.J. (2015). Late Quaternary megafaunal extinctions on the continents: a short review. *Geological Journal* 50: 338–363.

Zimov, S.A., Zimov, N.S., Tikhonov, A.N. et al. (2012). The Mammoth steppe: a high-productivity phenomenon. *Quaternary Science Reviews* 57: 26–45.

## Questions for Thought, Study, and Discussion

1 As you know, bison are quite large animals, and before they were slaughtered by modern rifles in the 1800s, they traveled in huge herds. Read again what "megaherbivores" are and discuss whether bison really are megaherbivores that survived the hunting of ancient Americans.

2 Elephants and rhinoceroses really are megaherbivores. How did they survive in Africa when megaherbivores were killed off on other continents?

3 Describe the science behind this limerick:
They fall from the branches to wait
But they are 12 000 summers too late
You can smell them for miles
They are rotting in piles
The fruits that the gomphotheres ate.

4 Describe the science behind this limerick:
The morning was hardly propitious
When sailors discovered Mauritius
They killed off the lot
Stewed them up in a pot
And pronounced them extinct, but delicious.

# Glossary

**absolute dating** establishing the exact age of a rock, and an exact time scale for geology, using mainly radiometric dating methods.

**acanthodian** a Paleozoic fish, usually small and present in huge numbers, and noted for bearing many spines on fins and body.

**Acheulean** the archaeological industry of stone tools and weapons associated especially with *Homo erectus* and *Homo heidelbergensis*.

**acid rain** a rain of dilute acid, deriving from carbon dioxide and other gases that mix with water, both from volcanic emissions and from pollution.

**acritarch** organic microfossils, known from early Precambrian to the present.

**aetosaur** an armored, plant-eating archosaur of the Late Triassic, often with a snub nose used for rooting out plants.

**Afrotheria** "African mammals"; the clade of elephants, sea cows, hyraxes, golden moles, and others that are unique to Africa.

**allantois** the membrane inside the cleidoic egg that encloses the waste products (cf. amnion, chorion).

**amino acid** basic building block of proteins; there are 21 amino acids, which may be strung together in a huge range of different sequences to produce all the multitudes of protein types.

**ammonite** a coiled cephalopod mollusk.

**amnion** the membrane inside the cleidoic egg that surrounds the embryo (cf. allantois, chorion).

**amniote** an animal that lays amniotic eggs, namely a member of the clade comprising reptiles, birds, and mammals.

**amniotic** of the amnion – referring to membranes, fluids, or eggs.

**anapsid** "no arches"; a tetrapod skull with no temporal openings (cf. diapsid, synapsid).

**angiosperm** a flowering plant.

**ankylosaur** a dinosaur with armor over the body and head, typically of the Cretaceous.

**anoxia** the condition where oxygen is absent.

**anoxic** oxygen poor or with no oxygen (cf. aerobic, anaerobic, dysoxic).

**anther** "flower"; the part of a flower that produces pollen.

**anthracosaur** mainly Carboniferous and Permian aquatic amphibians, close to the origin of amniotes.

**antiarch** a group of placoderm jawed fishes, mainly from the Devonian.

**Archaean** the first great division of the Precambrian in the geological time scale.

**archosaur** a diapsid reptile of the clade that includes crocodilians, dinosaurs, and birds.

**Archosauria** the clade of archosaurs.

**arms race** in evolutionary terms, a model of continuing evolution by ecological interaction, such as predator–prey or competitors for the same ecospace.

**arthrodire** a group of placoderm jawed fishes, mainly from the Devonian.

**asexual reproduction** reproduction in the absence of sex, sometimes called cloning.

**aspect ratio** the ratio of width to height of a wing or other structure.

**asteroid** a minor planet, and can include very large rocks that impact on the Earth.

**australopithecine** a group of many species of early humans, known especially from Africa, mostly from 5–1 million years ago.

**autotrophy** "self-feeding"; organisms that do not feed on others, so essentially photosynthesizing plants and microbes.

**Avemetatarsalia** the "dinosaur line" of archosaur evolution, including pterosaurs, dinosaurs, birds, and their ancestors.

**banded iron formation (BIF)** Precambrian rocks, which consisted of bands of iron-rich and iron-poor sediment, formed on the ocean floor in the absence of oxygen.

**batrachosaur** member of the "amphibian" branch of basal tetrapod evolution, including temnospondyls and other extinct groups, and modern frogs and salamanders.

**behavior**  the things animals do when they are feeding, moving, caring for their young, or interacting with each other.

**behavioral thermoregulation**  controlling body temperature by behavior, such as basking to warm up or hiding to cool down.

**belemnite**  a cephalopod mollusk of the Mesozoic, squid-like, and with an internal calcareous guard or shell.

**bilaterian**  triploblastic animal with bilateral symmetry, and a digestive system with a mouth and anus.

**biogeography**  the ways in which plants and animals occupy the land and sea; the distribution of plants and animals.

**biomarker**  an organic chemical that indicates the presence of life.

**bone histology**  the study of bone structure under the microscope.

**Boreoeutheria**  "northern mammals"; the great clade of mammals including horses, cattle, rodents, and primates that originated mainly in the northern hemisphere.

**braided stream**  a river that splits and criss-crosses, usually when flowing down steep gradients at high speed.

**bryophyte**  a moss or liverwort.

**Cambrian**  the first division of the Phanerozoic in the geological time scale.

**carapace**  a hard shell covering a turtle or other animal.

**carpel**  "fruit"; the specialized structure that encloses the ovule in an angiosperm flower.

**Carrier's Constraint**  the difficulty of breathing and running at the same time, especially in salamanders and lizards, which flex their bodies from side to side to walk, but at the same time they compress their lungs.

**caseid**  an early synapsid or pelycosaur of the Late Carboniferous or Early Permian.

**cell membrane**  the membrane that surrounds a cell, usually composed of a double wall.

**cephalochordate**  "head backbone"; a basal chordate with fish-like form, such as amphioxus.

**cephalopod**  "head foot"; mollusks such as ammonites, octopus, and squid, with advanced heads and eyes.

**ceratopsian**  a dinosaur with a frill at the back of the skull and thickenings or horns over the nose, from the Cretaceous.

**Ceratosauria**  the Jurassic and Cretaceous theropod group of large flesh-eaters, often with "horns" on their heads.

**chemical evolution**  the formation of complex organic molecules (cf. organic molecule) from simpler inorganic molecules through chemical reactions during the early history of the Earth.

**chemical fossil**  chemicals found in rocks that provide an organic signature of ancient life.

**chert gap**  the absence of chert, a form of silica, from deep oceans following the Permian-Triassic mass extinction.

**Chicxulub Crater**  the crater in Mexico created at a time that coincides with the extinction of the dinosaurs.

**chlorophyll**  the pigment that gives plants their green color, and is also key to the photosynthetic breakdown of carbon dioxide.

**choana**  a pit or opening in a vertebrate skull.

**choanocyte**  a collar cell in a sponge that moves water by beating its flagellum.

**choanoflagellate**  free-living single-celled eukaryote that may be the closest relative of animals.

**chorioallantoic placenta**  the placenta typical of placental mammals, found in a few marsupials, formed from the fusion of the chorion and allantois membranes.

**chorion**  the membrane inside the cleidoic egg that lines the eggshell (cf. allantois, amnion).

**circulation system**  a system in the body that transports fluids; usually refers to the blood circulation system.

**clade**  a monophyletic group.

**cladistics**  the classification of taxa, or of biogeographic regions, in terms of shared derived characters (synapomorphies).

**cladogram**  a dichotomously (two-way) branching diagram indicating the closeness of relationship of a number of taxa.

**classification**  the process of naming organisms and arranging them in a meaningful pattern; also, the end-result of such a procedure, a sequential list of organism names arranged in a way that reflects their postulated relationships.

**Clovis culture**  a prehistoric human culture from North America.

**cnidarian**  the animal clade including corals, sea anemones, jellyfish, and hydrozoans.

**coal gap**  the absence of coal from terrestrial sediments, following the Permian-Triassic mass extinction.

**coelom**  the principal body cavity in most animals, located between the body wall and the gut.

**Coelurosauria**  the large clade of theropod dinosaurs including tyrannosaurs and maniraptorans.

**colony**  (of corals, graptolites, bryozoans, etc.) living in fixed association with other individuals, and forming a unified "superorganism" (cf. solitary).

**community**  a group of plants and animals that live together and interact in a specified area.

**conducting strand**  a canal in a plant that conducts fluid flow, such as the xylem or phloem.

**convergent evolution**  evolution of different organisms or different body structures to look the same, usually because of a shared mode of life.

**coral gap** the absence of corals from the oceans, following the Permian-Triassic mass extinction.

**Crurotarsi** the "crocodilian line" of archosaur evolution, including crocodilians and their ancestors.

**Cryogenian** "cold making"; the time division of the Precambrian when climates were freezing worldwide.

**cuticle** horny protein outer covering in many plants and animals.

**cyanobacteria** a major clade of bacteria that gain their energy through photosynthesis.

**Deccan Traps** basalt lavas erupted over a large area of present-day India at the KPg boundary.

**dentary** "tooth-bearing"; the tooth-bearing bone in the lower jaw of a vertebrate.

**Deuterostomia** major clade of animals including echinoderms and chordates.

**diagnostic character** a distinguishing feature, unique to a particular clade.

**diaphragm** the muscular structure in mammals that separates lungs and digestive organs, and assists in pumping the lungs.

**diapsid** "two arches"; a tetrapod skull with two temporal openings (cf. anapsid, synapsid).

**digestive tract** the structure from mouth to anus that processes food.

**disaster species** species that become established in the disturbed times following an extinction event.

**diversity gradient** a gradual change in biodiversity along a geographic line, say from low to high values.

**dromaeosaurid** an active theropod dinosaur, close to the origin of birds.

**durophage** an organism that feeds on hard materials, for example crushing shells or bones.

**Ecdysozoa** the division of bilaterian animals that includes forms such as arthropods, nematodes, and priapulids that shed their external skeletons as they grow.

**ecomorphology** the relationship between morphology or an organism and its function.

**ecosystem** the combination of habitats and organisms in a particular place at a particular time.

**edaphosaur** a plant-eating synapsid, usually with a sail on its back, from Late Carboniferous to Early Permian in age.

**endemic** "in district"; restricted to a particular geographic area (cf. cosmopolitan).

**eutherian** "true therian": another term for placental mammal.

**eutriconodont** member of a group of Jurassic and Cretaceous mammals, including forms once called triconodont.

**evolutionary radiation** a time when a clade expands substantially, usually enabled by a novelty or triggered by some external event.

**evolution by adaptive change** description of the results of evolution by natural selection, that plants and animals adapt to the changing environments and so their adaptations change through time.

**extinction event** the disappearance of many species at a particular time, usually caused by a single set of environmental stresses.

**fermentation** the chemical breakdown of a substance by bacteria, yeasts, or other microorganisms, typically with production of heat and bubbling gases.

**fern spike** a sudden explosion of fern abundance, often following a crisis such as a major volcanic eruption or impact.

**fish kill** sudden death of many fish, often by poisoning or other crisis.

**flagellum (pl. flagella)** "whip"; hair-like organelle in a eukaryote cell that is used for swimming.

**flood basalt** basalt (black-colored igneous rock) that emerges from vent-style eruptions and can build great thicknesses of rock over large areas.

**food web** the complex feeding interactions among members of a community.

**fossil** "dug up"; the remains of a plant or animal that died in the distant past.

**founder population** a population from which later populations emerge; often in reference to the first arrivals on an island, for example.

**furcula** "small fork"; the fused clavicles of birds (the wishbone of a chicken or turkey), and also found in many theropod dinosaurs.

**gastrolith** "stomach stone"; a stone swallowed by a bird or dinosaur to help break up its food.

**gene regulatory network** a collection of molecular regulators that interact with each other and with other substances in the cell to govern protein production.

**genome** the "genetic code," the complete set of genes in any organism.

**geographic speciation** the formation of two or more species by splitting of founder species by geographic barriers.

**geological time scale** the internationally agreed scale of geological time.

**gill arch** bar of cartilage or bone that supports the gill slits.

**gliding** flying without flapping wings.

**gnathostome** "mouth-jaw"; a jawed vertebrate.

**Gondwana** the southern supercontinent, made up of modern South America, Africa, India, Australia, and Antarctica.

**gorgonopsian** a carnivorous synapsid, sometimes with saber-like teeth, from the Late Permian.

**gracile**  slender or delicate.

**Great Oxidation Event**  the time in the Precambrian when oxygen levels in the atmosphere stepped up from close to zero to measurable quantities.

**greenhouse gas**  a gas, such as methane or carbon dioxide, that promotes heating of the atmosphere.

**ground effect**  the increased lift and decreased aerodynamic drag of a wing close to a fixed surface.

**guard cell**  the cells on each side of the plant stoma that enable it to open and close.

**guild**  a broad ecological grouping of plants or animals, such as "top predator," that may be occupied by different species in different communities or at different times.

**habitat**  the environmental setting within which a species or community lives.

**hadrosaur**  a duck-billed dinosaur, a kind of ornithopod that lived in the Late Cretaceous.

**heterotrophy**  "different feeding"; organisms that feed on a variety of materials (cf. autotrophy).

**horizontal gene transfer**  transfer of genes, sometimes called "jumping genes," between simple organisms.

**hot spot**  (1) ecologically, a geographic location with high levels of something, usually in reference to biodiversity; (2) geologically, an area of the Earth's crust where magma from a tectonic plume breaks through to form volcanic features (see plume eruption).

**hydrothermal vent**  a fissure or crack in the deep sea from which fluids heated deep in the Earth emerge.

**hyperthermal**  an event of extreme heating, often in reference to certain mass extinctions driven by massive volcanic eruptions and global warming.

**hypsodont**  high-crowned, long-wearing teeth of mammals such as horses and cattle.

**Ichthyosauria**  marine reptiles with long snouts and dolphin-like bodies, that existed from the Early Triassic to mid-Cretaceous.

**iguanodont**  an ornithopod dinosaur, such as *Iguanodon*, mainly from the Early Cretaceous.

**incumbent effect**  the observation that existing species have a competitive advantage over incomers.

**intercellular gas transport system**  the movement of gases within an animal or plant across cell membranes.

**internal transport system**  any system for transport of food, fluids or gases inside a plant or animal.

**iridium**  a platinum group element found in deep-earth rocks, but arriving on the surface mainly in meteorites.

**isotope**  a variant of a chemical element that differs in the number of neutrons in the nucleus.

**lactation**  the production of milk.

**Laurasia**  the northern supercontinent, comprising present-day North America, Europe, and Asia.

**lepidosaur**  a lizard or snake or close relative.

**lignin**  "wood"; woody tissue.

**lipid**  fatty or waxy compound of the cell.

**lithotrophy**  "stone eating"; animals that can drill into rock usually using chemical means to dissolve calcium carbonate.

**living fossil**  a slowly evolving group of plants or animals that is still living but looks somehow ancient.

**Lophotrochozoa**  "crest or wheel animals"; the major clade of bilaterian animals that includes worms, mollusks, brachiopods, bryozoans, and others.

**Lucy**  the early hominid, a female example of *Australopithecus afarensis*, one of the oldest relatively complete human skeletons.

**Marginocephalia**  "margin-heads"; the group of ornithischian dinosaurs that includes ceratopsians and pachycephalosaurs.

**Maniraptora**  "hand hunters"; theropod dinosaurs with elongated arms and powerful hands, including dromaeosaurids, troodontids, and birds.

**marsupial**  a pouched mammal, such as an opossum or kangaroo.

**masseter**  one of the major jaw muscles in mammals.

**mass extinction**  a major extinction event, typically marked by the loss of 10% or more of families and 40% or more of species, in a short time.

**mass wasting**  the loss of soil and plants from the landscape, leaving bare rock.

**meandering river**  a river that is shaped in broad curves; typical of only slightly sloping ground.

**megaherbivore**  a "large" plant-eater, such as an elephant or sauropod dinosaur.

**melanin**  a pigment biomolecule that gives black, brown, and ginger colors.

**melanosome**  an organelle that contains melanin within the structure of a hair or feather.

**metabolism**  all aspects of the body that relate to food consumption, energy budget and body temperature.

**metatherian**  "like therian"; the marsupials and close relatives.

**methanogen**  "producing methane"; an anaerobic organism that absorbs carbon dioxide and hydrogen and produces methane as waste gas.

**microsaur**  a small amphibian, typically Carboniferous or Permian, often with quite terrestrial adaptations.

**molecular clock**  the assumption that each protein molecule has a constant rate of amino acid substitution; the amount of difference between two homologous molecules indicates distance of common ancestry, and hence closeness of relationship.

**molecular evidence**  evidence from molecular biology.

**monotreme**  "one hole"; the basal mammals such as platypus and echidna today, that have a single opening for anus, bladder and reproductive organs.

**morganucodont**  an early mammal, especially known from the Early Jurassic.

**Mousterian**  the archaeological industry of stone tools and weapons associated especially with Neanderthals.

**multituberculate**  "multiple-cusped"; the successful herbivorous mammals of the Cretaceous and Paleogene that had long cheek teeth bearing many cusps for chewing a variety of foods, including plants.

**mysticetes**  the whales, as opposed to the toothed whales, the odontocetes.

**naked gene**  a gene composed of DNA without any associated proteins, lipids or other molecules to help protect it.

**nectridean**  a small Carboniferous water-living amphibian, sometimes with unusual swimming abilities.

**nematocyst**  "thread bladder"; the sting within the cnidoblast of a cnidarian.

**neocortex**  an advanced part of the cortex of the brain in mammals, associated with sight and hearing.

**neopterygian**  "new wings"; the holostean and teleost fish, which diversified from the Triassic onwards.

**neoselachian**  "new sharks"; the modern shark clade, which diversified in the Mesozoic.

**niche**  lifestyle and ecological interactions of an organism.

**notochord**  "back string"; the flexible, rod-like structure that supports the body of basal chordates and is a precursor of the spinal cord within the backbone in vertebrates.

**novelty**  in evolutionary terms, a new character, especially one that enables a group to diversify into a new adaptive zone.

**nucleic acid**  a complex organic substance present in living cells, especially DNA or RNA, whose molecules consist of many nucleotides linked in a long chain.

**nucleobase**  a nitrogen-containing biological compound that forms one of the building blocks of nucleotides.

**nucleotide**  a nitrogen-containing biological compound that forms one of the building blocks of nucleic acids.

**ocean acidification**  lowering of the pH of the oceans; slightly elevated acid content can be harmful to animals with calcareous skeletons.

**odontocetes**  the toothed whales, as opposed to the whalebone whales, the mysticetes.

**ophiacodont**  an early synapsid, a pelycosaur of the Late Carboniferous and Early Permian.

**ornithomimid**  "bird mimic"; a theropod with slender body and long legs, and often a small head and toothless jaws.

**ornithopod**  an ornithischian dinosaur without armor and walking bipedally; includes iguanodontids and hadrosaurs.

**osmotrophy**  the uptake of dissolved organic compounds by osmosis for nutrition.

**ostracoderm**  "shell skin"; a Paleozoic jawless fish with heavy armor.

**Out-of-Africa hypothesis**  the model for human evolution suggesting that all modern humans evolved in the past 200 000 years from an African ancestor.

**ovary**  egg"; egg-producing organ in female animals; structure that contains the ovules in plants.

**overkill hypothesis**  a model for extinction of the Pleistocene megaherbivores mainly by human hunting.

**oviraptor**  a Cretaceous theropod with short jaws, no teeth, and adaptations apparently for feeding on tough plant foods.

**ovule**  an undeveloped (unfertilized) seed.

**pachycephalosaur**  "thick-headed reptile"; one of the plant-eating Cretaceous dinosaurs with a thickened skull roof.

**paleobiogeography**  "ancient biogeography"; how plants and animals were distributed in the geological past.

**Paleozoic**  the first era of the Phanerozoic in the geological time scale.

**Paleozoic Fauna**  the animals that dominated the Paleozoic, as opposed to the preceding Cambrian Fauna and the following Modern Fauna.

**Pangea**  "all world"; ancient supercontinent composed of all the modern continents.

**Panthalassa**  the ancient sea that surrounded Pangea.

**parachuting**  gliding and falling at the same time.

**pareiasaur**  a Middle and Late Permian herbivore with small head and massive body.

**pectoral fin**  the paired fin in fishes at the front; evolutionarily equivalent to the tetrapod arm.

**pelagornithid**  a bird of the mid-Cenozoic with roughened beak margins mimicking teeth.

**pelvic fin**  the paired fin in fishes at the back; evolutionarily equivalent to the tetrapod leg.

**pelycosaur**  an early synapsid, sometimes with a sail on the back, especially from Late Carboniferous and Early Permian.

**petal**  a single division of a flower, usually colored, and formed from a modifed leaf.

**Phanerozoic**  the time since the Precambrian, comprising Paleozoic, Mesozoic, and Cenozoic eras.

**phloem**  tissues in plants that conduct foods made in the leaves to all other parts of the plant.

**phorusrhacid**  sometimes called "terror birds", a group of Cenozoic birds in South America that reached considerable size and likely preyed on mammals.

**photosynthesis**  "light manufacture"; the breakdown of carbon dioxide and water in the presence of sunlight to produce sugars and oxygen.

**phototrophy** "light feeders"; green plants that acquire food through photosynthesis.

**phylum (pl. phyla)** a division in classification; contains one or more classes and is contained in a kingdom.

**phytolith** "plant stone"; microscopic silica structures inside plants such as grasses, used to identify the diet of herbivores between whose teeth they can become stuck.

**phytosaur** "plant reptile"; a Late Triassic archosaur that fed on fish.

**placental** mammals with a placenta or structure in the mother to nurture the developing embryo.

**plastron** the lower shell of the turtle, under its belly, and attached to the carapace.

**plate tectonics** the processes within the Earth's crust and mantle that drive continental drift.

**plesiosaur** a member of the Plesiosauria.

**Plesiosauria** a group of Mesozoic marine reptiles, typically with long necks and swimming with four expanded paddles.

**pliosaur** a type of plesiosaur with a large head and short neck.

**plume eruption** major volcanic eruptions over a magmatic hot spot.

**productivity** the production of food within an ecosystem.

**protein** a complex organic chemical composed of amino acids, the basic building block of organisms.

**protocell** "first cell"; a simple structure with some properties of a living cell but not all.

**province** a geographic region.

**proxy** a stand-in or representative; for example, oxygen isotopes provide a proxy for temperature.

**pterodactyloid** a pterosaur with a small tail, or none at all; the clade that dominated in the Late Jurassic and Cretaceous.

**radula** "scraper"; the rasping feeding organ of mollusks.

**rangeomorph** one of a variety of Ediacaran animals shaped like a branching frond.

**red bed** red-colored sediments, generally sandstones and mudstones, formed usually in hot conditions on land.

**regulatory gene** a gene involved in controlling the expression of one or more other genes, as opposed to a structural gene.

**relative time scale** determining simply what is older or younger, rather than absolute age.

**relative wing loading** the relative amount of body mass supported by a particular area of wings, in a bird, bat, pterosaur, or insect.

**replicate** a copy, or the process of copying.

**reproductive cycle** the cycle from egg to adult, but including the point of reproduction.

**reptiliomorph** member of the "reptile" branch of basal tetrapod evolution, including anthracosaurs and

other extinct groups, and modern reptiles, birds, and mammals.

**respiration** breathing.

**rhamphorhynchoid** a pterosaur with a long tail, the earlier forms that dominated in the Late Triassic to Late Jurassic.

**rhodopsin** a pigment in the rods of the eye which is extremely sensitive to light.

**Rhynchosauria** the clade of Triassic plant-eating diapsid reptiles with hooked snouts and digging claws.

**ribozyme** an RNA molecule that is capable of acting as an enzyme.

**robust** strong or heavily built.

**root** the structure of a plant that usually fixes the plant in the soil and draws in water and minerals.

**Sauropterygia** the clade of Mesozoic marine reptiles that include plesiosaurs, nothosaurs, and pachypleurosaurs.

**scleractinian coral** a coral of modern type, originating in the Triassic.

**seed** the first stage of most plants, the early-stage fertilized embryo and food supply.

**segment** a part or section.

**sepal** one of the outermost parts of a flower, lying outside the petals.

**sexual reproduction** reproduction involving a male and female.

**shocked quartz** a form of quartz with sets of parallel lamellae, indicating it has been subjected to high pressure, such as during an impact.

**soaring** flying high on rising air currents, and usually without flapping the wings.

**sphenacodont** a carnivorous pelycosaur of the Late Carboniferous or Early Permian.

**Sphenodontia** the living tuatara, relative of Squamata, and ancestral forms back to the Triassic.

**spherule** a rounded object.

**spike** a sharp peak, indicating sudden change if seen in the geological record.

**sponge** a member of the phylum Porifera, a marine organism composed of silica spicules.

**sporangium (pl. sporangia)** spore-bearing structure in a land plant.

**Squamata** the group of reptiles including lizards and snakes.

**squamosal** a skull bone at the upper posterior corner.

**stamen** "stand"; the pollen-producing structure of a flower.

**star phylogeny** a time of rapid or explosive splitting in an evolutionary tree.

**stegosaur** a plate-backed dinosaur of the Jurassic or Cretaceous.

**stigma** "point"; the part of the carpel in a flower that receives pollen.

**stoma (pl. stomata)** "mouth"; an opening on the underside of a leaf through which water vapor may pass.

**stromatolite** "bed/mattress rock"; a layered structure generally formed by alternating thin layers of cyanobacteria and lime mud, typically in shallow warm sea waters.

**structural gene** a gene that codes for any RNA or protein product other than a regulatory factor.

**style** "pen"; in a flower, the slender part above the carpels bearing the stigma.

**summer monsoon** heavy rainfall in summer, especially in a tropical area.

**superoxide dismutase** an enzyme that breaks down the superoxide radical into oxygen or hydrogen peroxide; an important antioxidant defense to prevent cell damage.

**supracoracoideus** muscle in the shoulder that swings the forelimb forward in reptiles, and raises the wing in birds.

**swim bladder** the structure in fishes used to maintain pressure at different water depths, and evolved from the lung of their ancestors.

**symbiogenesis** the model for eukaryotic origins by combination of several prokaryotes to form a single cell.

**synapsid** "joined arch"; a tetrapod skull with one (lower) temporal opening (cf. anapsid, diapsid).

**systematics** the study of relationships of organisms and of evolutionary processes.

**taxonomy** "arrangement"; the study of the morphology and relationships of organisms.

**tektite** a small meteorite or melt rock that falls from the sky.

**teleost** a bony fish with thin scales, symmetrical tail fin, and protrusible jaws; the vast majority of fishes today.

**temnospondyl** a broad-snouted amphibian of the Carboniferous to Cretaceous; mainly fish-eaters.

**teratorn** a large predatory bird known from the Oligocene to Pleistocene.

**test** "pot"; the skeleton of an echinoid, foraminifer or radiolarian.

**Tetanurae** the theropod clade that includes megalosaurs, allosaurids, and coelurosaurs.

**tetrapod** "four feet"; the clade of amphibians, reptiles, birds, and mammals.

**tetrapodomorph** "tetrapod-like"; the tetrapods and fishapods, or fishes with functioning legs.

**therapsid** an advanced synapsid, from the Middle Permian onwards.

**Theria** the clade including all marsupial and placental mammals.

**therizinosaur** a Cretaceous theropod with enlarged claws and shortened hind limbs, but a tiny head and teeth adapted for plant eating.

**Thyreophora** the dinosaurian clade including ankylosaurs and stegosaurs.

**time tree** a phylogenetic tree plotted against a geological time scale.

**trace fossil** remains of the activity of an ancient organism, such as a burrow or track (cf. body fossil).

**tracheid** "artery"; a water-conducting strand in a land plant.

**tribosphenic** the three-cusped molar of modern mammals.

**triconodont** "three-coned"; group of Jurassic and Cretaceous mammals.

**troodontid** Cretaceous theropods with slender, fast-moving bodies and enlarged eyes and brains.

**trophic pyramid** a food web, or representation of feeding relationships, with a large amount of plant material at the bottom, rising through smaller numbers of herbivores, and smallest number of carnivores at the top.

**tsunami** a massive tidal wave, often set off by an earthquake.

**tunicate** a nonvertebrate chordate, sometimes called a sea squirt.

**turbidite** a rock formed from turbidity flows, mass movements of sand and mud down a slope and into deep water.

**turtle** a reptile with carapace and plastron, protective shells above and below.

**tyrannosaur** a theropod dinosaur, often large, especially of the Cretaceous.

**upright** erect, as opposed to sprawling, of posture.

**uraninite** formerly called pitchblende, an ore of the radioactive mineral uranium.

**vascular plant** a plant with vessels for transport of fluids; includes all land plants.

**vesicle** "bladder"; a fluid-filled sac.

**viviparous** "live-bearing"; an animal that does not lay eggs but produces live young.

**waterproofing** chemicals that repel water and protect the outer cell layers of certain animals and plants.

**winter monsoon** sudden, heavy rainfall in winter.

**woodpecker guild** birds or mammals that feed as woodpeckers do, by pecking, digging or probing into wood for insect food.

**Xenarthra** "strange joints"; the South American mammals such as ant-eaters, sloths, and armadillos.

**xylem** the woody tissue in vascular plants in which tracheids conduct fluids and that also acts as a support.

**yolk** the nutritious, often yellow material inside an egg that provides the food supply for the developing embryo.

# Index

Page locators in *italics* indicate figures. This index uses letter-by-letter alphabetization.

*Cowen's History of Life*, Sixth Edition. Edited by Michael J. Benton.
© 2020 John Wiley & Sons Ltd. Published 2020 by John Wiley & Sons Ltd.